# Animal Welfare

## Volume 23

The Animal Welfare series has been designed to help contribute towards a culture of respect for animals and their welfare by producing academic texts addressing how best to provide for the welfare of the animal species that are managed and cared for by humans. Books in the series do not provide a detailed blue-print for the management of each species, but they do describe and discuss the major welfare concerns, often in relation to the wild progenitors of the managed animals. Welfare has been considered in relation to animals' needs, concentrating on nutrition, behaviour, reproduction and the physical and social environment. Economic effects of animal welfare provision were also considered where relevant, as were key areas where further research is required.

Marie Haskell
Editor

# Cattle Welfare in Dairy and Beef Systems

## A New Approach to Global Issues

Co-ordinating Editor: Clive Phillips

Springer

*Editor*
Marie Haskell
SRUC (Scotland's Rural College)
Edinburgh, UK

ISSN 1572-7408
Animal Welfare
ISBN 978-3-031-21019-8 ISBN 978-3-031-21020-4 (eBook)
https://doi.org/10.1007/978-3-031-21020-4



This Springer imprint is published by the registered company Springer Nature Switzerland AG
The registered company address is: Gewerbestrasse 11, 6330 Cham, Switzerland

*This book is dedicated to the memory of Fiona Rioja-Lang. Fiona was instrumental in the development of this book. She is sorely missed by the animal welfare community, but her contributions to the field live on and are a credit to her enthusiasm and dedication.*

# Foreword

Since the publication of *The Welfare of Cattle* in 2008, public and stakeholder interest in farmed animal welfare, including the welfare of dairy and beef cattle, has further increased in many parts of the world. Concurrently, animal welfare science has improved our understanding of welfare, extending its scope beyond impaired welfare resulting from factors such as behavioral restrictions and health problems to the consideration of positive welfare states. In addition, translating the knowledge into "real-world" improvements of cattle welfare has gained importance. This requires inter-disciplinary approaches and novel ways to communicate and collaborate with various stakeholders. Acknowledging the inextricable connection between animal welfare, human well-being and the physical as well as social environment, animal welfare also constitutes a fundamental dimension in the transition to more sustainable livestock farming practices. Although not explicitly mentioned in the United Nations Agenda 2030, aspects of animal health and welfare play a role in achieving the Sustainable Development Goals at a regional to global level.

Against this background, the present volume comprehensively addresses advances and challenges in important areas of dairy and beef cattle welfare, written by top experts in the field. It covers current conceptual approaches to cattle welfare and provides overviews of assessing cattle welfare using behavioral as well as physiological and immunological measures. This is followed by reviews of challenges to cattle welfare in a range of dairy and beef production systems, including intensive and extensive systems. More specific chapters address cattle welfare in relation to slaughter and human–animal relationship as well as the often under-represented area of welfare around calving and of growing cattle. Emphasis is also put on the potential of monitoring cattle welfare using the rapidly evolving precision livestock farming techniques and mitigation of welfare problems through genetic selection. The last part of the book puts animal welfare in the context of the above-mentioned broader context of sustainability of livestock farming and opens the perspective to animal welfare in the Global South by focusing on unwanted cows in India, new-entrant dairying and cattle welfare in sub-Saharan smallholder farming systems, thus acknowledging the global dimension of dairy and beef cattle welfare issues.

The topics covered in this book are highly relevant to different stakeholders including farmers, advisors, veterinary practitioners, farm assurance, retailers, policy makers, as well as scientists. Integrating different perspectives and application fields, it stimulates discussion with the ultimate aim of improving cattle welfare in different production systems and regions of the world.

Institute of Livestock Sciences (NUWI)
Vienna, Austria
Christoph Winckler,

# Series Preface

Animal welfare is attracting increasing interest worldwide, and the knowledge and resources are available to, at least potentially, provide better management systems for farm animals, as well as companion, zoo, laboratory, and performance animals. The key requirements for adequate food, water, a suitable environment, companionship, and health are important for animals kept for all of these purposes.

The attention given to animal welfare in recent years derives largely from the fact that the relentless pursuit of financial reward and efficiency, to satisfy market demands, has led to the development of intensive animal management systems that challenge the conscience of many consumers, particularly in the farm and laboratory animal sectors. Livestock are the world's biggest land users, and the farmed animal population is increasing rapidly to meet the needs of an expanding human population. This results in a tendency to allocate fewer resources to each animal and to value individual animals less than the group. In these circumstances, the importance of each individual's welfare is diminished.

Increased attention to welfare issues is just as evident for zoo, companion, sport, and wild animals. Of growing importance is the ethical management of breeding programs since genetic manipulation is now technically advanced. There is less public tolerance of the breeding of extreme animals if it comes at the expense of animal welfare (e.g., brachycephalic dogs). The quest for producing novel genotypes has fascinated breeders for centuries. Dog and cat breeders have produced a variety of deformities that have adverse effects on their welfare, but nowadays the breeders are just as active in the laboratory, where the mouse is genetically manipulated with equally profound effects.

In developing countries, human survival is still a daily uncertainty for many, so that provision for animal welfare has to be balanced against human welfare. Animal welfare is usually a priority only if it supports the output of the animal, be it food, work, clothing, sport, or companionship. However, in many situations the welfare of animals is synonymous with the welfare of the humans that look after them, because happy, healthy animals will be able to assist humans best in their struggle for survival. In principle, the welfare needs of both humans and animals can be provided for, in both developing and developed countries, if resources are properly husbanded. In reality, the inequitable division of the world's riches creates physical and psychological poverty for humans and animals alike in many parts of the world.

The intimate connection between animals and humans that was once so essential for good animal welfare is rare nowadays, having been superseded by technologically efficient production systems where animals on farms and in laboratories are tended by increasingly few humans in the drive to enhance labor efficiency. With today's busy lifestyles, companion animals too may suffer from reduced contact with humans, although their value in providing companionship, particularly for certain groups such as the elderly, is beginning to be recognized. Animal consumers also rarely have any contact with the animals that are kept for their benefit.

In this estranged, efficient world, people struggle to find the moral imperatives to determine the level of welfare that they should afford to animals within their charge. A few people, and in particular many companion animal owners, strive for what they believe to be the highest levels of welfare provision, while others, deliberately or through ignorance, keep animals in impoverished conditions in which their health and well-being can be extremely poor. Today's multiple moral codes for animal care and use are derived from a broad range of cultural influences, including media reports of animal abuse, guidelines on ethical consumption and campaigning and lobbying groups.

This series has been designed to contribute toward a culture of respect for animals and their welfare by producing learned treatises about the provision for the welfare of the animal species that are managed and cared for by humans. The early species-focused books were not detailed management blueprints; rather they described and considered the major welfare concerns, often with reference to the behavior of the wild progenitors of the managed animals. Welfare was specifically focused on animals' needs, concentrating on nutrition, behavior, reproduction, and the physical and social environment. Economic effects of animal welfare provision were also considered where relevant, as were key areas where further research is required.

This volume returns to the theme of single vertebrate species to address the welfare of cattle, for the second time in this Springer series. An initial volume, The Welfare of Cattle by J. Rushen, A.M. de Passillé, M.A.G. von Keyserlingk, and D.M. Weary was published in 2008. This volume acknowledges the major advances that have been made in our understanding of cattle welfare and the changes in industry practices since that time. Both beef and dairy cows are more likely to be kept in intensive systems, presenting challenges to their physical welfare, even though nutrition is more controllable than when cows are at pasture. However, extensive management systems are still common where pasture can be used economically. The use of technology in cattle management is a recent and wide-reaching innovation, which can be used to both promote and monitor welfare. An example is the use of electronic collars to control grazing, allowing the prospect of fenceless dairy farms in future. Welfare assessment has advanced dramatically during last few decades, even though we are still short of indicators of positive emotions—a major challenge for the next few decades. Welfare can also be improved by taking care of animals at key points, such as at the early stage of life, during routine handling and at slaughter. Welfare "harms" can be reduced by including welfare traits in genetic selection. In recognition of global climate change and the ever-increasing heat challenges for higher yielding dairy cows and beef cattle in feedlots, more attention is paid to heat stress, since this is being recognized now even in cooler parts of the world. Greater recognition too is being paid to the sustainability of cattle production systems, and the involvement of consumers, and non-consumers, in provision of social license for production. The models of silvopastoral systems in South America and retirement of cows to sanctuaries in India are both used as exemplars of how social license operates in different parts of the world. The book further recognizes the worldwide diversity of cattle production by including a focus on smallholder and transhumant systems, still the prevailing methods of cattle production in sub-Saharan Africa. The editor, Marie Haskell, is to be commended for her extensive coverage of the welfare of cattle in the diverse systems around the world, which I am sure will captivate readers for many years to come.

Perth, WA, Australia Clive Phillips

[illegible]

# Contents

# Part I
# Context and Measurement

# A Good Life for Cattle

# 1

Marina A. G. von Keyserlingk and Daniel M. Weary

## Contents



**Abstract**

In this first chapter, we summarize how the science and our thoughts have progressed since the publication of *The Welfare of Cattle,* (a predecessor to the present volume) 15 years ago. We discuss some of the advances made on cattle welfare issues over the last decade and also highlight areas where more work is needed. We also emphasize the importance of considering how human factors can have a profound influence on the welfare of cattle. We have highlighted the need to understand the views of different stakeholders, including the public, and call for focused work on understanding the barriers to adoption of practices that improve the welfare of cattle on farms. Lastly, we summarize recent conceptual innovations, including in how animal welfare is defined, and consider some of the challenges inherent to working in a field of science much affected by human values.

**Keywords**

Animal welfare · Well-being · Cow comfort · Affective state · Natural behavior · Pain

M. A. G. von Keyserlingk (✉) · D. M. Weary
Animal Welfare Program, Faculty of Land and Food Systems, The University of British Columbia, Vancouver, BC, Canada
e-mail: nina@mail.ubc.ca



M. Haskell (ed.), *Cattle Welfare in Dairy and Beef Systems*, Animal Welfare 23,
https://doi.org/10.1007/978-3-031-21020-4_1

## 1.1 Introduction

Concerns over the ethical treatment of farm animals have become prominent over the past century. Key events included the publication of The Jungle by Upton Sinclair in 1906 describing the conditions experienced by both the animals and the workers in the Chicago stockyards, and the publication of Ruth Harrison's Animal Machines (1964) describing the rearing conditions of farm animals kept on modern intensive farms. Since then, societal concerns about the quality of farm animals' lives and questions concerning the legitimacy of farm practices have continued to evolve.

The field of animal welfare science arose to address these issues. The field tackles questions that arise from society about how we care for animals, applying disciplinary approaches from sociology, economics, psychology, veterinary medicine, animal behavior, etc. (Appleby 2004; Lund et al. 2006). Thus, animal welfare science can be viewed as a socially mandated area of academic inquiry, opportunistically borrowing from other branches of science, unlike traditional disciplines within animal agriculture where scientists tend to engage in research focused on questions that sit within their own discipline, such as nutrition, genetics, reproduction, and physiology.

It has been 15 years since we (together with Jeff Rushen and Anne Marie de Passillé) wrote a book entitled *The Welfare of Cattle* (Rushen et al. 2008), which is a predecessor to this book. Our aims at the time were to (1) review some of the key methods used to assess animal welfare and (2) summarize the literature on a variety of welfare issues in beef and dairy cattle production. Looking back through our first book, there is no doubt that we know more about how to assess the welfare of cattle than we did at that time, especially how to assess the emotional states of cattle where we have seen the greatest progress since 2008 (see review by Ede et al. 2019). Another area where we have made great strides has been in the use of technology to identify animals at risk of poor welfare. In 2008, the notion of being able to track individual animals within a group was something that was discussed but mostly considered too costly to realistically integrate into research, let alone commercial practice. Now we have many examples of technologies that can provide real-time data for the assessment of cattle welfare at the cow level (Halachmi et al. 2019).

Since 2008, we have seen some progress on the issues we focused upon in *The Welfare of Cattle*. For instance, in 2008, the practice of tail docking dairy cattle was common, but now (in part due to the scientific work on the topic; reviewed by Sutherland and Tucker 2011) this practice is banned or at least no longer supported by dairy industries around the world. Similarly, the practice of disbudding dairy calves without pain mitigation was common, but research has helped identify effective multi-modal pain treatment (Herskin and Nielsen 2018). As a result, the use of pain relief for disbudding is now a requirement in many EU countries, including Sweden, Denmark, and The Netherlands. Industry-led initiatives to mitigate the effects of routine disbudding have also progressed, with many jurisdictions requiring both intraoperative and postoperative pain control for this procedure. For instance, the most recent Canadian Code for the Care and Handling of Dairy Cattle now states that horn bud removal must take place within the first 2 months of life and that "*When removing buds or horns, local anesthesia and systemic analgesia must be provided*" (DFC-NFACC 2023). In the beef industry, there is growing acceptance that this procedure is painful, and pain mitigation for disbudding is becoming more common. For example, The Canadian Code of Practice for the Care and Handling of Beef Cattle now requires at least some form of pain control (CCA-NFACC 2014), and more generally, we see a move toward the use of polled genetics, eliminating the need to disbud (see review by Prayaga 2007). A 2020 survey of the United States cow–calf sector reported that only 7.8% of calves were born with horns, verifying the widespread adoption of polled genetics (USDA 2020). Unfortunately, less progress has been made on pain mitigation for other painful procedures such as branding.

Other issues we identified 15 years ago, such as lameness, are still a topic of interest and research. The good news is that there is now widespread agreement in the dairy industry that lameness is a serious welfare concern (Roche et al. 2020), but the bad news is that despite the considerable volume of research on this topic (Oehm et al. 2019), we have seen little evidence of meaningful reductions in lameness prevalence (Griffiths et al. 2018). In 2008, our focus was on lameness in intensively housed dairy systems, but in the intervening years, we have also seen increased interest in lameness in grazing systems (Bran et al. 2018), as well as a worrisome increase in lameness in beef systems, especially in feedlots (Davis-Unger et al. 2019). The etiology of lameness varies greatly between and within these systems, and we suspect that 15 years hence, lameness will still be considered a difficult puzzle. That said, we are hopeful that the increase in longitudinal studies within both beef (Marti et al. 2021) and dairy (Randall et al. 2015) farming systems, documenting the development and resolution of cases over time, together with better detection methods (including the development of artificial intelligence–based measures that allow tracking of within-cow changes; Qiao et al. 2021), will lead to improved methods of preventing and treating lameness on farms.

Some issues that we only touched upon in 2008 have now become much more prominent and are becoming the focus of considerable research. A key example is the issue of early cow–calf separation (typically followed by individual rearing of calves) as practiced on many dairy farms. When we published *The Welfare of Cattle*, we recognized that early separation could provide some benefits in terms of reducing the emotional response to separation when this eventually occurs (Lidfors 1996), but little literature was available to address other types of benefit and harm associated with this procedure (see Beaver et al. 2019; Meagher et al. 2019). In recent years, we have seen an explosion of research interest in this topic (Johnsen et al. 2016), including work on how commercial farmers have found practical methods of making prolonged cow–calf contact systems work on their farms (i.e., Mutua and Haskell 2022).

On the related issue of how best to care for the newly separated calf, we have seen much progress, including feeding more milk (feeding approximately 10% of body weight per day was once common, but now many farms feed calves twice as much) and social housing (Khan et al. 2011; Costa et al. 2016). However, essentially all such developments have focused on the replacement heifer, the young female calves reared as eventual replacements for milking cows on the farm. The unpleasant reality is that many dairy production systems treat any remaining calves (typically all males and any females not needed as replacements) as surplus (Creutzinger et al. 2021). The fate of these surplus calves has received little research, and we see little evidence of meaningful improvement in the lives of these animals. That said, there is reason for hope, not least because technology that was still embryonic 15 years ago is now creating opportunity for change. Specifically, the widespread adoption of sexed semen on dairy farms allows farmers to deliberately select the genetics of their replacement heifers, which then allows remaining cows to be bred to beef sires (Balzani et al. 2021). The resulting hybrid calves are more valuable to the beef industry (Poock and Beckett 2022), and thus more likely to avoid the fate of early slaughter as bobby calves (Cave et al. 2005), or the rearing condition associated with milk-fed veal production (Renaud et al. 2018). Future research is required to better understand the welfare conditions for these hybrid calves reared for beef, including related issues such as how best to humanely castrate male calves on dairy farms. New work is also required to better understand the implications of this change on beef systems (Bolton and von Keyserlingk 2021), especially beef farms specializing in cow–calf rearing that may be threatened by these changes in the dairy sector.

### 1.1.1 Understanding the Human Dimensions of Cattle Welfare

We began this chapter by citing two key books that helped spur public interest in the welfare of farm animals and to some extent created the social mandate for later developments in the field of animal welfare science. However, when we published *The Welfare of Cattle*, there was little scholarship examining the human dimensions of cattle welfare, including public and farmer attitudes, and the types of interventions most likely to engage farmers and lead to meaningful changes on farms. In the years following the publication of the *Welfare of Cattle*, research in this area has expanded rapidly, and we are likely to see further important developments in the years to come.

Voices from within and outside animal agriculture calling for improvements in animal welfare continue to grow (Shields et al. 2017). Some working within agriculture may be tempted to dismiss the views of outsiders who are ignorant of the specifics of animal care practices, but in our view, dismissing these voices is a mistake for at least two reasons (Weary and von Keyserlingk 2017). One is that knowledge of farm practices is a poor predictor of public acceptance, as learning more about agricultural practices can make people more critical rather than less (Clark et al. 2016, 2017). More importantly, farm practices are increasingly viewed as operating under a social license (Rollin 2011). According to this view, farms that fail to adopt practices consistent with widely held (and sometimes changing) public views may lose their ability to operate. This can be seen in the case of public-led initiatives for changes in agriculture practice, such as the *End the Cage Age* European

Citizens Initiative (described by CWF 2022). Launched in 2018, this initiative resulted in 1.6 million signatures and culminated in 2021 in a decision by the European Commission to revise the current European Union legislation. The proposed legislation review was accompanied by a commitment to phase out the use of cages for all farmed animals, including calves, across Europe by 2027. Moreover, this initiative also came with an assurance that all imported farm animal products sold within the European Union would comply with the cage-free standard.

Such changes, driven by actors distant from the day-to-day care for farm animals, place increasing pressure upon the cattle industries to address contentious practices. The speed with which news travels, especially bad news via social media, means that increasingly it is no longer if the public will hear about contentious animal care practices, but rather when this will occur. Undercover videos showing poor animal handling, often interspersed with routine but unpleasant husbandry practices (such as cow–calf separation), are likely to continue to emerge. In response to such challenges, stakeholders in the supply chains (including processors and retailers) are increasingly demanding transparency (Bateman and Bonanni 2019) and calling for third-party audit programs to protect farm animals (and reduce reputational risk for the buyers) (Vizzier Thaxton et al. 2016).

Understanding the barriers to adoption of proven practices is also increasingly recognized as key to improving welfare on beef and dairy farms (Liu et al. 2019; Balzani and Hanlon 2020). For example, as we indicated above, there is a plethora of evidence indicating that dairy calves should be fed at least 20% body weight (BW) equivalent of milk during the milk feeding period (see review by Khan et al. 2011), but in the most recent US survey of dairy calf management procedures, over 50% of producers reported feeding just 10% BW equivalent of milk (3.8–4.7 L/d of whole milk or milk replacer; USDA 2016). The question that must be asked is why some farmers are quick to adopt improved practices while others do not.

At the most basic level, the day-to-day behaviors of individual farmers caring for cattle have a profound influence on the welfare of these animals (Adler et al. 2019). Even two decades ago, the importance of good handling practices on fearfulness (and even measures of production) in farm animals was well known (e.g., Hemsworth et al. 2000), and developments in the years that have followed continue to reinforce the importance of issues associated with human handling (Ceballos et al. 2018). One reason for this is that, as described above, instances of mistreatment by farmers are now commonly documented in video exposés. More work is now required to understand the situations in which mishandling is most likely, for example, assessing the role of worker training and workplace stress (Ramos et al. 2021).

The urgent need for such work is likely to increase due to the growth of farm size (and hence, also the number of employees working on farms) and perhaps also the increased use of labor-saving technologies such as automated milking systems (AMS). Such automation can reduce the need for some negative interactions (as can occur when moving cows to and from the milking parlor), but automation can also increase the requirement for workers with more specialized skills and may increase feelings of frustration for both workers and cows that fail to interact successfully with this technology. For instance, the adoption of AMS shifts interactions at milking from animal-human to animal-technology (Jacobs and Siegford 2012) and thus creates new challenges. On average 4% of cows fail to voluntarily enter the AMS unit, and these cows must instead be "fetched" by farm workers (Tse et al. 2018). The fetching process requires that farm workers find individual cows that failed to enter the AMS unit and then manually move them to the AMS. This extra level of intervention adds work (and likely frustration) for the farmer and can be a stressor for the individual cow as well as the group (e.g., by disrupting the flow of other cows that are trying to enter the AMS). For a single group of 60 cows (a typical number serviced by a single AMS), a 4% fetch rate equates to 876 fetch events per year, each adding stress and workload to the farm worker, to the individual cow and to the entire herd. This fetching can create unrest within the pen and can result in negative interactions between the stockperson, the cows, and the technology.

We believe that the human–animal–technology interface in cattle farming systems will promote efficiencies but also require that farmers develop systems to help ensure that these interactions are positive. To some extent, this may require that those working with cattle become more knowledgeable about how to consistently and positively train cattle, ensuring that animals are less fearful and likely improving the quality of life for cows and farmers (García Pinillos et al. 2016) (see Chap. 9 Waiblinger and Lurzel). The zoo animal literature provides some good examples of how to work positively and cooperatively with animals (Ward and Melfi 2013), and we see much opportunity for including such approaches when working with dairy and beef cattle in the years to come.

## 1.2 What Is Animal Welfare?

In 2008, we closely followed the three-sphere conception developed by Fraser et al. (1997), considering the animal's feelings (e.g., measures of pain and other affective states), biological functioning (e.g., measures of health, injury and perhaps production), and natural living (e.g., ability to express motivated natural behaviors and have access to some elements of natural environments such as pasture). Different people have different values, and this can correspond to different weightings placed on each sphere. For instance, some farmers may especially value good production, and some veterinarians may place particular emphasis on good health. Others (including some animal welfare scientists) may be most interested in affective states and thus focus especially on unpleasant or positive experiences. The three-sphere conception serves to remind us to look for blind spots. For example, when considering the housing or management systems, it is rarely adequate to just consider how this affects animal health (or either of the other components); instead, we should consider the effects in terms of all three aspects as this can allow us to identify multiple (and sometimes conflicting) effects on welfare.

Much of the recent research on cattle welfare has focused especially on concerns and measures associated with biological functioning. In contrast, a recent review by Laura Whalin and us (2021) highlighted the disconnect between how calves behave when allowed to roam on pasture compared to when housed intensively and suggested that a better understanding of how cows and calves behave in naturalistic settings can help us refine methods in more intensive housing systems, including providing opportunities for the calves to express some highly motivated natural behaviors, such as drinking milk from a teat rather than a bucket.

In addition to changes in the types of scientific approaches used, and the variety of issues considered important, our thinking on the broader issue of how animal welfare should be perceived has changed since the publication of *The Welfare of Cattle*. Animal welfare science is socially mandated; that is to say, it was created to address societal concerns about the appropriate care of animals. The three-sphere framework was intended to reflect the concerns identified from an *"informal content analysis of samples of script, combined with years of personal involvement in discussions of animal welfare"* (Fraser 2008, p. 71). We now know more about the range of concerns that people have about animal care. Many of these concerns can still be positioned within the three-sphere conception, but areas of tension have emerged.

One particularly interesting idea that has emerged from qualitative studies, examining people's views on animal welfare, is that farmers and others responsible for keeping animals have a positive duty of care which, for some at least, appears to be independent of the effects of care that can be assessed using the three-sphere conception. Specifically, when asked to comment on animal management practices, some members of the public argue that farmers have a moral responsibility to provide protection to their animals (Cardoso et al. 2018). For example, Ventura et al. (2016) found that people often called for *"…compassionate attention at the level of the individual animal, gentle handling techniques, and consistent and predictable management"* and even *"human kindness"* and *"love"* (p. 8). Of course, people may believe that well-intentioned care will translate to direct positive outcomes that fit within the three-sphere conception, but some at least may believe that well-intended care is important even if it fails to translate into improvements under the three spheres (Weary and Robbins 2019). This result suggests that future researchers may wish to directly assess and document the care farmers provide, and their intentions and motivations, as well as the outcomes of this care for the animals.

The developing literature on animal agency also challenges the boundaries of the three-sphere conception. The ability to act in ways that we want, to learn about the world around us, and to make informed choices are all crucially important to well-being in humans (Higgins 2012), and a rapidly developing literature also suggests that these aspects of agency are important to animals (Franks and Higgins 2012; Špinka 2019). Franks and Higgins (2012) used the term "effectiveness" and argued that animals could experience high welfare when different elements of effectiveness work together including value effectiveness (animals are successful in having desired results), truth effectiveness (animals are able to establish what is real; i.e., learn about the world around them), and control effectiveness (i.e., animals are able to use this information to better manage their behavior and experiences). We suggest that future research on cattle welfare focus more specifically on these different dimensions of agency, both as inherent welfare concerns and as factors that may contribute to improvements under the three-sphere conception (e.g., by reducing feelings of frustration in animals).

Finally, until now (and following the tone we set in *The Welfare of Cattle*) we have been rather egalitarian in our approach to the welfare concerns discussed, urging readers to consider a diversity of issues rather than simply focus on one welfare dimension or type of concern. The merit of this approach is that it encourages a breadth of consideration, reducing the risk of simply focusing on one issue due to our preexisting biases or lack of imagination. However, the reality is that a huge diversity of welfare concerns exists, far more than we can realistically study and address. Thus, we need to also discuss how issues should be prioritized, initially for study and ultimately for resolution on cattle farms. Although researchers may be tempted to focus on issues that are theoretically or methodologically most challenging (such as developing novel methods for assessing affective states in animals) or on responses and issues that are most pleasant to study (like play behaviors and positive affective states), we argue that our primary focus must remain on those issues most likely to result in true suffering, and those measures that most clearly indicate this state. Suffering is perhaps most likely when animals are least able to express agency and when a sequence of negative experiences occur in combination resulting in negative expectations regarding future experiences and may be indicated by measures of low mood and reduced cognitive functioning (Weary 2019). Lecorps et al. (2021) describe several examples in cattle production systems that likely qualify, and we call for more research that more specifically targets issues and measures that relate directly to animal suffering in the years to come.

Society will continue to need research on how cattle-rearing practices affect their welfare, how problematic practices can be improved, and how these changes are best implemented on farms. The remaining 15 chapters, written by animal welfare scientists from around the globe, highlight many important developments that have occurred in research on the welfare of cattle. The first two chapters review some of the advances made on the key behavioral, physiological, and health measures used to assess animal welfare. The second section consists of four chapters covering different aspects of dairy and beef cattle production, both extensive and intensive. This is followed by five chapters that each take a deeper look at the latest science in specific areas of concern, such as slaughter, the human animal bond, parturition and early life of the calf, the role of technology (precision livestock farming), and lastly, the role of genetics. The book ends with a series of case studies that bring to light some of the complexities of working in specific cattle system, reinforcing the important role of contextual factors that must be considered when identifying animal welfare solutions. For instance, small-scale beef and dairy farms in sub-Saharan Africa play a key role in supporting livelihoods in communities, but farmers working on these farms also face numerous challenges given low incomes and lack of resources (see Chap. 15). China provides a different example where dairy farming has undergone a rapid transformation from an industry made up of many small-scale farms to one that now embraces large-scale farms, making this country one of the top three milk-producing countries in the world (Fan et al. 2018; see Chap. 16). However, with rapid change come challenges. As such, the current volume helps identify ways of improving the lives of dairy and beef cattle that are culturally and regionally relevant and thus likely to be useful to researchers, those involved in cattle production, and more generally those in our society interested in tackling these important issues.

## References

Adler F, Christley R, Campe A (2019) Invited review: examining farmers' personalities and attitudes as possible risk factors for dairy cattle health, welfare, productivity, and farm management: a systematic scoping review. J Dairy Sci 102:3805–3824. https://doi.org/10.3168/jds.2018-15037

Appleby M (2004) Science is not enough: how do we increase implementation? Anim Welfare 13:159–162

Balzani A, Hanlon A (2020) Factors that influence farmers' views on farm animal welfare: a semi-systematic review and thematic analysis. Animals 10(9):1524. https://doi.org/10.3390/ani10091524

Balzani A, Aparacida Vaz do Amaral C, Hanlon A (2021) A perspective on the use of sexed semen to reduce the number of surplus male dairy calves in Ireland: a pilot study. Front Vet Sci 7:623128. https://doi.org/10.3389/fvets.2020.623128

Bateman A, Bonanni L (2019) What supply chain transparency really means. Harv Bus Rev. https://hbr.org/2019/08/what-supply-chain-transparency-really-means. Accessed 1 March

Beaver A, Meagher RK, von Keyserlingk MAG, Weary DM (2019) Invited review: a systematic review of the effects of early separation on dairy cow and calf health. J Dairy Sci 102:5784–5810. https://doi.org/10.3168/jds.2018-15603

Bolton SE, von Keyserlingk MAG (2021) The dispensable surplus dairy calf: is this issue a "wicked problem" and where do we go from here? Front Vet Sci 8:660934. https://doi.org/10.3389/fvets.2021.660934

Bran JA, Daros RR, von Keyserlingk MAG, LeBlanc SJ, Hötzel MJ (2018) Cow- and herd-level factors associated with lameness in small-scale grazing dairy herds in Brazil. Prev Vet Med 151:79–86. https://doi.org/10.1016/j.prevetmed.2018.01.006

Canadian Cattleman's Association (CCA) National Farm Animal Care Council (NFACC) (2014) Code of practice for the care and handling of beef cattle. https://www.nfacc.ca/pdfs/codes/beef_code_of_practice.pdf. Accessed 28 Feb 2022

Cardoso CS, von Keyserlingk MAG, Hötzel MJ, Robbins J, Weary DM (2018) Hot and bothered: public attitudes towards heat stress and outdoor access for dairy cows. PLoS ONE 13(10):e0205352. https://doi.org/10.1371/journal.pone.0205352

Cave JG, Callinan A, Woonton WK (2005) Mortalities in bobby calves associated with long distance transport. Aust Vet J 83:82–84. https://doi.org/10.1111/j.1751-0813.2005.tb12203.x

Ceballos MC, Sant'Anna AC, Boivin X, de Oliveira CF, de L. Carvalhal MV, Paranhos da Costa MJR (2018) Impact of good practices of handling training on beef cattle welfare and stockpeople attitudes and behaviors. Livest Sci 216:24–31. https://doi.org/10.1016/j.livsci.2018.06.019

Clark B, Stewart GB, Panzone LA, Kyriazakis I, Frewer LJ (2016) A systematic review of public attitudes, perceptions and behaviours towards production diseases associated with farm animal welfare. J Agric Environ Ethics 293(29):455–478. https://doi.org/10.1007/S10806-016-9615-X

Clark B, Stewart GB, Panzone LA, Kyriazakis I, Frewer LJ (2017) Citizens, consumers and farm animal welfare: a meta-analysis of willingness-to-pay studies. Food Policy 68:112–127. https://doi.org/10.1016/J.FOODPOL.2017.01.006

Compassion for World Farming (CWF) (2022) European Citizens Initiative (ECI) (2018) *'End the Cage Age'* see https://www.endthecageage.eu. Accessed 1 Mar 2022

Costa JHC, von Keyserlingk MAG, Weary DM (2016) Invited review: effects of group housing of dairy calves on behavior, cognition, performance, and health. J Dairy Sci:1–15. https://doi.org/10.3168/jds.2015-10144

Creutzinger K, Pempek J, Habing G, Proudfoot K, Locke S, Wilson D, Renaud D (2021) Perspectives on the management of surplus dairy calves in the United States and Canada. Front Vet Sci 8:661453. https://doi.org/10.3389/fvets.2021.661453

Dairy Farmers of Canada (DFC) National Farm Animal Care Council (NFACC) (2023) Code of practice for the care and handling of dairy cattle. https://www.nfacc.ca/codes-of-practice/dairy-cattle/code. Accessed 1 July, 2023

Davis-Unger J, Schwartzkopf-Genswein KSG, Pajor EA, Hendrick S, Marti S, Dorin C, Orsel K (2019) Prevalence and lameness-associated risk factors in Alberta feedlot cattle. Trans Anim Sci 3:595–606. https://doi.org/10.1093/tas/txz008

Ede T, Lecorps B, von Keyserlingk MAG, Weary DM (2019) Symposium review: scientific assessment of affective states in dairy cattle. J Dairy Sci 102:10677–10694. https://doi.org/10.3168/jds.2019-16325

Fan F, Chang J, Ren Y, Wu X, Du Y, Xu R, Liu D, Chang SX, Meyerson LA, Peng C, Ge Y (2018) Recoupling industrial dairy feedlots and industrial farmlands mitigates the environmental impacts of milk production in China. Environ Sci Tech 52:3917–3925. https://doi.org/10.1021/acs.est.7b04829

Franks B, Higgins ET (2012) Effectiveness in human and other animals. In: Olson KM, Zanna MP (eds) Advances in experimental social psychology, vol 45. Academic Press, New York, pp 285–346

Fraser D (2008) Understanding animal welfare: the science in its cultural context. Universities Federation of Animal Welfare (UFAW) Animal Welfare series, p 71

Fraser D, Weary DM, Pajor EA, Milligan BN (1997) A scientific conception of animal welfare that reflects ethical concerns. Anim Welfare 6:187–205

García Pinillos R, Appleby MC, Manteca X, Scott-Park F, Smith C, Velarde A (2016) One welfare – a platform for improving human and animal welfare. Vet Rec 179:412–413. https://doi.org/10.1136/vr.i5470

Griffiths BE, Grove White D, Oikonomou G (2018) A cross-sectional study into the prevalence of dairy cattle lameness and associated herd-level risk factors in England and Wales. Front Vet Sci 5:65. https://doi.org/10.3389/fvets.2018.00065

Halachmi I, Guarina M, Bewley J, Pastell M (2019) Smart animal agriculture: application of real-time sensors to improve animal well-being and production. Ann Rev Anim BioSci 7:403–425. https://doi.org/10.1146/annurev-animal-020518-114851

Harrison R (1964) Animal machines. Ballantine Books, London

Hemsworth PH, Coleman GJ, Barnett JL, Borg S (2000) Relationships between human-animal interactions and productivity of commercial dairy cows. J Anim Sci 78:2821–2831. https://doi.org/10.2527/2000.78112821x

Herskin MS, Nielsen BH (2018) Welfare effects of the use of a combination of local anesthesia and NSAID for disbudding analgesia in dairy calves — reviewed across different welfare concerns. Front Vet Sci 5:00117. https://doi.org/10.3389/fvets.2018.00117

Higgins ET (2012) Beyond pleasure and pain: how motivation works. Oxford University Press, New York

Jacobs JA, Siegford JM (2012) Invited review: the impact of automatic milking systems on dairy cow management, behavior, health, and welfare. J Dairy Sci 95:2227–2247. https://doi.org/10.3168/jds.2011-4943

Johnsen JF, Zipp KA, Kälber T, de Passillé AM, Knierim U, Barth K, Mejdell CM (2016) Is rearing calves with the dam a feasible option for dairy farms? — Current and future research. Appl Anim Behav Sci 181:1–11. https://doi.org/10.1016/j.applanim.2015.11.011

Khan MA, Weary DM, von Keyserlingk MAG (2011) Invited review: effects of milk ration on solid feed intake, weaning, and performance in dairy heifers. J Dairy Sci 94:1071–1081. https://doi.org/10.3168/jds.2010-3733

Lecorps B, Weary DM, von Keyserlingk MAG (2021) Negative expectations and vulnerability to stressors in animals. Neurosci Biobehav Rev 130:240–251. https://doi.org/10.1016/j.neubiorev.2021.08.025

Lidfors LM (1996) Behavioural effects of separating the dairy calf immediately or 4 days post-partum. Appl Anim Behav Sci 49:269–283. https://doi.org/10.1016/0168-1591(96)01053-2
Liu J, Toma L, Barnes AP, Stott A (2019) Farmers' uptake of animal health and welfare technological innovations. Implications for animal health policies. Front Vet Sci 6:410. https://doi.org/10.3389/fvets.2019.00410
Lund V, Coleman G, Gunnarsson S, Appleby MC, Karkinen K (2006) Animal welfare science—working at the interface between the natural and social sciences. Appl Anim Behav Sci 97:37–49. https://doi.org/10.1016/j.applanim.2005.11.017
Marti S, Jelinski MD, Janzen ED, Jelinski MJ, Dorin CL, Orsel K, Pajor EA, Shearer J, Millman ST, Schwartzkopf-Genswein KA (2021) A prospective longitudinal study of risk factors associated with cattle lameness in southern Alberta feedlots. Can J Anim Sci 101:647–654. https://doi.org/10.1139/cjas-2020-0128
Meagher RK, Beaver A, Weary DM, von Keyserlingk MAG (2019) Invited review: a systematic review of the effects of prolonged cow–calf contact on behavior, welfare, and productivity. Dairy Sci 102:5765–5783. https://doi.org/10.3168/jds.2018-16021
Mutua EK, Haskell MJ (2022) Factors contributing to milk yield variation among cows in a cow–calf contact system in early lactation. JDS Commun 3(1):55–58. https://doi.org/10.3168/jdsc.2021-0143
Oehm AW, Knubben-Schweizer G, Rieger A, Stoll A, Hartnack S (2019) A systematic review and meta-analyses of risk factors associated with lameness in dairy cows. BMC Vet Res 15:346. https://doi.org/10.1186/s12917-019-2095-2
Poock SE, Beckett JL (2022) Changing demographics of the commercial dairy calf industry: why use beef on dairy? Vet Clin Food Anim 38:1–15. https://doi.org/10.1016/j.cvfa.2021.11.001
Prayaga KC (2007) Genetic options to replace dehorning in beef cattle—a review. Aust J Agric Res (Crop Pasture Sci) 58:1–8. https://doi.org/10.1071/AR06044
Qiao Y, Kong H, Clark C, Lomax S, Su D, Eiffert S, Sukkarieh S (2021) Intelligent perception-based cattle lameness detection and behaviour recognition: a review. Animals 11:3033. https://doi.org/10.3390/ani11113033
Ramos AK, McGinley M, Carlo G (2021) The relations of workplace safety, perceived occupational stress, and adjustment among Latino/a immigrant cattle feedyard workers in the United States. Safety Sci 139:105262. https://doi.org/10.1016/j.ssci.2021.105262
Randall LV, Green MJ, Chagunda MGG, Mason C, Archer SC, Green LE, Huxley JN (2015) Low body condition predisposes cattle to lameness: an 8-year study of one dairy herd. J Dairy Sci 98:3766–3777. https://doi.org/10.3168/jds.2014-8863
Renaud DL, Kelton DF, LeBlanc SJ, Haley DB, Duffield DF (2018) Calf management risk factors on dairy farms associated with male calf mortality on veal farms. J Dairy Sci 101:1785–1794. https://doi.org/10.3168/jds.2017-13578
Roche SM, Renaud DL, Genore R, Shock DA, Bauman C, Croyle S, Barkema HW, Dubuc J, Keefe GP, Kelton DF (2020) Canadian National Dairy Study: describing Canadian dairy producer practices and perceptions surrounding cull cow management. J Dairy Sci 103:3414–3421. https://doi.org/10.3168/jds.2019-17390
Rollin BE (2011) Animal right as a mainstream phenomenon. Animals 1:102–115. https://doi.org/10.3390/ani1010102
Rushen J, de Passillé AM, von Keyserlingk MAG, Weary DM (2008) The welfare of cattle. Springer, Dordrecht
Shields S, Shapiro P, Rowan A (2017) Decade of progress toward ending the intensive confinement of farm animals in the United States. Animals 7(5):40. https://doi.org/10.3390/ani7050040
Špinka M (2019) Animal agency, animal awareness and animal welfare. Anim Welfare 28(1):11–20. https://doi.org/10.7120/09627286.28.1.011
Sutherland MA, Tucker CB (2011) The long and short of it: a review of tail docking in farm animals. Appl Anim Behav Sci 135:179–191. https://doi.org/10.1016/j.applanim.2011.10.015
Tse C, Barkema HW, DeVries TJ, Rushen J, Pajor EA (2018) Impact of automatic milking systems on dairy cattle producers' reports of milking labour management, milk production and milk quality. Animal 12:2649–2656. https://doi.org/10.1017/S1751731118000654

United States Department of Agriculture (USDA) (2016) Dairy 2014, Dairy Cattle Management Practices in the United States, 2014 USDA–APHIS–VS–CEAH–NAHMS. Fort Collins, CO #692.0216

United States Department of Agriculture (USDA) (2020) Beef 2017, Beef Cow-calf Management Practices in the United States, 2017, report 1. USDA-APHIS-VS-CEAH-NAHMS. Fort Collins, CO #.782.0520

Upton Sinclair (1906) The Jungle. Doubleday, Page & Co.

Ventura BA, von Keyserlingk MAG, Wittman H, Weary DM (2016) What difference does a visit make? Changes in animal welfare perceptions after interested citizens tour a dairy farm. PLoS One 11(5):e0154733. https://doi.org/10.1371/journal.pone.0154733

Vizzier Thaxton Y, Christensen KA, Mench JA, Rumley ER, Daugherty C, Feinberg B, Parker M, Siegel P, Scanes CG (2016) Symposium: animal welfare challenges for today and tomorrow. Poultry Sci 95:2198–2207. https://doi.org/10.3382/ps/pew099

Ward SJ, Melfi V (2013) The implications of husbandry training on zoo animal response rates. Appl Anim Behav Sci 147:179–185. https://doi.org/10.1016/j.applanim.2013.05.008

Weary DM (2019) Suffering, agency and the Bayesian mind. In: McMillan FD (ed) Mental health and well-being in animals. CAB eBooks, pp 1–35. https://doi.org/10.1079/9781786393401.0156

Weary DM, Robbins JA (2019) Understanding the multiple conceptions of animal welfare. Anim Welfare 28:33–40. https://doi.org/10.7120/09627286.28.1.033

Weary DM, von Keyserlingk MAG (2017) Invited review: public concerns about dairy cow welfare — how should the industry respond? Anim Prod Sci 57:1201–1209. https://doi.org/10.1071/AN16680

Whalin L, Weary DM, von Keyserlingk MAG (2021) Understanding behavioural development of calves in natural settings to inform calf management. Animals 11(8):2446. https://doi.org/10.3390/ani11082446

# Using Behaviour to Understand and Assess Welfare in Cattle

# 2

Marie J. Haskell and Fritha M. Langford

## Contents



**Abstract**

Animals show behavioural responses to their environments based on the integration of their underlying emotional, cognitive and motivational processes, with

M. J. Haskell (✉)
SRUC, Edinburgh, UK
e-mail: marie.haskell@sruc.ac.uk

F. M. Langford
SRUC, Edinburgh, UK

University of Newcastle, Newcastle-upon-Tyne, UK



M. Haskell (ed.), *Cattle Welfare in Dairy and Beef Systems*, Animal Welfare 23,
https://doi.org/10.1007/978-3-031-21020-4_2

the ultimate goal of fulfilling needs that promote survival. Attempts to improve welfare involve understanding these needs and the factors affecting them. Assessing behaviour is one of the major methods that can be used to improve welfare. This chapter discusses the different methods to use animal behaviour to investigate welfare in cattle. The two main methods are direct observations of spontaneous behaviour and behavioural tests. Observations of spontaneous behaviour capture the undisturbed behaviour of cattle in their 'home' surroundings. This can be used to document the natural or 'baseline' patterns of behaviour. Behavioural tests typically require placing an animal in an experimental pen with stimuli or resources that we want to understand the animal's response to. Behavioural tests allow us to investigate animal preference, motivation and emotional responses in more depth. Welfare assessment protocols often include behavioural indicators of positive and negative welfare but are typically observations of ongoing behaviour rather than behavioural tests. Finally, this chapter will examine what behavioural assessments cannot tell us and where further developments in this science are necessary.

**Keywords**

Animal behaviour · Animal welfare · Motivation · Behavioural test · Affective state

## 2.1 Introduction

Cattle are kept in many different management systems across the world, from indoor housing systems to outdoor feedlots and pasture. The sizes of the herds and scale of the farms also differ, from small-holdings or landless farmers with one or two cattle in Asia and Africa, to stations in Australia and feedlots in America with thousands of animals. Welfare issues have been identified in many of these systems and can arise from poor handling, housing or management or a mismatch between the genotype of the animal and its current environment. There are a number of stakeholders that have an interest in improving the welfare of cattle in these systems. Consumers are becoming increasingly interested in the welfare status of the animals used to produce the food that they buy (Ventura et al. 2013; Cornish et al. 2016). Because of this, private companies and governmental groups involved in selling food products derived from cattle are particularly interested in welfare assessment protocols as they give them a means of providing evidence to their customers on welfare standards on farms (e.g., Veissier et al. 2008). Good farmers view high standards of animal welfare as a cornerstone of their enterprises, but they may need guidance on how it can be achieved. Clearly, the animals themselves have a stake in their own welfare. The need to provide information on how to promote and improve animal welfare has necessitated research into the welfare of cattle, which has been funded by governmental, retailer, levy boards and non-governmental bodies interested in improving welfare. This research has allowed us to learn a lot about the resources, social and environmental variables that cattle need to achieve good standards of welfare. The role of researchers and those monitoring and managing cattle welfare in commercial settings is to develop and use the appropriate methods to achieve these different aims.

To assess animal welfare in the most comprehensive way, aspects of health, physiology and behaviour should all be measured. In this chapter, we discuss the ways that welfare can be assessed using behaviour while Chap. 3 will consider health and physiology. There are a number of ways that behaviour can be used to assess welfare. The evaluation of behaviour can be used in research settings to investigate the biological background underlying the welfare issue in question and work towards a solution. A great deal of work has been done to understand how to deliver better standards of welfare by investigating what cattle want or what resources should be provided, allowing for basic functioning such as lying, feeding, drinking and thermal comfort. The effect of management strategies on welfare, including the quality of human–animal interactions and the application of painful procedures, is another important area of research. The extent of disease and pain can be assessed using behaviour, for example, at calving and during disbudding and dehorning of calves. Recently, there has been increasing interest in how to recognise and provide for positive experience in a move to create 'positive welfare' or a life that includes positive experiences, rather than just the lack of negative experiences.

Behaviour can also be used as one indicator of welfare in a welfare assessment protocol. These protocols aim to quantify the welfare status of groups of animals in a short space of time (i.e., capturing a 'snapshot' of the welfare status of the animals). This assessment can be done using audit-type protocols, as part of a government or veterinary inspection or as part of a certification scheme that provides consumers with an indication of the level of welfare of the animals on that farm. Welfare indicators can also be used as a day-to-day measure by cattle owners to regularly assess individual animals or groups of animals to check their welfare status. Scoring cattle for their level of lameness is an example of a behavioural indicator of welfare that is widely used.

An in-depth analysis of the welfare issues affecting cattle in production systems across the world is provided in other chapters in this book. The aim of this chapter is to review the use of behavioural methods of assessing welfare in cattle in experimental and welfare assessment settings. Examples from published studies will be used to illustrate the use of these methods and the pros and cons of the methods assessed. The possibilities for future advances in methodology will also be discussed.

## 2.2 What Is the Relationship Between Behaviour and Animal Welfare?

Welfare is concerned with the animal's state in terms of its emotional and physical well-being (Dawkins 2008). The science of animal welfare involves testing hypotheses about what we can do about the environment and management of cattle to reduce harms and promote positive effects. This is where observation of behaviour plays a major role. Evolution has produced animals that have goals that promote successful survival and reproduction (such as caring for offspring, eating food that provides sufficient nutrients or seeking shade in hot sunny weather). The performance of a particular pattern of behaviour in a particular environment and the elements of that environment which the animal chooses to avoid or interact with are an observable outward expression of the animal's motivation to achieve these goals. This behavioural outcome is the result of a conscious or sub-conscious integration of all influencing motivational, cognitive and emotional factors. The process of domestication and selective breeding has changed the genetic make-up of the livestock seen today. Although it may have altered some aspects of the goals or the intensity of the behaviours expressed (e.g., lower levels of maternal behaviour shown in dairy cattle), it has not altered the fact that behaviour represents an external expression of the animal's needs. The motivational factors include internal factors (e.g., neural and hormonal factors) and external factors (e.g., environmental stimuli), which are influenced by what the animal has learned about the positive or negative consequences of its interaction with these stimuli in the past. Thus, a cow deciding to move to a feed-trough and eat may have done so in response to physiological factors that indicate low blood glucose and other metabolic demands, as well as her knowledge about the presence of feed in the trough and social characteristics of the other cows present there. Thus, behaviour is an integrated output that can tell us about animal needs.

Much of the investigative work done to date has been aimed at understanding and preventing states of poor welfare, such as pain, stress and discomfort. In recent years, animal welfare science has also encompassed efforts aimed at trying to understand positive welfare states in cattle and other livestock, to find ways to allow animals to have a 'good life' (FAWC 2009). While measuring physiology can tell us that the animal has responded to the stimuli, it is only by observing the quality of the behavioural response that we can understand the valence (positive or negative quality) of the experience of interacting with that stimulus. For example, when we measure cortisol concentrations in saliva, we can gain an insight into the changes in activity in the hypothalamic–pituitary–adrenal axis, but only by using this measure in conjunction with behavioural signs (such as approach or avoidance behaviour or expressions of anxiety) can we tell whether the increase in cortisol has been brought about by a negative or positive experience. The way the behaviour is performed (the intensity or the behavioural expression) or whether the action is repeated (the performance of the behaviour has been reinforced) can also tell us about the quality of the emotional experience.

While this all makes sense at a conceptual level, assessing needs and feelings is not straightforward. The application of appropriate methods and careful interpretation are required to understand these behavioural outputs, and there are some limitations to the methodologies to be aware of. This will be discussed in the following sections.

## 2.3 How Do We Decide What Type of Behavioural Assessment to Use?

There are different methods available to assess and record the behaviour of an individual animal or a group of animals. We often divide the observation of behavioural responses into 'spontaneous behaviour' and 'behavioural tests'. In the former, we are often looking at the animal in its 'normal' environment, such as in the natural wild habitat, or in its home pen, barn or field. The animal is free to perform whatever behaviour pattern it chooses within the constraints of the environment. Behavioural tests typically involve observing behaviour in a custom-built test arena or involve an animal interacting with an object or person presented in a standardised manner (such as in a human approach test or novel object test—see below for further details). In terms of time, behavioural tests can be relatively short in duration if they aim to capture a specific response, while observations of spontaneous behaviour may be carried out across many hours or days depending on the situation and the type and frequency of the performance of the behaviour(s) of interest. In observations of spontaneous behaviour, the ethogram (see Box 2.1 for a definition) may be quite comprehensive and take in more of the total behavioural repertoire of the animal, while the ethogram used for a typical behavioural test may just focus on the specific behaviours or responses that are relevant for the test situation.

**Box 2.1 Key Methodological Concepts.**

Definitions of the basic elements of behavioural assessments are defined here. Please see bateson and martin (2021) for more detail.

Behavioural repertoire: The animal's behavioural repertoire is a 'list' of all the possible behaviours that it could perform. It is rare that every possible behaviour that an animal performs is recorded in detail in any single set of observations or in a specific situation. Observational studies are a good way to determine what the typical repertoire consists of for an experimental context and can be used to guide the construction of the ethogram.

Ethogram: An ethogram is a list of behaviour patterns and definition for each behaviour pattern and is typically compiled at the start of any experiment that involves the recording of animal behaviour. An important element in the construction of an ethogram is the consideration of what behaviours are likely to occur in the experimental situation. Typically, an experiment will focus on behaviours that are relevant and important for the hypothesis being tests and in the situation being studied.

Time budget: The term 'time budget' refers to the time that the animal spends in each behaviour in its repertoire across a pre-determined time period (e.g., a day). In animal welfare studies, we are often trying to identify whether a factor has affected the overall time budget or the specific behaviours within it. If we imagine that the performance of each behaviour is the result of a 'decision' that relates the behaviour to the overall set of needs of the animal, the time spent in each behaviour would provide insight into the animals' needs.

Preference test: A preference test is an experimental method in which an animal is presented with two or more options of resources (e.g., food types or lying surfaces) or experiences (e.g., handling quality) in an experimental setting, and it must choose between them. Typically, this involves the animal moving towards the preferred option, but may also involve making an operant response (such as pressing a lever or touch-screen button) to gain access to that option.

Motivational test: The aim of a test of motivation is to determine how much 'value' the subject animal places on access to a particular resource. Typically, the experimenter imposes an increasing 'cost' or 'price' for access to a resource. This involves the animal performing a task that takes time or energy to complete or prevents it performing other competing behaviours. This may include walking increasing distances or pushing against increasing weights on a gate. The point at which the animal gives up performing the task or stops accessing the resource is often used as a measure of the motivational strength or 'value' of the resource to the animal. Inter-observer reliability: This is the extent to which different individual observers agree on the identity of the behaviours being observed. Inter-observer reliability can be assessed by having each observer conducting a trial to observe the same set of animals performing behaviour and using statistical methods to assess the degree of agreement.

Intra-observer reliability: This term refers to the extent to which a single observer shows agreement in identifying behavioural elements compared to what they had identified previously when watching the same recording. Intra-observer reliability is important to assess, as interpretations of behaviour can change over time as the observer becomes more practised at identifying behaviour, or with fatigue or boredom. Intra-observer reliability is typically tested by video-recording a period of behaviour and recoding that section of video on a later date.

Which type of behavioural assessment is used in any research trial or welfare audit depends on the question being asked. If we are interested in understanding how an animal or group of animals behaves in a particular environment, then observing ongoing undisturbed spontaneous behaviour will address that question. If we are more interested in getting a greater understanding of the precise characteristics of the behavioural or emotional response to a stimulus or environment, or of factors affecting that response, a behavioural test is often better suited.

## 2.4 Using Behaviour to Understand and Assess Welfare in Cattle

### 2.4.1 Direct Observations of Spontaneous Behaviour

#### 2.4.1.1 Observations of Behaviour in the Home Environment to Provide Background Information

Direct observations of behaviour take place in the animal's home environment and involve observing and recording the subject animal's (or animals') behaviour in that environment. The study may involve observing a particular behaviour pattern of interest (e.g., grazing behaviour in a natural pasture; e.g., Braghieri et al. 2011), a behaviour at a key period in life (e.g., cow behaviour before calving; e.g., Miedema et al. 2011) or documenting the full range of behaviours performed in a specific context such as a housing or management system (e.g., calves housed in calf hutches; Ugwu et al. 2021). Most studies have a stated aim or hypothesis that leads to a focus on a particular type of behaviour that is relevant to the hypothesis (e.g., social or feeding behaviour), and the ethogram used is limited to these focal behaviours. In some cases, however, the study may be more observational and aim to describe the repertoire of behaviours of animals in a particular situation or context. In a classic study, Hall (1989) studied a group of free-living, relatively unmanaged White Park cattle on the Chillingham Estate in northern England. The aim of the study was to assess the social and maintenance behaviours in this undisturbed group, but the study also documented grazing and lying times across the different seasons. This made the study a valuable foundation source of information on the behaviour of free-living cattle. In a similar wide-ranging study, Kiley-Worthington and de la Plain (1983) dedicated a whole book to detailing the behaviour of beef suckler cattle in herds in southern England with observations lasting 2 years. The observations recorded social and maternal behaviour and communication, thereby providing background information that has been referenced in a number of subsequent studies (e.g., Lidfors et al. 1994; Rørvang et al. 2019).

Observations of spontaneous behaviour of cattle in a particular management context are often a good way to start a programme of work on a welfare issue. As welfare problems may be detected by a deviation away from normal behaviour, we need to know what 'normal' looks like (Nielsen 2020). Spontaneous observations of behaviour can help with this, as records of undisturbed behaviour can help to establish what behaviours are present in the repertoire of cattle in that context and to define aspects such as time budgets and circadian rhythms for that system. There are a number of examples of where this type of background research has been done. For instance, to understand the normal behaviour of beef cattle at pasture as a basis for future studies on grazing cattle, Kilgour et al. (2012) observed six herds of cattle grazing on commercial farms. They established that although they displayed a diverse repertoire of behaviours, the cattle spent 95% of their time performing just two major behaviours (grazing and resting/ruminating). Vitale et al. (1986) used extensive focal observations to characterise the early social interactions between cow and calf in an undisturbed herd of an ancient breed of cattle and showed that calves hide for the first few days of life and then establish diurnal feeding and activity patterns. These in-depth studies of undisturbed behaviour can also be made in more intensive systems to provide background information on behavioural patterns. Lidfors et al. (1994) assessed behaviour at calving and choice of calving site in beef and dairy cattle in a range of systems. The study found that all cows isolated themselves from other cows to some extent prior to calving, but there was high variation between individuals. Ugwu et al. (2021) defined the behavioural repertoire of calves in a hutch rearing system. The observations made in these studies have been used as a kind of 'baseline' for other studies investigating the influence of management or housing interventions on these natural behaviours.

#### 2.4.1.2 Specific Behaviours

There are some behaviour patterns that can be used as clear indicators of an animal's welfare. Some provide information on the animal's emotional experience. The performance of play is an indicator of a positive emotional state, whereas certain vocalisations and the performance of oral stereotypies are indicators of pain or other negative states. Many of these specific behaviours are easily identified visually, which may explain why they have been frequently used in research.

Vocalisations are an important class of specific behaviour because they are seen as an 'honest indicator' of emotional state, particularly of emotional distress (Watts and Stookey 2000). Quantifying vocalisations offers a potential method of assessing how the animal is 'coping' with its environment (Green et al. 2018). The frequency of vocalisations is typically higher in animals experiencing negative states. Valizaheh et al. (2008) found that during the dry-off process, dairy cows increased their vocalisation rate when transferred from the energy-rich lactation diet to a hay-based diet. Similarly, cows on a restricted ration vocalised more than cows fed more of the same ration (Tucker et al. 2009). Grooming is a behaviour that appears to occur when the animal is in a good state of welfare but declines when ill health or stressful events are experienced. Studies by Mandel et al. (2013, 2018) showed that lameness, high heat load and disruption due to artificial insemination procedures reduced the use of a grooming brush. This was particularly evident when the brush was distant from the feeding areas and access presented some sort of 'cost' to the animal.

Cows and calves vocalise more frequently in response to separation than cows and calves kept together (Lidfors 1996). The time the cow and calf spend together after birth, and hence their level of bonding, has been shown to affect their vocal responses. Spectrographic analysis of the quality of the post-separation calls of groups of cows separated from their calves after different periods of bonding showed that specific acoustic characteristics of the calls differed (Weary and Chua 2000). The possibility of gaining more information on emotional state by detailed acoustic analysis of vocalisations is an interesting new avenue of research (e.g., Manteuffel et al. 2004; Green et al. 2018; see Sect. 2.7, Future Directions).

Other specific behaviours have been used as indicators of negative emotional states. In a study on the calving process in dairy cows, Miedema et al. (2011) showed that an increased frequency of tail-raising was a reliable indicator that the birth of the calf was imminent. This particular behaviour is likely to be an indicator of the pain of the parturition process. The occurrence of stereotypic behaviours (abnormal, repetitive behaviours) is used as an indicator of a severe compromise in welfare in cattle, as well as in other species such as pigs and horses. Tongue-rolling and bar-biting are common stereotypies in cattle. In a study in cattle that were housed in tie-stalls in the winter and grazed during the summer, Corazzin et al. (2010) found that fewer cattle performed tongue playing during the summer grazing period than during the winter, indicating that the confinement of the tie-stall had adverse effects on well-being.

Play is a behavioural pattern that is considered to be a good indicator of a positive emotional experience (Boissy et al. 2007) and is performed mostly by young animals. A study by Jensen et al. (1998) described the behaviours shown by calves during locomotor play in detail. This has been the foundation of studies that have shown that play is more frequent in positive situations, such as when calves have high milk allowances and are free from pain (e.g., Rushen and de Passillé 2012; Grössbacher et al. 2020). These findings validate the hypothesis that the occurrence of play can be used as an indicator of a positive state.

#### 2.4.1.3 Direct Observations of Behaviour with 'Treatments'

Welfare improvements for cattle have often involved the addition of resources or improving the quality of structural elements of their environment. Giving the cattle new or improved environmental elements (e.g., different types of bedding or cubicle formations), a resource (e.g., access to shade) or a management intervention (e.g., different social grouping strategies) are typical examples. There are two major ways of running these trials. In one format, different groups of animals are used: some live in the experimental housing with access to the resource and some without. In the other format, animals live in the experimental housing or conditions across periods of many days or weeks, and behaviour in 'treatment' periods (i.e., with access) is compared with 'no treatment' periods without access. The 'use' of the resource is assessed by recording time spent with the resource, interactions with the resource or alterations in time budget associated with use of the resource. As this type of study involves a comparison between behaviour with the resource present and when it is absent, a key element is to establish 'baseline' levels of the key behaviours without the resource or under standard management conditions. An ethogram is used that reflects this focus. This type of study can be used on commercial farms, as well as on experimental research farms, as the modifications can be easy to implement. The distinction between 'spontaneous' observations of behaviour and 'behavioural test' may seem a bit artificial for this type of test regime. However, key assessments are made of the spontaneous behaviour of the animal(s) when the resource is present, and then when it is not. This is unlike the preference test paradigm which is an experimental method in which options being tested are presented in an experimental set-up (see Box 2.1 and Sect. 2.4.2.1).

This format of 'live-in' test has been useful in understanding the preferences and needs of cattle. In a comparison of groups design, a study by Park et al. (2020) assessed the effect of a grooming brush on the behaviour of beef cattle in feedlots. Groups of cattle were assigned to pens with or without a grooming brush. In pens with brushes, cattle performed fewer aggressive and stereotypic behaviours. Brush use was steady and did not decline once the animals have used it. Similarly, Palacio et al. (2015) assessed the effect of the portable shade devices on the use of this shade and on lying and grazing behaviour in groups of cows with and without the shade devices. Cows with shade spent 40% of their time in the shade. These cows also drank less, lay down more and grazed more than the cows in pastures without shade.

Another design that is frequently used allows each test group of animals to experience all options in sequence and compares the behaviour between the periods when the resource is and is not present. Fregonesi et al. (2007) investigated the effect of overstocking on the lying time of dairy cattle. They housed groups of 12 cows in pens with differing numbers of cubicles to create 100, 109, 120, 133 and

150% stocking density. Cows were observed for 1 week in each of the stocking density treatments and then moved back to a 'control' level of 100% stocking density for 1 week between each treatment. The results showed that cows lie down less and compete more for cubicles as stocking density increases. It is preferable to use a study design where each animal experiences each treatment, as it allows individual animals to act as their own control, which accounts for individual differences in the statistical analysis. However, this may not always be possible when working with commercial farms.

This crossover experimental design has also been used to assess the effect of management strategies. Schirmann et al. (2011) assessed the effect of regrouping on the feeding, social, lying and rumination behaviour of dairy cows in the dry period or non-lactating period. Cows were monitored before and after the regrouping event. The study showed that regrouping resulted in decreases in feed intake, rumination and lying and increases in aggression compared to maintaining stable groups.

#### 2.4.1.4 Qualitative Behavioural Assessment

While most research using observation of spontaneously occurring behaviour describes the behavioural action (e.g., standing, feeding, and head-butting another animal), we can also look more closely at the way in which the behaviour is performed.

Qualitative behavioural assessment (QBA) is method of assessing the 'quality' of behavioural expression. It differs from conventional methods of behavioural assessment as it requires the observer to assess 'how' the animal is performing the behaviour, rather than quantifying the frequency and duration of the behaviours observed using the traditional ethogram-based approach It is described as an assessment of animal body language or of the expressive qualities of the way the animal is behaving and is seen as an indicator of the animal's emotional experience (Wemelsfelder 2007). In this method, observers are asked to create their own qualitative descriptors of how the animals are behaving, such as 'calm' or 'agitated', or to use a previously created descriptor list. A QBA approach has been used to assess various aspects of management and animal welfare status. For instance, the approach was used to show that calm behaviour by stockpersons is associated with more positive states in calves than elicited by the behaviour of nervous handlers (Ellingsen et al. 2014). QBA was also used to show that behavioural expression differs between cows with and without mastitis (De Boyer des Roches et al. 2018). It is one of the few methods that are currently capable of capturing the expressive qualities of behaviour, particularly for positive emotional states.

#### 2.4.1.5 Facial Expression

The analysis of facial expressions as an indicator of emotional state is a rapidly advancing area of research. The so-called 'grimace scales' were first developed to assess pain in rodents (e.g., Langford et al. 2010) and showed that pain was expressed as tightness in facial features such as the cheeks and eyes. This approach has also been applied to cattle. Gleerup et al. (2015) found that pain was evident in the facial expression of cows, as well as in other behavioural aspects such as their posture and attention to their surroundings. Other aspects of facial expression have been assessed. The degree to which the white sclera of the eye is visible is under autonomic control and may be used to assess emotional state (Sandem et al. 2002). Proctor and Carder (2015) found that relaxed cows showed less eye-white than cows in a neutral state.

### 2.4.2 Behavioural Tests

Behavioural tests typically involve asking the animal specific 'questions', typically in a purpose-built pen or arena. Behavioural tests are often used to assess the animal's response to stimuli or resources of interest (e.g., to a novel object or person, a new type of cubicle format or to a procedure such as milking). The specific behaviours recorded depend on the aim and type of test, are often related to the animal's engagement with the stimuli being tested, such as latency to approach, time spent in contact with the stimuli offered or energy expended in reaching the stimulus. This is in contrast to observations of spontaneous behaviour where durations and frequencies of different behaviours are typically measured. For instance, in a novel object test (see below for detail), we typically record time to approach or touch the object and occurrence of other behaviours indicative of fear or curiosity. In cognitive bias tests (see below for a full description), approach or response to the different options is assessed.

Any behavioural response is the result of a conscious or unconscious decision by the animal, based on its internal state, the presence of relevant stimuli in the environment and its prior knowledge of the utility of those stimuli. Emotions serve to add 'meaning' to the memories of the stimuli and their consequences. Because of this, behavioural tests can be used to tell us much about the motivational, cognitive and emotional factors that underlie the behavioural responses that we see in cattle. Specific tests of emotion are discussed in Sect. 2.4.3 below.

#### 2.4.2.1 Choice or Preference Tests

##### 2.4.2.1.1 Introduction and Concepts

The aim of a choice or preference test is to determine what an animal wants; we are asking the animal to 'vote with its feet', which ought to be a good indication of how it views the options presented to it in the test (Dawkins 1990, 2017). In practice, choice or preference tests involve offering the animal access to two or more resources or options and recording which one they make the most use of (also see also Box 2.1 for a short definition). The type of resource and the frequency and duration of the bouts of behaviour performed by the animal to engage with that resource dictate the type of preference test to be used. In a 'classical' test set-up, subject animals are brought individually from their home environment to the test pen, and choice is assessed within the period of the test (typically a few minutes). This testing regime works well for resources that can be used within a short period of time, such as assessing preferences for different food types. However, a 'live-in' version of the choice test is preferable for resources that take time to experience, such as lying areas, or are used sporadically over time, such as grooming brushes or drinking water. In both formats, a period of training is necessary at the start of the experiment, where animals have access to each option, to ensure they have experienced all aspects of the options and are thus making an informed choice.

In cattle, many of the welfare issues that have been identified involve aspects of the environment, such as poor lying areas, lack of shade or shelter or restricted access to feed. Choice tests have been used to identify resources that better meet the needs of cattle. Many studies have been done that assess preference for physical aspects of the environment, such as different types of housing, feed, sprinklers, presence of bedding or lying areas. More rarely, they are used to ask about preferences for different types of experience (e.g., quality of handling: Pajor et al. 2003; exposure to cold air: Bell et al. 2019). Some examples are discussed below.

#### 2.4.2.1.2 Preference Tests Using a Choice Point

One important way of assessing preference is to present the animal with two options in physically separate locations or areas that are accessible from a single 'choice point'. In some cases, laneways lead to the different options, while in others, the choice point is located at the intersection of a Y- or T-shaped pen or 'maze'. Preference is indicated by the number of times that the animal chooses to enter each of the areas or the time it spends in each choice area. Charlton et al. (2011) assessed the preference for dairy cattle for indoor cubicle housing versus access to pasture. Twice a day, the cows were brought to a choice point where they could choose to go out to pasture or go into a typical cow house. Preference was assessed by the number of choices directed towards each environment and time spent there. The cows chose the indoor housing more often than the pasture. Using a similar choice-point methodology, Pajor et al. (2003) tested the preference of dairy cattle for different types of handling. They used a Y-maze set-up and ran a series of trials in which a control (a handler standing beside one arm of the maze) was offered against options of a handler offering feed or hitting and shouting at the animal. The heifers quickly learnt to approach the feeding handler and to avoid the hitting and shouting handler in over 80% of the trials. However, in a second series of trials, when negative options were presented together, no preference was shown between use of a cattle prod and shouting or hitting and shouting, suggesting all options were equally aversive.

#### 2.4.2.1.3 Preference Tests with a Test Pen

In many preference test set-ups, the options are simply presented in an open pen. The 'choice' is less clearly indicated by the animal's movement in this set-up than in a Y-maze or T-maze, so direct contact with each option is typically measured. In an experiment assessing the preferences of cattle for heights of water trough and volume of water, cattle were deprived of water for 5 h. They were tested individually in a small paddock which contained the two options (Experiment 1: high vs low height and Experiment 2: big vs small volume) being studied. The choice of trough for drinking was recorded (Pinheiro Machado Filho et al. 2004). Preferences for feed types have also been tested this way. Meagher et al. (2017) used a test pen system to determine whether heifers preferred a constant, familiar feed or a varied feed, showing that most heifers preferred the constant feed, but there was a lot of variation between animals.

#### 2.4.2.1.4 Assessing Preference for Resources in a 'Live-in' Context

A number of studies have assessed the preference in cattle by presenting animals with various options within a 'live-in' housing or grazing system. For instance, Parola et al. (2012) investigated preference for sprinklers in hot climatic conditions by housing beef cattle in two connected pens: one with sprinklers above the feed bunks and one without. The steers spent more time in the pen with the sprinklers and within the area covered by the sprinklers. A similar study by Lee et al. (2013) assessed the preference of beef cattle for feedlots or pasture. Groups of beef cattle were habituated to a feedlot environment and adjacent pasture paddocks before being offered free access to both. The cattle spent 75% of their time at pasture. Schütz et al. (2009) offered cows shading structures that reduced the solar radiation to various degrees (Fig. 2.1) and found that they preferred those that provided most shade.

### 2.4.2.2 Strength of Motivation Tests

#### 2.4.2.2.1 Introduction and Concepts

Motivation can be defined as the strength of an animal's need for access to a stimulus or environmental or social resource. It is influenced by internal factors such as

**Fig. 2.1** Dairy cows in a 'live in' preference test. Cows were given access to wooden structures with shade netting fitted as a roof. Different types of netting that blocked different levels of solar radiation (25, 50 or 99%) were used, and cow preference was assessed (Schütz et al. 2009). Photo courtesy of Karin Schütz (AgResearch, New Zealand)

hormones and the presence of relevant stimuli and the animal's knowledge of these stimuli. For instance, the motivation of a beef steer to seek shade on a hot day at pasture will be influenced by internal factors such as core body temperature and current state of water balance and external factors such as the presence of shade-providing features of the pasture and previous knowledge of the effectiveness of the shade resources at reducing heat load. Strength of motivation is assessed by asking the animal to spend energy or time to access the resource or forego access to another valued resource. The 'price' that the animal is willing to pay is an indication of the value or need for that resource and can tell us about how welfare is impacted by the absence of the resource (e.g., Matthews and Ladewig 1994). Therefore, the strength of motivation for any particular resource is important to understand when we are trying to improve the physical environment of the animal. Additionally, if we want farmers keeping cattle to spend money on altering the facilities offered in the cow housing or at pasture, providing evidence of a strong need is an important justification. There are a number of different techniques that can be used to assess motivation that are discussed below with examples of studies that have used these methods.

#### 2.4.2.2.2 Speed, Time and Usage as Measures of Motivation

Motivation can be assessed by quantifying approach or avoidance behaviours, using measures such as the speed of approach, latency to respond or use of the resource. To assess the dietary preferences of calves, Webb et al. (2014) first habituated groups of calves to five different feed types and then gave them simultaneous access to all feeds. Preference, and thus motivation, was assessed by time spent eating, feed intake and visit frequency for each type of feed. Similarly, Schütz et al. (2021) investigating willingness to consume water contaminated with small concentrations of manure by assessing volume of water consumed. Cows drank more clean water than contaminated water at any manure concentration. Similarly, social motivation can be assessed by recording the time taken by an individual to return to the group of pen-mates. Typically, the social group of animals is held at one end of a runway, and one by one, each group member is separated and taken to the other end of the runway. The time taken to move back to the group and the time spent in proximity to the group are a good measure of social motivation, which correlates well with other measures of sociality (Gibbons et al. 2010).

Motivation to avoid an aversive stimulus can be assessed by measuring the latency to move away from the stimulus. To determine what climatic conditions contribute to cold stress in calves, Bell et al. (2019) exposed calves to three wind speeds (0 as a control, 1 and 3.3 m/s). Calves were tested in a two-part test pen, with wind applied only in one half (see Fig. 2.2). The latency to move from the 'windy' to the 'shelter' pen was used as a measure of aversion to the wind. Calves showed decreasing latencies to move to the shelter pen as wind speed increased.

#### 2.4.2.2.3 Expending Energy

Determining how much energy the animal is prepared to expend to gain access to a resource is a common and effective method of motivational testing in cattle. This can be done by asking the animal to push on a weighted gate or to walk increasing

**Fig. 2.2** Response to aversive stimuli. In this study, cool air was blown into the pen where the calf was located (note the fans at the top left of the photo). The calf was able to move into the pen on the right of the photo to avoid the draught. The response to the cool air movement and the latency to move away from it was used to assess the thermal comfort zone under a range of temperatures and wind speeds (Bell et al. 2019). Photo courtesy of David Bell, SRUC, UK

distances. There are a number of techniques that are used with cattle to assess the amount of energy that an animal is willing to expend.

#### 2.4.2.2.3.1 Increasing Operant Responses

Asking an animal to perform an operant response, such as pushing a panel or pressing a lever, is a classic way of quantifying the effort an animal is prepared to make to access a resource (Matthews and Ladewig 1994). The advantage of using this approach is that each response is short and distinct, so a range of responses is possible. This granularity aids in distinguishing between levels of the same resource and assess differences between individual animals. Jensen et al. (2005) used a press panel in a consumer demand trial to assess the demand (or need) for lying in dairy heifers. The heifers showed a high demand for lying equivalent to 12–13 h per day.

#### 2.4.2.2.3.2 Increasing Distances and Increasing Weight

To assess the motivation of cows for access to pasture, a study by Charlton et al. (2013) asked cows to walk distances of 60, 140 and 260 m to access a field. They showed that during the day, cows used the pasture less when they had to walk the longest distance, but at night there was no drop in the use of pasture even when cows were asked to walk the longest distance. This suggests that pasture access is more valuable to cows at night. In a similar vein, Shewbridge-Carter et al. (2021) used increasing walking distances (34.5, 80.5 and 126.5 m) to determine whether cows value access to an open lying area within the housing compared to cubicles (Fig. 2.3). Cows maintained a high level of usage of the open surfaces over the three test

**Fig. 2.3** Using increasing distance to assess motivation. Cow walking through a 'passport queue' style raceway to access an open straw-bedded lying area (Shewbridge-Carter et al. in press). Photo courtesy of Laura Shewbridge-Carter (HAU and SRUC, UK)

distances. Schütz et al. (2006) assessed the relationship between feeding motivation and feed deprivation in a study in which dairy cows that were food-deprived to various levels had to walk increasing distances to feed. In lactating cows, there was a linear relationship between distance walked and feed deprivation levels, indicating that feed deprivation causes hunger in a linear manner.

Tucker et al. (2018) developed a pneumatic push-gate to assess the motivation of cows to access a bedded area (Fig. 2.4). The push-gate consisted of two hinged panels attached to a frame, allowing the cow to push through the gap between the panels with her shoulders. The mechanical resistance required to open the gate could be increased until the cow stopped using the gate. Another type of weighted gate was used by von Keyserlingk et al. (2017) to show that cows are prepared to 'work' as hard for access to pasture as they are for fresh feed.

It is important to note that there are some limitations of motivational tests. One major issue is the so-called 'ceiling effect'. When animals are asked to push against weighted gates or walk long distances, they may reach the limits of their physical ability to perform the task (e.g., push even greater weights) before they have expressed their motivation for the resource. It is also important to ensure that other motivations are met so that the response seen is truly representative of the value the animal places on that resource. For instance, thirst may interfere with an animal's motivation for food.

#### 2.4.2.3 Motivational Priorities

As well as assessing motivation for a single resource at a time, we can also assess motivational priorities. This refers to whether the performance of one behaviour or use of a resource is more important than others. We may wish to know this to

**Fig. 2.4** Cow pushing through the pneumatic gate as a measure of her motivation to access a bedded area. Technique used in Tucker et al. (2018). Photo credit: Aarhus University, Denmark

understand the 'ranking' of animal needs, so that the most important can be prioritised in farm building design or in managing the animals.

Depriving animals of access to resources such as feed and lying areas and then allowing them access to the resources to perform all of these behaviours, but only for a limited time, force them to prioritise the most important behaviours. In a study with lactating dairy cows, Metz (1985) prevented cows from lying down after milking. When released from restriction, the cows prioritised lying over other behaviours. In another study of motivational priorities, Munksgaard et al. (2005) held the cows in a pen which did not allow them to eat, lie down or have social contact for increasing lengths of time, before allowing them a test period in which all behaviours were possible. As the deprivation time increased, the cows showed a higher proportion of lying in the unrestricted period, while the proportions of feeding and social behaviour remained constant, further indicating the importance of lying behaviour.

In some senses, this type of test is asking the animal to 'trade off' or choose between the behaviours that they might perform. As well the priority for resources, the social priorities of cattle can be assessed. The willingness of cattle to feed close to more dominant animals was tested by Rioja-Lang et al. (2009). Test cows were trained that both a bin of high-quality feed and one of low-quality feed were present in a test arena. Another cow that was dominant to the test cow was then allowed to feed from the high-quality feed bin, and the test cow allowed to choose whether she would stand next to the dominant to feed or 'trade off' access to high quality feed with maintaining distance to the dominant cow by feeding from the low-quality feed bin (Fig. 2.5). The majority of the test cows chose to avoid the dominant cow and eat the low-quality food, indicating an aversion to close proximity to dominant cows when feeding. In a trade-off test that assessed the importance of shade for dairy cattle in hot conditions, Schütz et al. (2008) deprived cows of lying for 0, 3 and 12 h and then offered them the choice between lying down and seeking shade. This scenario was repeated over a range of climatic conditions. It was not until the temperature rose to over 30 °C that cows gave up lying down to seek out the shade. Considering the high priority given to lying behaviour shown by previous experiments, this result indicates cows give high priority to seeking shade in hot conditions.

### 2.4.3 Tests of Emotional State or Emotional Responsiveness

As discussed previously, tests of preference and motivation have primarily been used to ask what cattle 'want' or 'need'. While it is assumed that providing them with what they want will improve their welfare and therefore promote a more positive (or at least, less negative) emotional state, these tests do not directly assess how the animal feels when it has access to the resources. There is increasing emphasis placed on understanding animal emotional states, not least because welfare is about feelings (Dawkins 1990) but also because, increasingly, consumers are more aware of animal sentience and experience. Additionally, recent developments in methodology have given us better tools to assess emotion.

There are a range of tests that are designed to assess the emotional state of the animals more directly. Many of these tests are tests of fear. Fear has been described as an emotional state induced by the perception of actual danger (Boissy 1995). Fear has received a lot of research attention in cattle, as there are a number of situations in farming and husbandry that have the potential to induce fear in cattle. These include the proximity to humans and the exposure to unfamiliar pens, procedures and facilities. As well as the animal welfare element, fearful and reactive animals can be dangerous to human handlers due to their size and strength. Fearfulness can also affect productivity, which means that there is a commercial incentive to reduce fear in cattle (Breuer et al. 2000).

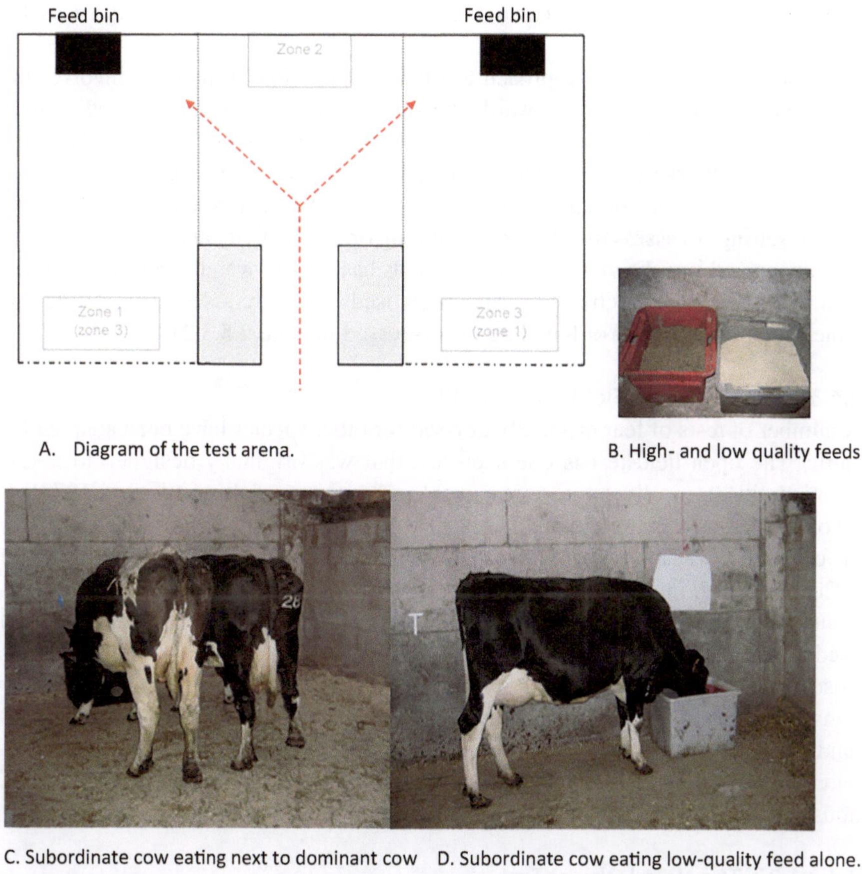


**Fig. 2.5** Example of a trade-off motivational test (Rioja-Lang et al. 2009). Subordinate cows were trained to receive high-quality feed in either the left or right feed bin (**a** and **b**). A dominant animal was then introduced that ate at the high-quality feed bin. The subordinate cow had to decide whether to eat the high-quality feed but stand in close proximity to the dominant cow (**c**) or to 'trade-off' access to high-quality feed, but eat without the close proximity to the dominant cow (**d**) (images and diagrams courtesy of Fiona Lang). **a** Diagram of the test arena. **b** High- and low-quality feeds. **c** Subordinate cow eating next to dominant cow. **d** Subordinate cow eating low-quality feed alone.

#### 2.4.3.1 Tests of Fear

##### 2.4.3.1.1 Human Approach Tests

As mentioned above, the fear of humans has been an important area of study for economic and animal welfare reasons. Breuer et al. (2000, 2003) investigated the fear of humans in dairy cows by assessing approach behaviour (time to approach, time in proximity to and interactions with the experimenter) towards an experimenter seated on one side of an open arena. The studies showed that the experience of poor-quality handling increases fear in cattle with negative consequences for productivity and welfare.

Another type of human approach test that assesses fear of humans involves the experimenter moving slowly towards the subject animal in a standardised manner (Waiblinger et al. 2002; Gibbons et al. 2009 and see Chap. 9). The point at which the animal withdraws is recorded, with animals that withdraw when the person is the furthest away considered to be the most fearful. The test has been used in experimental settings to assess the effects of handling on productivity and longer-term fear of humans and has shown that calmer animals have better weight gain (e.g., Lürzel et al. 2015). The approach test has also been used in welfare assessment settings, as a measure of the quality of handling (as discussed in Sect. 2.5.3.2).

#### 2.4.3.1.2 The Open-Field (or Arena) Test

A number of tests of fear originally devised for other species have been adapted for cattle. The open-field test is one such test that was originally designed to assess emotionality (typically the fear response) in laboratory rodents (Archer 1973). It involves placing the animal into an open arena and recording behaviours indicative of fear such as escape attempts, vocalisations or urination and defaecation (Nielsen 2020). Research into the effects of housing or management systems on fearfulness of novelty in adult cattle and calves has used open-field tests. Boivin et al. (1992) used an open-field test to assess the response of indoor- and outdoor-reared animals to social isolation and the novelty of an unfamiliar arena. They found that indoor housed animals showed more indications of fear than animals from extensive rangeland management system. Jensen et al. (1997) also used an open-field test to compare fear responses in calves housed individually or in groups of four calves. Calves housed individually behaved more fearfully than the calves housed in groups.

#### 2.4.3.1.3 The Novel Object Test

The novel object test is also adapted from a test first used with laboratory rodents. In the typical test, the animal is moved into an open test arena which contains an unfamiliar object (a traffic cone is often used for cattle). The latency to first contact with the object and the time spent interacting with it are assessed. Some studies use an unfamiliar animal as a novel 'object' where the aim is to assess social fear or motivation (e.g., Wagner et al. 2013). Other tests of neophobia have assessed cattle in more 'cow-relevant' situations like response to novel objects, food and people within the home pen (Herskin et al. 2004). More commonly, novel object tests are carried out in conjunction with open-field tests and human approach tests to determine whether fear of different stimuli (of humans, isolation and objects) is related to other personality traits or states. It has been shown that responses to the different fear-inducing stimuli are unrelated, suggesting that the response is mediated by independent underlying mechanisms (e.g., Gibbons et al. 2009; van Reenan et al. 2013), but that calves that are fearful are also likely to be pessimistic (LeCorps et al. 2018; see below for a discussion of cognitive bias tests).

#### 2.4.3.1.4 Crush or Chute Tests

Crush or chute tests have been used a great deal to assess fear in beef cattle. They typically involve scoring the response of an animal to the handling procedure, which involves confinement in a handling crush or chute, a procedure such as drenching or vaccination as well as close proximity to humans. The response of the animal is typically scored on a scale from calm to very agitated (Grandin 1993). The response can also be assessed by 'flight speed', which is the speed at which the animal runs away from this handling situation (e.g., Burrow 1997). Although many stressors are involved in this situation (humans, confinement, novelty of the situation), it is a widely used test, which has been used to understand underlying personality in animals and responses to other stressors (e.g., Turner et al. 2011).

### 2.4.3.2 Laterality

The two hemispheres of the brain are specialised to process the information coming from the environment in different ways. The right hemisphere is specialised to process inputs from new or challenging stimuli while the left side specialises in information from positive or predictable inputs. Because of a crossover in the optic nerves, the left side of the brain processes information from the right eye, while the right side of the brain processes information from the left eye. Therefore, observing whether an animal chooses to look at a stimulus with its left or right eye can indicate whether it regards that stimulus as being threatening or not (Rogers 2010). A number of studies have used this emotional-driven laterality in eye use to determine how threatening cattle perceive stimuli such as other cows, novel objects and human handlers to be. Subordinate cows viewed other cows predominantly with their left eye, indicating that this situation was challenging to them. Unfamiliar humans were viewed through the left eye, whilst familiar humans were viewed through the right eye (Phillips et al. 2015). Kappel et al. (2017) hung two identical novel objects on either side of a passageway. Most cows did not show a one-sided approach, but cows approached some of the objects on the right, which given the link between this eye and the left side of the brain suggests that they found the objects positive to engage with. These studies involve little or no training of the subjects and can be implemented relatively easily, showing that this method may be used to assess emotional responses.

### 2.4.3.3 Aversion Learning

Following the same principle as using the speed of approach or consumption to assess motivation for a resource, we can assess the strength of an animal's negative perception of an experience using aversion learning techniques. If an animal has learnt that a specific location is associated with negative experiences, it will be more reluctant to approach that area than if it had been associated with a more positive experience. For instance, in cattle, the handling area or crush is often a site where injections and other treatments are administered, so the animal may be reluctant to approach it. This is known formally as conditioned place preference or conditioned place avoidance/aversion depending on whether an approach towards or an avoidance of a location is being tested. It is useful when assessing emotional states in response to a discrete event that can be administered in a discrete location, as the experimenter needs the animal to learn this event: location association. In cattle, these techniques have been used particularly to assess responses to unpleasant or painful events.

Aversion learning was used by Pajor et al. (2000) to investigate what handling techniques dairy cattle find aversive. In the first study, a test raceway was used. Groups of test cows were used, with groups receiving food, being brushed, being hit and shouted at and no interaction as a control. The cows that received the hit/shout treatment moved more slowly and required more forceful handling to get them to move down the race than the group that received food. When different types of handling treatment (hitting, shouting, electric prod, tail twist) were applied, the group receiving the electric prod appeared to be slower and require more force on some trials, but overall, the technique was not able to differentiate between the aversive treatments. It was concluded that this test was able to differentiate between treatments that differed greatly in their aversiveness but lacked the sensitivity to detect differences between similar treatments. As discussed above, Pajor et al. (2003) used a preference test set-up with different options placed in a Y-maze to further investigate response to different types of handling technique. The preference test paradigm was able to differentiate between treatments more readily.

Ede et al. (2019) used the conditioned place aversion methodology to assess pain during disbudding. In a 'within calf' trial design, calves had one bud removed with local anaesthetic plus additional pain relief (an anti-inflammatory drug) applied and one bud removed with only anaesthetic applied. The procedures took place in distinctive test pens. When the calves were returned 2 days later, calves avoided the pen in which the disbudding without additional pain relief was done.

Similarly, Adcock and Tucker (2020) used conditioned place preference to assess the effectiveness of nerve-block treatments in reducing pain following disbudding. Learning of associations in calves that were disbudded and injected with lidocaine was compared with calves that experienced sham-disbudding with lidocaine. Stimuli associated with the use of the lidocaine were approached preferentially by calves that were disbudded, indicating that they found the relief from pain to be positive. Sham-disbudded calves on high doses of lidocaine, however, avoided stimuli associated with the drug, which suggests that they found the injection to be painful, which is in agreement with human experience. Disbudded calves appear to be trading-off short-term discomfort with longer-term pain relief. In terms of methodology, what is interesting is that the conditioned place preference technique could identify these effects.

#### 2.4.3.4 Judgement Bias Tests

There is increasing evidence in humans and other animals that emotion and cognition are interlinked (Paul et al. 2005). Positive and/or negative emotions experienced whilst interacting with a stimulus are stored in the memory and influence future interactions. There is also evidence that ongoing mood states and personality affect response to novel stimuli in animals. Animals and humans in low mood states are more likely to treat new stimuli with suspicion and have more negative expectations of the outcomes of interacting with these new stimuli than animals in more positive mood states. This is known as 'judgement bias'. Judgement bias tests typically involve training the animal that positive and negative (or less positive) outcomes are associated with two contrasting test stimuli (Paul et al. 2005).

The methodology to assess judgement bias was first developed in an experiment with rats (Harding et al. 2004), but since then, it has been used a great deal to assess the effects of barren housing and poor handling on the mood states of animals. In cattle, the effect of various management practices on emotional state has been investigated, including effects of pain and other negative emotional states. Neave et al. (2013) assessed the effect of pain due to hot-iron disbudding on judgement bias in calves. They trained calves to use a touch screen to gain access to milk and then trained them to associate a particular colour of screen with a milk reward (e.g., red) and another (e.g., white) with non-reward. Once the calves were trained to these contingencies, intermediate colours were presented. Calves were tested with these intermediates before and after disbudding. It was shown that the calves were more likely to show pessimistic responses after disbudding than before the procedure. Daros et al. (2014) used a similar technique to assess the response of calves to abrupt weaning from their dams and demonstrated a negative judgement bias in calves after weaning compared to before. Bučková et al. (2019) used location rather than colour to distinguish the reward and non-reward (e.g., left side of arena = food; right side = non-reward, and vice versa for a second group) and assessed the effect on the emotional state of calves of being pair- or singly housed. Pair-housed calves showed more responses when the intermediate positions were tested. This technique is not easy to implement, as it requires time to train the animals, but holds great promise in understanding the emotional impact of many other husbandry and management procedures.

Attention bias is another method of assessing judgement bias (e.g., Monk et al. 2018). Rather than training animals to associate particular rewards or punishments with particular stimuli or locations, this test method requires no training. Animals are presented with an alarming stimulus (such as a dog) whilst engaging in a motivated behaviour such as feeding. It has been shown that anxious animals pay more attention to the alarming stimuli than calm animals. This is a good method to determine whether housing or management practices has increased fear or anxiety in animals, and it has the advantage of not requiring the extensive training needed for judgement bias tests. However, it has only recently been used in cattle, with one study showing that beef cattle experiencing repeated housing and management stressors responded differently to a perceived threat than animals that had not experienced these stressors (Sommariba et al. 2019).

### 2.4.4 Direct Observations of Spontaneous Behaviour Versus Behavioural Tests: Pros, Cons and Pitfalls

Having discussed the different types of tests in the previous sections, the pros and cons of each approach in understanding welfare should be considered. Firstly, there is much to be learned from the observation of spontaneously occurring behaviour. The observations can take place in an undisturbed setting in which animals are likely to behave in a normal manner. This is useful in gaining an understanding of the extent of the behavioural repertoire and the daily and seasonal patterns. Cattle can also be observed in their social groups, which can allow the effects of social hierarchy to be assessed. This is not the case for behavioural tests, which are typically carried out with individual animals tested one at a time.

However, because of this 'hands-off' approach, we do not always get a deep understanding of factors such as the emotional state of the animal or the motivation and cognitive factors involved. To delve deeper into these aspects, behavioural tests must be used. Additionally, our conclusions about the animals' behaviour in their home environment, and consequently, their welfare are somewhat limited to the specific context that the animals were observed in. So, when we compare the results of a study that has recorded spontaneous behaviour in cattle with cattle in a different context, we have to be aware that animal characteristics (e.g., age, breed, reproductive status) and environmental factors (e.g., climate, feed availability, light/dark cycles) will affect the behaviour of the animals. Each of the 'foundation' studies discussed above had specific combinations of animal and environmental factors that may make the 'generalisability' to other situations limited unless the system we want to compare to is similar. The specific elements present in the environment may also limit the range of behaviours that have may be performed, so the behaviour repertoire observed may also be specific to the situation. Some important behaviours may not occur. However, these studies provide broad parameters to allow comparisons of behaviour and are the best understanding we often can obtain of 'normal' behaviour and thus can be used as long as these differences and limitations are acknowledged.

Behavioural tests are useful as they can facilitate a better understanding of the motivational, cognitive and emotional aspects underlying the behaviour. However, because we must make decisions about every aspect of the test (the options presented, the size of the arena, the time the animals have in contact with the options, etc.), the results may depend on the way the test was carried out. There are many literature reviews focussing on the methodological issues. The best way to design preference tests is discussed thoroughly by Duncan and others (e.g., Dawkins 1983; Duncan 1992) and includes careful choice of offered resources (such that both are not bad for welfare or so good that the choice is meaningless), attention to previous experience with the options and care in interpretation of results. It is important to remember that as choice is relative, short-term choice does not necessarily reflect long-term welfare.

Tests of motivation also need to be set up carefully. As with preference tests, aspects of the test set-up, the previous experience of the animal and the relative attractiveness of the options the animal must work for must be considered. It is also difficult to create a test scenario where only one motivation is being assessed. The social isolation of a test arena and the handling that may be involved to move the animal in and out of the test arena are elements that may affect its response to the context in question. Additionally, the amount of resource or access to the resource must be appropriate (i.e., offering access to a lying area for a period that is shorter than the average lying bout is not appropriate) and the response required (to expend energy or walk increasing distances) must be appropriate. The 'ceiling effect' refers to the situation in which the animal reaches the limit of their physical ability to 'pay the cost' (e.g., push even greater weights) before they have fully expressed their motivation for the resource (e.g., a cow cannot physically push open a weighted gate allowing access to the outdoors even though she is still motivated to go there). The value of each choice and the consequences of making the choice must be understood by the animals. Many of these aspects are considered in reviews (e.g., Fraser and Matthews 1997; Kirkden and Pajor 2006).

When emotional responses are assessed, it is useful to be aware that there may be strong emotional responses, and that the measure of this intensity of response, or the experimental set-up must be designed to capture the full range of response if possible. For instance, there may be a ceiling effect in the 'crush score' used to assess responsiveness of beef cattle to handling in yards (Grandin 1993) if the scoring method used is not suitable to distinguish between responses on the extreme end of the scale. 'Floor' effects are also possible if no animal experiences an emotional response to the stimuli or context being tested. The use of pilot trials and appropriate measurement methods can help to resolve these issues.

One further issue is that not all animals cope well in a test situation. Behavioural tests typically assess one animal at a time. While this allows each animal to express their choices and range of behaviour without the presence of competitors, it does mean that the animal is tested in isolation from pen-mates. This can lead to some element of fearfulness of the experimental set-up. Because of this social isolation and the handling involved in testing, it is common that not all of the animals available to be tested habituate to the test procedure as they respond in an overtly fearful manner and have to be removed from the study. Tests that involved the animals having to successfully learn a task or association before they can be tested are particularly prone to this problem. Judgement bias testing suffers from this 'test amenability' issue, as animals must learn a discrimination task before their response to a housing manipulation can be tested. There are a reasonably high percentage of animals in many trials that fail to learn the discrimination and are not tested further. This may mean that the results are biased towards the responses of test-amenable animals. Using 'habituation' or 'training' periods at the start of the experiment that allow animals to habituate to the handling and isolation is often helpful.

## 2.5 Using Behaviour in Welfare Assessment Protocols

### 2.5.1 Introduction

Welfare assessment protocols consist of a set of measures that tell us something about animal welfare. The measures (or indicators) can be a direct assessment of the state of welfare of an animal, such as body condition or number of injuries. These are known as 'animal-based' measures. Measures of aspects of the environment and management which affect welfare are also used and are known as 'resource-based' measures, such as length of feeder space available per cow or cleanliness of the water (Johnsen et al. 2001; Main et al. 2003). Animal behaviour is a key animal-based measure. However, unlike the previous methods discussed, welfare assessment protocols (WAPs) are carried out in commercial settings on working farms, which places some limitations on what behavioural methodologies can be used.

There are a number of criteria that any measure, behavioural or otherwise, must fulfil to be useful in a WAP. Most importantly, the indicator must be valid. This means that there must be evidence to show that the indicator measures something real and relevant about the animals' welfare. Fortunately, there is a wide body of evidence from research in animal welfare science and veterinary medicine to draw upon to decide upon which indicators are valid. The indicator must also be reliable. Reliability is a measure of whether the same score is recorded when the measurement is repeated. Thus, if the underlying welfare state has not changed, any indicator used must give the same set of results if: (1) assessed by different assessors (i.e., the inter-observer reliability must be high); (2) assessed by the same assessor on a different day (i.e., the intra-observer reliability must be high); (3) used on different farms (i.e., it must not be too situation specific); and (4) used at different times of year or even different times of day (i.e., it must be stable over time). Although for most indicators inter- and intra-observer reliability issues can be reduced with effective and repeated training of assessors, the problems caused by poor reliability between farms are more difficult to overcome and may be due to the diversity in characteristics of the cattle, buildings and facilities across farms. Finally, the feasibility of each indicator needs to be taken into account. This means that regardless of the indicator's validity and reliability, if it cannot be undertaken over a short period of time by a single person, needs bulky or fragile equipment or is impossible to measure in all types of farms, then it should not be included in the WAP.

### 2.5.2 Types of Welfare Assessment Protocol

There are two main types of welfare assessment protocols currently in use in commercial systems. Firstly, there are protocols that are used by external organisations to assess the welfare of the whole herd of cattle on a farm. These use an 'audit-type' system that incorporates a suite of welfare indicators, including behavioural indicators of welfare. Audit-type protocols are most frequently used by organisations such as food retailers, who aim to certify that the farm reaches particular standards that are communicated to their customers. Some organisations only assess animal welfare (e.g., Global Animal Partnership 2021) while others assess welfare alongside other measures on the farm, such as that for human safety and environmental protection (e.g., Red Tractor, UK 2022). This type of assessment protocol is also used by government veterinarians or other official bodies to ensure compliance with legislation on animal welfare. These protocols give a snap-shot of the welfare conditions for the cattle on farm on that day and can be used to benchmark one farm against a group of others. However, these audits are often carried out relatively infrequently (yearly or twice yearly in many cases) due to the expense of sending an auditor to all farms in any scheme or group.

The second type of welfare assessment protocol is the 'monitor-type' protocol that can be used more frequently (monthly, weekly or even daily) to provide ongoing information about the welfare of individual animals, groups of animals or the whole herd. The monitor-type assessment protocols are most frequently used by cattle owners themselves or in conjunction with their veterinarian or retail body to provide information that the farmer themselves can act on. In most cases these protocols have a few key indicators that are relatively quick to observe and that can be then compared over time to monitor welfare and changes to welfare.

### 2.5.3 Behavioural Indicators of Welfare

While the protocols differ in the number of indicators used, the indicators themselves are often scored in the same way. There are a number of behavioural indicators of welfare that are used in welfare assessment protocols (e.g., Winckler 2018), and these will be discussed below.

#### 2.5.3.1 Gait Scoring

Gait scoring (also known as lameness scoring) is the most commonly used behavioural measurement in cattle welfare protocols, especially for dairy cattle. Alterations in gait are the result of pain resulting from pathological or conformational foot or leg conditions. Gait alteration as an indicator of pain has been validated through a comparison of gait before and after use of pain relief (e.g., Rushen et al. 2007). Gait scoring typically involves watching animals as they walk past on a suitable flat surface. The score focuses on the quality of the steps made (e.g., the length of the stride, if the stride is straight or contains abduction or if there are differences in weight bearing between the four feet) and may also assess aspects of the body movement (e.g., the straightness or arching of the back or the lowering or bobbing of the head on walking). The assessor carefully watches the animal as it walks a few strides and must observe at least one step from all four feet. The assessor puts together the qualities of the movement to give the animal a single numerical score. There are a number of gait or locomotion scoring systems that have been developed (Sprecher et al. 1997; Flower and Weary 2006; AHDB 2020) and are reasonably similar, varying only in the size of scale (0–3 or 0–5) and the emphasis placed on the different aspects of the movement of the body (e.g., stride length and back arch).

Although assessors do require training, gait scoring is not a difficult assessment and does not need specialist knowledge, so can be carried out by lay assessors, farmers and/or veterinarians. Good inter-observer reliability can be achieved (Winckler and Willen 2001). Gait scoring is also likely to yield similar results in different context, so that the same animal scored in the parlour or walking along a path is likely to get a similar score. There can be time-of-day problems, since cows with full udders walk differently from cows with 'empty' udders, and so it is always best to score after milking or a few hours before milking in lactating cattle (Flower et al. 2006). Gait scoring is also feasible on most farms but may take some time to complete on a big herd. Some protocols therefore recommend only sampling a proportion of the animals in larger herds (e.g., Welfare Quality® 2009; Global Animal Partnership 2021).

However, gait scoring is somewhat of a 'blunt' tool. Gait scoring is not necessarily the best indicator of severity of pain or specific disease state in the feet and legs (e.g., Schlageter-Tello et al. 2014). Even when pain relief or treatment has been administered, cattle can still show some gait alterations (due to a curved spine or joint alterations) if the lameness has been untreated for a prolonged period. Additionally, we do not know if individual differences in tolerance to pain affect the extent to which the cow shows signs of lameness. However, even with these considerations, it is likely to remain a useful and commonly used method of behavioural observation.

#### 2.5.3.2 Tests of the Responsiveness to Humans

As discussed above and more comprehensively in Chap. 9, 'human approach' tests can be used to evaluate fear of humans in cattle. There are two types of human approach test, and both can be used in welfare assessment protocols in commercial farm settings. In the first type of test, the human approaches the animal, while in the second type of tests, the voluntary approach of the animal to the human is recorded.

The human approach test is more commonly used to assess fear/friendliness in cattle and calves (e.g., Welfare Quality® 2009). This test can be applied when the animal is at the feeder or standing in the body of the barn or pen. When at the feed-trough, the assessor will start the test standing a set distance away from the animal and will slowly move forward in a standardised manner. When the test is carried out in the home pen, the assessor should allow the animal a short period to adjust to their presence and then slowly move towards the animal. The distance from the animal that the assessor reaches when the animal starts to move away, or whether the animal can be touched, is recorded (e.g., Welfare Quality® 2009; Andreasen et al. 2014 (using the Danish Cattle Federation protocol)). Again, this type of test scores well in terms of observer reliability (Rousing and Waiblinger 2004; Winckler et al. 2007). However, factors such as motivation to feed or ability to move freely around other animals and structures will affect response, and every attempt should be made to standardise conditions when comparing scores across farms. Using this test in pastured animals requires careful interpretation, and results cannot directly be compared with housed animals.

Voluntary approach tests are used to assess the fear/friendliness of calves towards humans in both dairy and beef settings. In this test, the assessor will enter the calf accommodation and stand in one place at the front of a pen for a standard period of time, while monitoring the behaviour of the calves (e.g., orientation of calves towards the person) and the latency within the test time for the calves to approach. These measurements are also reliable both between/within observers and in most types of situations where calves are group housed. Special care needs to be taken in interpretation if the calves are at pasture (the outside environment might be more interesting than the new person). Other factors will affect the response to humans such as hunger or tiredness, or recent experience of painful procedures such as disbudding, and these need to be taken into account.

#### 2.5.3.3 Lying and Transitions into Lying Posture

Lying is a behaviour that is given a high priority in cattle (see Sect. 2.4.2.3 on Motivational Priorities). Because of this, the amount of time that cows are able to spend lying down on a farm is seen as a key measure of cow welfare on that farm. Lying time is also a measure of the quantity of lying area available per cow and comfort of these lying areas (e.g., Tucker et al. 2021). The 'Cow Comfort Index' was one of the first to be used in commercial systems and records the proportion of cows touching a cubicle that is lying down (described in Cook et al. 2005). While this index has been widely used in the dairy industry, it really only provides an indication of cubicle use and not of lying time. As the use of pedometers or activity monitors in dairying has become more widespread, data on daily lying times in individual cows have become easier to quantify. Studies have shown that dairy cows in tie stall and cubicle systems lie down for between 10 and 12 h/d on average, while cows at pasture and open packs and feedlots lie down for 9 h/d (Tucker et al. 2021). However, as discussed in this review by Tucker et al. (2021), lying time can be reduced by the time needed for other activities such as feeding and milking and poor-quality cubicle configuration and bedding (see Chap. 4 also). However, lying time is higher in lame cows, suggesting that high lying times are not always an indicator of good welfare. As the use of technology on farms increases, the use of lying times may be increasingly used as part of welfare audits, but these provisos should be considered.

The ability of cattle to move between standing and lying is also used as a measure of mobility but also of the quality of the lying area (e.g., Welfare Quality® 2009). The amount of cushioning for the body that the bed base and bedding of the lying area provide as the animal lies down and the frictional properties of the surface that prevent the feet from sliding during this movement likely affect the process of lying down (Zambelis et al. 2019). Typically, the time taken for the complete transition from standing to lying is recorded (Welfare Quality® 2009), and this measure is reliable and feasible to assess in commercial conditions (Plesch et al. 2010).

#### 2.5.3.4 Aggressive Behaviour

Cows may experience aggressive or competitive behaviour when there is competition for resources such as access to feed, shade or lying spaces. Recipients of aggression may experience stress and reduced access to these resources, and there is a higher risk of injury when animals must move quickly to avoid aggression. Aggressive behaviour is typically higher in housed systems than at pasture due to the higher stocking densities involved. There are a few WAPs that include agonistic behaviour (e.g., Welfare Quality® 2009) for both beef and dairy cattle. Aggressive behaviour is most frequently seen at the feeding area, and observations typically focus on this area (Welfare Quality® 2009). However, these observations are time-consuming and so are not typically included in WAPs run by commercial companies. Despite this, aggressive encounters cause acute stress and fear in animals and are an important element to consider.

### 2.5.4 Qualitative Behavioural Assessment (QBA)

Qualitative behavioural assessment (as discussed above in Sect. 2.4.1.4) is a technique in which observers quantify the expressive quality of animal's interactions with each other and with the environment. QBA can be used in on-farm WAPs. At present, it is one of the few measures available that is currently capable of assessing positive welfare states in an on-farm setting. Welfare Quality® (2009) is one of the few WAPs to include QBA. In this context, it involves the use of a 'fixed list' of 20 qualitative terms that are measured using a visual analogue scale. Up to eight observations points are chosen on each farm that represents the whole farm structure.

### 2.5.5 Summary

In summary, a great deal of research has gone into understanding what behaviour patterns can be used as welfare indicators in welfare assessment tools, when a snapshot of the state of the farm is all that can be gathered in the few hours necessary for welfare assessment. Assessing the validity, reliability and feasibility of each measure has been important in the process of incorporating these indicators into welfare assessment tools. However, the range of behavioural measures used in protocols like Welfare Quality is much larger than those used by commercial companies. This is because many of the measures are time-consuming to assess, rather than the indicators being considered unimportant. A range of 1–2 h is considered to be the ideal length of an assessment visit (Winckler 2018). One of the solutions to this may be automated data collection. Some of these measures may be captured using technology, and research underway at present may facilitate this goal.

## 2.6 Cautionary Notes: What Assessing Behaviour Cannot Do

Precisely because a behavioural 'output' represents the integrated outcome of the internal and external factors modified by experience (as suggested in Sect. 2.2), it does not tell us about the constituent parts. Studies using neurobiological, immunological and physiological methods are required to do this. Likewise, the genetic regulation of behavioural and physiological responses can tell us about a more fundamental influence on animal welfare. Understanding these influences is important in efforts to understand animal states and improve welfare, and alongside assessments of behaviour, are tools in our toolbox in animal welfare science. These types of studies require different methods and will be discussed in other chapters of this book. In most cases, we need a combination of approaches.

However, we should not entirely rely on behavioural outcomes, especially where a behavioural response is likely to be highly influenced by physiological states. One of the earliest experiments that alerted researchers to a lack of correspondence between behavioural and physiological responses was shown in an experiment that assessed behaviour and heart rate responses to fear in two strains of chicken (Duncan and Filshie 1980). One strain showed an active escape response to the stimuli, and a sharp increase in heart rate that soon returned to normal levels. The other strain did not show such a marked behavioural response, but the heart rate increased and did not return to normal levels, indicating a greater stress response, despite the lack of an overt behavioural response.

An example in cattle where behaviour is not always reliable is in using lying postures to distinguish sleep from resting states in cows. Sleep is a behaviour ubiquitous among mammals and has been shown to have both a restorative function and be involved in learning and memory consolidation. In general, good-quality sleep of an appropriate duration for the species improves emotional mood and cognitive performance while poor-quality sleep has a detrimental effect on mood, health outcomes and cognitive abilities. Because of this, there has been research interest in determining whether housing facilities and procedures such as transport deprive cattle of sufficient good-quality sleep. This has been most successful in dairy calves, with Hänninen et al. (2008) using electrophysiological methods (including non-invasive electroencephalogram (EEG)) and behaviour to show that when calves lay down with their heads resting on a substrate, it was predictive of Rapid-Eye-Movement (REM) sleep, while lying still with their heads off the ground was predictive of non-REM sleep. However, studies by Ternman et al. (2014) and Hunter et al. (2021a) showed that the same postures were not as reliable for predicting sleep measured by EEG in adult cows. Hunter et al. (2021b) suggest that it may be possible to use other physiological measures (such as heart-rate features) to accurately predict sleep in cows even in differing housing conditions. This example serves to remind us that relying entirely on behavioural measures of welfare may be ill-advised, especially as a proxy for an underlying physiological state. Using the most relevant behavioural and physiological indicators to address a welfare issue is typically the most advisable method.

## 2.7 Future Directions

Our ability to use behaviour to understand factors affecting welfare and assess welfare status in animals is only as good as our ability to come up with good methodologies. As Franz de Waal argues with respect to animal cognitive abilities in his book 'Are We Smart Enough to Know How Smart Animals Are?' (2016), our ability to understand animals is limited only by our ability to come up with good methodologies.

Where these advances are needed is in the assessment of emotion, both positive and negative. Advances in cognitive bias test methods and the interpretation of facial and behavioural expression have been important, but more work is needed in this area. Studies that have identified different behavioural outcomes with different emotional states in sheep may be useful to follow-up in cattle. Welfare scientists have often 'borrowed' methods from human psychology, and there may be more to learn there. Additionally, as studies have shown that behavioural expression in the face and body of animals can indicate emotional state, computer image recognition and analysis systems will allow pain, stress, fear and positive states to be identified.

Further investigation into the emotional expression contained in vocalisations is warranted (Briefer 2012). Spectrographic analysis of vocalisations to assess the different parameters (such as amplitude, the lowest frequency sound produced and range of frequencies) may allow us to interpret the emotional 'message' of the vocalisations (Manteuffel et al. 2004; Briefer 2012; Green et al. 2018). As vocalisations are short in duration and occur spontaneously in response to a stressor, continuous automated monitoring may be necessary to capture vocalisations. Methodology to distinguish the vocalisations from background noise is necessary and to identify the individual emitting the sound. We also need to determine vocalisations that are always 'honest' (i.e., are reliable signals of the animal's state (Laidre and Johnstone 2013)), but if this is possible, analysis of vocalisations offers the possibility of a direct and meaningful way of capturing this key indicator of emotional state.

It may also be important to determine how to take into account individual variation in the expression of emotion to allow us to infer the presence of positive and negative states. We know that cows respond to alarming stimuli by any combination of running, freezing, snorting or jumping (Gibbons et al. 2009). Individual variation in emotional response may also be present for many other experiences that evoke an emotional state.

It is likely that the future will see further development of automated methods of assessing behaviour (see Chap. 11). Automated methods allow for continuous, non-invasive monitoring of animal behaviour and will release researchers from endless hours of video analysis. This will be important for enabling better welfare assessment protocols, as well as allowing easier assessment of behaviour in experimental settings. As shown above, changes in ongoing behaviour are an important method of detecting problems with welfare. Changes in behaviour such as feeding and resting are currently known to indicate disease (Gonzalez et al. 2008; Svensson and Jensen 2007 and see Chap. 11), but it is likely that changes in other behaviours, such as play or social behaviour, will be explored as indicators of compromise in health or welfare. This is particularly relevant for behaviours such as play, which are clear indicators of welfare status, but are infrequent and therefore difficult to assess using real-time observation. Recent studies have shown that periods of play can be detected using accelerometers (Luu et al. 2013; Gladden et al. 2020), opening up the opportunity to use the occurrence of play as a welfare indicator. Other important, but short-lived and infrequent behaviours, such as grooming and social behaviours, may also be addressed in this way.

## 2.8 Conclusion

Analysis of behaviour is an important method of assessing and understanding welfare in cattle. Various test methodologies have been established including observation of spontaneously occurring behaviour and a range of behavioural tests that assess motivation and emotional aspects of cattle welfare. Assessment of emotional states is a growing and important area of research, with new methodologies being developed. The use of behaviour as an indicator of welfare has advanced greatly in recent decades and is likely to become more important as more consumers and food retailers pay more attention to animal welfare. Future advances are likely to use technology, both in welfare assessment and to facilitate the more traditional methods of behavioural assessment.

**Acknowledgements** We would like to thank Kenny Rutherford and Simon Turner for their valuable comments on drafts of this chapter. Major thanks is due to past and present students for their academic interest and enthusiasm for animal welfare science that has vastly contributed to our own knowledge and experience of the range and complexities of behavioural testing methods and indeed, positively contributed to the scientific community. We are also grateful to our families for their support during the writing phase.

## References

Adcock SJJ, Tucker CB (2020) Conditioned place preference reveals ongoing pain in calves 3 weeks after disbudding. Nat Sci Rep 10:3849

Agricultural and Horticultural Development Board (AHDB) (2020) Dairy mobility scoresheet. https://ahdb.org.uk/knowledge-library/dairy-mobility-scoresheet. ©Agricultural and Horticultural Development Board, 2020

Andreasen SN, Sande P, Forkman B (2014) Can animal-based welfare assessment be simplified? A comparison of the Welfare Quality® protocol for dairy cattle and the simpler and less time-consuming protocol developed by the Danish Cattle Federation. Anim Welf 23:81–94. https://doi.org/10.7120/09627286.23.1.081

Archer J (1973) Tests for emotionality in rats and mice: a review. Anim Behav 21:205–235

Bateson M, Martin P (2021) Measuring behaviour. An introductory guide, 4th edn. Cambridge University Press

Bell D, Haskell M, Macrae A et al (2019) Behavioural responses of dairy calves to temperature and wind. In: Proceedings of the 53rd Congress of the ISAE, Bergen, Norway, 5–9 August, 2019, p 318

Boissy A (1995) Fear and fearfulness in animals. Q Rev Biol 70:165–191

Boissy A, Manteuffel G, Jensen MB et al (2007) Assessment of positive emotions in animals to improve their welfare. Physiol Behav 92:375–397

Boivin X, Le Neindre P, Chupin JM et al (1992) Influence of breed and early management on ease of handling and open-field behaviour of cattle. Appl Anim Behav Sci 32:313–323

Braghieri A, Pacelli C, Girolami A et al (2011) Time budget, social and ingestive behaviours expressed by native beef cows in Mediterranean conditions. Livest Sci 141:47–52. https://doi.org/10.1016/j.livsci.2011.05.001

Breuer K, Hemsworth PH, Barnett JL et al (2000) Behavioural response to humans and the productivity of commercial dairy cows. Appl Anim Behav Sci 66:273–288

Breuer K, Hemsworth PH, Coleman GJ (2003) The effect of positive or negative handling on the behavioural and physiological responses of nonlactating heifers. Appl Anim Behav Sci 84:3–22

Briefer EF (2012) Vocal expression of emotions in mammals: mechanisms of production and evidence. J Zool 288:1–20

Bučková K, Špinka M, Hintze S (2019) Pair housing makes calves more optimistic. Sci Rep 9:20246. https://doi.org/10.1038/s41598-019-56798-w

Burrow HM (1997) Measurements of temperament and their relationships with performance traits of beef cattle. Anim Breed Abstr 65:477–495

Charlton GL, Rutter SM, East M et al (2011) Preference of dairy cows: indoor cubicle housing with access to a total mixed ration vs. access to pasture. Appl Anim Behav Sci 130:1–9

Charlton GL, Rutter SM, East M, Sinclair LA (2013) The motivation of dairy cows for access to pasture. J Dairy Sci 96:4387–4396. https://doi.org/10.3168/jds.2012-6421

Cook NB, Bennett TB, Nordlund KV (2005) Monitoring indices of cow comfort in free-stall-housed dairy herds. J Dairy Sci 88:3876–3885

Corazzin M, Piasentier E, Dovier S, Stefano Bovolenta S (2010) Effect of summer grazing on welfare of dairy cows reared in mountain tie-stall barns. Ital J Anim Sci 9(3):e59. https://doi.org/10.4081/ijas.2010.e59

Cornish A, Raubenheimer D, McGreevy P (2016) What we know about the public's level of concern for farm animal welfare in food production in developed countries. Animals 6:74. https://doi.org/10.3390/ani6110074

Daros RR, Costa JHC, von Keyserlingk MAG et al (2014) Separation from the dam causes negative judgement bias in dairy calves. PLoS ONE 9(5):e98429. https://doi.org/10.1371/journal.pone.0098429

Dawkins MS (1983) Battery hens name their price: consumer demand theory and the measurement of ethological 'needs'. Anim Behav 31:1195–1205

Dawkins MS (1990) From an animal's point of view: motivation, fitness and animal welfare. Behav Brain 13:1–61

Dawkins MS (2008) The science of animal suffering. Ethology 114:937–945. https://doi.org/10.1111/j.1439-0310.2008.01557.x

Dawkins MS (2017) Animal welfare with and without consciousness. J Zool 301:1–10

De Boyer des Roches A, Lussert A, Faure M et al (2018) Dairy cows under experimentally-induced Escherichia coli mastitis show negative emotional states assessed through Qualitative Behaviour Assessment. Appl Anim Behav Sci 206:1–11

De Waal F (2016) Are we smart enough to know how smart animals are? WW Norton & Company, New York

Duncan IJH (1992) Measuring preferences and the strength of preferences. Poult Sci 71:658–663

Duncan IJH, Filshie JH (1980) The use of radio telemetry devices to measure temperature and heart rate in domestic fowl. A handbook on bioltelemetry and radio tracking. In: Proceedings on an International Conference of Telemetry and Radio Tracking in Biology and Medicine, Oxford, 20–22 March 1979

Ede T, Von Keyserlingk MAG, Weary DM (2019) Assessing the affective component of pain, and the efficacy of pain control, using conditioned place aversion in calves. Biol Lett 15:20190642

Ellingsen K, Coleman GJ, Lund V et al (2014) Using qualitative behaviour assessment to explore the link between stockperson behaviour and dairy calf behaviour. Appl Anim Behav Sci 153:10–17

FAWC (2009) Farm Animal Welfare in Great Britain: past, present and future. Farm Animal Welfare Council, London

Flower FC, Weary DM (2006) Effect of hoof pathologies on subjective assessments of dairy cow gait. J Dairy Sci 89:139–146

Flower FC, Sanderson DJ, Weary DM (2006) Effects of milking on dairy cow gait. J Dairy Sci 89:2084–2089

Fraser D, Matthews LR (1997) Preference and motivation testing. In: Appleby MC, Hughes BO (eds) Animal welfare. CAB International, New York, pp 159–173

Fregonesi JA, Tucker CB, Weary DM (2007) Overstocking reduces lying time in dairy cows. J Dairy Sci 90:3349–3354. https://doi.org/10.3168/jds.2006-794

Gibbons JM, Lawrence AB, Haskell MJ (2009) Responsiveness of dairy cows to human approach and novel stimuli. Appl Anim Behav Sci 116:163–173. 2008.08.009

Gibbons JM, Lawrence AB, Haskell MJ (2010) Measuring sociability in dairy cows. Appl Anim Behav Sci 122:84–91. https://doi.org/10.1016/j.applanim.2009.11.011

Gladden N, Cuthbert E, Ellis K et al (2020) Use of a tri-axial accelerometer can reliably detect play behaviour in newborn calves. Animals 10(7):1137

Gleerup KB, Andersen PH, Munksgaard L et al (2015) Pain evaluation in cattle. Appl Anim Behav Sci 171:25–32

Global Animal Partnership (2021) Global Animal Partnership's 5-Step® Animal Welfare Pilot Standards for Dairy Cattle. © Global Animal Partnership

Gonzalez LA, Tolkamp BJ, Coffey MP et al (2008) Changes in feeding behavior as possible indicators for the automatic monitoring of health disorders in dairy cows. J Dairy Sci 91:1017–1028

Grandin T (1993) Behavioral agitation during handling of cattle is persistent over time. Appl Anim Behav Sci 36:1–9

Green AC, Johnston IN, Clark CEF (2018) Invited review: the evolution of cattle bioacoustics and application for advanced dairy systems. Animal 12(6):1250–1259

Grössbacher V, Lawrence AB, Winckler CG et al (2020) Negative play contagion in calves. Sci Rep 10:21699. https://doi.org/10.1038/s41598-020-78748-7

Hall SJG (1989) Chillingham cattle: social and maintenance behaviour in an ungulate that breeds all year round. Anim Behav 38:215–225

Hänninen L, Mäkelä JP, Rushen J et al (2008) Assessing sleep state in calves through electrophysiological and behavioural recordings: a preliminary study. Appl Anim Behav Sci 111:235–250

Harding EJ, Paul ES, Mendl M (2004) Animal behavior – cognitive bias and affective state. Nature 427:312

Herskin MS, Kristensen A-M, Munksgaard L (2004) Behavioural responses of dairy cows toward novel stimuli presented in the home environment. Appl Anim Behav Sci 89:27–40. https://doi.org/10.1016/j.applanim.2004.06.006

Hunter LB, Haskell MJ, Langford FM et al (2021a) Heart rate and heart rate variability change with sleep stage in dairy cows. Animals 11(7):2095. https://doi.org/10.3390/ani11072095

Hunter LB, O'Connor C, Langford FM et al (2021b) Machine learning prediction of sleep stages in dairy cows from heart rate and muscle activity measures. Sci Rep 11(1):10938. https://doi.org/10.1038/s41598-021-90416-y

Jensen MB, Vestergaard KS, Krohn CC et al (1997) Effect of single versus group housing and space allowance on responses of calves during open-field tests. Appl Anim Behav Sci 54:109–121

Jensen MB, Vestergaard KS, Krohn CC (1998) Play behaviour in dairy calves kept in pens: the effect of social contact and space allowance. Appl Anim Behav Sci 56:97–108. https://doi.org/10.1016/S0168-1591(97)00106-8

Jensen MB, Pedersen LL, Munksgaard L (2005) The effect of reward duration on demand functions for rest in dairy heifers and lying requirements as measured by demand functions. Appl Anim Behav Sci 90:207–217

Johnsen PF, Johannesson T, Sandoe P (2001) Assessment of farm animal welfare at herd level. Acta Agric Scand Sect A Anim Sci (Suppl 30):26–33

Kappel S, Mendl MT, Barrett DC et al (2017) Lateralized behaviour as indicator of affective state in dairy cows. PLoS ONE 12:e0184933

Kiley-Worthington M, de la Plain S (1983) The behaviour of suckler beef cattle (*Bos Taurus*). Tierhaltung: BD 14. Birkhäuser Verlag Basel

Kilgour RJ, Uetake K, Ishiwata T et al (2012) The behaviour of beef cattle at pasture. Appl Anim Behav Sci 138:12–17. https://doi.org/10.1016/j.applanim.2011.12.001

Kirkden R, Pajor E (2006) Using preference, motivation and aversion tests to ask questions about animals' feelings. Appl Anim Behav Sci 100:29–47

Laidre ME, Johnstone RA (2013) Animal signals. Curr Biol 23:R829–R833. https://doi.org/10.1016/j.cub.2013.07.070

Langford DJ, Bailey A, Chanda ML et al (2010) Coding of facial expressions of pain in the laboratory mouse. Nat Methods. https://doi.org/10.1038/nmMeth.1455

LeCorps B, Weary DM, von Keyserlingk MAG (2018) Pessimism and fearfulness in dairy calves. Sci Rep 8:1421. https://doi.org/10.1038/s41598-017-17214-3
Lee C, Fisher AD, Colditz IG et al (2013) Preference of beef cattle for feedlot or pasture environments. Appl Anim Behav Sci 145:53–59
Lidfors LM (1996) Behavioural effects of separating the dairy calf immediately or 4 days postpartum. Appl Anim Behav Sci 49:260–283
Lidfors LM, Moran D, Jung J et al (1994) Behaviour at calving and choice of calving place in cattle kept in different environments. Appl Anim Behav Sci 42:11–28. https://doi.org/10.1016/0168-1591(94)90003-5
Lürzel S, Münsch C, Windschnurer I et al (2015) The influence of gentle interactions on avoidance distance towards humans, weight gain and physiological parameters in group-housed dairy calves. Appl Anim Behav Sci 172:9–16
Luu J, Føske Johnsen J, de Passillé AM et al (2013) Which measures of acceleration best estimate the duration of locomotor play by dairy calves? Appl Anim Behav Sci 148:21–27
Main DCJ, Kent JP, Wemelsfelder F et al (2003) Applications for methods of on-farm welfare assessment. Anim Welf 12:523–528
Mandel R, Whay HR, Nicol CJ et al (2013) The effect of food location, heat load, and intrusive medical procedures on brushing activity in dairy cows. J Dairy Sci 96:6506–6513
Mandel R, Harazy H, Gygax L et al (2018) Short communication: detection of lameness in dairy cows using a grooming device. J Dairy Sci 101:1511–1517
Manteuffel G, Puppe B, Schon PC (2004) Vocalization of farm animals as a measure of welfare. Appl Anim Behav Sci 88:163–182
Matthews LR, Ladewig J (1994) Environmental requirements of pigs measured by behavioural demand functions. Anim Behav 47:713–719
Meagher RK, Weary DM, von Keyserlingk MAG (2017) Some like it varied: individual differences in preference for feed variety in dairy heifers. Appl Anim Behav Sci 195:8–14
Metz JHM (1985) The reaction of cows to a short-term deprivation of lying. Appl Anim Behav Sci 13:301–307
Miedema HM, Cockram MS, Dwyer CM et al (2011) Changes in the behaviour of dairy cows during the 24 h before normal calving compared with behaviour during late pregnancy. Appl Anim Behav Sci 131(2011):8–14. https://doi.org/10.1016/j.applanim.2011.01.012
Monk JE, Doyle RE, Colditz IG et al (2018) Towards a more practical attention bias test to assess affective state in sheep. PLoS ONE 13(1):e0190404. https://doi.org/10.1371/journal.pone.0190404
Munksgaard L, Jensen MB, Pedersen LJ et al (2005) Quantifying behavioural priorities—effects of time constraints on behaviour of dairy cows, Bos taurus. Appl Anim Behav Sci 92(2005):3–14
Neave HW, Daros RR, Costa JHC et al (2013) Pain and pessimism: dairy calves exhibit negative judgement bias following hot-iron disbudding. PLoS ONE 8(12):e80556. https://doi.org/10.1371/journal.pone.0080556
Nielsen BL (2020) Asking animals: an introduction to animal behaviour testing. CABI, Wallingford
Pajor EA, Rushen J, de Passillé AMB (2000) Aversion learning techniques to evaluate dairy cattle handling practices. Appl Anim Behav Sci 69:89–102
Pajor EA, Rushen J, de Passillé AMB (2003) Dairy cattle's choice of handling treatments in a Y-maze. Appl Anim Behav Sci 80:93–107
Palacio S, Bergeron R, Lachance S et al (2015) The effects of providing portable shade at pasture on dairy cow behavior and physiology. J Dairy Sci 98:6085–6093. https://doi.org/10.3168/jds.2014-8932
Park RM, Schubach KM, Cooke RF et al (2020) Impact of a cattle brush on feedlot steer behavior, productivity and stress physiology. Appl Anim Behav Sci 228:104995. https://doi.org/10.1016/j.applanim.2020.104995
Parola F, Hillmann E, Schütz KE et al (2012) Preferences for overhead sprinklers by naïve beef steers: test of two nozzle types. Appl Anim Behav Sci 137:13–22. https://doi.org/10.1016/j.applanim.2011.12.010

Paul ES, Harding EJ, Mendl M (2005) Measuring emotional processes in animals: the utility of a cognitive approach. Neurosci Biobehav Rev 29:469–491

Phillips CJC, Oevermans H, Syrett KL et al (2015) Lateralization of behavior in dairy cows in response to conspecifics and novel persons. J Dairy Sci 98:2389–2400

Pinheiro Machado Filho LC, Teixeira DL, Weary DM et al (2004) Designing better water troughs: dairy cows prefer and drink more from larger troughs. Appl Anim Behav Sci 89:185–193

Plesch G, Broerkens N, Laister S et al (2010) Reliability and feasibility of selected measures concerning resting behaviour for the on-farm welfare assessment in dairy cows. Appl Anim Behav Sci 126:19–26

Proctor HS, Carder G (2015) Measuring positive emotions in cows: do visible eye whites tell us anything? Physiol Behav 147:1–6

Red Tractor, UK (2022). https://redtractor.org.uk/our-standards/dairy/

Rioja-Lang FC, Roberts DJ, Healy SD et al (2009) Dairy cows trade-off feed quality with proximity to a dominant individual in Y-maze choice tests. Appl Anim Behav Sci 117:159–164. https://doi.org/10.1016/j.applanim.2008.12.003

Rogers LJ (2010) Relevance of brain and behavioural lateralization to animal welfare. Appl Anim Behav Sci 127:1–11

Rørvang MV, Nielsen BL, Herskin MS et al (2019) Prepartum maternal behavior of domesticated cattle: a comparison with managed, feral, and wild ungulates. Front Vet Sci 12 March 2018. https://doi.org/10.3389/fvets.2018.00045

Rousing T, Waiblinger S (2004) Evaluation of on-farm methods for testing the human–animal relationship in dairy herds with cubicle loose housing systems—test–retest and inter-observer reliability and consistency to familiarity of test person. Appl Anim Behav Sci 85:215–231

Rushen J, de Passillé AM (2012) Automated measurement of acceleration can detect effects of age, dehorning and weaning on locomotor play of calves. Appl Anim Behav Sci 139:169–174

Rushen J, Pombourcq E, de Passillé AM (2007) Validation of two measures of lameness in dairy cows. Appl Anim Behav Sci 106:173–177

Sandem I, Braastad BO, Bøe KE (2002) Eye white may indicate emotional state on a frustration-contentedness axis in dairy cows. Appl Anim Behav Sci 79:1–10

Schirmann K, Chapinal N, Weary DM et al (2011) Short-term effects of regrouping on behavior of prepartum dairy cows. J Dairy Sci 94:2312–2319

Schlageter-Tello A, Bokkers EAM, Groot Koerkamp PWG et al (2014) Manual and automatic locomotion scoring systems in dairy cows: a review. Prev Vet Med 116:12–25

Schütz K, Davison D, Matthews L (2006) Do different levels of moderate feed deprivation in dairy cows affect feeding motivation? Appl Anim Behav Sci 101:253–263. https://doi.org/10.1016/j.applanim.2006.02.008

Schütz KE, Cox NR, Matthews LR (2008) How important is shade to dairy cattle? Choice between shade or lying following different levels of lying deprivation. Appl Anim Behav Sci 114:307–318. https://doi.org/10.1016/j.applanim.2008.04.001

Schütz KE, Rogers AR, Cox NR et al (2009) Dairy cows prefer shade that offers greater protection against solar radiation in summer: shade use, behaviour, and body temperature. Appl Anim Behav Sci 116:28–34

Schütz KE, Huddart FJ, Cox NR (2021) Effects of short-term exposure to drinking water contaminated with manure on water and feed intake, production and lying behaviour in dairy cattle. Appl Anim Behav Sci 238:105322

Shewbridge-Carter L, Rutter SM, Ball D et al (2021) Dairy cow trade-off preference for two different lying qualities: lying surface vs. lying space. J Dairy Sci 104:862–873. https://doi.org/10.3168/jds.2020-18781

Sommariba M, Lonis W, Roehe R et al (2019) The effects of a composite chronic stress treatment on fear responses and attention bias in beef cattle. In: Proceedings of the 53rd Congress of the International Society for Applied Ethology, p 142. 5–9 August 2019, Bergen, Norway

Sprecher DJ, Hostetler DE, Kaneene JB (1997) A lameness scoring system that uses posture and gait to predict dairy cattle reproductive performance. Theriogenology 47:1179–1187

Svensson C, Jensen MB (2007) Identification of diseased calves by use of data from automatic milk feeders. J Dairy Sci 90:994–997

Ternman E, Pastell M, Agenäs S et al (2014) Agreement between different sleep states and behaviour indicators in dairy cows. Appl Anim Behav Sci 160:12–18. https://doi.org/10.1016/j.applanim.2014.08.014

Tucker CB, Lacy-Hulbert SJ, Webster JR (2009) Effect of milking frequency and feeding level before and after dry off on dairy cattle behavior and udder characteristics. J Dairy Sci 92:3194–3203

Tucker CB, Munksgaard L, Mintline EM et al (2018) Use of a pneumatic push gate to measure dairy cattle motivation to lie down in a deep-bedded area. Appl Anim Behav Sci 201:15–24. https://doi.org/10.1016/j.applanim.2017.12.018

Tucker CB, Jensen MB, de Passillé AM et al (2021) Lying time and the welfare of dairy cows. J Dairy Sci 104:20–46. https://doi.org/10.3168/jds.2019-18074

Turner SP, Navajas EA, Hyslop JJ et al (2011) Associations between response to handling and growth and meat quality in frequently handled Bos taurus beef cattle. J Anim Sci 89:4239–4248. https://doi.org/10.2527/jas.2010-3790

Ugwu N, Love E, Murrell J et al (2021) Development of an ethogram for hutch-housed dairy calves and determination of factors influencing their behaviour. Appl Anim Behav Sci 234:105165. https://doi.org/10.1016/j.applanim.2020.105165

Valizaheh R, Veira DM, von Keyserlingk MAG (2008) Behavioural responses by dairy cows provided two hays of contrasting quality at dry-off. Appl Anim Behav Sci 109:190–200

Van Reenan CG, van der Werf JTN, O'Connell NE et al (2013) Behavioural and physiological responses of heifer calves to acute stressors: long-term consistency and relationship with adult reactivity to milking. Appl Anim Behav Sci 147:55–68

Veissier I, Butterworth A, Buck B et al (2008) European approaches to ensure good animal welfare. Appl Anim Behav Sci 113:279–297

Ventura BA, von Keyserlingk MAG, Schuppli CA et al (2013) Views on contentious practices in dairy farming: the case of early cow-calf separation. J Dairy Sci 96:6105–6116. https://doi.org/10.3168/jds.2012-6040

Vitale AF, Tenucci M, Papini M et al (1986) Social behaviour of the calves of semi-Wild Maremma cattle, *Bos Primigenius Taurus*. Appl Anim Behav Sci 16:217–231

von Keyserlingk MAG, Cestari AA, Franks B et al (2017) Dairy cows value access to pasture as highly as fresh feed. Sci Rep 7:44953 l. https://doi.org/10.1038/srep44953

Wagner K, Barth K, Hillman E, Palme R, Futschik A, Waiblinger S (2013) Mother rearing of dairy calves: reactions to isolation and to confrontation with an unfamiliar conspecific in a new environment. Appl Anim Behav Sci 147:43–54

Waiblinger S, Menke C, Coleman GJ (2002) The relationship between attitudes, personal characteristics and behaviour of stockpeople and subsequent behaviour and production of dairy cows. Appl Anim Behav Sci 79:195–219

Watts JM, Stookey JM (2000) Vocal behaviour in cattle: the animal's commentary on its biological processes and welfare. Appl Anim Behav Sci 67:15–33

Weary DM, Chua B (2000) Effects of early separation on the dairy cow and calf 1. Separation at 6 h, 1 day and 4 days after birth. Appl Anim Behav Sci 69:177–188

Webb LE, Engel B, Berends H et al (2014) What do calves choose to eat and how do preferences affect behaviour? Appl Anim Behav Sci 161:7–19. https://doi.org/10.1016/j.applanim.2014.09.016

Welfare Quality® (2009) Welfare Quality assessment protocol for cattle. Welfare Quality® Consortium, Lelystad

Wemelsfelder F (2007) How animals communicate quality of life: the qualitative assessment of behaviour. Anim Welf 16:25–31

Winckler C (2018) Assessment of cattle welfare: approaches, goals, and next steps on farms. Advances in Cattle Welfare, pp 55–69. Woodhead Publishing Series in Food Science, Technology and Nutrition. https://doi.org/10.1016/B978-0-08-100938-3.00003-6. Copyright © 2018 Elsevier Ltd.

Winckler C, Willen S (2001) The reliability and repeatability of a lameness scoring system for use as an indicator or welfare in dairy cattle. Acta Agric Scand Sect A Anim Sci Supplementum 30:103–107

Winckler C, Brinkmann J, Glatz J (2007) Long-term consistency of selected animal-related welfare parameters in dairy farms. Anim Welf 16:197–199

Zambelis A, Gagnon-Barbin M, St John J et al (2019) Development of scoring systems for abnormal rising and lying down by dairy cattle, and their relationship with other welfare outcome measures. Appl Anim Behav Sci 220:104858. https://doi.org/10.1016/j.applanim.2019.104858

# Physiological and Immunological Tools and Techniques for the Assessment of Cattle Welfare

# 3

Bernadette Earley, Kelly Buckham-Sporer, Aran O'Loughlin, and Dayle Johnston

## Contents




**Abstract**

Cattle are exposed to many different types of stressors, for example, the housing of animals during winter, castration, dehorning, weaning, transport, handling and slaughter, each of which have the potential to cause stress, pain and injury if not managed correctly. An increase in stress levels in cattle has the potential to lower



B. Earley (✉) · K. Buckham-Sporer · A. O'Loughlin · D. Johnston
Animal and Bioscience Research Department, AGRIC, Teagasc, Grange, Dunsany, Co. Meath, Ireland
e-mail: bernadette.earley@teagasc.ie







M. Haskell (ed.), *Cattle Welfare in Dairy and Beef Systems*, Animal Welfare 23,
https://doi.org/10.1007/978-3-031-21020-4_3

immunity, impair growth performance and increase susceptibility to disease. All of these potential responses contribute to, and result in, reduced animal welfare.

The aim of this chapter is to present the main actions of the major stress axes and describe the relationships between the stress and immune systems. The underlying mechanisms will be highlighted using examples relating to castration, transport, housing and weaning management practices of beef cattle. The type of stress response depends on the nature (acute or chronic) of the stressful event and stressors commonly encountered by cattle fall into both these categories. Immunocompetence is clearly important in resistance against disease and may be compromised by welfare challenges. The immune response of an animal to a stressor may be influenced by the type or duration of the stressor, genetics, age, social status, the time of sampling in relation to both the onset of stress and the time of day, pathogen exposure, health status and the functioning of the immune system There are a number of immune measures available to us to quantify the response, including the relative populations of white blood cells, laboratory-based assays of white blood cell function and measurement of physiological biomarkers produced by or relevant to the immune system and disease. The application of these measures in assessment of welfare will be discussed in this chapter.

**Keywords**

Stress · Cortisol · Immune function · Biomarkers

## Abbreviations

| | |
|---|---|
| ACTH | Adrenocorticotropic hormone |
| ADP | Adenosine diphosphate |
| AMP | Adenosine monophosphate |
| AP-1 | Activation protein 1 |
| APC | Antigen presenting cell |
| APP | Acute-phase protein |
| APR | Acute-phase response |
| AVP | Arginine-vasopressin |
| BRD | Bovine respiratory disease |
| BRDC | Bovine respiratory disease complex (BRDC) |
| CD | Cluster of differentiation |
| CD62L | L-selectin |
| cDNA | Complementary DNA |
| CNS | Central nervous system |
| CRF | Corticotrophin-releasing factor |
| CRH | Corticotrophin-releasing hormone |
| DEG | Differentially expressed gene |
| DHEA | Dehydroepiandrosterone |

| | |
|---|---|
| DNA | Deoxyribonucleic acid |
| GC | Glucocorticoid |
| GR | Glucocorticoid receptor |
| GRE | Glucocorticoid response element |
| HPA | Hypothalamic–pituitary–adrenal axis |
| HSP | Heat shock protein |
| IBR | Infectious bovine rhinotracheitis |
| IFN | Interferon |
| IL | Interleukin |
| LPS | Lipopolysaccharide |
| MHC | Major histocompatibility complex |
| MMP | Matrix metalloproteinase |
| MR | Mineralocorticoid receptor |
| mRNA | Messenger RNA |
| N:L ratio | Neutrophil:lymphocyte ratio |
| NFκB | Nuclear factor kappa B |
| NGS | Next-generation sequencing |
| NK | Natural killer cell |
| PBMC | Peripheral blood mononuclear cell |
| PVN | Paraventricular nucleus |
| RBC | Red blood cell |
| RNA | Ribonucleic acid |
| ROS | Reactive oxygen species |
| RT | Reverse transcriptase |
| SAA | Serum amyloid A |
| SAM | Sympathetic adrenomedullary |
| Th1 | T helper cell type 1 |
| Th2 | T helper cell type 2 |
| TLR | Toll-like receptor |
| TNF | Tumour necrosis factor |

## 3.1 Introduction

Cattle are exposed to many different types of husbandry management procedures, all of which have the potential to induce stress. In cattle, research measuring stress-related immune function has focused on a number of husbandry management practices including castration, housing, transport and weaning. Exposure to stress is inevitable, but healthy animals with access to appropriate resources are able to cope with mild stressors. However, severe or prolonged stress can ultimately affect health as prolonged stress affects the immune systems (Thornton et al. 2022). To improve welfare in these situations, we need to understand the stress and immune function systems and the interactions between the two systems. In this chapter, we will firstly define what is meant by the term stress and discuss the physiological response to stressors.

## 3.2 Stress

### 3.2.1 Definition of Stress

A comprehensive model of stress was developed and outlined by Moberg (1985, 1987; Moberg and Mench 2000) which used the terms "stressful event" resulting in a "stress response" leading to the "consequences of stress". A stress response was considered not only in terms of adrenocortical activity, but as encompassing a range of behavioural, neuroendocrine and autonomic nervous system reactions. The recognition of the threat to homeostasis posed by the stressful event occurs within the central nervous system, which then organises the biological response. The resulting change in biological function may progress to a pre-pathological state, where changes are seen in behaviour, the immune system and reproductive function. This may further progress to a pathological state (Moberg 1985, 1987; Moberg and Mench 2000; Herman et al. 2020; Niu et al. 2022). Von Borell (2001) defined stress as a condition in an animal that results from the action of one or more stressors that may be of either external or internal origin. A stressor could be considered harmful depending on how the organism is able to cope with the insult as it regains homeostasis. The expression of the stress can be measured using physiological, immunological and behavioural approaches.

### 3.2.2 Physiological Responses of Animals to Stress

The physiological response to a stressor involves a complex interaction between the nervous, endocrine and immune systems. When a stressor is first encountered, two systems respond to it. The most immediate is the "flight or fight response" that may allow the situation to be resolved by the animal repelling the threat or removing itself from the threatening situation. The sympathetic–adrenal medullary (SAM) axis is involved in this response. The term "fight or flight" was first described by Cannon (1932) when he stated that "animals in threatening situations show adaptive responses in which they fight or retreat". The second is a more general stress response that alters body function to in response to the stressor. This is the hypothalamic–pituitary–adrenal (HPA) axis (Carroll and Burdick Sanchez 2014; Brown and Vosloo 2017). These axes will be discussed in more detail below.

### 3.2.3 The Structure and Function of the SAM Axis

The SAM axis controls the "fight or flight" response (Griffin 1989) by controlling physiological homeostasis and the coping mechanism of an animal in response to a stressor (Schommer et al. 2003). Following the description of the "fight or flight" response, it was Cannon's work that highlighted the SAM axis as one of the main components initially responding to a threat but also involved in reinstating homeostasis after the threatening encounter. The SAM axis contains the sympathetic nervous system and the adrenal glands. When a stressor is present, the SAM axis will trigger the production of sympathetic neurotransmitters (known as catecholamines and includes epinephrine, norepinephrine and dopamine (Sapolsky et al. 2000; Sejian et al. 2018)). These neurotransmitters have different functions that facilitate the animal's immediate attempts to deal with the stressor. Norepinephrine is released from the sympathetic nerve endings resulting in greater mental acuity, and epinephrine released from the medullae of the adrenal glands changes metabolic function and activities of the cardiovascular and respiratory systems. This facilitates oxygenation of blood and raises plasma glucose concentration, thereby delivering enriched blood selectively to tissues and organs upon which the stressors are making most demands.

### 3.2.4 The Structure and Function of the HPA Axis

The HPA axis is made up of the hypothalamus, the anterior pituitary gland and the adrenal glands (Fig. 3.1). The HPA axis response to stress commences within the brain. It is initiated by hypophysiotropic neurons within the paraventricular nucleus (PVN). These neurons release neuropeptides into the hypophyseal portal blood system and synthesise and release corticotrophin-releasing hormone (CRH) and arginine vasopressin (AVP). Corticotrophin-releasing hormone acts synergistically with AVP to act on corticotrophin cells in the anterior pituitary which stimulates the secretion of adrenocorticotrophic hormone (ACTH). ACTH then acts upon the outer adrenal cortex of the adrenal gland which regulates glucocorticoid (GC) and adrenal androgen secretion by the zona fasiculata and reticularis, respectively.

The levels of these GCs in the blood peak between 10 minutes and 1 hour after the initiation of the stress response (Sapolsky et al. 2000) and act upon a wide range of cells, tissues and organs throughout the body including those of the immune system (Chrousos 2009). While GCs are beneficial to the short-term survival of cattle, prolonged exposure can lead to serious metabolic, immune and psychological dysfunction (McEwen and Stellar 1993; Burton et al. 2005). To combat this, GCs exert a rapid negative feedback on the secretion of CRH from the hypothalamus by inhibiting gene transcription (Herman 1992; Herman et al. 2020) and ACTH secretion from the pituitary thus preventing further secretion of GCs from the adrenal gland (Elenkov and Chrousos 2002). The stress response can be terminated in one of three ways, (i) a rate-sensitive feedback which is rapid and occurs within minutes of elevated GC levels, (ii) an intermediate feedback which is slightly slower or (iii) a delayed feedback which occurs at a transcriptional level over the course of a few hours.

The main active GC in cattle is cortisol, a cholesterol-derived steroid (Mormède et al. 2007; Burdick et al. 2011). Only 10% of cortisol is found in its unbound state, as 80% is bound by corticosterone and the other 10% is bound by albumin. Glucocorticoids have a wide range of functions in the body such as anti-inflammatory actions, gluconeogenesis, metabolism of proteins (Vegiopoulos and Herzig 2007), reproduction (Tempel and Leibowitz 1994; Burton et al. 2005), immune function

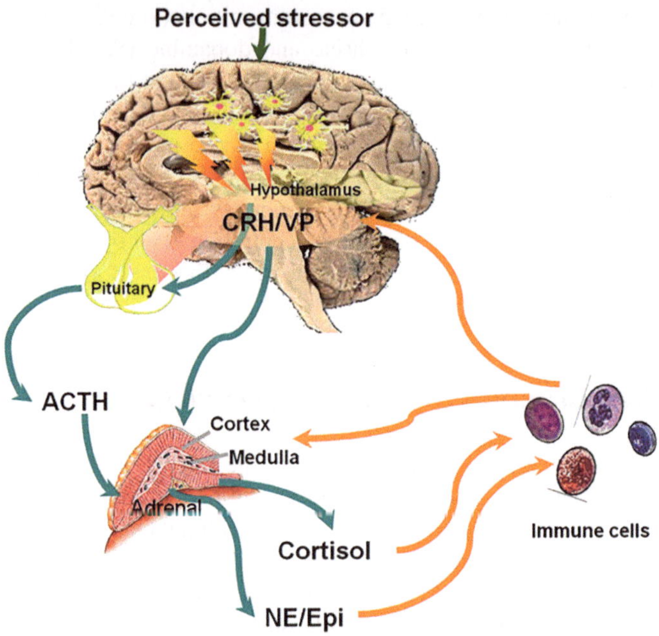


**Fig. 3.1** Stress regulation by the hypothalamic–pituitary–adrenal (HPA) axis. The HPA axis is the primary pathway by which glucocorticoid (primarily cortisol in most mammals including cattle) secretion is regulated. The HPA axis is primarily regulated via two neurohormones, corticotropin-releasing hormone (CRH) and vasopressin (VP), which are released from the hypothalamus when cattle are exposed to or perceive a stressor. Release of these neurohormones stimulates the release of ACTH from the anterior portion of the pituitary gland. Once released into circulation, ACTH stimulates the release of cortisol from the adrenal cortex. Cortisol then acts as an endocrine hormone travelling throughout the body where it initiates various stimulatory and inhibitory actions on target tissues, including immune cells. In addition to cortisol, the catecholamines epinephrine (EPI) and norepinephrine (NE) are also typically released by the adrenal medulla during times of stress and/or immunological insult (Source: Carroll and Burdick Sanchez 2014; Image reproduced with permission from American Journal of Animal Science, Oxford University Press)

(Carroll and Forsberg 2007), growth regulation (Sartin et al. 1998), cardiovascular output (Brotman et al. 2007) and regulation of the stress response.

There are connections between the SAM axis and the HPA axis. If the SAM axis fails to resolve the stressor, the HPA axis is activated. Epinephrine and norepinephrine influence the HPA axis by regulating the release of hormones such as corticotrophin-releasing hormone (CRH) from the PVN of the hypothalamus, adrenocorticotrophic hormone (ACTH) from the anterior pituitary and cortisol from the adrenal cortex (Chrousos and Gold 1992). The sympathetic component of the SAM axis contains preganglionic neurons, which lie within the spinal cord, which release the neurotransmitter acetylcholine. Acetylcholine then triggers postganglionic neurons to discharge norepinephrine directly into target tissue.

Although improved assays have been developed for the measurement of plasma catecholamines, circulating cortisol may still be a more reliable stress marker and more easily measured than plasma epinephrine. This is mainly due to the difficulty in scheduling blood collection around the very short half-life of epinephrine in the blood following a stressful event (Minton 1994). In addition, alterations of measures of sympathetic nervous activity occur in humans in conjunction with daily activities such as eating or movement, suggesting that catecholamines may not be a reliable and quantifiable indicator of a stress response in farm animals. Heart rate and respiratory rate have been used as possible indicators of SAM activity (Andrade et al. 2001). A final drawback to measuring endocrine activity is that chronically stressed animals may not exhibit any change in endocrine response (Dantzer and Mormède 1983; Dickens and Romero 2013).

## 3.2.5 Biomarkers of Stress in Cattle/Hormones of the Neuroendocrine System

### 3.2.5.1 Measures of Stress in Cattle

The endocrine activity that is part of the stress response process is commonly measured as an indicator of stress (Collier et al. 2017). As discussed above, it is generally agreed that activation of the HPA axis signals a stress response; therefore, analysis of circulating glucocorticoids, cortisol being the most predominant, has been measured in many cattle stress studies (Chen et al. 2015). These studies have assessed conditions and contexts including lameness (Almeida et al. 2007), transportation (Blecha et al. 1984; Yagi et al. 2004; Earley and O'Riordan 2006; Gupta et al. 2007a, b, Meléndez et al. 2021), branding (Lay et al. 1992), castration (Pang et al. 2006), weaning (Hickey et al. 2003a), regrouping and relocation (Gupta et al. 2008) and restraint (reviewed by Grandin (1997)). Less frequently, indicators of SAM axis activation may be monitored. Measurement of the circulating catecholamines epinephrine and norephinephrine is considered a marker of very acute stress (Griffin 1989; Chen et al. 2015) and has also been utilised in bovine stress studies investigating weaning, branding and housing quality (Hickey et al. 2003a, b; Lay et al. 1992). Elevations in heart rate have also been used as a physiological marker of stress in cattle (Lay et al. 1992).

Cortisol is perhaps the best-known biomarker of stress and can be rapidly measured (Sapolsky et al. 2000). However, there are some issues with using plasma cortisol concentrations to assess stress, and these are discussed further below. However, numerous bovine studies have successfully used cortisol measurements as an adjunct to other measurements in order to assess welfare in cattle. These include studies examining weaning (Hickey et al. 2003a, b; Blanco et al. 2009; Lynch et al. 2010a, 2011), transport (Yagi et al. 2004; Buckham Sporer et al. 2008; Sporer et al. 2008; Gupta et al. 2007b), castration (Pang et al. 2006, 2009a, b, 2010) and regrouping (Gupta et al. 2008). The ratio of cortisol to dehydroepiandrosterone (DHEA) is also sometimes used as a biomarker of stress in weaning (Lynch et al. 2011) and transport (Buckham Sporer et al. 2008) studies. DHEA, a cortisol precursor, has been implicated in the stress response (Zinder and Dar 1999) and acts to counteract the assumed immunosuppressive effects of cortisol (Saccò et al. 2002;

Bauer 2005). A reduction in DHEA concentration has been shown to occur with an increase in cortisol concentration in a number of studies (Buckham Sporer et al. 2008). The pituitary-derived stress hormone ACTH is responsible for glucocorticoid secretion from the adrenal gland. ACTH has been shown to increase during transport stress, highlighting its potential use as a biomarker in future studies of stress (Dixit et al. 2001; Knights and Smith 2007).

Unlike basal cortisol concentrations, the cortisol response induced by exogenous ACTH administration may provide an independent index of adrenocortical sensitivity (Moberg and Mench 2000). Given the difficulties in interpretation of a lack of stress response in situations of chronic stress, other methodologies have been developed to assess acute stress responses when chronic stress may be involved. The use of exogenous administration of glucocorticoids is one such method. Through dynamic testing methodologies, the HPA axis can be pharmacologically stimulated with the use of exogenous CRH or ACTH, and the response at both the pituitary and adrenal levels can be evaluated. Indeed, HPA axis challenges by way of exogenous CRH or ACTH stimulation have been shown to be appropriate for investigation of the bovine stress response (Fig. 3.2) (Fisher et al. 1997, 2002; Gupta et al. 2007a, b, 2008).

In a study by Gupta et al. (2008) involving the regrouping and relocation (R&R) of cattle while on an indoor housing study, the response of the adrenal cortex to

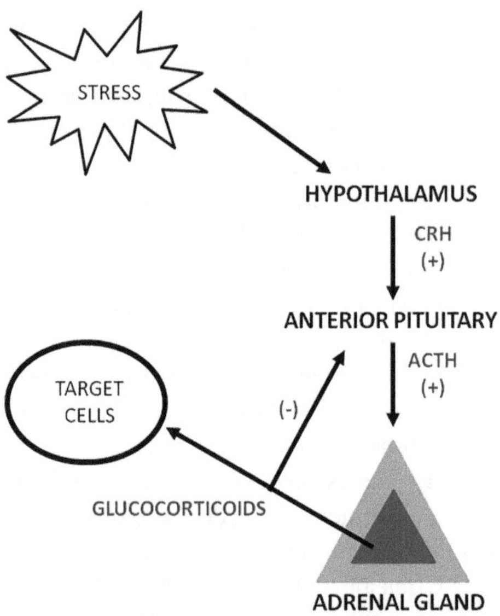


**Fig. 3.2** Representation of HPA axis activation. Perceived stressors stimulate the release of corticotropin-releasing hormone (CRH) from the hypothalamus in the brain. This triggers the anterior pituitary to secrete adrenocorticotropic hormone (ACTH) into the circulation, stimulating the release of glucocorticoids, cortisol included, by the adrenal cortex of the adrenal glands. Glucocorticoids bind to their receptors (GR) in target cells and also signal the anterior pituitary to slow its release of ACTH in a negative feedback loop

exogenous ACTH was tested over six time-points. Steers were implanted with a jugular catheter 24 h before the exogenous ACTH challenge to facilitate serial blood sample collections. Dexamethasone (20 μg/kg BW; Faulding Pharmaceuticals Plc, UK) was administered (i.m.) at −12 h to steers undergoing the ACTH challenge. The purpose of the administration of the dexamethasone prior to the ACTH challenge was to equalise systemic concentrations of cortisol in animals across the treatment groups, in an effort to facilitate a more equitable examination of ACTH on cortisol release. The authors reported a decreased responsivity of the adrenal gland to ACTH following acute stress stimuli (first R&R) coupled with the chronic effect of subsequent repeated R&R (first, second and third) that down-regulated the pituitary–adrenal axis or increased the sensitivity of the pituitary to cortisol negative feedback. Furthermore, following repeated exposure (i.e. after the sixth R&R), there was no change in cortisol response to ACTH challenge, which indicates a decrease in the adrenal gland responsiveness to ACTH. The effect could be occurring either via ACTH receptor concentrations in the adrenal gland, or in the synthesis, release or clearance of cortisol.

### 3.2.6 Limitations to Measurements of Stress in Cattle

Although the assessment of circulating cortisol levels is the most predominant measure of stress studied in cattle, there are limitations to relying solely on this measure as an indicator of the extent of stress that an animal experiences. Firstly, it has been shown that the blood sampling necessary to obtain plasma or serum for the assay of cortisol can itself induce HPA activation (Möstl and Palme 2002). Blood GC concentrations are also influenced by factors such as diet, time of day and GR sensitivity (Sapolsky et al. 2000; Nader et al. 2010). A study using intensive sampling illustrated the circadian rhythm of plasma cortisol in the absence of stress. The study sampled sexually mature bulls at 30 min intervals for 24 h and demonstrated rapid and frequent cortisol fluctuations throughout the day with lowest concentrations in the evening and highest concentrations in the morning (Thun et al. 1981). Interestingly, a blind bull included in the same study did not exhibit a circadian rhythm, implying that the light–dark cycle of each day is in part responsible for this rhythm. Protocols for assaying glucocorticoids in faeces or saliva have been developed (Palme 2019) in order to avoid confounding results due to stress incurred by handling and blood sampling (Loerch and Fluharty 1999; Möstl and Palme 2002).

Difficulties in quantifying circulating cortisol may also arise in circumstances of chronic stress (von Borell 2001). While an increase in cortisol is typically viewed as evidence of stress in cattle, high levels of basal cortisol may already be present in an animal in a chronic stress situation, so it may not be able to show an additional response to an added stressor. For example, animals subjected to heat stress for very long periods may actually have higher basal cortisol concentrations; therefore, any observation of an increase may be non-existent or misleading.

## 3.3 Stress and the Immune System

### 3.3.1 Effect of Stress on the Immune System

Stress may have beneficial and detrimental effects on the immune system depending on the duration of the stressor and the underlying state of the animal. According to Kelley (1985), the phenomenon that stressed animals are more susceptible to disease was first reported by Louis Pasteur in 1878 when he demonstrated that chickens exposed to cold stress were far more likely to die when infected with the anthrax pathogen compared to chickens not exposed to cold stress, which were able to suppress the anthrax pathogen. In reference in farmed livestock, Moberg (1985) stated that animals under severe stress are likely to succumb to disease, fail to reproduce or fail to develop properly. In cattle, research has shown that the levels of stress experienced in some of the common husbandry practices such as abrupt weaning and transport can adversely affect the immune system.

#### 3.3.1.1 Acute Versus Chronic Stress

Suppression of the immune system is usually attributed to chronic stress (stressors lasting days to months) (Carroll and Forsberg 2007; Salak-Johnson and McGlone 2007). Conversely, the immune system is usually elevated and enhanced in response to acute stressors (stressors lasting minutes to days) (Carroll and Forsberg 2007). Responses to acute stress are designed to prime the immune system for invading pathogens and infections (Carroll and Forsberg 2007).

Chronic stress can cause the immune system to switch from preparing for infection to the suppression of immune function. Chronic stress causes continued glucocorticoid stimulation from immune cells. Pro-inflammatory cytokines are secreted by activated immune cells during chronic stress, and they stimulate the further release of glucocorticoids. As immune cells down-regulate under continued glucocorticoid stimulation to prevent damage to the host, there is a danger that they could become tolerant of glucocorticoids and lose their ability to respond (Carroll and Forsberg 2007). A dysfunctional neuroendocrine–immune interface with abnormal anti-inflammatory feedback and hyperactive pro-inflammatory responses may play a role in the pathogenesis of many diseases (Elenkov et al. 2005).

The effect of chronic stress on the immune system can be illustrated with the example of bovine respiratory disease. Bovine respiratory disease typically involves infection with a primary viral or mycoplasma agent, followed by a secondary bacterial infection. Immune suppression that is caused by stress allows the primary viral or mycoplasma infection to establish itself in the host animal. These microorganisms will further compromise its immune defence by allowing a secondary infection by bacteria to occur (Griffin et al. 2010; Glass et al. 2012). In cattle, these bacteria are generally strains of the *Mannheimia haemolytica, Pasteurella multocida, Histophilus somnus, Mycoplasma bovis* and occasionally *Trueperella pyogenes* bacteria which are the bacteria that make up the bovine respiratory disease complex (BRDC) (Griffin et al. 2010; Klima et al. 2014). These bacteria are present in both healthy and sick cattle; however, when cattle are stressed, the immune system fails to suppress them from initiating pathogenesis (Griffin et al. 2010). Stressed cattle are also more susceptible to coccidiosis (Daugschies and Najdrowski 2005).

## 3.3.2 Glucocorticoid Actions on Specific Elements of the Immune System

Chronic stress–induced cortisol secretion has been associated with immune suppression (Ju et al. 2014), which results in the animal becoming more susceptible to disease and immune challenges. Pharmacological levels of glucocorticoids (i.e., levels resulting from an exogenous source, such as a challenge dose of cortisol) generally cause immunosuppression, but physiological levels can be immunomodulatory, immune-enhancing or immune-suppressive, depending on the conditions experienced by the animal (Chen et al. 2015). The concentration of the glucocorticoid hormone, the effects of cytokines, hormones and neurotransmitters and the activation state of the leukocytes, all influence the direction in which glucocorticoids drive the immune response (Dhabhar 2002).

Understanding the pathways and chemical interactions involved in regulating the effect of stress on the immune system can help us to identify the appropriate biomarkers to assess effects of stress on cattle health and welfare. Thus, the extent of the effects can be detected by assessment of a number of different aspects of the immune system. The glucocorticoids produced as part of the stress response adversely affect the functioning of white blood cells such as lymphocytes, monocytes and neutrophils (Griffin 1989). The role of the white blood cell types, the effects that stress has on these cells and their use as biomarkers of stress will be discussed in the section below. Glucocorticoids also influence the expression of genes controlling growth, reproduction, metabolism and resource allocation which also adversely affect the immune response (Maciel et al. 2001), and a later section will discuss this aspect.

### 3.3.2.1 Effects of Stress on Cell Signalling Across the Immune System

#### 3.3.2.1.1 Stress Effects on the Glucocorticoid Receptors

Glucocorticoid receptors (GR), found in the cytoplasm of the target cell, enable GCs to stimulate the alteration in immune function. There are two types of GR, high-affinity mineralocorticoid receptor (MR) (Pascual-Le Tallec and Lombes 2005) and low-affinity GR. Mineralocorticoid receptors bind GCs when the hormone is at normal levels and the MR-GC are responsible for numerous physiological processes, such as mediating non-stress-related circadian fluctuations in GCs and autonomic outflow (Van den Berg et al. 1994). Mineralocorticoid receptors are responsible for the negative feedback mechanism during basal activity, and GR receptors are responsible for feedback mechanisms during basal and stress-induced GC levels (Tilbrook et al. 1999). Glucocorticoid affinity for MR is ten times greater than low-binding-affinity GR (Kino and Chrousos 2001); therefore, large-scale binding of GCs to GRs only happens when GC hormone levels are increased, such as during a stress response. Approximately 80–90% of GCs circulating in the blood are bound by protein, mainly corticosteroid-binding globulin (CBG) and to a lesser extent, albumin (Breuner et al. 2013). When GCs are bound to proteins, they are biologically inactive and are not able to migrate from blood into cells. Corticosteroid-binding globulin and albumin are necessary for the transportation and protection of GCs (Sapolsky et al. 2000).

The generally accepted model (Fig. 3.3) is that GC binding to the cytoplasmic GR complex leads to dissociation of the GR protein from the complex, rapid translocation of the GR to the nucleus and binding to GR response elements leading to increased or decreased transcription (Sternberg 2006).

Glucocorticoids that are bound to proteins can become available for physiological actions when they reach a target cell. Corticosteroid-binding globulin (CBG) is synthesised in the liver, and it has been shown to increase in response to elevated levels of GCs (Ralph and Tilbrook 2016). However, this does not mean that free GC concentration will be dependent on the synthesis of CBG. Corticosteroid-binding globulin binding affinity can be influenced by temperature and pH. Hence, if an animal is under heat stress, GCs will be not be bound to CBG; therefore, glucocorticoids may not be transported to the necessary tissue (Breuner et al. 2013). Glucocorticoids enter a cell and bind with the GR which triggers a change that releases the heat shock protein 90 complex (HSP90) (Deb et al. 2014). On ligand binding to GR, heat shock protein 90 (HSP90) and other proteins dissociate from the oligomeric complex that constitutes the inactive GR, allowing the GR to form

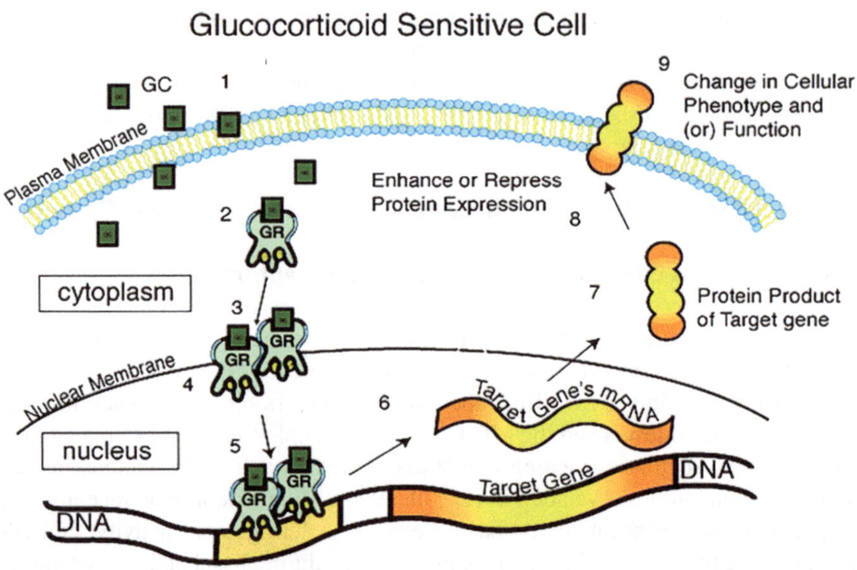


**Fig. 3.3** Glucocorticoids (GC) activate cytosolic receptors (GR), which then act as ligand-activated transcription regulators of GC-sensitive target genes. Steps 1–8 depict the chain of events (adapted from Burton et al. 2005. Reproduced with permission of Elsevier)

homodimers and translocate to the nucleus where they can bind to glucocorticoid response elements (GRE), resulting in transactivation of gene expression (Boumpas et al. 1993). The role of HSP90 in the untransformed GR heterocomplex functioning is linked to proper folding of the receptor's steroid-binding domain and to maintaining the receptor in untransformed, that is, non-DNA-binding state. The GR can then regulate gene expression by up-regulating the expression of anti-inflammatory proteins and repressing the expression of pro-inflammatory proteins. Heat stress produces a typical cellular stress response characterised by elevated cellular HSPs level, which in turn produces functional alterations of the GR, such as reduction of the hormone-binding capacity, decline in the GR protein cytoplasmic level as well as enhancement of the receptor activity in transcriptional activation (Mishra 2021). The most commonly studied HSPs in farm animals are HSP70, HSP90 and HSP27. Of all these HSPs studied, HSP70 is identified to be the ideal biological marker for heat stress in cattle.

#### 3.3.2.1.2 Acute-Phase Response

Acute-phase proteins are used in the assessment of stress in cattle due to their response to glucocorticoids. The APPs are proteins whose plasma concentrations increase or decrease in response to inflammation. This response that is generated is called the acute-phase reaction (APR). Abrupt weaning, social disruption and transport stress have been shown to induce increase in the APPs serum amyloid A (SAA), fibrinogen, haptoglobin and ceruloplasmin in cattle (Arthington et al. 2005; Hickey et al. 2003b; Murata 2007; Qiu et al. 2007; Aich et al. 2009; Carroll et al. 2009; Lynch et al. 2011). However, there are studies that show an inconsistent pattern in the APR following a stressor. For instance, haptoglobin and fibrinogen were reported to decrease following 9 h of transport (Buckham Sporer et al. 2008) while Arthington et al. (2003) demonstrated that haptoglobin decreased while SAA, fibrinogen and ceruloplasmin increased following transport and comingling. Haptoglobin was demonstrated to be a good APP for distinguishing differences in clinical health stasis within groups of calves (Carter et al. 2002) and was previously considered to be a useful measure of inflammatory and stress responses in cattle (Murata et al. 2004). It has been reported to increase following abrupt weaning in beef calves (Lynch et al. 2012). Recently, however, variations in haptoglobin concentrations following weaning have been considered to be an inconsistent and unreliable biomarker of an abrupt weaning stress response in suckled beef calves (O'Loughlin et al. 2014). These findings suggest that while an acute-phase response (APR) appears to be detectable following exposure to stress in cattle, the direction of change is not always consistent, although on the whole, an increase is APP is observed. Therefore, caution must be adhered to when using APPs as biomarkers, and it is necessary to use them in unison with other, well-characterised biomarkers.

#### 3.3.2.1.3 Cytokines

Cytokines are small soluble proteins that covey instructions and mediate communication among immune and non-immune cells. Cytokines are secreted as part of the immune response. They stimulate the HPA axis to increase the level of glucocorticoids, which in turn feedback to inhibit the synthesis of cytokines (Salak-Johnson and McGlone 2007; Hulbert and Moisa 2016). The TNFα, IL-1, IL-12, IL-4, IL-5, IL-6, IL-12 and IFNγ cytokines are inhibited by glucocorticoids, but IL-10 is increased in response to glucocorticoids (Carroll and Forsberg 2007; Salak-Johnson and McGlone 2007) (Fig. 3.4).

Glucocorticoids can inhibit gene expression by activating glucocorticoid receptors (GR) on immune cells (Carroll and Forsberg 2007; Chen et al. 2015). Immune cells (e.g. lymphocytes, macrophages, granulocytes) process the GR and respond to high levels of glucocorticoid during stressful periods (Carroll and Forsberg 2007). Nuclear factor-κB (NF-κB) represents a family of inducible transcription factors, which regulates a large array of genes involved in different processes of the immune

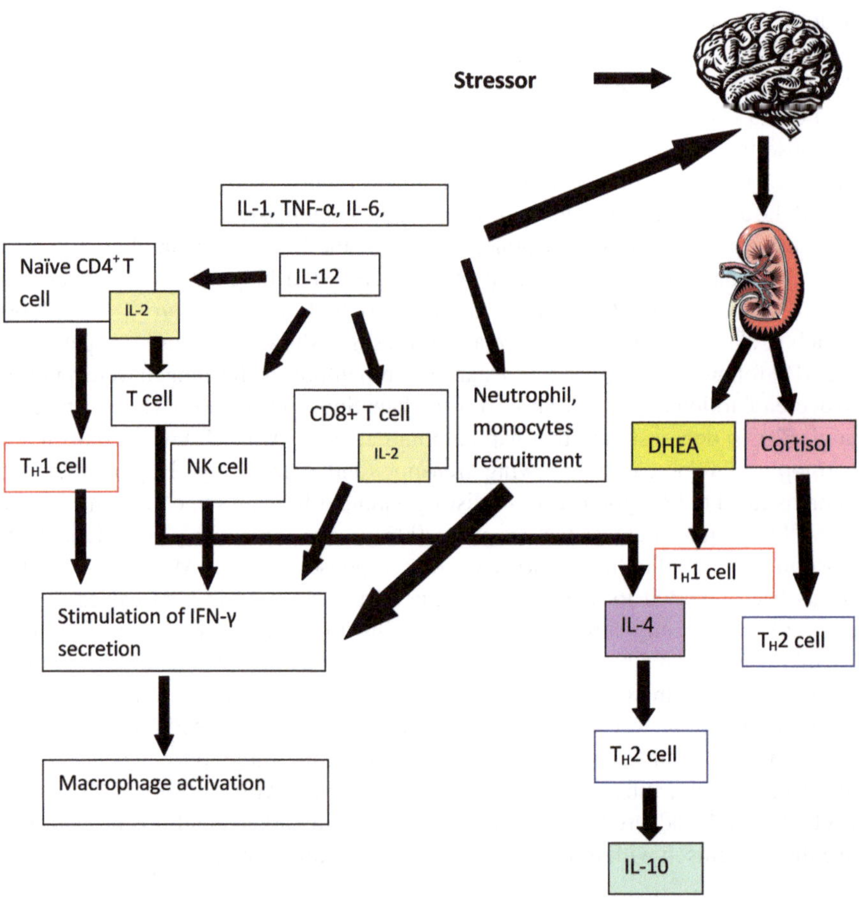


**Fig. 3.4** Tentative model to explain the physiological and immunological changes associated with a stressor. Thick arrows indicate enhancement or positive stimulus, while thin arrows indicate inhibition. Dehydroepiandrosterone (DHEA); Interferon-Y (IFN-Y); Interleukin (IL); Natural killer cell (NL) cell

and inflammatory responses. Activated GR can bind to NF-kβ and prevent it from migrating to the nucleus of the cell and consequently interfere with the production of cytokines by macrophages and Th cells (Carroll and Forsberg 2007; Chen et al. 2015). Cytokine inhibition may be a mechanism to prevent the body from an over-reaction of the immune response as generally glucocorticoids stimulate anti-inflammatory cytokines and suppress pro-inflammatory cytokines (Salak-Johnson and McGlone 2007).

### 3.3.2.2 Effects of Stress on White Blood Cells

Research in cattle has shown that stress has a number of effects on the major white blood cell types involved in immune function. The following sections will describe the basic functioning of the cell types and discuss effects of stress as shown in cattle.

#### 3.3.2.2.1 Leukocytes in General

##### 3.3.2.2.1.1 Leukocyte Distribution

White blood cells (WBCs) are also called leukocytes or leucocytes. They are the cells of the immune system that are involved in protecting the body against both infectious disease and foreign invaders. Leukocyte subsets (or sub-populations), which are predominantly responsible for the surveillance and clearance of pathogens, are targets of stress hormones derived from the HPA axis. This makes them highly suitable as biomarkers of stress in cattle, with particular emphasis on neutrophils, lymphocytes and the neutrophil:lymphocyte (N:L) ratio (Hickey et al. 2003a, b; Yagi et al. 2004; Gupta et al. 2004; Pang et al. 2006; Sporer et al. 2008; Blanco et al. 2009; Earley and Murray 2010; Lynch et al. 2010a, b, 2011) (Table 3.1). Following a stressor, a concurrent increase in total circulating neutrophil number

**Table 3.1** Normal complete blood cell count (CBC) for cattle

| Variable | Range | Average | |
|---|---|---|---|
| Erythron | | | |
| Erythrocytes (×10⁶/μL) | 5.0–10.0 | 7.0 | |
| Haemoglobin (g/dL) | 8.0–15.0 | 11.0 | |
| Haematocrit (%) | 24.0–46.0 | 35.0 | |
| Mean corpuscular volume (MCV) (fL) | 40.0–60.0 | 52.0 | |
| Mean corpuscular haemoglobin conc. (MCHC) (pg) | 11.0–17.0 | 14.0 | |
| Leukocytes (μL) | | | Percent (%) |
| Total leukocytes | 4000–12,000 | 8000 | |
| Neutrophils | 600–4000 | 2000 | 15–45 |
| Lymphocytes | 2500–7500 | 4500 | 45–75 |
| Monocytes | 25–840 | 400 | 2–7 |
| Eosinophils | 0–2400 | 700 | 0–20 |
| Basophils | 0–200 | 5 | 0–2 |
| Platelets (×10³/μL) | 100–800 | | |

Adapted from Jones and Allison (2007)

and decrease in total circulating lymphocyte number in the periphery are one of the more visible biomarkers of the stress response in cattle, resulting in a two-fold increase in the N:L ratio.

### 3.3.2.2.2 Neutrophils

#### 3.3.2.2.2.1 Neutrophil Structure and Function

The neutrophil, characterised by a polymorphic, segmented nucleus, cytoplasmic granules and large glycogen stores, is the first line of cellular defence against pathogens (Paape et al. 2003). The segmented nucleus is an important feature of the neutrophil as it allows rapid transendothelial migration via a mechanism that arranges the nuclear lobes into a straight line. This enables the neutrophil to arrive at sites of infection before macrophages, which have a larger, horseshoe-shaped nucleus. The bovine neutrophil contains three main types of bactericidal granules. The primary granules (also known as azurophilic granules) produce peroxidase which is an important antibacterial substance. Very low concentrations of lysozyme are present in the bovine azurophilic granules. Secondary granules outnumber the primary granules in mature neutrophils, but the tertiary granules are the predominant granule in cattle and, like the secondary granule, are peroxidase-negative. These tertiary granules contain powerful, oxygen-independent bactericidal proteins called β-defensins. These β-defensins inhibit both Gram-positive and -negative bacteria in addition to fungi and viruses. Certain inflammatory mediators, such as platelet-activating factor (PAF) and CXCL8 (formally IL-8), cause the secretion of lactoferrin from neutrophil secondary and tertiary granules into the phagosomes and extracellular environment (Swain et al. 2000). The high affinity of lactoferrin for iron results in the sequestration of iron, preventing its availability to Gram-negative bacteria. This technique in itself may not be enough to prevent bacterial uptake of iron as bacteria contain a number of iron-binding molecules known as siderophores that efficiently transport iron to the pathogen. To counter these siderophores, neutrophils secrete lipocalin-2 which binds bacterial siderophores and prevents them from returning iron to the bacteria. Neutrophils become activated upon receptor binding by cytokines, complement components and immunoglobulins. This results in the activation of the oxidative burst reaction, where neutrophils release large amounts of superoxide ($O_2^-$) and hydrogen peroxide ($H_2O_2$) (O'Loughlin 2011). Myeloperoxidase, located in the azurophilic granules, is released into phagosomes containing pathogens and acts to form hypochlorite from $H_2O_2$ and $Cl^-$, aiding in the destruction of pathogens (Paape et al. 2003; Janeway et al. 2005) (Fig. 3.5).

Oxygen radicals and proteases, released by neutrophils in the process of inflammation, serve to clear infection but are also cytotoxic to host tissues (Ledbetter et al. 2001). While neutrophils play an essential role in the clearance of pathogens by the immune system, prolonged exposure to activated neutrophils can result in severe tissue damage via prolonged inflammation (Boaventura et al. 2010). In order to regulate this tissue damage, rapid phagocytosis of apoptotic neutrophils by macrophages is required, reducing systemic tissue damage from inflammation (Chin et al. 2000). Sladek and Rysanek (2001) demonstrated that neutrophil apoptosis plays a

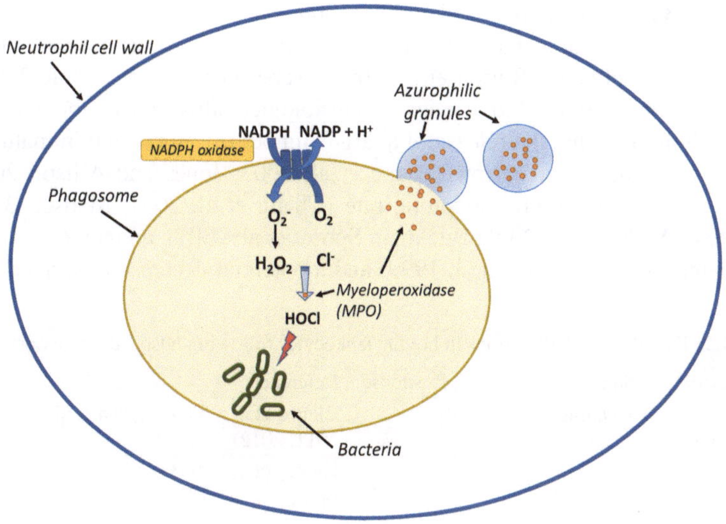


**Fig. 3.5** Bacterial killing and inflammatory tissue damage by neutrophils. The neutrophil plays an important role in host defence and is equipped with various weapons. Bacteria are ingested into the neutrophil and incorporated into a phagosome. Myeloperoxidase, located in the azurophilic granules, is released into phagosomes containing pathogens (bacteria) and acts to form hypochlorite from $H_2O_2$ and $Cl^-$, aiding in the destruction of pathogens

primary role in the resolution of inflammation in cattle. During this resolution of inflammation, neutrophils do not re-circulate as occurs with lymphocytes, rather they undergo apoptosis (Janeway et al. 2005). Apoptosis, also known as programmed cell death, is morphologically characterised by chromatin condensation, nuclear fragmentation, cell shrinkage and blebbing of the plasma membrane (Reed 2000) and is a protective mechanism that plays an essential role in the resolution of the inflammatory response (Haslett 1999). The small, membrane-bound bodies that result from this process can be rapidly cleared by phagocytic cells such as macrophages. Numerous pro-inflammatory cytokines, including IL-1β, IL-2, IL-6, CXCL8, IL-15 and IFN-γ, slow the apoptotic rate, extending neutrophil survival (Akgul et al. 2001; Paape et al. 2003). Pro-inflammatory TNF-α has the opposite effect, inducing apoptosis in bovine neutrophils (Van Oostveldt et al. 2002). However, the induction of apoptosis by TNF-α can be halted by exposure to lipopolysaccharide (LPS) (Hachiya et al. 1995), indicating a role for TNF-α in removing activated neutrophils from sites of inflammation only after bacterial clearance.

#### 3.3.2.2.2.2 Neutrophil Response to Stress

The bovine neutrophil is a direct target for stress hormones, as in normal conditions, it contains both glucocorticoid (Chang et al. 2004) and functional adrenergic receptors (LaBranche et al. 2010). Neutrophilia (increased number of neutrophils) has been frequently reported in response to either endogenous GC (produced in a response to the stress of weaning (Hickey et al. 2003a; Lynch et al. 2010a, 2011), transport (Buckham Sporer et al. 2007, 2008; Riondato et al. 2008), castration (Fisher et al. 1997; Ting et al. 2003; Pang et al. 2009b) or exogenous GC following experimental challenge (Burton et al. 2005; Weber et al. 2006) (Table 3.2). This response can be attributed to a series of physiological alterations in bovine neutrophil function including the release of a large number of reserve and immature neutrophils from the bone marrow (Paape et al. 2003; Jones and Allison 2007), a reduction in the neutrophil apoptotic rate (Chang et al. 2004; Madsen-Bouterse et al. 2006; Weber et al. 2006; Buckham Sporer et al. 2007), an increase in neutrophil chemotaxis (Anderson et al. 1999) and a potential decrease in surface marker

**Table 3.2** Distributional alterations in bovine leukocytes following husbandry stressors

| Leukocyte | Stress | Response | Reference |
|---|---|---|---|
| Total leukocytes | Weaning | ↑ | Blanco et al. (2009), Lynch et al. (2010a, 2011, 2012) |
| | | ↔ | Hickey et al. (2003a) |
| | | \| | Phillips et al. (1989) |
| | Transportation | ↑ | Yagi et al. (2004), Buckham Sporer et al. (2007), Riondato et al. (2008) |
| | Castration | ↑ | Fisher et al. (1997), Earley and Crowe (2002), Ting et al. (2003), Pang et al. (2006, 2009b) |
| | Dexamethasone challenge | ↑ | Menge and Dean-Nystrom (1999) |
| Neutrophils | Weaning | ↑ | Hickey et al. (2003b), Blanco et al. (2009), Lynch et al. (2010a, 2011) |
| | Transportation | ↑ | Yagi et al. (2004), Earley and O'Riordan (2006), Earley et al. (2006), Buckham Sporer et al. (2007), Gupta et al. (2007b), Riondato et al. (2008) |
| | Castration | ↑ | Fisher et al. (1997), Ting et al. (2003), Pang et al. (2009b) |
| | Dexamethasone challenge | ↑ | Burton et al. (2005), Weber (2006), Menge and Dean-Nystrom (1999) |
| Lymphocytes | Weaning | ↓ | Hickey et al. (2003a), Blanco et al. (2009), Lynch et al. (2010a, 2011) |
| | Transportation | ↓ | Earley and O'Riordan (2006), Earley et al. (2006), Buckham Sporer et al. (2007), Gupta et al. (2007b) |
| | Castration | ↓ | Ting et al. (2003) |
| Monocytes | Weaning | ↔ | Lynch et al. (2010a, 2011) |
| | Transportation | ↔ | Riondato et al. (2008) |
| | Castration | ↑ | Ting et al. (2003) |
| Eosinophils | Weaning | ↔ | Lynch et al. (2010a) |
| | | ↑ | Lynch et al. (2011) |
| | Transportation | ↑ | Gupta et al. (2007a, b) |
| | | ↓ | Riondato et al. (2008) |
| | Castration | ↔ | Ting et al. (2003) |

↑ = increase, ↓ = decrease, ↔ = no significant change

CD62L expression, reducing neutrophil margination (transit of neutrophils) along endothelial walls (Weber et al. 2004; Burton et al. 2005; Lynch et al. 2011). In contrast, Pang et al. (2009b) reported no alterations to surface expression of CD62L following either band or Burdizzo castration or following hydrocortisone infusion.

The GR appears to be responsible for the vast majority of these GC-induced effects on neutrophils (Chang et al. 2004; Weber et al. 2006). However, the GC-bound GR also appears to play a role in enhancing phagocytosis of apoptotic neutrophils by macrophages which suggests that GC binding to the GR of macrophages plays a role in resolving the inflammatory process (Liu et al. 1999). Additionally, two of the above studies identified increased neutrophil expression of genes involved in tissue remodelling and wound healing following *in vitro* dexamethasone treatment (Burton et al. 2005; Weber et al. 2006) with a subsequent transport stress study identifying the same effect *in vivo* (Buckham Sporer et al. 2007).

Weaning can act as a stressor for beef calves, and it creates a short-lived, acute response which enhances cellular mobilisation and creates a pro-inflammatory response (O'Loughlin et al. 2011, 2012). There are many reports in the literature of disruptions in immune function due to the onset of abrupt weaning in beef cattle. For example, the *in-vitro* secretion of IFNγ in abruptly weaned beef calves was decreased after stimulation with both concanavalin A and keyhole limpet hemocyanin (Hickey et al. 2003a). An increase in neutrophil number (neutrophilia) has been observed in response to abrupt weaning (Hickey et al. 2003a; Lynch et al. 2010a; O'Loughlin et al. 2011; Lynch et al. 2012; O'Loughlin et al. 2012). Neutrophilia usually occurs due to the down-regulation of surface adhesion molecules and the up-regulation of anti-adhesion molecules (Burton et al. 2005). Additionally, abruptly weaned beef calves often display an increased neutrophil:lymphocyte ratio (Hickey et al. 2003a; O'Loughlin et al. 2011, 2012) and a decreased lymphocyte number (Hickey et al. 2003a, b; Lynch et al. 2010a, b; O'Loughlin et al. 2011; Lynch et al. 2012; O'Loughlin et al. 2012) (Fig. 3.6 and Table 3.2).

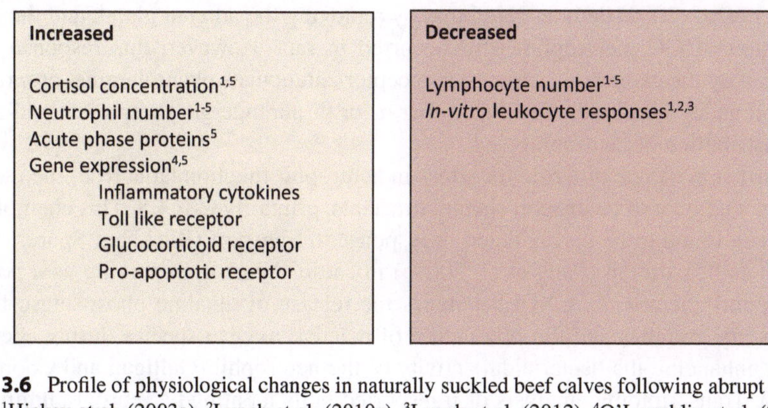


**Fig. 3.6** Profile of physiological changes in naturally suckled beef calves following abrupt weaning. [1]Hickey et al. (2003a), [2]Lynch et al. (2010a), [3]Lynch et al. (2012), [4]O'Loughlin et al. (2011), [5]O'Loughlin et al. (2012)

Furthermore, the stress hormones, cortisol (Hickey et al. 2003a; O'Loughlin et al. 2012) and noradrenaline (Hickey et al. 2003a), were increased due to abrupt weaning in beef calves while after weaning, the number of neutrophils undergoing phagocytosis decreased (Lynch et al. 2010a, 2012).

The redistribution of immune cells during stress may serve to direct them to specific target organs and enhance the speed and efficacy of the immune response. Immune cells travel into the bloodstream under the influence of epinephrine and norepinephrine. Epinephrine and glucocorticoids subsequently cause them to move from the circulatory system to tissue surveillance pathways, lymphoid tissues and sites of infection.

As discussed, many studies have reported changes in the immune response of beef calves to abrupt weaning including alterations in haematological cell numbers, APP concentrations, gene expression and *in vitro* immune responses (Hickey et al. 2003a; Lynch et al. 2010a, b; O'Loughlin et al. 2011, 2012). There is a paucity of literature available on the dairy calves' immune response to weaning, but studies by Kim et al. (2011) and Hulbert et al. (2011) provided evidence of a stress response associated with gradual weaning of dairy calves. Cortisol and TNF within the serum increased following gradual weaning in dairy calves, while serum IFNγ and the percentage of CD25+ T cells in the peripheral blood decreased (Kim et al. 2011). Additionally, this study also found that the APPs, haptoglobin and serum amyloid A increased significantly, while lactoferrin (an iron-binding protein) decreased in response to gradual weaning in dairy calves Furthermore, an increase in neutrophil number and in the neutrophil:lymphocyte ratio was also observed. Additionally, neutrophil oxidative burst and phagocytic activity was decreased folowing gradual weaning of dairy calves (Hulbert et al. 2011). However, changes in the transcriptome of dairy calves in response to gradual weaning have yet to be investigated.

Despite numerous studies indicating that GC is the main effector of neutrophilia following a stress response in cattle (Chang et al. 2004; Burton et al. 2005; Madsen-Bouterse et al. 2006; Weber et al. 2006; Buckham Sporer et al. 2007; Jones and Allison 2007; Lynch et al. 2010a, 2011, 2012), it is also likely that catecholamines play a role via the adrenergic receptor (Stevenson et al. 2001; Engler et al. 2004). Engler et al. (2004) demonstrated that by removing the adrenal gland, and thus the secretion of GC, neutrophila still occurred in rats. However, this response was blocked by the exogenous adrenergic receptor antagonists phentolamine, propranolol and nadolol and indicated that either α- or β- adrengergic receptors could alter the distribution of neutrophils.

Further evidence of a role for stress in neutrophil function hails from the induction of a number of neutrophil chemoattractants, primarily CXCL8. This chemokine has been found to be up-regulated in response to transport (Buckham Sporer et al. 2007) and castration (Pang et al. 2009a) in cattle. Apart from serving as a potent neutrophil chemokine, CXCL8 induces the release of alkaline phosphatase from secondary granules and the production of reactive oxygen species during chemotaxis, enhancing the bactericidal activity of the neutrophil (Galligan and Coomber 2000). Transcriptomic analysis of transported bulls identified a reprogramming of neutrophil gene expression for greater bactericidal activity (Buckham Sporer et al.

2007; Sporer et al. 2008) although bactericidal activity has been shown to be unaffected by weaning (Lynch et al. 2010a). Another cytokine induced by weaning stress, TNF-α (Carroll et al. 2009), works synergistically with the complement component C5a at sites of inflammation to increase the rate of neutrophil phagocytosis. This suggests that stress enhances neutrophil function by not only increasing circulating neutrophils, but also through enhanced phagocytosis and chemotaxis as evidenced by increased chemokine expression. Nevertheless, not all studies have reported this effect, with Lynch et al. (2010a) identifying a decrease in the phagocytic capacity of peripheral neutrophils following weaning. This is potentially due to the measurement of peripheral neutrophils rather than activated neutrophils at the site of infection. It has also been reported that endogenous glucocorticoids do not adversely impact upon oxidative burst activity in bovine neutrophils, even at supraphysiological concentrations (Pang et al. 2009a; Lynch et al. 2010a). In fact, the oxidative burst activity of bovine neutrophils is actually enhanced by exposure to IFN-γ and TNF-α, both of which may be involved in a typical stress response (Hoeben et al. 1998).

#### 3.3.2.2.3 Lymphocytes

##### 3.3.2.2.3.1 Lymphocyte Function

Lymphocytes are typically considered to be part of the adaptive immune response, which consists of the humoral (antibody mediated) and cellular systems. Humoral immunity is the part of immune system that is mediated by macromolecules found in extracellular fluids such as secreted antibodies, complement proteins and certain antimicrobial peptides. Common lymphoid progenitors differentiate into four major populations of mature lymphocytes: B cells, T cells, NK cells and NK-T cells (Chaplin 2010). The B cells play a vital role in humoral immunity by producing antibodies against foreign antigens and operating as antigen-presenting cells (APC).

This process occurs in unison with the work of T cells which are responsible for cell-mediated immunity. T cells possess T cell receptors (TCR) on their surface membrane which enables the identification of peptide antigens presented by either class I or class II major histocompatibility complex (MHC) proteins on infected cells. These T cells can be further subdivided into $CD4^+$ and $CD8^+$ T cells. $CD4^+$ T cells are associated with MHC class II molecules and function in the activation of the humoral response, stimulating B cells while $CD8^+$ cells interact with MHC class I molecules, reacting with cytotoxic effects upon identification of infected cells.

##### 3.3.2.2.3.2 Lymphocyte Response to Stress

Stress can alter lymphocyte function and distribution in a number of ways, as identified by numerous bovine stress studies (Van Kampen and Mallard 1997; Dixit et al. 2001; Odore et al. 2004, 2011; Lynch et al. 2010a, Mallard et al. 2018) (Table 3.2). Similar to neutrophils, bovine lymphocytes express both glucocorticoid (Preisler et al. 2000; Odore et al. 2004) and adrenergic (Abraham et al. 2004; Odore et al. 2004) receptors which act to regulate lymphocyte activity based on neuroendocrine signalling. Glucocorticoids have been shown to result in significant reductions in lymphocyte number (Hickey et al. 2003a, b; Lynch et al. 2010a). However, it also seems that catecholamines may contribute to the reduction in circulating lymphocytes (Hickey et al. 2003a; Engler et al. 2004). Contrary to being immunosuppressive, the depletion of lymphocytes from peripheral circulation actually reflects increased extravasation to susceptible tissues rather than apoptosis or necrosis, indicating an increased capacity for immune surveillance (Viswanathan and Dhabhar 2005).

The proportion of $CD4^+$ and $CD8^+$ T cells in peripheral circulation has been shown to decrease in calves following weaning (Lynch et al. 2010a) or 14 h of truck transportation (Riondato et al. 2008). Additionally, lymphocytes presenting MHC class II molecules increased considerably following either weaning (Lynch et al. 2010a) or transportation (Riondato et al. 2008). This is likely due to GC action as demonstrated by a pharmacological challenge of Holstein bulls with dexamethasone, resulting in an increase in MCH class II presenting lymphocytes (Saama et al. 2004). Additionally, lymphocytes possess receptors for ACTH. Stefano and Smith (1996) demonstrated that ACTH can regulate the immune system through direct effects on cellular targets independent of its capacity to stimulate GC secretion from the adrenal gland. In fact, lymphocytes can actually synthesise ACTH, particularly in response to stress, and this autocrine action may be an important part of this process (Dixit et al. 2001). This has most clearly been highlighted following transportation stress, whereby ACTH secretion from non-stimulated lymphocytes increased in cows transported for 14 h (Dixit et al. 2001). Lymphocyte-derived ACTH remained elevated in cows left on the truck for 24 h following transportation while cows that were off-loaded had baseline ACTH concentrations following 24 h of rest. This suggests that lymphocyte-derived ACTH may not follow the cyclical nature of pituitary derived ACTH and may be a more reliable biomarker of stress.

### 3.3.3 Effect of Glucocorticoids on Gene Expression

#### 3.3.3.1 DNA Binding

As well as directly affecting the signalling and white blood cell functioning, stress can also affect the expression of genes associated with the immune system. For instance, the presence of glucocorticoids can affect the gene expression pathways involved in the production of cytokines. In most mammalian cells, including bovine neutrophils, GR exists in two main isoforms, GRα and GRβ (Kino and Chrousos 2001; Burton et al. 2005). However, only GRα has glucocorticoid-binding capacity. In cells not exposed to stress or therapeutic levels of glucocorticoids, most GRα are located in the cytoplasm as complexes with accessory proteins that maintain the receptors in a high affinity hormone-binding state. When glucocorticoid concentrations rise inside the cell due to elevated extracellular concentrations, hormone binding to GRα activates the receptor causing it to dissociate from its complex. This allows GRα to translocate into the cell's nucleus. In the nucleus, hormone-bound GRα employs multiple mechanisms to change the expression of hormone sensitive genes, including binding to DNA and other transcription factors to influence rates of gene transcription and to mRNAs already existing in the cytoplasm to affect their stability (Burton et al. 2005).

Many genes of the immune system contain the glucocorticoid response element (GRE) in their promoter regions. The GRE is a specific sequence of DNA that allows GR binding, allowing GR to directly induce transcription (Fig. 3.7). The GRE is a 15 bp sequence encompassing two half-sites separated by three base pairs

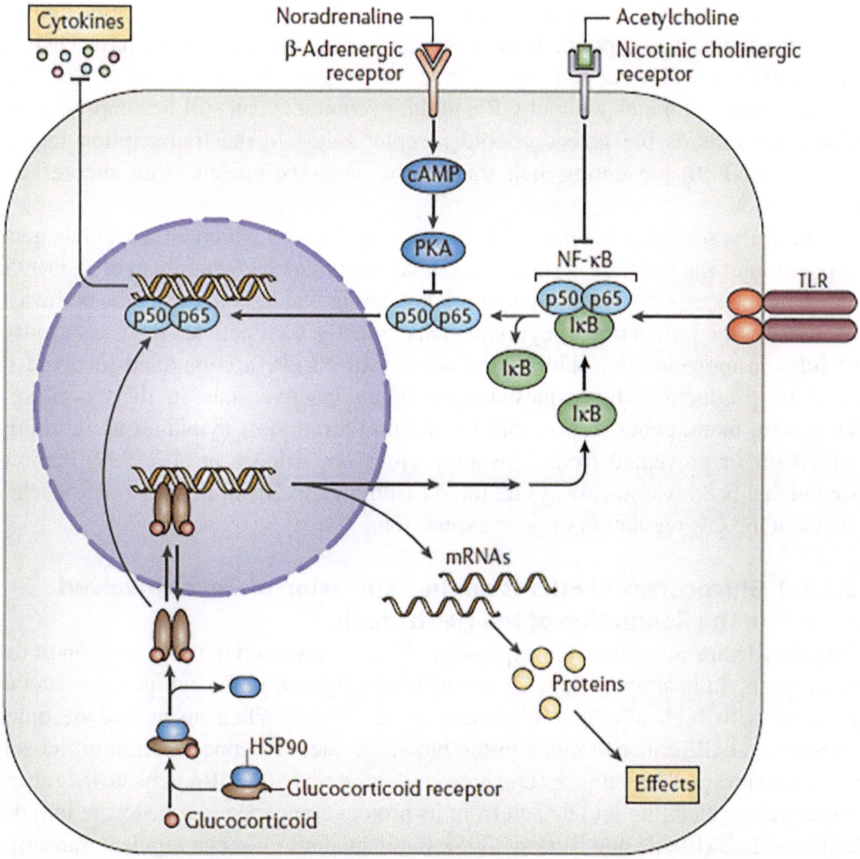


**Fig. 3.7** Molecular regulation of cytokine gene expression by glucocorticoids and catecholamines. Glucocorticoids bind to glucocorticoid receptors in the cytoplasm of the cell, resulting in the dissociation of HSP90 and the formation of a glucocorticoid-bound glucocorticoid receptor homodimer. This glucocorticoid homodimer translocates to the nucleus of the cell where it acts as a transcription factor, binding to GRE and increasing the transcription of a number of genes. One such gene is IκBα which sequesters NFκB in an inactive state, preventing its activation of pro-inflammatory cytokines. Norepinephrine (noradrenaline) binds with the β-adrenergic receptor to induce activation of cAMP while acetylcholine binds to the nicotinic cholinergic receptor, both of which lead to the inhibition of NFκB (Adapted from Sternberg 2006. Reproduced with permission of Springer Nature)

that can consist of any arrangement of nucleotides. The binding of the GR to the GRE can also allow other transcription factors to bind to previously unavailable regions of DNA as a result of the induction of chromatin remodelling. GR can also bind to a "negative" glucocorticoid response element (nGRE) to inhibit transcription. This process involves the binding of three GR molecules instead of the typical formation of a GR homodimer. GR binds to the nGRE to suppress activation through transactivating factors. One key gene inhibited by the GR-nGRE complex is POMC reducing local synthesis of ACTH. However, GR also results in the transrepression of a number of inflammatory cytokines. Glucocorticoids indirectly regulate expression of these cytokines through binding to a number of transcription factors, inhibiting subsequent transcription. Many inflammatory genes that do not have GRE in their promoter regions contain sites for activation protein (AP) -1 and NFκB. This explains how cytokines without GRE in their promoters can still be suppressed by glucocorticoids as the glucocorticoid receptor binds to the transcription factors AP-1 and NFκB, preventing their transmigration to the nucleus, thus suppressing transcription.

Currently, several other mechanisms are suspected by which GCs regulate gene expression of the immune system. Two separate studies by Scheinman et al. (1995) and Auphan et al. (1995) presented work indicating that GCs interrupt the pathways involved in the production of cytokines. Specifically, CGs activate the transcription of IκBα, a molecule that inhibits the activity of NFκB (a compound involved in cytokine production) by sequestering it in an inactive state in the cytoplasm. Therefore, many genes responsible for the proliferation of cytokines are abruptly turned off or prevented from activation. However, Adcock et al. (1999) demonstrated that IκBα was not always required to inhibit NFκB, implicating other mechanisms in the GC regulation of gene expression.

#### 3.3.3.2 Glucocorticoid Effects on the Expression of Genes Involved in the Regulation of the Neutrophils

Glucocorticoids also affect the expression of genes involved in the regulation of the neutrophils. In healthy cattle, GRs are generally located in the cytoplasm bound to proteins with high affinity (Bamberger et al. 1996). When an animal becomes stressed and GC concentrations in the blood elevate, GC concentration in the cell will also elevate. When this occurs, hormones binding to the GRα activate the receptor, therefore enabling it to detach from its protein complex and translocate into the cell's nucleus (Eicher and Burton 2005). In the nucleus GRα can regulate transcription of target genes by binding to DNA and mRNAs already in the cytoplasm (Bamberger et al. 1996). Glucocorticoid action is dependent on GR-mediated transcriptional regulation of specific target genes as a result of sequence-specific DNA binding which, in turn, inhibits the promoter regions of genes such as nuclear factor kappa B (NF-B) and activator protein-1 (AP-1) which are potent transcription factors for many pro-inflammatory cytokines and adhesion genes. Central to the anti-inflammatory action of glucocorticoids is the induction of inhibitor kappa B alpha (IκBα) which binds to and inhibits NF-B by sequestering it in the cytoplasm. The interaction of GR with the proinflammatory transcription factors, AP1 and NF-κB, antagonises their activity and is considered to be a primary mechanism by which glucocorticoids suppress inflammation. Therefore, the actions of GCs within the cell will vary depending on the sensitivity of the GC receptor. This is influenced by nutrition, immune status, circadian rhythm and reproductive status (Bamberger et al. 1996; Chrousos 2009). During mid and late lactation in dairy cows, Preisler et al. (2000) demonstrated that when cortisol concentration in the blood was elevated, GRα proteins translocate into the nucleus of blood neutrophils. When GRα is lost from the cytosol of neutrophils, it leads to a noticeable change in the expression of two genes which regulate neutrophil behaviour: CD62L which enables adhesion of neutrophils to the site of the infection on surface epithelium (Weber et al. 2004) and *Fas* which promotes apoptosis (Chang et al. 2004). The outcome of these processes means that altered GC concentrations will not decrease the number of neutrophils in the blood; however, it will inhibit some of their functions. This will be of major significance should a stressed animal's immune system be challenged.

## 3.4 New Techniques in Molecular Biology and Their Application in Cattle

### 3.4.1 The Transcriptome

Understanding the transcriptome can provide a great deal of insight into the functioning of an organism. Given recent advances in genome sequencing technologies (Marioni et al. 1999; Marioni et al. 2008), the field of transcriptomics currently provides unparalleled potential to understand key physiological processes and identify biomarkers. Numerous genes that play a role in the stress response have already been identified (Chang et al. 2004; Weber et al. 2004, 2006; Madsen-Bouterse et al. 2006; Buckham Sporer et al. 2007, 2008; O'Loughlin et al. 2012; Gao et al. 2019; Cassar-Malek et al. 2022). Future work will contribute to the identification of multiple genes which can be incorporated into pathway analysis tools or multivariate analysis to detect gene signatures for disease. One of the shortcomings of using the transcriptome for biomarker discovery is that the presence of a transcript does not necessarily imply the presence of its respective protein. Nevertheless, so long as a gene biomarker presence is sensitive and accurate, it does not necessarily have to translate to a protein level (Aronson 2005). This is because the increased expression of a gene may be in response to a physiological process of interest, even if the transcript itself does not have a physiological effect.

O'Loughlin et al. (2011, 2012) demonstrated that there are also molecular biomarkers of weaning stress in beef calves. The cytokine signalling, transmembrane transport, haemostasis and G-protein-coupled receptor pathways underwent transcriptomic alterations following the abrupt weaning of beef calves (O'Loughlin et al. 2012). Leukocyte gene expression was altered for at least 7 days after weaning and simultaneous housing (O'Loughlin et al. 2012). Gene expressions of pro-inflammatory cytokines including *IL-1β, CXCL8, IFNγ* and *TNFα,* were up-regulated in response to weaning in single-suckled beef calves. Furthermore, the glucocorticoid receptor, *GRα*, the pro-apoptotic gene, *Fas*, the Gram-negative pattern recognition receptors *TLR4, CD62L* and *NFkβ1*, were also increased in response to weaning, while *BPI* expression decreased following weaning (O'Loughlin et al. 2011). Therefore, the stress associated with abrupt weaning of beef calves causes some immune cells to be transcriptionally enhanced to better enable them to locate and destroy pathogens, evidenced by the up-regulation of chemokines, cytokines and integrins following abrupt weaning (O'Loughlin et al. 2012).

### 3.4.2 The Proteome

Proteomics is a science that is at the forefront of biomarker discovery as proteins have a massive potential to serve as biomarkers for stress as well as numerous disease states (Rifai and Gerszten 2006). Effectively, the use of these biomarkers allows detection of the actual transcribed message, eliminating the ambiguity that can occur at the transcriptomic level where transcripts can be degraded or modified prior to translation and alternate splicing can result in multiple protein products. The plasma proteome can be accessed from a sample of blood. Surprisingly, it represents the expression of almost all of the proteins from tissues throughout the body (Anderson et al. 2004). While the proteomic approach has yet to be aggressively applied to the detection of biomarkers in cattle, several proteomic-based approaches have been used to identify biomarkers of weaning stress in cattle (Herzog 2007), bovine spongiform encephalopathy (BSE) from urine samples (Simon et al. 2008), distinguish between subclinical mycobacterial infections (Seth et al. 2009), identify proteomic biomarkers for BRD fatality when challenged with BHV-1 and *Mannheimia haemolytica* following transport (Aich et al. 2009) and also for muscle growth in cattle (Ikegami et al. 2008; Keady et al. 2011). Unfortunately, proteomic technology lags behind that of genomics, lacking both the sensitivity and the large dynamic range that is required to accurately identify and characterise proteomic biomarkers, although this is rapidly changing (Diamandis and van der Merwe 2005; Koomen et al. 2005; Rifai and Gerszten 2006). Therefore, the search for biomarkers has focused on genomic studies, principally using transcriptomics to identify genes that are responsible for the regulation of physiological processes of interest (O'Loughlin 2011).

## 3.5 Conclusion About Effects of Stress on Immune Function and How to Measure It

Stress has a number of major effects on the immune system. In acute stress, the immune system is primed to resist disease. However, chronic stress has a number of detrimental effects on immune function by influencing gene expression pathways, by impacting on inflammatory responses, inducing neutrophilia (increase in neutrophil number) and lymphopaenia (reduction in lymphocyte number). The stress hormones such as glucocorticoids and catecholamines play a key role in the host following exposure to stress, having both anti-inflammatory and immunosuppressive effects. Aspects of the activation of these pathways can be used as biomarkers, but require an in-depth knowledge of the mechanisms involved.

## References

Abraham G, Gottschalk JJ, Ungemach FR (2004) Possible role of dexamethasone in sensitizing the beta-2-adrenergic receptor system in vivo in calves during concomitant treatment with clenbuterol. Pharmacology 72:196–204

Adcock JM, Nasuhara Y, Stevens DA, Barnes PJ (1999) Ligand-induced differentiation of glucocorticoid receptor (GR) trans-repression and transactivation: preferential targetting of NF-κB and lack of I-κB involvement. Br J Pharmacol 127:1003–1011

Aich P, Babiuk LA, Potter AA et al (2009) Biomarkers for prediction of bovine respiratory disease outcome omics. J Integr Biol 13:199–209

Akgul C, Moulding DA, Edwards SW (2001) Molecular control of neutrophil apoptosis. FEBS Lett 487:318–322

Almeida PE, Weber PSD, Burton JL et al (2007) Depressed DHEA and increased sickness response behaviors in lame dairy cows with inflammatory foot lesions. Domest Anim Endocrinol 34:89–99

Anderson BH, Watson DL, Colditz IG (1999) The effect of dexamethasone on some immunological parameters in cattle. Vet Res Commun 23:399–413

Anderson NL, Polanski M, Pieper R et al (2004) The human plasma proteome. Mol Cell Proteomics 3:311–326

Andrade O, Orihuela A, Solano J et al (2001) Some effects of repeated handling and the use of a mask on stress responses in Zebu cattle during restraint. Appl Anim Behav Sci 71:175–181

Aronson JK (2005) Biomarker and surrogate endpoints. Br J Clin Pharmacol 59:491–494

Arthington JD, Eicher SD, Kunkle WE et al (2003) Effect of transportation and commingling on the acute-phase protein response growth and feed intake of newly weaned beef calves. J Anim Sci 81:1120–1125

Arthington JD, Spears JW, Miller DC (2005) The effect of early weaning on feedlot performance and measures of stress in beef calves. J Anim Sci 83:933–939

Auphan N, Didonato JA, Rosette C, Helmberg A, Karin M (1995) Immunosuppression by glucocorticoids: inhibition of NF-κB activity through induction of IκB synthesis. Science 270:286–289

Bamberger CM, Schulte HM, Chrousos GP (1996) Molecular determinants of glucocorticoid receptor function and tissue sensitivity to glucocorticoids. Endocr Rev 17:245–261

Bauer ME (2005) Stress glucocorticoids and ageing of the immune system. Stress 8:69–83

Blanco M, Casasús I, Palacio J (2009) Effect of age at weaning on the physiological stress response and temperament of two beef cattle breeds. Animal 3:108–117

Blecha FS, Boyles L, Riley JG (1984) Shipping suppresses lymphocyte blastogenic responses in Angus and Brahman x Angus feeder calves. J Anim Sci 59:576–583

Boaventura VS, Santos CS, Cardoso CR et al (2010) Human mucosal leishmaniasis: neutrophils infiltrate areas of tissue damage that express high levels of Th17-related cytokines. Eur J Immunol 40:2830–2836

Boumpas DT, Chrousos GP, Wilder RL et al (1993) Glucocorticoid therapy for immune-mediated diseases: basic and clinical correlates. Ann Inter Med 119:1198–1208

Breuner CW, Delehanty B, Boonstra R (2013) Evaluating stress in natural populations of vertebrates: total CORT is not good enough. Funct Ecol 27:24–36

Brotman DJ, Golden SH, Wittstein IS (2007) The cardiovascular toll of stress. Lancet 370:1089–1100

Brown EJ, Vosloo A (2017) The involvement of the hypothalamo-pituitary-adrenocortical axis in stress physiology and its significance in the assessment of animal welfare in cattle. Onderstepoort J Vet Res 84(1):a1398
Buckham Sporer KR, Burton JL, Earley B et al (2007) Transportation stress in young bulls alters expression of neutrophil genes important for the regulation of apoptosis tissue remodeling margination and anti-bacterial function. Vet Immunol Immunopathol 118:19–29
Buckham Sporer KR, Weber PSD, Burton JL et al (2008) Transportation of young beef bulls alters circulating physiological parameters that may be effective biomarkers of stress. J Anim Sci 86:1325–1334
Burdick N, Randel R, Carroll J et al (2011) Interactions between temperament stress and immune function in cattle. Int J Zool:1–9. https://doi.org/10.1155/(2011)/373197
Burton JL, Madsen SA, Chang LC et al (2005) Gene expression signatures in neutrophils exposed to glucocorticoids: a new paradigm to help explain "neutrophil dysfunction" in parturient dairy cows. Vet Immunol Immunopathol 105:197–219
Cannon WB (1932) The wisdom of the body. W.W. Norton & Company, Inc., New York
Carroll JA, Burdick Sanchez NC (2014) Bill E. Kunkle interdisciplinary beef symposium: overlapping physiological responses and endocrine biomarkers that are indicative of stress responsiveness and immune function in beef. cattle. J Anim Sci 92:5311–5318
Carroll JA, Forsberg NE (2007) Influence of stress and nutrition on cattle immunity. Vet Clin N Am: Food Anim Pract 23:105–149
Carroll JA, Arthington JD, Chase CC Jr (2009) Early weaning alters the acute-phase reaction to an endotoxin challenge in beef calves. J Anim Sci 87:4167–4172
Carter JN, Meredith GL, Montelongo M et al (2002) Relationship of vitamin E supplementation and antimicrobial treatment with acute-phase protein responses in cattle affected by naturally acquired respiratory tract disease. Am J Vet Res 63:1111–1117
Cassar-Malek I, Pomiès L, de la Foye A et al (2022) Transcriptome profiling reveals stress-responsive gene networks in cattle muscles. Peer J 6(10):e13150
Chang LC, Madsen SA, Toelboell T, Weber PSD et al (2004) Effects of glucocorticoids on Fas gene expression in bovine blood neutrophils. J Endocrinol 183:569–583
Chaplin DD (2010) Overview of the immune response. J Allergy Clin Immunol 125:S3–S23
Chen Y, Arsenault R, Napper S et al (2015) Models and methods to investigate acute stress responses in cattle. Animals 5:1268–1295
Chin AC, Lee WD, Murrin KA et al (2000) Tilmicosin induces apoptosis in bovine peripheral neutrophils in the presence or in the absence of pasteurella haemolytica and promotes neutrophil phagocytosis by macrophages. Antimicrob Agents Chemother 44:2465–2470
Chrousos GP (2009) Stress and disorders of the stress system. Nat Rev Endocrinol 5:374–381
Chrousos GP, Gold PW (1992) The concepts of stress and stress system disorders: overview of physical and behavioral homeostasis. J Am Med Assoc 267:1244–1252
Collier RJ, Renquist BJ, Xiao Y (2017) A 100-year review: stress physiology including heat stress. J Dairy Sci 100(12):10367–10380
Dantzer R, Mormède P (1983) Stress in farm animals: a need for reevaluation. J Anim Sci 57:6–18
Daugschies A, Najdrowski M (2005) Eimeriosis in cattle: current understanding. J Vet Med B 52:417–427
Deb R, Sajjanar B, Singh U et al (2014) Effect of heat stress on the expression profile of Hsp90 among Sahiwal (Bos indicus) and Frieswal (Bos indicus×Bos taurus) breed of cattle: a comparative study. Gene 536(2):435–440
Dhabhar FS (2002) Stress-induced augmentation of immune function- the role of stress hormones leukocyte trafficking and cytokines. Brain Behav Immun 16:785–798
Diamandis EP, Van Der Merwe DE (2005) Plasma protein profiling by mass spectrometry for cancer diagnosis: opportunities and limitations. Clin Cancer Res 11:963–965
Dickens MJ, Romero LM (2013) A consensus endocrine profile for chronically stressed wild animals does not exist. Gen Comp Endocrinol 15(191):177–189
Dixit VD, Marahrens M, Parvizi N (2001) Transport stress modulates adrenocorticotropin secretion from peripheral bovine lymphocytes. J Anim Sci 79:729–734

Earley B, Crowe MA (2002) Effects of ketoprofen alone or in combination with local anesthesia during castration of bull calves on plasma haptoglobin, in-vitro interferon-γ production, white blood cell numbers and animal performance. J Anim Sci 80:1044–1052

Earley B, Murray M (2010) The effect of road and sea transport on inflammatory adrenocortical metabolic and behavioural responses of weanling heifers. BMC Vet Res 6:36

Earley B, O'Riordan EG (2006) Effects on transporting bulls at different space allowances on physiological haematological and immunological responses to a 12-h journey by road. Irish J Agric Food Res 45(3):9–50

Earley B, Fisher AD, O'Riordan EG (2006) Effects of pre-transport fasting on the physiological responses of young cattle to 8-hour road transport. Irish J Agric Food Res 45:51–60

Eicher SD, Burton JL (2005) Immune system: stress effects. In: Pond WC, Bell AW (eds) Encyclopedia of animal science. Marcel Dekker, New York, pp 544–547

Elenkov IJ, Chrousos GP (2002) Stress hormones pro-inflammatory and anti-inflammatory cytokines and autoimmunity. Ann N Y Acad Sci 966:290–303

Elenkov IJ, Iezzoni DG, Daly A et al (2005) Cytokine dysregulation inflammation and well-being. Neuroimmunomod 12:255–269

Engler H, Dawils L, Hoves S et al (2004) Effects of social stress on blood leukocyte distribution: the role of [alpha]- and [beta]-adrenergic mechanisms. J Neuroimmunol 156:153–162

Fisher A, Crowe M, O'Kiely P et al (1997) Growth behaviour adrenal and immune responses of finishing beef heifers housed on slatted floors at 15 20 25 or 30 $m^2$ space allowance. Livest Prod Sci 51:245–254

Fisher A, Verkerk G, Morrow C et al (2002) The effects of feed restriction and lying deprivation on pituitary–adrenal axis regulation in lactating cows. Livest Prod Sci 73:255–263

Galligan CL, Coomber BL (2000) Effects of human IL-8 isoforms on bovine neutrophil function *in vitro*. Vet Immunol Immunopathol 74:71–85

Gao ST, Ma L, Zhou Z et al (2019) Heat stress negatively affects the transcriptome related to overall metabolism and milk protein synthesis in mammary tissue of midlactating dairy cows. Physiol Genom 51(8):400–409

Glass EJ, Baxter R, Leach RJ et al (2012) Genes controlling vaccine responses and disease resistance to respiratory viral pathogens in cattle. Vet Immunol Immunopathol 148:90–99

Grandin T (1997) Assessment of stress during handling and transport. J Anim Sci 75:249–257

Griffin JFT (1989) Stress and immunity: a unifying concept. Vet Immunol Immunopathol 20:263–312

Griffin D, Chengappa M, Kuszak J et al (2010) Bacterial pathogens of the bovine respiratory disease complex. Vet Clin N Am: Food Anim Pract 26:381–394

Gupta S, Earley B, Ting S et al (2004) Technical note: effect of corticotropin-releasing hormone on adrenocorticotropic hormone and cortisol in steers. J Anim Sci 82:1952–1956

Gupta S, Earley B, Crowe MA (2007a) Pituitary adrenal immune and performance responses of mature Holstein × Friesian bulls housed on slatted floors at various space allowances. Vet J 173:594–604

Gupta S, Earley B, Crowe MA (2007b) Effect of 12-hour road transportation on physiological immunological and haematological parameters in bulls housed at different space allowances. Vet J 173:605–616

Gupta S, Earley B, Nolan M et al (2008) Effect of repeated regrouping and relocation on behaviour of steers. Appl Anim Behav Sci 110:229–243

Hachiya O, Takeda Y, Miyata H et al (1995) Inhibition by bacterial lipopolysaccharide of spontaneous and tnf-α-induced human neutrophil apoptosis *in vitro*. Microbiol Immunol 39:715–723

Haslett C (1999) Granulocyte apoptosis and its role in the resolution and control of lung inflammation. Am J Respir Crit Care Med 160:S5–S11

Herman JL (1992) Complex PTSD: a syndrome in survivors of prolonged and repeated trauma. J Traum Stress 5:377–391

Herman JP, Nawreen N, Smail MA et al (2020) Brain mechanisms of HPA axis regulation: neurocircuitry and feedback in context Richard Kvetnansky lecture. Stress 23:617–632

Herzog K (2007) The effects of weaning stress on the serum protein profile of calves: a proteomic analysis. Ph.D Thesis, University of Saskatchewan, Saskatchewan, Canada

Hickey MC, Drennan M, Earley B (2003a) The effect of abrupt weaning of suckler calves on the plasma concentrations of cortisol catecholamines leukocytes acute phase proteins and in vitro interferon-gamma production. J Anim Sci 81:2847–2855

Hickey MC, Earley B, Fisher A (2003b) The effect of floor type and space allowance on welfare indicators of finishing steers. Irish J Agric Food Res 42:89–100

Hoeben D, Burvenich C, Massart-Leën AM (1998) Glucocorticosteroids and in vitro effects on chemiluminescence of isolated bovine blood granulocytes. Eur J Pharmacol 354:197–203

Hulbert LE, Moisa SJ (2016) Stress, immunity, and the management of calves. J Dairy Sci 99:3199–3216

Hulbert LE, Cobb CJ, Carroll JA et al (2011) Effects of changing milk replacer feedings from twice to once daily on Holstein calf innate immune responses before and after weaning. J Dairy Sci 94(5):2557–2565

Ikegami H, Sono Y, Nagai K et al (2008) Discovery of a protein biomarker candidate related to carcass weight in Japanese black beef cattle (Wagyu). J Prot Bioinfor S2

Janeway CA, Travers M, Walport M et al (2005) Immunobiology the immune system in health and disease, 6th edn. Garland Science Publication, New York

Jones ML, Allison RW (2007) Evaluation of the ruminant complete blood cell count. Vet Clin North Am Food Anim Pract 23:377–402

Ju XH, Xu HJ, Yong YH et al (2014) Heat stress up regulation of toll-like receptors 2/4 and acute inflammatory cytokines in peripheral blood mononuclear cell (PBMC) of Bama miniature pigs: an in vivo and in vitro study. Animal 8:1462–1468

Keady SM, Kenny DA, Keane MG et al (2011) Effect of sire breed and genetic merit for carcass weight on the transcriptional regulation of the somatotropic axis in M. longissimus dorsi of crossbred steers. J Anim Sci 89:4007–4016

Kelley KW (1985) Immunological consequences of changing environmental stimuli. Anim Stress:193–223

Kim M-H, Yang J-Y, Upadhaya SD et al (2011) The stress of weaning influences serum levels of acute-phase proteins iron-binding proteins inflammatory cytokines cortisol and leukocyte subsets in Holstein calves. J Vet Sci 12:151–157

Kino T, Chrousos GP (2001) Glucocorticoid and mineralocorticoid resistance/hypersensitivity syndromes. J Endocrinol 169:437–445

Klima CL, Alexander TW, Hendrick S et al (2014) Characterization of Mannheimia haemolytica isolated from feedlot cattle that were healthy or treated for bovine respiratory disease. Can J Vet Res 78:38–45

Knights M, Smith GW (2007) Decreased ACTH secretion during prolonged transportation stress is associated with reduced pituitary responsiveness to tropic hormone stimulation in cattle. Dom Anim Endocrinol 33:442–450

Koomen JM, Li D, Xiao L-C et al (2005) Direct tandem mass spectrometry reveals limitations in protein profiling experiments for plasma biomarker discovery. J Proteome Res 4:972–981

Labranche TP, Ehrich MF, Eyre P (2010) Characterization of bovine neutrophil β2-adrenergic receptor function. J Vet Pharmacol Ther 33:323–331

Lay DC Jr, Friend TH, Randel RD et al (1992) Behavioral and physiological effects of freeze or hot-iron branding on crossbred cattle. J Anim Sci 70:330–336

Ledbetter TK, Paape MJ, Douglass LW (2001) Cytotoxic effects of peroxynitrite polymorphonuclear neutrophils free-radical scavengers inhibitors of myeloperoxidase and inhibitors of nitric oxide synthase on bovine mammary secretory epithelial cells. Am J Vet Res 62:286–293

Liu YJ, Cousin M, Hughes J et al (1999) Glucocorticoids promote nonphlogistic phagocytosis of apoptotic leukocytes. J Immunol 162:3639–3646

Loerch SC, Fluharty FL (1999) Physiological changes and digestive capabilities of newly received feedlot cattle. J Anim Sci 77:1113–1119

Lynch EM, Earley B, McGee M et al (2010a) Effect of abrupt weaning at housing on leukocyte distribution functional activity of neutrophils and acute phase protein response of beef calves. BMC Vet Res 6:39

Lynch E, Earley B, McGee M et al (2010b) Characterisation of physiological and immunological responses in beef cows to abrupt weaning and subsequent housing. BMC Vet Res 6:37

Lynch EM, Earley B, McGee M et al (2011) Effect of post-weaning management practices on physiological and immunological responses of weaned beef calves. Irish J Agric Food Res 50(2):161–174

Lynch EM, Earley B, McGee M et al (2012) Effect of pre-weaning concentrate supplementation on the peripheral distribution of leukocytes the functional activity of neutrophils and the acute phase protein responses of abruptly weaned beef calves. BMC Vet Res 8:1

Maciel S, Chamberlain C, Wettemann R et al (2001) Dexamethasone influences endocrine and ovarian function in dairy cattle. J Dairy Sci 84:1998–2009

Madsen-Bouterse SA, Rosa GJM, Burton JL (2006) Glucocorticoid modulation of Bcl-2 family members A1 and Bak during delayed spontaneous apoptosis of bovine blood neutrophils. Endocrinology 147:3826–3834

Mallard B, Cartwright S, Nayeri S et al (2018) Genome-wide association and functional annotation of positional candidate genes for immune response in Canadian Holstein Cattle. In: Emam M, Sargolzaei M (eds) Proceedings of the World Congress on Genetics Applied to Livestock Production, p 711

Marioni JC, Mason CE, Mane SM et al (1999) RNA-seq: an assessment of technical reproducibility and comparison with gene expression arrays. Genome Res 18:1509–1517

Marioni JC, Mason CE, Mane SM, Stephens M, Gilad Y (2008) RNA-seq: an assessment of technical reproducibility and comparison with gene expression arrays. Gen Res 18:1509–1517

McEwen BS, Stellar E (1993) Stress and the individual: mechanisms leading to disease. Arch Inter Med 153:2093–2101

Meléndez DM, Marti S, Haley DB, Schwinghamer TD, Schwartzkopf-Genswein KS (2021) Effects of conditioning, source, and rest on indicators of stress in beef cattle transported by road. PLoS ONE 16(1):e0244854

Menge C, Dean-Nystrom EA (1999) Dexamethasone depletes [gamma][delta]T cells and alters the activation state and responsiveness of bovine peripheral blood lymphocyte subpopulations. J Dairy Sci 91:2284–2298

Minton JE (1994) Function of the hypothalamus-pituitary-adrenal axis and the sympathetic nervous system in models of acute stress in domestic farm animals. J Anim Sci 72:1891–1898

Mishra SR (2021) Behavioural, physiological, neuro-endocrine and molecular responses of cattle against heat stress: an updated review. Trop Anim Health Prod 53

Moberg GP (1985) Biological response to stress: key assessment of animal well-being. In: Moberg GP (ed) Animal stress. Am Physiol Soc, Bethasda, pp 27–49

Moberg GP (1987) Problems in defining stress and distress in animals. J Am Vet Med Assoc 191:1207–1211

Moberg GP, Mench JA (2000) The biology of animal stress: basic principles and applications for animal welfare. CABI Publishing, Wallingford, Oxon. ISBN 9780851993591, pp 1–22. https://doi.org/10.1079/9780851993591.0000

Mormède P, Andanson S, Aupérin B et al (2007) Exploration of the hypothalamic–pituitary–adrenal function as a tool to evaluate animal welfare. Physiol Behav 92:317–339

Möstl E, Palme R (2002) Hormones as indicators of stress. Dom Anim Endocrinol 23:67–74

Murata H (2007) Stress and acute phase protein response: an inconspicuous but essential linkage. Vet J 173:473–474

Murata H, Shimada N, Yoshioka M (2004) Current research on acute phase proteins in veterinary diagnosis: an overview. Vet J 168:28–40

Nader N, Chrousos GP, Kino T (2010) Interactions of the circadian CLOCK system and the HPA axis. Trends Endocrinol Metab 21:277–286

Niu X, Ding Y, Chen S, Gooneratne R, Ju X (2022) Effect of immune stress on growth performance and immune functions of livestock: mechanisms and prevention. Animals 12(7):909

O'Loughlin A (2011) Examination of the effect of weaning stress on the physiological and molecular regulation of immune function in circulating bovine leukocytes. PhD Thesis, University of Ireland, Maynooth

O'Loughlin A, McGee M, Waters S et al (2011) Examination of the bovine leukocyte environment using immunogenetic biomarkers to assess immunocompetence following exposure to weaning stress. BMC Vet Res 7:45

O'Loughlin A, Lynn DJ, McGee M et al (2012) Transcriptomic analysis of the stress response to weaning at housing in bovine leukocytes using RNA-seq technology. BMC Genom 13:250

O'Loughlin A, McGee M, Doyle S, Earley B (2014) Biomarker responses to weaning stress in beef calves. Res Vet Sci 97:458–463

Odore R, D'angelo A, Badino P et al (2004) Road transportation affects blood hormone levels and lymphocyte glucocorticoid and [beta]-adrenergic receptor concentrations in calves. Vet J 168:297–303

Odore R, Badino P, Re G et al (2011) Effects of housing and short-term transportation on hormone and lymphocyte receptor concentrations in beef cattle. Res Vet Sci 90:341–345

Paape MJ, Bannerman DD, Zhao X et al (2003) The bovine neutrophil: structure and function in blood and milk. Vet Res 34:597–627

Palme R (2019) Non-invasive measurement of glucocorticoids: advances and problems. Physiol Behav 199:229–243

Pang WY, Earley B, Sweeney T et al (2006) Effect of carprofen administration during banding or burdizzo castration of bulls on plasma cortisol in vitro interferon- production acute-phase proteins feed intake and growth. J Anim Sci 84(35):1–359

Pang WY, Earley B, Sweeney T et al (2009a) Temporal patterns of inflammatory gene expression in local tissues after banding or burdizzo castration in cattle. BMC Vet Res 5:36

Pang WY, Earley B, Sweeney T et al (2009b) Effects of banding or burdizzo castration of bulls on neutrophil phagocytosis and respiratory burst CD62-L expression and serum interleukin-8 concentration. J Anim Sci 87:3187–3195

Pang WY, Earley B, Murray M et al (2010) Banding or Burdizzo castration and carprofen administration on peripheral leukocyte inflammatory cytokine transcripts. Res Vet Sci 90(1):127–132

Pascual-Le Tallec L, Lombes M (2005) The mineralocorticoid receptor: a journey exploring its diversity and specificity of action. Mol Endocrinol 19:2211–2221

Phillips WA, Juniewicz PE, Zavy MT et al (1989) The effect of the stress of weaning and transport on white blood cell patterns and fibrinogen concentration of beef calves on different genotypes. Can J Anim Sci 69:333–340

Preisler MT, Weber PSD, Tempelman RJ et al (2000) Glucocorticoid receptor down-regulation in neutrophils of periparturient cows. Am J Vet Res 61:14–19

Qiu X, Arthington JD, Riley DG, Chase CC, Phillips WA, Coleman SW, Olson TA (2007) Genetic effects on acute phase protein response to the stresses of weaning and transportation in beef calves. J Anim Sci 85:2367–2374

Ralph C, Tilbrook A (2016) Invited review: the usefulness of measuring glucocorticoids for assessing animal welfare. J Anim Sci 94:457–470

Reed, JC (2000) Mechanisms of Apoptosis. Am J Pathol 157: 1415-1430

Rifai N, Gerszten RE (2006) Biomarker discovery and validation. Clin Chem 52:1635–1637

Riondato F, D'Angelo A, Miniscalco B et al (2008) Effects of road transportation on lymphocyte subsets in calves. Vet J 175:364–368

Saama PM, Jacob JB, Kehrli ME et al (2004) Genetic variation in bovine mononuclear leukocyte responses to dexamethasone. J Dairy Sci 87:3928–3937

Saccò M, Valenti G, Corvi Mora P et al (2002) DHEA a selective glucocorticoid receptor antagonist: its role in immune system regulation and metabolism. J Endocrinol Invest 25:81–82

Salak-Johnson JL, Mcglone JJ (2007) Making sense of conflicting data: stress and immunity in swine and cattle. J Anim Sci 85:E81–E88

Sapolsky RM, Romero LM, Munck AU (2000) How do glucocorticoids influence stress responses? Integrating permissive suppressive stimulatory and preparative actions. Endocr Rev 21:55–89

Sartin J, Elsasser T, Gunter D et al (1998) Endocrine modulation of physiological responses to catabolic disease. Dom Anim Endocrinol 15:423–429

Scheinman RI, Cogswell PC, Lofquist AK et al (1995) Role of transcriptional activation of IkappaBalpha in mediation of immunosuppression by glucocorticoids. Science 270:283–286

Schommer NC, Hellhammer DH, Kirschbaum C (2003) Dissociation between reactivity of the hypothalamus-pituitary-adrenal axis and the sympathetic-adrenal-medullary system to repeated psychosocial stress. Psychosomatic Med 65:450–460

Sejian V, Bhatta R, Gaughan JB et al (2018) Review: adaptation of animals to heat stress. Animal 12:S431–S444

Seth M, Lamont EA, Janagama HK et al (2009) Biomarker discovery in subclinical mycobacterial infections of cattle. PLoS ONE 4:e5478

Simon S, Lamoureux L, Plews M et al (2008) The identification of disease-induced biomarkers in the urine of BSE infected cattle. Prot Sci 6:23

Sladek Z, Rysanek D (2001) Neutrophil apoptosis during the resolution of bovine mammary gland injury. Res Vet Sci 70:41–46

Sporer KR, Xiao L, Tempelman RJ et al (2008) Transportation stress alters the circulating steroid environment and neutrophil gene expression in beef bulls. Vet Immunol Immunopathol 121(3–4):300–320

Stefano GB, Smith EM (1996) Adrenocorticotropin - A central trigger in immune responsiveness: tonal inhibition of immune activation. Med Hypotheses 46:471–478

Sternberg EM (2006) Neural regulation of innate immunity: a coordinated nonspecific host response to pathogens. Nat Rev Immunol 6:318–328

Stevenson JR, Westermann J, Liebmann PM et al (2001) Prolonged alpha-adrenergic stimulation causes changes in leukocyte distribution and lymphocyte apoptosis in the rat. J Neuroimmunol 120:50–57

Swain SD, Jutila KL, Quinn MT (2000) Cell-surface lactoferrin as a marker for degranulation of specific granules in bovine neutrophils. Am J Vet Res 61:29–37

Tempel DL, Leibowitz SF (1994) Adrenal steroid receptors: interactions with brain neuropeptide systems in relation to nutrient intake and metabolism. J Neuroendocrinol 6:479–501

Thornton P, Nelson G, Mayberry D et al (2022) Impacts of heat stress on global cattle production during the 21st century: a modelling study. Lancet Planet Health 6(3):e192–e201

Thun R, Eggenberger E, Zerobin K et al (1981) Twenty-four-hour secretory pattern of cortisol in the bull: evidence of episodic secretion and circadian rhythm. Endocrinology 109:2208–2212

Tilbrook AJ, Rivalland EA, Turner AI et al (1999) Responses of the hypothalamopituitary adrenal axis and the sympathoadrenal system to isolation/restraint stress in sheep of different adiposity. Neuroendocrinology 87:193–205

Ting STL, Earley B, Hughes JML et al (2003) Effect of ketoprofen lidocaine local anesthesia and combined xylazine and lidocaine caudal epidural anesthesia during castration of beef cattle on stress responses immunity growth and behavior. J Anim Sci 81:1281–1293

Van Den Berg D, De Jong W, De Kloet E (1994) Mineralocorticoid antagonist inhibits stress-induced blood pressure response after repeated daily warming. Am J Physiol Endocrinol Metab 267:921–926

Van Kampen C, Mallard BA (1997) Effects of peripartum stress and health on circulating bovine lymphocyte subsets. Vet Immunol Immunopathol 59:79–91

Van Oostveldt K, Paape MJ, Dosogne H et al (2002) Effect of apoptosis on phagocytosis respiratory burst and CD18 adhesion receptor expression of bovine neutrophils. Dom Anim Endocrinol 22:37–50

Vegiopoulos A, Herzig S (2007) Glucocorticoids, metabolism and metabolic diseases. Mol Cell Endocrinol 275:43–61

Viswanathan K, Dhabhar FS (2005) Stress-induced enhancement of leukocyte trafficking into sites of surgery or immune activation. Proc Nat Acad Sci 102:5808–5813

Von Borell E (2001) The biology of stress and its application to livestock housing and transportation assessment. J Anim Sci 79:260–267

Weber PS, Toelboell T, Chang LC et al (2004) Mechanisms of glucocorticoid-induced down-regulation of neutrophil L-selectin in cattle: evidence for effects at the gene-expression level and primarily on blood neutrophils. J Leukocyte Biol 75:815–827

Weber PSD, Madsen-Bouterse SA, Rosa GJM et al (2006) Analysis of the bovine neutrophil transcriptome during glucocorticoid treatment. Physiol Genomics 28:97–112

Yagi YH, Shiono Y, Chikayama A et al (2004) Transport stress increases somatic cell counts in milk and enhances the migration capacity of peripheral blood neutrophils of dairy cows. J Vet Med Sci 66:381–387

Zinder O, Dar DE (1999) Neuroactive steroids: their mechanisms of action and their function in the stress response. Acta Physiol Scand 167:181–188

# Part II

# Challenges and Solutions in Different Management Systems for Cattle

# Housing of Dairy Cattle: Enhancing Movement Opportunity in Housing Systems

4

Véronique Boyer, Elise Shepley, Sarah E. McPherson, Jessica St John, and Elsa Vasseur

## Contents




**Abstract**

The intensification of dairy production over the course of the last decades has led to substantial changes in the housing and the management of dairy cattle, and increased human control over the cow and her environment. These changes have been accompanied by an increase in cow productivity, but also by a sustained decline in cow longevity, and raised new concerns regarding the health and the welfare of dairy cows reared in intensive, highly productive environments. At the heart of this, housing systems were found to impact the cow's ability to move comfortably in her environment, and as such, the quality of the housing is thought of as critical elements in ensuring good welfare of dairy cows. This chapter discusses the impact of such intensive housing systems—more specifically stall-based systems—on the opportunity for movement and on the dairy cow's ability



V. Boyer · E. Shepley · S. E. McPherson · J. St John · E. Vasseur (✉)
Department of Animal Science, McGill University, Sainte-Anne-de-Bellevue, QC, Canada
e-mail: elsa.vasseur@mcgill.ca







M. Haskell (ed.), *Cattle Welfare in Dairy and Beef Systems*, Animal Welfare 23,
https://doi.org/10.1007/978-3-031-21020-4_4

to fulfill her needs. Such impacts are traditionally assessed through an array of outcome measures, which reflect the effects of inadequate housing and management on the cows. Yet, although such indicators reflecting the negative impacts of housing systems on the cow's welfare are now common, the scientific literature remains scarce on indicators of positive welfare status to assess the effectiveness of improvements in housing conditions that go beyond the minimum standards.

**Keywords**

Cow behavior · Movement opportunities · Housing systems · Outcome measures · Cow welfare

## 4.1 Introduction

Dairy production has intensified over the course of the last century. While progress in areas of nutrition, genetic selection, and management of dairy cows has led to substantial increases in productivity, it has also led to an increase in human control over the cow's environment. Consequently, while cows have always been dependent on their human caregivers to some extent, the intensification of housing and management has also brought an increase in the dependence of dairy cows upon humans when it comes to fulfilling their various needs such as proper nutrition, meaningful social contacts, high-quality rest, and natural movement, or when it comes to be provided with opportunities to fulfill said needs on their own. As intensification of dairy production has progressed, a concurrent decrease in cow longevity has been observed across all countries with highly productive dairy sectors (Dallago et al. 2021), leading to research efforts being organized to identify the different causes and create solutions to reduce or resolve this decline in longevity. Among the questions tackled, the concept of the opportunity for movement, or the ability of the cow to move with ease in her environment appears to be one promising avenue; indeed, an environment restricting the cow's ability to move could well be a factor implicated in multiple welfare issues identified by research, such as levels of body injuries and lameness among other issues. The physical outcomes of these welfare issues, such as alterations in locomotion or lameness and injuries to the legs, are used as indicators in assessment tools that are used to assess the effects of housing. This chapter will discuss the impact of intensive, stall-based housing systems on the opportunity of movement and on the ability of cows to fulfill their needs, as reflected in routinely collected outcome measures of welfare.

## 4.2 The Concept of Opportunity for Movement

Housing systems are thought to play a central role in this phenomenon, as they represent the key to the human's control over the dairy cow's environment. As dairy production intensified, it increasingly turned away from more extensive production systems such as pasture-based systems to move toward indoor-based, total confinement housing systems. Housing systems vary in the level of human control on the cow's environment, and conversely, in the potential restriction imposed upon the ability of cows to satisfy their needs. Natural environments, or production systems allowing cows to access such environments, are usually considered as the least restrictive, as they allow cows to express their behavioral repertoire with reduced (or even better, minimal) constraints. On the opposite side of the spectrum, stall-based systems (tie-stalls and free-stalls) are considered as the most restrictive of all dairy cattle housing designs, especially in aspects relating to opportunities for movement (Fig. 4.1). However, these types of systems are also present in most farms, particularly in countries with highly productive dairy sectors.

The intended role of the stall is to be the main resting area for the cow. Thus, given the importance of rest in a dairy cow's daily schedule (8–13 h/d; Tucker et al. 2021), it can be expected that the cow will experience a certain level of restriction to her movement in any stall-based housing system, since the stall properly accommodates only a limited use of different resting postures, all involving a sternal body recumbence with a tucked position of the hind legs relative to the body. This can explain why furthering our understanding of the role that the stall and its components play in restricting movement opportunities for dairy cows has become an important topic of research (Fig. 4.2).

The space within a stall that the cow can use can be subdivided in a multitude of ways, according to the function or the needs of the cow. All of the different stall structure elements play a role—albeit to a varying degree—in providing the cow

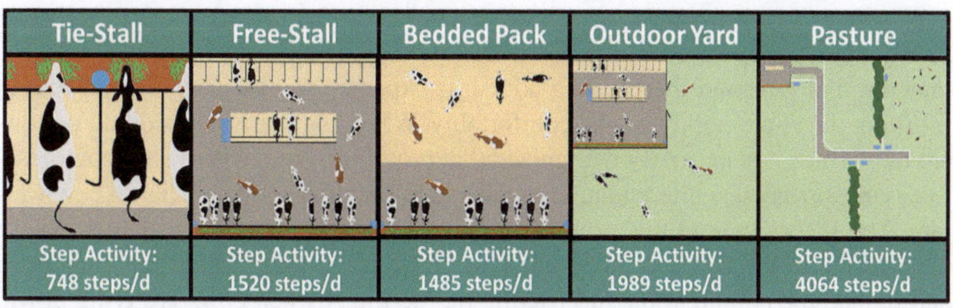


**Fig. 4.1** An overview of the different types of housing systems for dairy cows, along with the average daily step count per cow according to scientific literature (see review in Shepley et al. 2020). One should note that the locomotor activity represents only one type of movement opportunity offered to dairy cows in each housing system and by no means represents a full picture on its own. Other opportunities provided besides locomotor activity, such as opportunities for social interactions and opportunities to engage in a diversity of resting postures, can also be impacted by the type and configuration of housing systems cows are found in.

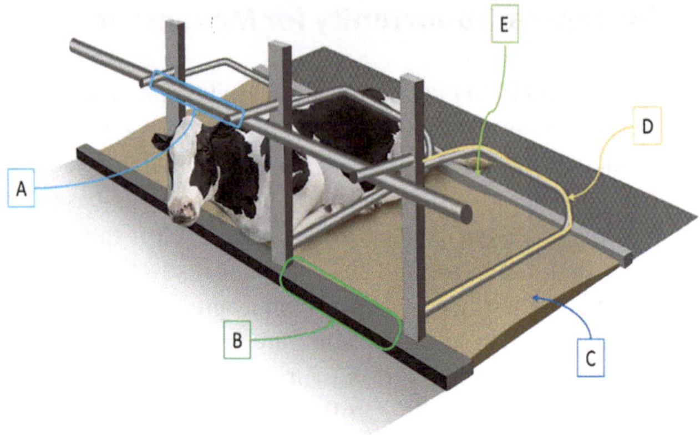


**Fig. 4.2** An overview of the principal structural components of the stall: A-Tie-rail (tie-stall) or neck-rail (free-stall). The recommended position of the tie-rail is illustrated; neck-rails are most commonly positioned a certain distance behind the stall divider posts, closer to the rear end of the stall compared to this image. The tie chain (not illustrated here) is usually tied to the middle of the tie-rail and is a structure unique to tie-stalls; B-Brisket board (free-stall) or manger wall (tie-stall); C-Stall surface; D-Side divider; E-Rear curb. Beyond the dimensions of the stall bed, there are a number of structural elements such as the configuration of the side dividers or the size of the front curb, which impact the cow's positioning and ability to use the stall for lying down. Stalls are also used for standing, in which case the position of the rail and the dimensions of the stall will have an impact on the cow.

with the opportunity to rest or to move comfortably. Thus, the impact of each aspect of stall design will be different depending on which need we are assessing, and stall components are often divided into a small number of groups based on their role or function in the system.

## 4.3 The Role of the Stall and Its Components

Firstly, the stall surface, which comprises the combination of both the stall base material and type and the bedding added, is expected to play a significant role in fulfilling the cow's ability to rest comfortably. While stall base and bedding options are numerous, it is generally recognized that providing cows with a soft, compressible lying surface—while maintaining adequate traction—is beneficial to cow welfare. The theory behind this principle is logical: softer stall surfaces better absorb the pressure or the shock imposed upon the cows' joints upon rising and lying down and reduce the compression and friction on the limbs when cows are lying down. This objective is the basis of recommendations made for stall surfaces across the world. However, it must be noted that data on compliance with these recommendations are not readily available.

The surface area made available to the cow, namely the stall bed, is defined by the various stall elements, which define its limits at the front, on the sides, and at the back (see Fig. 4.2). The front of the stall bed is usually delimited by a brisket board (free-stall) or manger wall (tie-stall) made of wood, metal, plastic, or concrete, which, in tie-stalls, also serves to keep feed from being transferred into the stall from the manger in front of it. In all stall-based systems, the stall comprises side dividers, which define the lateral space it provides to the cow. Finally, the back of the stall is delineated either by another short wall called the curb or by a gutter positioned right behind the stall. The gutter may or may not be covered by a grid. When the cow is standing in the stall, the presence of another optional rail—the neck rail in free-stalls, and the tie-rail in tie-stalls—also acts to regulate the amount of space available for standing and moving in the stall. In tie-stalls, the chain attached to the tie-rail represents the element keeping the cow from leaving the stall confines.

Recommendations related to the stall dimensions, that is, the positioning of the dividers, neck rail, and divider wall are typically based on cow dimensions, which can be based on cow size or weight. Basing the dimensions of the stall elements on the size or weight of the cow should take into account the fact that not all cows are of the same size, and therefore, that cows differ in their need for space. However, the recommendations cannot account for individual differences between cows in terms of the extent to which they can fulfill their needs within the constraints imposed by the structural components of the stall. For instance, the preferred lying posture of an individual cow (e.g., degree of tucking of the legs, resting of the head, position of the body) may be more or less accommodated by the configuration of the stall in which she is housed.

Stall systems were first and foremost designed to facilitate more intensive management of cows. The advent of stall systems was meant to facilitate feces and urine being deposited in specific areas while having the cows rest in others, easing excreta management and maintaining cow cleanliness. Stall systems also represent the most efficient types of housing system with regard to housing the maximum number of animals in a given space. Space is saved by imposing a predetermined lying-down orientation on all animals.

Traditional assessments of the welfare impact of a housing system's features mostly focused on negative outcomes, which are considered to reflect issues, which could result from the lack of opportunity for the cows to fulfill key needs. The outcome measures employed to identify problems and find solutions are thought to reflect the consequences of suboptimal environments on cows. Common outcomes used in current on-farm assessment programs (e.g., Welfare Quality® Assessment Protocol for Cattle, Welfare Quality Consortium 2009; Farmers Assuring Responsible Management (FARM) Animal Care Reference Manual, NMPF 2020; proAction® Reference Manual, DFC 2021) mostly use visual scoring following a pre-established chart (e.g., body condition score, body injuries, cleanliness, lameness, lying-down and rising movements). Other avenues for welfare assessments in the future may rely on various automatic recording devices (e.g., collars, leg-mounted accelerometers) to collect data on indicators such as lying time metrics, activity levels, and frequency of visits to automatic milking systems (Vasseur 2017) to assess the welfare of cows.

## 4.4 Impact of Stall Bed Surface

Literature pertaining to the stall bed surface type and bedding shows that bedding depth appears to be the most influential material component of the stall bed. Stall bedding has the greatest ability to compensate for shortcomings of the stall base type and/or of the bedding type that are detrimental to cow comfort (e.g., hardness, abrasiveness; see review by McPherson and Vasseur 2021a). Increasing bedding depth in the stall increases lying time, and therefore, cow comfort, regardless of the stall base type or bedding type. Hock injuries result from bed surface abrasiveness and/or lack of compressibility, thus the combination of bedding depth and stall base characteristics can play a major role in decreasing the likelihood of a cow developing hock injuries. This role was confirmed by results from an experiment, which found improved lying time and reduced body injuries in stalls with deep bedding (≥7.5 cm; McPherson and Vasseur 2021b). Bedding is known to contribute to stall bed compressibility, especially with harder stall bases such as concrete or rubber mats (Villettaz Robichaud et al. 2020). As such, deeper bedding in stalls could compensate, up to a certain level, for shortcomings in other areas of stall design, improving comfort to acceptable levels. Increased bedding depth is also associated with reduced prevalence of lameness on farms. However, more research needs to be conducted to provide further insight into this interaction.

While no specific bedding system or type of stall base is known to yield cleaner cows, increased bedding depth can increase cow cleanliness when combined with proper stall management (see review by McPherson and Vasseur 2021a). Bedding quality is another key aspect of bedding and is characterized by how soft and dry the bedding is. Cows appear to show a clear preference for dry lying surfaces and will spend more time standing when only wet bedding is available. Thus, wet and soiled lying surfaces can negatively impact the welfare of the animals by affecting the quantity and quality of rest and can result in poorer hygiene, which can in turn affect health and productivity.

## 4.5 Impact of Stall Bed Size

The length and width of the stall bed are thought to contribute to the comfort of cows, particularly their comfort while resting. Bed length is defined as the space between the front manger and the gutter, thus defining the longitudinal space allowance that the cow lies into, and is an important factor to consider, as it affects welfare outcomes. Stall bed length affects transition movements (lying down and rising) but also impacts the cow's use of the stall for standing. Other factors such as the position of the neck rail or of the tie-rail and the height of the manger in tie-stalls will affect the response of the cow to stall length as well, by modifying her lying postures, her position in the stall, and concurrently, her ability to perform lying down and rising movements with ease. Stall width can be defined as the space between the 2 dividers, which act as the right and left side limits of each stall, thus the lateral space available in the stall bed. The greatest need for lateral space in cows comes at the time of lying down (Ceballos et al. 2004) as it is intrinsically linked to the ease of movement of cows when lying down and rising and to the comfort of the bed. Stall width is also associated with the use of the stall for standing; cows spend more time standing in narrow stalls but less time lying down (Tucker et al. 2004).

Increasing stall bed length has been demonstrated to increase lying time and to decrease injury and lameness prevalence (see review by McPherson and Vasseur 2021a) but is often not utilized by producers due to concerns about cleanliness. Longer bed lengths allow the cow to stand and defecate within the stall rather than into the rear passageway, while shorter stalls force the cow to stand close to the rear edge of the stall and defecate into the passageway. Because of this, longer stalls are often anecdotally associated with being dirtier and therefore making the cows dirtier. However, relatively few studies have been conducted to experimentally or epidemiologically investigate the relationship between stall or bed length and cleanliness. The limited research available (see McPherson and Vasseur 2021b) suggests that longer stalls lead to slightly dirtier cows and stalls, suggesting that management practices (i.e., additional cleaning) may need to be adapted when using longer stalls in order to maintain the same standards of cleanliness.

Increasing stall width is associated with longer lying times and increased ease of movement (see review by Boyer and Vasseur 2021). There are demonstrated links between increasing stall width and a decreased risk for hock, knee, and neck injuries, although different studies published on the matter present contradicting results. However, the width of stalls in many of the epidemiological studies falls well below the dimensions recommended for cow body size in current national guidelines, which could have played a role in the results obtained. The same situation (contradicting results) applies to the link between lameness and stall width. Regarding cow cleanliness, increasing stall width has been linked in different studies with decreased cleanliness, increased cleanliness, or as having no significant impact on the cleanliness of cows.

## 4.6 Impact of Height of the Manger Wall and of the Brisket Board

This structural element is meant to define the front limit of the stall bed, thus it is hypothesized to primarily impact the cow's ability to use the stall for resting in conjunction with other elements defining the size of the stall bed. The impact of the manger wall height on cow welfare has not been researched extensively, with only a handful of studies presenting results on the matter (McPherson and Vasseur 2021a). Lower, less restrictive manger walls and brisket boards have been associated with a decrease in lameness prevalence, an increase in lying time, but also with an increase in the likelihood of dirty udders. Brisket boards also influence where larger cows lie down in the stall. While lower appears to be better, reducing the height of the manger wall has also been associated with a reduced ability to rise and lie down in cows, likely due to the fact that although longitudinal space was increased by the lower manger wall, the tie-rail position in the particular study remained too restrictive for the cows to fully benefit from the increase in the space made available to them (McPherson and Vasseur 2021b).

## 4.7 Impact of the Rail Position

Neck-rails and tie-rails define the dynamic space available to the cow at the front of the stall. As such, improperly positioned rails may hinder lying down and rising movements and impair the ability of the cow to utilize the stall for standing if needed. Most studies agree that rails positioned at mid-range height increase the risk of neck and hock injuries (see review by St John 2019). Other than this evidence, however, there is no clear link between rail height and outcome measures, especially with increasing rail height. This suggests that increasing tie-rail or neck-rail height in stalls can only increase comfort up to a certain point; the interaction between this factor, other aspects of stall design, and cow factors likely explains the lack of conclusive results regarding tie-rail/neck-rail height.

Height is not the only aspect relating to the positioning of the rail in stalls, however. The forward position, measured either in terms of distance from the rear end of the stall or in distance from the manger wall or brisket board, also plays a role in the ability of the cow to use the space of the stall. While the rail position appears to particularly affect the cow's ability to use the stall for standing (Tucker et al. 2005), it may also have an impact on the cow's ability to use the stall to fulfill her need for rest. This is reflected in various welfare outcome measures, such that moving the position of the rail (tie-rail or neck-rail) toward the front of the stall may decrease the incidence of body injuries, sole lesions, digital dermatitis, lameness, and may increase the number of lying bouts (see review in St John 2019). A reduction in cow and stall cleanliness was, however, found with increased forward position of the rail. This could represent a compromise between cow resting comfort and cleanliness to be considered by producers, as adaptations to management can compensate for the increased risk associated with decreased cleanliness levels.

## 4.8 Intrinsic Limitations to Investigating the Impact of the Stall and Its Components

While the current body of evidence suggests that several different welfare outcomes can be improved through the modification of stall configuration and features, the relationships between individual features cannot be fully understood if these interrelated elements are evaluated separately, as one can compensate for another, and as it is their combination in the stall that creates the overall level of comfort for the cow. Many of these relationships are derived from results of epidemiological studies, which have examined the links between those outcomes and stall design elements without any predetermined focus on specific aspects. While such studies may lead to the identification of unsuspected factors contributing to aspects of cow comfort, their reliance on commercial farm installations represents a weakness.

Compliance to recommendations for stall dimensions was often found to be lacking for multiple aspects of the stalls within the same farm (e.g., Bouffard et al. 2017). This is likely due to the age of the buildings. As cow housing is expensive, producers are often reluctant to replace the stalls or the housing, which means that the stalls will often not accommodate the growth in size of modern dairy cows. Experimental studies tend to use a slightly different approach, focusing on only one, or a few related stall components and evaluating their impacts on the cow need of interest and on the corresponding health and welfare outcomes. For instance, researchers may assess the effect of stall length on lying time, but not consider the associated issue of how a longer stall length affects where the neck rail and brisket board should be positioned. The manipulations conducted as part of those studies allow researchers to properly isolate the impact of specific stall design factors and investigate dimensions or configurations not employed on farms, which represents an interesting avenue to better our understanding of each factor's role in the comfort and welfare of housed dairy cows. However, the design of these studies often takes away our ability to assess the multiple interactions between different factors and their impact on cow welfare, because they typically focus on one or two stall elements at a time.

One should also keep in mind that most outcome measures currently used are indicators of welfare problems, and that as such, the absence of negative outcomes does not guarantee a good welfare status per se. For one, the scope of issues measured by commonly used outcomes may not capture certain types of problems frequently encountered in housing systems. An example of this would be the occurrence of contacts with the stall partitions, which has only been included in the scope of measures collected in a few studies. Cows were found to bump into or lean on the dividers and rails of their stalls up to 12,000 times per day, even in stalls of recommended dimensions (Freinberg et al. 2020). Such findings have important implications regarding the welfare of dairy cows, as it signals that the current dimensions of stalls, although theoretically allowing sufficient space to accommodate the cows' movements, do not provide enough of a margin of error for cows to avoid hitting the different stall partitions. Such partitions are thus thought to play the role of a physical barrier or hindrance to movement instead of acting as they should—as visual indicators of the stall limits. While the contacts with the front rail could eventually result in neck injuries—a commonly measured outcome—contacts with the side dividers, while still occurring at an abnormally high rate, may pass unmeasured. This is because the main areas of contact, namely the flanks and the hip bones, are often not included in injury assessment protocols. Research has found that provision of additional lateral space for cows can help reduce the number of contacts during lying-down movements, although it did not reduce these contacts completely (Boyer et al. 2021). The magnitude of this reduction has to be considered in a context where individual cows do not adapt their behavior in the same manner when provided with additional space, much like individual cows will choose to prioritize needs at different levels when the environment they live in forces such decisions (Zambelis and Vasseur 2022). Figure 4.3 shows an example of differences in the use of a double stall by two different individuals, and as such, is a good illustration of these

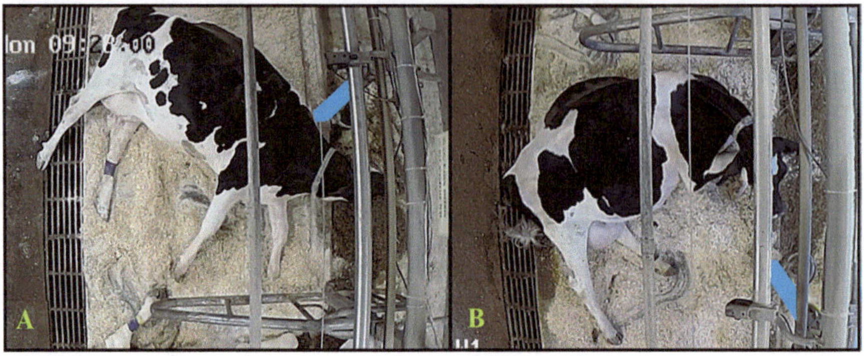


**Fig. 4.3** Differences in the use of additional lateral space for lying down by individual cows, as tested in Boyer et al. (2021). The lying space was doubled by removing dividers between two adjacent stalls. The cow in A has substantially increased her use of the additional space to engage in lying postures different from ones commonly seen in stall-based systems. The cow in B has also expanded her use of space while lying down, albeit to a lesser degree than the cow in A, as only her hind legs can be found in the second stall. The structural post marking where the additional space starts is highlighted in blue

individual differences. The impact of stall design on cow welfare is the result of the interaction between multiple aspects of stall design and multiple factors related to each cow, and her lying behavior and conformational characteristics, and such interactions have yet to be investigated.

## 4.9 Other Aspects of the Free-Stall Housing System Affecting the Cow's Ability to Move and to Access Resources

The free-stall-housed cow has to move around within her housing to access the different resources provided to her within the system. These resources principally consist of stalls for resting, feeding area(s) for eating, water troughs or bowls for drinking, but may also include the automatic milking system (AMS) to be milked, and brushes for grooming. Cows may have to compete with others for access to these physical resources.

Competition between cows for these resources may occur due to the need for synchronous access, poor alleyway configuration, high stocking densities, or an interaction between these factors. Cattle are a social species and prefer to perform behaviors such as feeding and lying in a synchronized manner. Additionally, feeding and lying are behaviors that take up a substantial amount of the cow's daily time budget, which may further increase competition for the stalls and feeding areas. Therefore, the housing should ideally be designed to allow all cows to access to feeding and lying areas at the same time, and the appropriate dimensions of alleyways to allow them to access these resources easily. Competition may occur even when simultaneous access is available to all cows, as cows compete for favored aspects of these resources, such as access to stalls that are the closest to the feeding area or the larger or cleaner stalls. Preference for these aspects varies between individuals (Val-Laillet et al. 2008). While access to water, grooming brushes, and the AMS are important, the behaviors involved are not known to be performed in a socially synchronized way, so providing high-volume access is less critical.

Two important interconnected factors affect the cow's ability to access the resources provided to her in a free-stall system: the physical layout of the facility and the stocking density. Stocking density is a key aspect of management under the producer's control. The quality of the flooring is a key feature of the physical layout of the housing. It affects the cows' ease of movement and can be assessed using locomotion metrics such as stride length and slip (Telezhenko and Bergsten 2005). The provision of flooring with good traction and good compressibility (e.g., the use of grooving in the concrete flooring or using rubber matting over the concrete flooring), which is closer to the natural characteristics of a pasture environment than the traditional concrete flooring, has been shown to enhance the cow's locomotion (Telezhenko and Bergsten 2005). Access to resources will be facilitated by designing the housing to promote a continuous flow of movement by providing large passageways and configuring the rows of free-stalls so that there are no or few dead-ends. Incorporating these features into housing needs to be considered at the design stage of the housing. Failure to do so may result in increased competition or forcing cows to have to stand and wait for access to feed and other resources. Too much time spent idle or waiting, especially on hard surfaces, increases the risk of lameness and decreased cow welfare (Cook and Nordlund 2009).

Stocking density also plays an important role in regulating access to resources and interacts with the physical layout of the barn. In most instances, increasing the stocking rate of the barn negatively impacts the ability of cows to access the different resources, particularly ones of primary interest such as stalls and feed bunk space. Overstocking the resting space is known to reduce the time the cows spend lying down, particularly so for subordinate or less competitive individuals (Winckler et al. 2015). Overstocking also has a number of effects on feeding behavior and access to the feed bunk space. High stocking densities lead to competition for access to the feed bunk and results in high levels of displacements, where one animal forces another from the feed-bunk, particularly when fresh feed is delivered (DeVries et al. 2004; Proudfoot et al. 2009). This results in alterations in feeding behavior, particularly in subordinate or younger animals, with higher feeding rates and longer/shorter feeding bouts typically observed (DeVries et al. 2004; Huzzey et al. 2006; Collings et al. 2011). Partitions at the feed face that create individual feeding spaces reduce the rate of displacement (Huzzey et al. 2006). In addition to this, overstocking affects the overall time budget as cows spend more time standing in the alleyway behind the feeding area, waiting for an opportunity to feed (Huzzey et al. 2006). This increased waiting time negatively affects lying time, but when coupled with poor access to feed, can lead to increased lameness and body injuries, decreased reproductive success, and decreased milk production. It is generally the least competitive or subordinate cows that are the most affected (Val-Laillet et al. 2008).

Attention to cleanliness and maintenance of the stalls, floors, and feeding areas is also important. Poorly distributed or dirty bedding in the stalls reduces the motivation for cows to lie down (Weary 2017) and soiled or wet floors can become slippery, reducing the ease of movement of the cows and increasing the risk of injuries. Overall, proper management of facilities and of the herd (e.g., appropriate grouping strategies) is important and will contribute to mitigate (or exacerbate, in the case of poor or inadequate management) the impacts of the environment and resources access constraints on the welfare of the cows.

## 4.10 Conclusion: Moving Beyond Meeting Minimal Requirements

Research in animal welfare has traditionally tackled questions with approaches centered on negative outcomes for welfare, with the improvement of the animals' situation confirmed by the absence or reduction of such outcomes. This has certainly been the case for studies of the effect of adjustments to housing conditions, where the outcome measures have primarily centered on injury, lameness, and cleanliness. The idea of studying positive outcomes or positive indicators of welfare, such as positive animal–animal or human–animal interactions and positive affective states, has slowly started making its way into the farm animal-related scientific literature and could represent a way to assess the efficacy of improvements to housing conditions, by going beyond the minimum standards required to obtain an absence of negative outcomes (Vigors and Lawrence 2019).

The provision of enrichment to dairy cows could hold an important place in future welfare improvement schemes, especially as it extends beyond the physical needs of the dairy cows to include the fulfilling of some of their psychological needs as well. An example of practices studied is the provision of outdoor access for cows housed in a movement-restrictive environment (e.g., Aigueperse and Vasseur 2021), which represents one avenue to not only provide housed cows with the physical benefits of a greater opportunity to express locomotion, but also increased opportunities to fulfill their need for socialization with conspecifics, and provide exposure to novel environments and stimuli likely to further contribute to their well-being. Future research endeavors should help to further document the benefits of improvements to housing systems and management practices. This new knowledge will help bring forward a more comprehensive perspective on cow welfare, which encompasses all aspects of the cow's physical and psychological experience, giving new keys to the dairy industry to ensure its sustainability and societal acceptability.

## References

Aigueperse N, Vasseur E (2021) Providing and outdoor exercise area affects tie-stall cow reactivity and human-cow relations. Front Vet Sci 7

Bouffard V, de Passillé AM, Rushen J et al (2017) Effect of following recommendations for tiestall configuration on neck and leg lesions, lameness, cleanliness, and lying time in dairy cows. J Dairy Sci 100(4):2935–2943
Boyer V, Vasseur E (2021) Graduate student literature review: the effect of chain length and stall width on common outcome measures of dairy cow welfare in stall-based housing systems. J Dairy Sci 104(3):3751–3760
Boyer V, Edwards E, Guiso MF et al (2021) Making tiestalls more comfortable: III. Providing additional lateral space to improve the resting capacity and comfort of dairy cows. J Dairy Sci 104(3):3327–3338
Ceballos A, Sanderson D, Rushen J, Weary DM (2004) Improving stall design: use of 3-d kinematics to measure space use by dairy cows when lying down. J Dairy Sci 87(7):2042–2050
Collings LKM, Weary DM, Chapinal N, von Keyserlingk MAG (2011) Temporal feed restriction and overstocking increase competition for feed by dairy cattle. J Dairy Sci 94:5480–5486
Cook NB, Nordlund KV (2009) The influence of the environment on dairy cow behavior, claw health and herd lameness dynamics. Vet J 179(3):360–369
Dairy Farmers of Canada (2021) proAction® Reference Manual – July 2021. Dairy Farmers of Canada, Ottawa
Dallago GM, Wade KM, Cue RI et al (2021) Keeping dairy cows for longer: a critical literature review on dairy cow longevity in high milk-producing countries. Animals 11(3)
DeVries TJ, von Keyserlingk MAG, Weary DM (2004) Effect of feeding space on the inter-cow distance, aggression, and feeding behavior of free-stall housed lactating dairy cows. J Dairy Sci 87(5):1432–1438
Freinberg C, Zambelis A, Vasseur E (2020) Validation of contact mats to monitor dairy cow contact with stall partitions. Animals 10(6)
Huzzey JM, DeVries TJ, Valois P, von Keyserlingk MAG (2006) Stocking density and feed barrier design affect the feeding and social behavior of dairy cattle. J Dairy Sci 89:126–133
McPherson SE, Vasseur E (2021a) Graduate student literature review: the effects of bedding, stall length, and manger wall height on common outcome measures of dairy cow welfare in stall-based housing systems. J Dairy Sci 103(11):10940–10950
McPherson SE, Vasseur E (2021b) Making tiestalls more comfortable: IV. Increasing stall bed length and decreasing manger wall height to heal injuries and increase lying time in dairy cows housed in deep-bedded tiestalls. J Dairy Sci 104(3):3339–3352
National Dairy FARM Program (2020) Farmers Assuring Responsible Management Animal Care Reference Manual Version 4: 2020-2022. National Milk Producers Federation, Arlington
Proudfoot KL, Veira DM, Weary DM, von Keyserlingk MAG (2009) Competition at the feed bunk changes the feeding, standing, and social behavior of transition dairy cows. J Dairy Sci 92:3116–3123. https://doi.org/10.3168/jds.2008-1718
Shepley E, Lensink J, Vasseur E (2020) Cow in motion: a review of the impact of housing systems on movement opportunity of dairy cows and implications on locomotor activity. Appl Anim Behav Sci 230:105026
St John J (2019) Effect of positioning the tie-rail to follow the natural neck line of cows when eating and rising on the welfare of dairy cows housed in tie-stall barns. Dissertation, McGill University
Telezhenko E, Bergsten C (2005) Influence of floor type on the locomotion of dairy cows. Appl Anim Behav Sci 93(3):183–197
Tucker CB, Weary DM, Fraser D (2004) Free-stall dimensions: effects on preference and stall usage. J Dairy Sci 87(5):1208–1216
Tucker CB, Weary DM, Fraser D (2005) Influence of neck-rail placement on free-stall preference, use, and cleanliness. J Dairy Sci 88(8):2730–2737
Tucker CB, Jensen MB, de Passillé AM et al (2021) Invited review: lying time and the welfare of dairy cows. J Dairy Sci 104(1):20–46
Val-Laillet D, de Passillé AM, Rushen J et al (2008) The concept of social dominance and the social distribution of feeding-related displacements between cows. Appl Anim Behav Sci 111(1–2):158–172

Vasseur E (2017) Animal behavior and well-being symposium: optimizing outcome measures of welfare in dairy cattle assessment. J Anim Sci 95(3):1365–1371

Vigors B, Lawrence A (2019) What are the positives? Exploring positive welfare indicators in a qualitative interview study with livestock farmers. Animals 9(9)

Villettaz Robichaud M, Pic A, Delgado H, Adam S, Lacroix R, Pellerin D, Vasseur E (2020) Short communication: use of the Clegg hammer measure as an indicator of stall-surface compressibility in tie-stall housing and its relationship with stall comfort. J of Dairy Sci 103(1):871–876

Weary DM (2017) Getting stocking density right for your cows. WCDS Adv Dairy Technol 29:243–253

Welfare Quality Consortium (2009) Welfare Quality Assessment Protocol for Cattle. Welfare Quality® Consortium, Lelystad

Winckler C, Tucker CB, Weary DM (2015) Effects of under- and overstocking freestalls on dairy cattle behaviour. Appl Anim Behav Sci 170:14–19

Zambelis A, Vasseur E (2022) Testing the ability of contact mats to identify problematic stall configurations. Can J Anim Sci 102(2): 317-324 https://doi.org/10.1139/cjas-2021-0109

# Welfare of Dairy Cows in Pasture-Based Systems

# 5

Cheryl O'Connor and Jim Webster

## Contents



**Abstract**

Dairy cattle are particularly good at converting forage into metabolic energy, making them well suited for milk production from grazing. Major efforts have been put into improving pasture production to suit the grazing characteristics of dairy cows. While this is a very efficient way to produce milk and has the inherent naturalness of outdoor grazing, the system is more complex to manage than concentrate-based systems. Many aspects of the welfare of the cows can be compromised.

Many of the welfare challenges arise from the seasonal nature of pasture production and the limitations of energy supply from pasture. This can cause periods of significant negative energy balance for cows, leading to hunger and low body condition. Genetic selection for high milk production exacerbates this issue. Additional welfare challenges come from the climate, particularly wet and muddy conditions, whether on or off pasture, which cows distinctly dislike. Heat

C. O'Connor (✉) · J. Webster
AgResearch Ltd., Ruakura Research Centre, Hamilton, New Zealand
e-mail: Cheryl.OConnor@agresearch.co.nz





M. Haskell (ed.), *Cattle Welfare in Dairy and Beef Systems*, Animal Welfare 23,
https://doi.org/10.1007/978-3-031-21020-4_5

stress is another common welfare issue and is exacerbated by high-producing cows and a lack of provision of shade. Skilled management is required to maintain acceptable welfare.

This chapter describes the welfare issues of dairy cows in pasture-based systems, providing an understanding of the risks that lie alongside the positive aspects of grazing systems.

**Keywords**
Cow welfare · Pasture · Body condition · Shade and shelter · Natural behaviour

## 5.1 Introduction to Pasture-Based Systems

There are large differences in dairy production systems globally due to physical, climatic and financial factors (Kay et al. 2015). Grazed pasture is the major component of the total diet for only 10% of the world's cows and pasture-based systems are based on different operating principles compared with housed systems (Holmes et al. 2007).

The objective of most pasture-based milk production systems across the world is to maximise the amount of milk produced by the cows from grazed grass (O'Donovan et al. 2021; Cahill et al. 2019). In many countries operating a pasture-based system, grass growth is seasonal with peak growth occurring between spring and autumn. In a successful grazing system, the cow's feed requirements must be synchronised with the rate of supply of feed from pasture, which typically varies seasonally. To achieve this, each cow must calve once every 365 days, and all cows must calve within 8–10 weeks in late winter/early spring so that peak feed demand matches peak pasture growth in late spring (Dillon et al. 1995; Macdonald et al. 2008). Grass has a more variable nutritional value than the concentrate feed used in housed systems, and this influences which breeds are used in these systems for best results. Cows selected for high milk production, such as North American or Dutch Holstein Friesian strains, perform well when fed a high-energy total mixed ration (TMR), whereas Jersey, Ayrshire and crosses with Holsteins such as the 'Kiwi-cross' (a commonly used Holstein/Jersey cross; Fig. 5.1) are more suitable for a pasture-based system as they are able to maintain body condition and milk production from grazed grass (Kolver et al. 2002; Horan et al. 2006; Roche et al. 2006).

**Fig. 5.1** Kiwicross cows (a cross between the Holstein Friesian and Jersey breeds) (Photo courtesy of Laura B Hunter)

## 5.2 Benefits of Pasture-Based Systems

There are a number of benefits for the welfare of dairy cattle managed at pasture, particularly in regard to behaviour. Outdoor farming allows animals to perform more natural behaviours, such as feeding, movement, exploration and social behaviours than is possible in intensive, indoor-based systems. Naturalness is a fundamental component of good animal welfare and therefore promotes a more positive welfare image (Hemsworth et al. 1995). Pasture provides grazing opportunities and allows cows to have more choice how they spend their time (Fig. 5.2). There is evidence that aspects of a pasture environment are attractive to cows; for instance, housed cows have a strong preference to spend time outside on pasture especially at night (Charlton et al. 2011; Charlton and Rutter 2017; Legrand et al. 2009). What the exact basis for this attraction is not yet determined.

A beneficial aspect of pasture environments compared to indoor systems is freedom of movement (Phillips et al. 2013; Arnott et al. 2017). There is evidence of reduced frustration and improved physical health and well-being (Gonyou 1996). Cows in grazing systems had better locomotion, straighter rear legs, steeper foot angle, flatter and more refined bones and better legs and feet (all of which are considered beneficial in terms of functionality) compared with cows in housed systems (Onyiro and Brotherstone 2008) and were less susceptible to digital dermatitis (Onyiro et al. 2008). A study on housed cows showed that a period on pasture was beneficial in promoting recovery from lameness. This effect was not due to the cows

**Fig. 5.2** A typical pasture-based system in New Zealand (Photo courtesy of Laura B Hunter)

at pasture lying down for longer, but was probably due to pasture being a more comfortable surface to stand on, helping cows to recover from hoof and leg injuries (Hernandez-Mendo et al. 2007).

Pasture environments have complexity and stimulus value which promotes desirable behaviours like foraging, exploration and social play (Wilson et al. 2002). Dairy cows with access to an outdoor paddock will spend time exploring the environment (Loberg et al. 2004). Cows at pasture have also been shown to be less restless and have an undisrupted lying and rumination pattern, more social interactions and less agonistic behaviour than cows indoors (O'Connell et al. 1989). Grazing cattle, if given the opportunity, select mixed diets, with a preference for legumes (e.g., clover), which optimise their own efficiency of nutrient capture (Rutter 2006). Pasture routinely has excessive macronutrients, especially nitrogen, but on mixed-species swards, animals are able to choose appropriately and manage nutrient intake themselves, balancing the supply of energy and nitrogen to the rumen (Manteca et al. 2008).

## 5.3 Challenges of Pasture-Based Systems

There are a number of aspects of pasture-based systems that can lead to poor welfare if not managed well, and arguments around these aspects are often raised when comparisons with housed systems are made. These include climatic exposure (heat, cold, mud), body condition and hunger, animal health and calf management.

### 5.3.1 Climatic Conditions

#### 5.3.1.1 Cold

Cows are markedly resilient to cold weather. Healthy cattle are robust in their ability to tolerate the cold conditions encountered in most pasture-based climates including its wet and changeable nature. This resistance is due to the fermentation of roughage that results in considerable heat production, which in turn reduces the net energy required for maintenance of body temperature (Hahn 1985).

The opportunity for cows to graze outdoors allows for the expression of natural behaviours but because the management of grazing systems tends to focus on maximising dry matter intake (DMI) and maximising the available land area for growing pasture, cows can often lack access to shelter because trees and hedges have been removed or reduced. Cold winter temperatures in combination with wind and rain cause cattle to lose heat to the environment and increases metabolic requirements (Tucker et al. 2007). Because cold stress requires an animal to raise its regulatory heat production to maintain a normal body temperature (Christopherson 1985), it will lead to increased feed intake where feed is available. When additional feed is not obtainable, this will negatively affect weight gain and milk production (Young 1981; Bryant et al. 2007), reduced reproductive function (Gwazdauskas 1985) and in extreme circumstances can lead to death (Martin et al. 1975; Mader 2003; Stull et al. 2008). Production and welfare benefits of shelter in cold, wet and windy conditions have been demonstrated in outdoor pasture conditions (Holmes et al. 1978; Tucker et al. 2007; Webster et al. 2008; Schütz et al. 2010a). Schütz et al. (2010a) suggested that protection from rain and the combination of rain and wind is most important for dairy cattle welfare.

A number of management methods, including the provision of additional feed, are used to improve protection during cold conditions. In terms of absolute temperatures, well-fed lactating cows can withstand temperatures as cold as −30 °C under dry conditions with no air movement while non-lactating cows can withstand conditions to −20 °C. Young growing stock and calves are more sensitive and can experience cold stress from 5 °C (Kadzere et al. 2002). Wind chill and rain have a significant impact on the animal's ability to maintain body temperature. Rainfall greater than 1 mm/h will reduce the insulation value of the coat by 30% (Turnpenny et al. 2000) while air speed of 2 m/s increases the air temperature at which lower critical limits are met from −10 °C to 0 °C (Higgins and Dodd 1989). Thus, in cold, wet and windy conditions, young or non-lactating animals may be particularly vulnerable to cold stress.

The changeable conditions in temperate pasture environments include large swings in temperature, wind, rain and poor underfoot conditions, which make it not only difficult for cows to adapt thermally, but also to adapt behaviourally. For example, cows on an outdoor woodchip pad without shelter from wind and rain and fed to appetite on good quality silage showed an increase in levels of plasma and faecal cortisol and serum total thyroxine compared to cows housed indoors (Webster et al. 2008). Despite the ability to maintain body temperature, cows also spent less time lying and adopted more heat-conservative postures, such as lying with their front legs bent and the hind legs touching their body to reduce heat loss, as well as shivering (Tucker et al. 2007). While these behaviours were observed in an experimental setting, it is likely that they indicate that welfare was affected, as the behaviours indicate the animals' awareness of the cold conditions and illustrate their efforts to avoid hyperthermia. However, it is likely these behavioural strategies would have limited success for cows exposed to these conditions on commercial farms for extended periods of time.

#### 5.3.1.2 Winter Grazing Management

Rainfall inevitably makes conditions wet underfoot, and managing those winter conditions is particularly challenging for animals in pasture-based systems. Intensive winter grazing is used in pastoral farming to manage feed supply at a time of year when pasture growth is limited by cool temperatures and short day length. Intensive winter grazing on a crop or at pasture involves a system of herd and pasture management where animals are held on a restricted area of pasture or crop (a break or strip) at a high stocking density. When one section of forage is eaten, animals are given access to a new break. They may also be given supplementary feed such as silage or hay at the same time. Relatively large amounts of surplus rainfall occur in the winter months in temperate regions, and these intensive winter grazing systems help to preserve soil structure and pasture quality on other parts of the farm. However, these areas become wet and muddy very quickly, and this is a big concern for cow welfare. Cows are reluctant to lie down on wet, muddy surfaces (O'Connor et al. 2019), and this results in cows not getting the sleep and rest they need to support good mental and physiological function (Seigel 2005). Heat loss will occur if cows do lie down on cold and wet surfaces. Convection of body heat to the colder surface can lead to cold stress (Morrison et al. 1970; Holmes et al. 1978; Muller et al. 1996; Fisher et al. 2003). Thin cows will be more susceptible to cold stress (Thompson et al. 1983) and Tucker et al. (2007) found that both better body condition and provision of shelter helped mitigate negative effects of cold, wet and windy weather. When outside, cows can choose sheltered areas or change their lying and standing postures to reduce the amount of surface area exposed to rain and wind. In these conditions, cattle adopt more tucked lying postures (front and hind legs tucked close to the body), which could be an attempt to preserve body temperature and reduce heat loss (Tucker et al. 2007).

In pasture-based systems, the temporary use of uncovered off-paddock facilities, particularly 'stand-off' or 'out-wintering' pads, is becoming more commonplace (Botha and O'Connor 2015; Fig. 5.3). A stand-off pad is a purpose-built, drained loafing area where animals can lie down. In most cases, they are not a place to feed animals but may be next to a feeding area. They are typically used to reduce environmental impacts of cow presence on pasture during the winter, such as pugging (poaching) of the soil and nitrogen leaching from urine and dung deposition. In New Zealand, Australia, and Ireland, these facilities are predominantly used during winter when the cows are not lactating. Cows will often be kept for part of the day on the pad and spend the other part on pasture. However, if the surface is not maintained in these facilities, there are serious consequences for welfare. O'Connor et al.

**Fig. 5.3** Stand-off pad with post peelings (a long thin wood-chip product) as bedding (Photo courtesy of Cheryl O'Connor)

(2019) found that the daily lying time of non-lactating dairy cows declined during a 5-week period in a typical New Zealand winter stand-off pad/pasture hybrid system. The management of an uncovered woodchip pad in this manner (where there is no refreshing of bedding) produced a surface that cows were reluctant to lie on after 3 weeks and resulted in cows being severely deprived of lying for the following 3 weeks. Cows on a pad with refreshed woodchips achieved lying times of 10.5 h/ a day, which is typical for cows on pasture (Fisher et al. 2008; Schütz et al. 2013). This illustrates the importance of frequent refreshing of the bedding to achieve adequate lying times. Many studies have shown that cows prefer a dry surface to lie on (Fregonesi et al. 2007; Reich et al. 2010). O'Connor et al. (2019) found that refreshing the woodchip bedding maintained moisture levels at less than 65%, however, after 3–4 weeks the moisture content of the woodchips on the stand-off pad exceeded 75%. The cows also reduced their lying times on the pad after 3–4 weeks, confirming that a surface with a moisture level above 75% is too wet to be a comfortable lying surface. For long-term effectiveness and the well-being of cows, good design, maintenance and management of stand-off pads are essential. In particular, famers need to consider the level of rainfall in their area, and how many cows will be stood off, how often and for how long when designing and using a stand-off pad.

#### 5.3.1.3 Heat

In most countries with pasture-based systems, the weather conditions that can generate heat stress for dairy cattle occur more frequently than those that generate cold stress. Cattle rely on evaporation from the lungs to a large extent and evaporation from the skin to a lesser extent to lose excess heat. Increased respiratory rate is a key indicator of increasing heat load in cattle and is used as an indicator that core body temperatures are under threat of excessive heat loading (Ominski et al. 2002; Schütz et al. 2010a). Behavioural responses to high heat load include an increase in shade use (Tucker et al. 2008; Schütz et al. 2009, 2010b), time spent near water (Schütz et al. 2010b; Legrand et al. 2011), decreased feed intake (Hahn 1999; Ominski et al. 2002), panting (Schütz et al. 2014) and a reduction in lying time (Tucker et al. 2008; Schütz et al. 2010b).

The natural response of cows to hot conditions is to seek and use shade to reduce the impact of heat (Fisher et al. 2008; Schütz et al. 2009). As ambient temperatures rise, the motivation to find shade is strong relative to other valued behaviours such as resting (Schütz et al. 2008). Cows will compete for shade (Schütz et al. 2010b), and use of shade is positively related to solar radiation levels and high air temperatures (Kendall et al. 2006; Tucker et al. 2008; Schütz et al. 2009). Access to shade reduces respiration rate and body temperature (Kendall et al. 2007; Schütz et al. 2010b, 2011), and these cooling benefits are greater if the source of shade blocks out more solar radiation (Tucker et al. 2008). Choice tests have shown that dairy cows also have a preference for types of shade that are more effective at blocking solar radiation (Schütz et al. 2009) and will increase the use of this shade as temperatures increase (Tucker et al. 2008). When there is insufficient shade to accommodate all cows and their natural shade-seeking behaviour is thwarted, there are signs of frustration and increased aggressive interactions between cows (Schütz et al. 2010b). Simply reducing body temperature with sprinklers does not satisfy the natural behavioural drive to seek shade, as cows preferred shade over sprinklers when given the choice (Schütz et al. 2011).

### 5.3.2 Body Condition and Hunger

Cows on pasture can face nutritional challenges caused by seasonal pasture availability and quality. It is often stated that it is unlikely that high-yielding lactating cows can meet their nutritional requirements from grazing alone (e.g., Charlton et al. 2011). However, but a lower production system does not necessarily indicate a poorer welfare system (Mee and Boyle 2020). Although herbage growth patterns and rates will vary with climate conditions, the basic principles of a pasture-based system are consistent—utilise herbage grown as the primary feed, conserve herbage grown surplus to cow requirements as silage or hay and introduce either home-produced or purchased supplements when herbage growth is not sufficient for cow nutrient requirements (Roche et al. 2011). In both temperate and Mediterranean-type climates, with or without irrigation, herbage production is characterised by a spring peak that typically exceeds DMI requirements of the herd (Roche et al. 2009a). Generally, cool temperatures result in less than required herbage growth in winter and early spring, and insufficient moisture and excessive temperatures often result in reduced summer herbage growth and less than desired herbage quality. It is therefore unsurprising that seasonal increases and decreases in body condition need to be carefully managed in most pasture-based systems.

In most pasture-based systems, the reproductive cycle is based around seasonal pasture growth and ability to turn the flush of pasture growth in spring and summer into milk and milk solids. Climatic variation that causes inconsistent and unpredictable quantity and quality of feed can have a negative effect on the welfare of adult cows in grazing systems (Roche et al. 2009b). However, to offset poor grass growth in winter, strategic supplementation from conserved forages or purchased supplements can be used to maintain production and cow body condition. Peak body condition is aimed for at calving time (i.e. typically early spring to ensure that peak lactation coincides with peak pasture growth and quality) and then generally declines through lactation as summer pasture quality falls.

The importance of managing body condition is highlighted by Roche et al. (2013), who found that health and metabolic risks were elevated in cows with both excessively high and low body condition scores (BCS) at the time of calving. Despite the strong focus on the impact of body condition on productivity, there are few publications that have addressed the welfare impacts of body condition. In fact, despite the obvious relevance of the question, the association between BCS and hunger has not been elucidated for dairy cows. Some have gone as far as to suggest that low body condition in itself may not be a welfare concern (Roche et al. 2009c).

However, it is known that cows with low BCS typically eat more if it is available (Hayirli et al. 2002; Tucker et al. 2007). Using competition levels as an indicator, Schütz et al. (2013) found little evidence of hunger in low BCS cows; nonetheless these cows were less aggressive than those with higher BCS and therefore might have been less successful at competing for limited feed in this type of test. Matthews et al. (2012) found that thinner cows spent more time foraging and less time lying; however, the cows' cognitive impact of this difference in time budget was not quantified. The mental consequences of thwarting grazing behaviour in a hungry cow are largely unknown, although it may be presumed to cause some level of discomfort (Gregory 2004). Feelings of hunger in pasture grazing situations are likely to differ from those in other feeding systems because pasture has a low overall dry matter content and is bulky by its very nature; thus intake may be limited by physical restrictions and palatability (Butris and Phillips 1987) or rumen capacity even if dietary energy requirements for production are not met (Kolver and Muller 1998). It is possible that the rate of decrease in BCS might be linked more closely to hunger and welfare compromise.

High BCS in dairy cows has also been associated with impaired health through immunosuppression (Lacetera et al. 2005), greater risk of metabolic disease especially at calving (Roche et al. 2009c) and 'fat cow syndrome' where excessive mobilisation of fat depots results in liver disease (Morrow 1976). Reports of the relationships between BCS and calving-related problems are equivocal. Markusfeld et al. (1997) reported that poor body condition was associated with a risk of retained placenta and uterine infection after calving while Berry et al. (2007) could find no relationship between body condition and dystocia or still births. Good body condition will assist an animal as a buffer against periods of cold stress. Cows with higher

BCS (9 out of 10) were able to maintain a more even body temperature during cold conditions while thinner cows (BCS 4 out of 10) had lower body temperatures and were more likely to shiver and lie in postures that conserve body heat (Tucker et al. 2007).

### 5.3.3 Animal Health

There are numerous health challenges faced by pasture-based cows. While some are similar to those in indoor systems, there are a unique set of challenges that farm management must address. Lameness is an issue but stems from different causes compared to indoor systems. Similarly, body condition and metabolic issues are a concern, but they may be triggered by climatic conditions, and also require different diagnostic tools. Other issues, like nutrient deficiencies and intolerances of specific fungi, are unique to pasture systems.

#### 5.3.3.1 Lameness

The main lameness-causing condition seen in New Zealand and Ireland is white line disease (Chesterton et al. 2008; Somers and O'Grady 2015) The white line is a weak point in the hoof where the sole and the wall of the hoof are joined, which is easily damaged. At calving or during other periods of stress, the white line becomes more susceptible to damage. Cows in pasture-based systems do a lot of walking and twisting and turning of the feet on tracks and yards. These movements can cause the wall and sole to separate (Ranjbar et al. 2016). Stones may be forced upward into the white line which results in further separation of the wall from the sole of the hoof. If this continues, the stone and/or bacteria will reach the sensitive tissues of the hoof causing pain and infection. Most dairy industry programmes in countries with pasture-based systems focus on prevention of lameness on dairy farms caused by these physical factors, particularly focussing on increasing awareness of the disease and the impact of cow flow and movement on lameness by the people who work with the cows, along with improving the handling yards of the milking parlour and physical environment especially of the tracks. For example, a good track is well designed, constructed and maintained and uses the appropriate track width for the herd size, is distraction-free and with good drainage and walking surface.

Bovine digital dermatitis (BDD) is a common infectious cause of lameness in housed cattle (Laven and Lawrence 2006; Solano et al. 2016). In contrast, BDD is significantly less common in pasture-based systems (Holzhauer et al. 2012; Laven and Lawrence 2006), but despite this, has been reported in many countries where cattle are permanently kept at pasture including Chile (Rodriguez-Lainz et al. 1998) and Australia (Milinovich et al. 2004). In New Zealand, BDD was only identified in 2011; however, a cross-sectional survey (Yang et al. 2019) found that in the last 10 years, it had become more common and found in more herds, however, the prevalence on the affected farms was very low.

The pH in the rumen of grazing cows fluctuates throughout the day, but the low pH phases are driven by rapid production of volatile fatty acids that are quickly absorbed from the rumen and therefore have no detrimental effect. Unlike indoor housed cows, adding a fibre source, such as hay or straw, to cows primarily grazing pasture will not alter rumen pH nor impact on lameness.

In pasture-based systems, lame cows are routinely separated from the main herd and milked once daily. This management routine, combined with being on pasture rather than in housing, may reduce recovery time (Hernandez-Mendo et al. 2007; Laven et al. 2008). Improving farmer diagnosis/identification of lameness is likely to result in more prompt treatment of cows and thus also improve animal welfare outcomes. To that end, Australian (Dairy Australia 2015) New Zealand (DairyNZ 2017) and Irish (Teagasc 2016) dairy industries have developed extension programmes promoting the formal lameness scoring of dairy herds on a regular basis.

#### 5.3.3.2 Mastitis

Mastitis is divided into two types: cow-associated and environmental. The bacteria causing cow-associated mastitis usually live in the udder tissue and on teat skin and most commonly spread at milking. The bacteria causing environmental mastitis survive in the cow's environment (including soil, manure, bedding, calving pads, water, or on body sites of the cow other than the udder). Although milking may enable their entry through the teat canal, the environment is the primary source of infection. These bacteria include *Streptococcus uberis* and the *coliforms*. *Strep. uberis* has become the major cause of mastitis in Australia (Shum et al. 2009) and New Zealand (McDougall 2003) and the second most prevalent in Ireland (Keane et al. 2013). Infection of the udder with these organisms is often opportunistic, taking advantage of circumstances that favour environmental contamination and changes in the mammary gland's susceptibility to infection. Most cases of environmental mastitis occur within a few weeks of calving, when the cows' natural immune defence mechanisms are low and their teats have been in contact with mud and manure during calving.

Increasing antimicrobial resistance is an issue for human and animal health, and antimicrobial stewardship is of growing importance. The greatest use of antimicrobials in pasture-based dairy farming is in the treatment and control of mastitis (Bates et al. 2020; More et al. 2017). More selective use of dry cow therapy (i.e. treating only those cows with infections rather than treating all of the cows) (Regan et al. 2021) and the benefits of non-antibiotic internal teat sealants (Rabiee and Lean 2013) as an alternative are being actively promoted.

#### 5.3.3.3 Facial Eczema

Facial eczema (FE) is caused by a toxin (sporidesmin) produced by the spores of the fungus *Pithomyces chartarum* that grows on pasture. The fungus grows in the dead litter at the base of pasture plants in warm moist conditions (Brook 1963). When ingested by cattle, sporidesmin damages the liver and bile ducts. The damaged liver cannot rid the body of wastes and a breakdown product of chlorophyll builds up in the blood causing sensitivity to sunlight, which in turn causes inflammation of the skin (Clare 1944). Any exposed unpigmented or thin skin (such as the skin on the udder) may redden, thicken and peel (Morris et al. 2004). Cows suffering from FE will have a drop in milk production (Towers and Smith 1978), be restless, seek shade and lick their udder (Morris et al. 2004). Not all animals affected with FE show physical signs although liver damage will have occurred and a resulting impact on milk production has recently been demonstrated (Cuttance et al. 2021) with sub-clinical FE. There is no cure for FE, so prevention is the only way of protecting animals. Preventative measures include monitoring pasture spore count and either dosing animals with zinc or spraying pastures with a fungicide.

#### 5.3.3.4 Ketosis

Ketosis is a metabolic disease that occurs when the cow is in severe negative energy balance and cannot efficiently use mobilised body fat for energy. A common cause is a sudden drop in energy intake due to underfeeding or adverse weather events (e.g. snowstorms) that prevent the cows from eating sufficient amounts of dry matter. Ketosis can also occur post-calving, when the cow is mobilising excess body fat to meet the demands of milk production (Daros et al. 2017). Cows that are too fat at calving or cows that have been overfed pre-calving are particularly at risk (Roche et al. 2009c, 2013). Ketosis is often diagnosed when blood levels of beta-hydroxybutyrate (BHBA) are high, but pasture-based cows naturally have a greater basal concentration of BHBA than those fed a high proportion of starch-based supplements or a TMR (Roche et al. 2010). Therefore, in pasture-based systems, ketosis should not be diagnosed based on BHBA concentrations alone. Additional indicators of energy balance, in particular NEFA (and glucose if possible) should also be measured to allow ketosis to be properly diagnosed. Preventative measures include careful management of body condition around calving and providing supplementary feed during adverse weather events.

#### 5.3.3.5 Acidosis

Acidosis is a metabolic disease that occurs when rumen pH levels fall below normal. Research in pasture-based systems shows that neither clinical nor sub-acute rumen acidosis occurs in cows grazing high-quality pasture (Kolver and De Veth 2002). However, it can occur in these systems when cows are not properly transitioned onto high sugar/starch feeds (commonly brassicas or fodder beet) in winter or when large quantities of high sugar/starch feeds are included in the diet (Waghorn et al. 2018). Careful management is required during transition to allow the rumen microbes to adapt to a change in feed type. In particular, it is important to ensure that all cows have equal access to the crop, whether it is grazed or spread in the pasture paddock, and that they have sufficient time to adapt to it.

#### 5.3.3.6 Copper Deficiency

Copper deficiency in cattle is a common and complex problem and occurs in many countries using pasture-based dairy systems. In general, dairy cows in early lactation will not be getting sufficient copper from a pasture diet; therefore, supplementation is likely to be beneficial (Grace and Knowles 2012). Dairy cows are most likely to be deficient in winter/early spring, coinciding with higher demands over this period due to pregnancy and early lactation. Fast-growing calves over 6 months of age are also likely to be deficient (Givens et al. 1981).

#### 5.3.3.7 Grass Staggers

Grass staggers (tetany or hypomagnesaemia) is a metabolic disease caused by magnesium deficiency (Storry and Rook 1962). Magnesium plays an important role in grass staggers prevention, and the cow is dependent on what magnesium is supplied in their diet and from supplements. Magnesium requirements are also affected by the levels of potassium and calcium in the diet (Mann et al. 2019). Some farms with very high potassium levels in pasture will require high rates of magnesium supplementation.

#### 5.3.3.8 Milk Fever

Milk fever is a metabolic disorder caused by insufficient calcium availability. Magnesium also plays an important role in milk fever prevention as magnesium is required for the production of hormones that are important for the absorption of calcium from the gut and the mobilisation of calcium from the bones (Mann et al. 2019). Supplementation with magnesium for 2–3 weeks pre-calving will reduce the risk of milk fever; however, the cow will not build up a store of magnesium, and so continued supplementation is required during early lactation.

### 5.3.4 Young Calf Management

This section will briefly describe the key differences for calves raised in pasture-based systems compared to those described in Chap. 10 (by Jensen and Proudfoot) for indoor-housed calves. The ability to interact with and modulate their own environments through performance of specific natural behaviours (e.g., sheltering and social contact) is important to reduce negative affective states (e.g. frustration) for both calves and cows. Calves in pasture-based systems are usually born outdoors to their dams at pasture in winter and early spring and are generally moved into groups indoors within 12–24 h to reduce exposure to cold and wet weather. From an affective state perspective, it is unknown whether the enhanced social contact in groups makes up for the lack of maternal contact and natural calf behaviour (but see Chap. 10 for further discussion).

Calves born outdoors in spring are particularly sensitive to cold stress. For example, calf mortality in California was significantly related to cold, wet, windy weather in winter (Stull et al. 2008), and in New Zealand, the risk of perinatal mortality was greater on days with high rainfall (Cuttance et al. 2017). Calves born during cold (<10 °C), wet and windy weather conditions have lower rectal temperatures and take a longer time to stand up than calves born during warm and dry weather (Diesch et al. 2004). If able to, calves will change their behaviour to conserve heat and avoid cold temperatures, for instance, by lying more (Schrama et al. 1993) and lying with their limbs and tail tucked in close to the body (Brunsvold et al. 1985). In cold conditions, the immune system of young calves may be compromised as energy is directed to maintaining thermoneutrality (Hemsworth et al. 1995). This immunosuppresion will contribute to increased morbidity and mortality. Cold conditions at birth and thereafter have also been shown to have detrimental effects on health and growth performance (Scibilia et al. 1987; Nabenishi and Yamazaki 2017).

It is important that the calf receives an adequate amount of high-quality colostrum as quickly as possible after birth. Failure of passive transfer (FPT) is common in most dairy systems. In New Zealand, the proportion of calves with FPT has been reported as between 25 and 49% (Vermunt et al. 1995; Wesselink et al. 1999; Lawrence et al. 2017; Cuttance et al. 2017), and in Australia it is 38% (Vogels et al. 2013). Failure of passive transfer is recognised as one of the leading causes of calf mortality and morbidity, with 65–70% of post-mortem samples in Ireland reporting FPT (e.g. Anon 2017). In most dairy systems, strategies to reduce the risk of FPT focus on ensuring that all calves ingest sufficient good quality colostrum within 4 h of birth (Beam et al. 2009). However, in pasture-based systems, it is not feasible or practical to ensure that all calves are fed within 4 h of birth because on most farms, calves are only collected from calving paddocks once or twice daily. It is possible that variables identified in indoor-housed studies are not as important in pasture-based management systems as calves are often left on the dam for longer than reported in other dairy management systems (Beam et al. 2009). Cuttance et al. (2018) found that variables which influence how well a cow can feed its calf in the first 12–24 h have a larger influence on successful passive transfer than the collection and management of calves once they reach the rearing shed.

Although the same welfare issues occur as with any system where the calf is removed from the dam soon after birth, most calves from pasture systems are raised on grass. This may be throughout the entire calf rearing period or after 4–6 weeks of being housed indoors. Calves are usually group reared whether indoors or outside. As with all livestock, a comfortable lying surface is critical for calves, and although most are reared on sawdust or woodchips, other surfaces such as river stones are being used. Although a study assessing different types of bedding materials for calves found no effect of bedding type on weight gain, cleanliness or health (Sutherland et al. 2013), there were reduced skin temperatures and lying and playing times for calves reared on stones (Sutherland et al. 2014).

Play behaviour has been suggested as an indicator of positive emotions and welfare (reviewed by Boissy et al. 2007). Young animals are highly motivated to play when all their basic needs, such as health and thermal comfort, are met (Jensen et al. 1998). The expression of play behaviour in calves is reduced after a negative experience, such as weaning (Krachun et al. 2010) or disbudding (Mintline et al. 2013), or when housed on an uncomfortable surface (Sutherland et al. 2013). Therefore, investigating play behaviour in addition to lying times and physiological parameters may provide a more comprehensive assessment of the welfare of calves reared in different systems.

## 5.4 Conclusion

Dairy production on pasture has many potential welfare benefits but also many welfare risks, as discussed above. Pasture-based dairy systems are very complex systems to manage, and a great deal of skill and knowledge are required, particularly as the increasing variability of climate change requires a very reactive approach. Add to this a drive to both increase production and reduce environmental impact, and pasture-based systems can start to run into serious problems. Under these circumstances, the welfare benefits of grazing systems, of a better environment and opportunity for natural behaviours, can be quickly lost if not expertly managed.

## References

Anon (2017) All island animal disease surveillance report. A joint AFBI/DAFM Veterinary Laboratories publication. Available at https://www.afbini.gov.uk/sites/afbini.gov.uk/files/publications/AFBI.pdf

Arnott G, Ferris CP, O'Connell NE (2017) Review: welfare of dairy cows in continuously housed and pasture-based production systems. Animal 11:261–273

Bates A, Laven R, Bork O et al (2020) Selective and deferred treatment of clinical mastitis in seven New Zealand dairy herds. Prev Vet Med 176:104915

Beam AL, Lombard JE, Kopral CA et al (2009) Prevalence of failure of passive transfer of immunity in newborn heifer calves and associated management practices on US dairy operations. J Dairy Sci 92:3973–3980

Berry DP, Lee JM, Macdonald KA, Roche JR (2007) Body condition score and body weight effects on dystocia and still births and consequent effects on postcalving performance. J Dairy Sci 90:4201–4211

Boissy A, Manteuffel G, Jensen MB, Moe RO, Spruijt B, Keeling LJ, Wincker C, Forkman B, Dimitrov I, Langbein J, Bakken M, Veissier I, Aubert A (2007) Assessment of positive emotions in animals to improve their welfare. Physiol Behav 92:375–397

Botha N, O'Connor C (2015) Off-paddock dairy systems farmer survey report. AgResearch Client Report, 2526. AgResearch, New Zealand, 36 pp

Brook PJ (1963) Ecology of the fungus *Pithomyces chartarum* (Berk & Curt) MB Ellis in pasture in relation to facial eczema disease in sheep. N Z J Agric Res 6:147–228

Brunsvold RE, Cramer CO, Larsen HJ (1985) Behavior of dairy calves reared in hutches as affected by temperature. Trans ASAE 28:1265–1268

Bryant JR, López-Villalobos N, Pryce JE et al (2007) Quantifying the effect of thermal environment on production traits in three breeds of dairy cattle in New Zealand. N Z J Agric Res 50:327–338

Butris GY, Phillips CJC (1987) The effect of herbage surface water and the provision of supplementary forage on the intake and feeding behaviour of cattle. Grass Forage Sci 42:259–264

Cahill L, Reilly B, Patton D et al (2019) The effect of grazing season length and stocking rate on milk production and supplementary feed requirements within spring calving dairy systems in the north-east of Ireland. Paper presented at the 9th European Conference on Precision Livestock Farming, Cork, Ireland, August 26–29 2019, pp 803–806

Charlton GL, Rutter SM (2017) The behaviour of housed dairy cattle with and without pasture access: a review. Appl Anim Behav Sci 192:2–9

Charlton GL, Rutter SM, East M, Sinclair LA (2011) Effects of providing total mixed rations indoors and on pasture on the behavior of lactating dairy cattle and their preference to be indoors or on pasture. J Dairy Sci 94:3875–3884

Chesterton RN, Lawrence KE, Laven RA (2008) A descriptive analysis of the foot lesions identified during veterinary treatment for lameness on dairy farms in north Taranaki. N Z Vet J 56:130–138

Christopherson RJ (1985) Management and housing of animals in cold environments. In: Yousef MK (ed) Stress physiology in livestock, vol 2. CRC Press, Boca Raton

Clare NT (1944) Photosensitivity diseases in New Zealand. III The photosensitizing agent in facial eczema. N Z J Sci Technol Sect A 25:202–220

Cuttance EL, Mason WA, Laven RA et al (2017) Prevalence and calf-level risk factors for failure of passive transfer in dairy calves in New Zealand. N Z Vet J 65:297–304

Cuttance EL, Mason WA, Laven RA, Denholm KS, Yang D (2018) Calf and colostrum management practices on New Zealand dairy farms and their associations with concentrations of total protein in calf serum. N Z Vet J 66:126–131

Cuttance EL, Mason WA, Laven RA (2021) The association of milk-solid production during the current lactation with liver damage due to presumptive ingestion of spores from *Pithomyces chatarum* by dairy cattle. N Z Vet J 69:201–210

Dairy Australia (2015) Healthy hooves investigation pack healthy hooves: investigation pack, 1st edn. Dairy Australia Ltd, Southbank

DairyNZ (2017) Lameness field guide–a farmer's guide to treating lameness, 2nd edn. DairyNZ Ltd., Hamilton

Daros RR, Hötzel MJ, Bran JA et al (2017) Prevalence and risk factors for transition period diseases in grazing dairy cows in Brazil. Prev Vet Med 145:16–22

Diesch TJ, Mellor DJ, Stafford KJ, Ward RN (2004) The physiological and physical status of single calves at birth in a dairy herd in New Zealand. N Z Vet J 52:250–255

Dillon P, Crosse S, Stakelum G, Flynn F (1995) The effect of calving date and stocking rate on the performance of spring-calving dairy cows. Grass Forage Sci 50:286–299

Fisher A, Stewart M, Verkerk GA et al (2003) The effects of surface type on lying behaviour and stress responses of dairy cows during periodic weather-induced removal from pasture. Appl Anim Behav Sci 81:1–11. https://doi.org/10.1016/S0168-1591(02)00240-X

Fisher AD, Roberts N, Bluett SJ et al (2008) Effects of shade provision on the behaviour, body temperature and milk production of grazing dairy cows during a New Zealand summer. N Z J Agric Res 51:99–105

Fregonesi JA, Veira DM, von Keyserlingk MAG, Weary DM (2007) Effects of bedding quality on lying behavior of dairy cows. J Dairy Sci 90:5468–5472. https://doi.org/10.3168/jds.2007-0494

Givens DI, Hopkins JR, Brown ME, Walsh WA (1981) The effect of copper therapy on the growth rate and blood composition of young growing cattle. J Agric Sci 97:497–505

Gonyou HW (1996) Design criteria: Should freedom of movement be retained. Acta Agric Scand A Anim Sci 27:36–39

Grace ND, Knowles SO (2012) Trace element supplementation of livestock in New Zealand: meeting the challenges of free-range grazing systems. Vet Med Int 2012:639472

Gregory NG (2004) Physiology and behaviour of animal suffering. Blackwell Science, Oxford

Gwazdauskas FC (1985) Effects of climate on reproduction in cattle. J Dairy Sci 68:1568–1578

Hahn GL (1985) Management and housing of farm animals in hot environments. In: Yousef MK (ed) Stress physiology in livestock, vol 2. CRC Press, Boca Raton, pp 151–174

Hahn GL (1999) Dynamic responses of cattle to thermal heat loads. J Anim Sci 77:10–20

Hayirli A, Grummer RR, Norheim EV, Crump PM (2002) Animal and dietary factors affecting feed intake during the pre-fresh transition period in Holsteins. J Dairy Sci 85:3430–3443

Hemsworth PH, Barnett JL, Beveridge L, Matthews LR (1995) The welfare of extensively managed dairy cattle: a review. Appl Anim Behav Sci 42:161–182

Hernandez-Mendo O, von Keyserlingk MAG, Veira DM, Weary DM (2007) Effects of pasture on lameness in dairy cows. J Dairy Sci 90:1209–1214

Higgins KP, Dodd VA (1989) A model of the bioclimatic value of shelter to beef cattle. J Agric Eng Res 42:149–164

Holmes CW, Christensen R, Mclean NA, Lockyer J (1978) Effects of winter weather on the growth-rate and heat-production of dairy-cattle. N Z J Agric Res 21:549–556

Holmes C, Brookes IM, Garrick DJ et al (2007) Milk production from pasture. Massey University, Palmerston North
Holzhauer M, Brummelman B, Frankena K, Lam T (2012) A longitudinal study into the effect of grazing on claw disorders in female calves and young dairy cows. Vet J 193:633–638
Horan B, Faverdin P, Delaby L, Buckley F, Rath M, Dillon P (2006) The effect of strain of Holstein-Friesian dairy cow and pasture-based system on grass intake and milk production. Anim Sci 82:435–444
Jensen MB, Vestergaard KS, Krohn CC (1998) Play behaviour in dairy calves kept in pens: the effect of social contact and space allowance. Appl Anim Behav Sci 56:97–108
Kadzere CT, Murphy MR, Silanikove N, Maltz E (2002) Heat stress in lactating dairy cows: a review. Livest Prod Sci 77:59–91
Kay JK, Loor JJ, Heiser A, McGowan J, Roche JR (2015) Managing the grazing dairy cow through the transition period: a review. Anim Prod Sci 55:936–942
Keane OM, Budd KE, Flynn J, McCoy F (2013) Pathogen profile of clinical mastitis in Irish milk-recording herds reveals a complex aetiology. Vet Rec 173:2–28
Kendall PE, Nielsen PP, Webster JR (2006) The effects of providing shade to lactating cows in a temperate climate. Livest Sci 103:148–157
Kendall PE, Verkerk GA, Webster JR, Tucker CB (2007) Sprinklers and shade cool cows and reduce insect avoidance behaviour in pasture-based dairy systems. J Dairy Sci 90:3671–3680
Kolver ES, De Veth MJ (2002) Prediction of ruminal pH from pasture-based diets. J Dairy Sci 85:1255–1266
Kolver ES, Muller LD (1998) Performance and nutrient intake of high-producing Holstein cows consuming pasture or a total mixed ration. J Dairy Sci 81:1403–1411
Kolver ES, Roche JR, de Veth MJ, Thorne P, Napper AR (2002) Total mixed ration versus pasture diets. Evidence of a genotype × diet interaction. Proc N Z Soc Anim Prod 6:246–251
Krachun C, Rushen J, de Passillé AM (2010) Play behaviour in dairy calves is reduced by weaning and by a low energy intake. Appl Anim Behav Sci 122:71–76
Lacetera N, Scalia D, Berbabucci U, Riocnchi B, Pirazzi D, Nardone A (2005) Lymphocyte functions in over-conditioned cows around parturition. J Dairy Sci 88:2010–2016
Laven RA, Lawrence KR (2006) An evaluation of the seasonality of veterinary treatments for lameness in UK dairy cattle. J Dairy Sci 89:3858–3865
Laven RA, Lawrence KE, Weston JF, Dowson KR, Stafford KJ (2008) Assessment of the duration of the pain response associated with lameness in dairy cows, and the influence of treatment. N Z Vet J 56:210–217
Lawrence K, Broerse N, Hine L, Yapura J, Tulley WJ (2017) Prevalence of failure of passive transfer of maternal antibodies in dairy calves in the Manawatu region of New Zealand. N Z Vet J 65:1–5
Legrand AL, von Keyserlingk MAG, Weary DM (2009) Preference and usage of pasture versus free-stall housing by lactating dairy cattle. J Dairy Sci 92:3651–3658
Legrand A, Schütz KE, Tucker CB (2011) Using water to cool cattle: behavioral and physiological changes associated with voluntary use of cow showers. J Dairy Sci 94:3376–3386
Loberg J, Telezhenko E, Bergsten C, Lidfors L (2004) Behaviour and claw health in tied dairy cows with varying access to exercise in an outdoor paddock. Appl Anim Behav Sci 89:1–16
Macdonald KA, Penno JW, Lancaster JAS, Roche JR (2008) Effect of stocking rate on pasture production, milk production, and reproduction of dairy cows in pasture-based systems. J Dairy Sci 91:2151–2163
Mader TL (2003) Environmental stress in confined beef cattle. J Anim Sci 81:E110–E119
Mann S, McArt J, Abuelo A (2019) Production-related metabolic disorders of cattle: ketosis, milk fever and grass staggers. In Pract 41:205–219
Manteca X, Villalba JJ, Atwood SB, Dziba L, Provenza FD (2008) Is dietary choice important to animal welfare? J Vet Behav 3:229–239
Markusfeld O, Galon N, Ezra E (1997) Body condition score, health, yield and fertility in dairy cows. Vet Rec 141:67–72

Martin SW, Schwabe CW, Franti CE (1975) Dairy calf mortality rate: influence of meteorologic factors on calf mortality rate in Tulare County, California. Am J Vet Res 36:1105–1109

Matthews LR, Cameron C, Sheahan AJ, Kolver ES, Roche JR (2012) Associations among dairy cow body condition and welfare-associated behavioral traits. J Dairy Sci 95:2595–2601

McDougall S (2003) Intramammary treatment of clinical mastitis of dairy cows with a combination of lincomycin and neomycin, or penicillin and dihydrostreptomycin. N Z Vet J 51:111–116

Mee JF, Boyle LA (2020) Assessing whether dairy cow welfare is "better" in pasture-based than in confinement-based management systems. N Z Vet J 68:168–177

Milinovich GJ, Turner SA, McLennan MW, Trott DJ (2004) Survey for papillomatous digital dermatitis in Australian dairy cattle. Aust Vet J 82:223–227

Mintline EM, Stewart M, Roger AR, Cox NR, Verkerk GA, Stookey JM, Webster JR, Tucker CB (2013) Play behavior as an indicator of animal welfare: disbudding in dairy calves. Appl Anim Behav Sci 144:22–30

More JS, Clegg TA, McCoy F (2017) The use of national-level data to describe trends in intramammary antimicrobial usage on Irish dairy farms from 2003 to 2015. J Dairy Sci 100:6400–6413

Morris CA, Towers NR, Hohenboken WD, Maqbool N, Smith BL, Phua SH (2004) Inheritance of resistance to facial eczema: a review of research findings from sheep and cattle in New Zealand. N Z Vet J 52:205–215

Morrison SR, Givens RL, Garrett WN, Bond TE (1970) Effects of mud-wind-rain on beef cattle performance in feed lot. Calif Agric (Berkeley) 24:6–7

Morrow D (1976) Fat cow syndrome. J Dairy Sci 59(9):1625–1629

Muller CJC, Botha JA, Smith WA (1996) Effect of confinement area on production, physiological parameters and behaviour of Friesian cows during winter in a temperate climate. S Afr J Anim Sci 26:1–5

Nabenishi H, Yamazaki A (2017) Effects of temperature–humidity index on health and growth performance in Japanese black calves. Tropic Anim Health Prod 49:397–402

O'Connell J, Giller PS, Meany W (1989) A comparison of dairy cattle behavioral patterns at pasture and during confinement. Ir J Agric Res 28:65–72

O'Connor C, Dowling S, Cave V, Webster J (2019) Cow lying behaviour and bedding quality changes during five weeks on a stand-off pad. Animals 9(5):257. https://doi.org/10.3390/ani9050257

O'Donovan M, Hennessy D, Creighton P (2021) Ruminant grassland production systems in Ireland. Ir J Agric Food Res 59:225–232

Ominski KH, Kennedy AD, Wittenberg KM, Moshtaghi Nia SA (2002) Physiological and production responses to feeding schedule in lactating dairy cows exposed to short-term, moderate heat stress. J Dairy Sci 85:730–737

Onyiro OM, Brotherstone S (2008) Genetic analysis of locomotion and associated conformation traits of Holstein-Friesian dairy cows managed in different housing systems. J Dairy Sci 91:322–328

Onyiro OM, Andrews LJ, Brotherstone S (2008) Genetic parameters for digital dermatitis and correlations with locomotion, production, fertility traits, and longevity in Holstein-Friesian dairy cows. J Dairy Sci 91:4037–4046

Phillips CJC, Beerda B, Knierim U et al (2013) A review of the impact of housing on dairy cow behaviour, health and welfare. In: Aland A, Banhazi T (eds) Livestock housing: modern management to ensure optimal health and welfare of farm animals. Wageningen Academic Publishers, Wageningen, pp 37–54

Rabiee AR, Lean IJ (2013) The effect of internal teat sealant products (Teatseal and Orbeseal) on intramammary infection, clinical mastitis, and somatic cell counts in lactating dairy cows: a meta-analysis. J Dairy Sci 96:6915–6931

Ranjbar S, Rabiee AR, Gunn A, House JK (2016) Identifying risk factors associated with lameness in pasture-based dairy herds. J Dairy Sci 99:7495–7505

Regan Á, Clifford S, Burrell AMG, Balaine L, Dillon E (2021) Exploring the relationship between mastitis risk perceptions and farmers' readiness to engage in milk recording. Prev Vet Med 193:105393

Reich LJ, Weary DM, Veira DM, von Keyserlingk MAG (2010) Effects of sawdust bedding dry matter on lying behavior of dairy cows: a dose-dependent response. J Dairy Sci 93:1561–1565. https://doi.org/10.3168/jds.2009-2713

Roche JR, Berry DP, Kolver ES (2006) Holstein-Friesian strain and feed effects on milk production, body weight and body condition score profiles in grazing dairy cows. J Dairy Sci 89:3532–3543

Roche JR, Turner LR, Lee JM et al (2009a) Weather, herbage quality and milk production in pastoral systems. 3. Inter-relationships and associations between weather variables and herbage growth rate, quality and mineral concentration. Anim Prod Sci 49:211–221

Roche JR, Turner LR, Lee JM et al (2009b) Weather, herbage quality and milk production in pastoral systems. 4. Effects on dairy cattle production. Anim Prod Sci 49:222–232

Roche JR, Friggens NC, Kay JK et al (2009c) Invited review: body condition score and its association with dairy cow productivity, health, and welfare. J Dairy Sci 92:5769–5801. https://doi.org/10.3168/jds.2009-2431

Roche JR, Kay JK, Phyn CVC et al (2010) Dietary structural to nonfiber carbohydrate concentration during the transition period in grazing dairy cows. J Dairy Sci 93:3671–3683

Roche JR, Burke CR, Meier S, Walker CG (2011) Nutrition reproduction interaction in pasture-based systems: is nutrition a factor in reproductive failure? Anim Prod Sci 51:1045–1066

Roche JR, Macdonald KA, Schütz KE et al (2013) Calving body condition score affects indicators of health in grazing dairy cows. J Dairy Sci 96:5811–5825

Rodriguez-Lainz A, Melendez-Retamal P, Hird DW, Read DH (1998) Papillomatous digital dermatitis in Chilean dairies and evaluation of a screening method. Prev Vet Med 37:197–207

Rutter SM (2006) Diet preference for grass and legumes in free-ranging domestic sheep and cattle: current theory and future application. Appl Anim Behav Sci 97:17–35

Schrama JW, Arieli A, Brandsma HA et al (1993) Thermal requirements of young calves during standing and lying. J Anim Sci 71:3285–3292

Schütz KE, Cox NR, Matthews LR (2008) How important is shade to dairy cattle? Choice between shade or lying following different levels of lying deprivation. Appl Anim Behav Sci 114:307–318

Schütz KE, Rogers AR, Cox NR, Tucker CB (2009) Dairy cows prefer shade that offers greater protection against solar radiation in summer: shade use, behaviour, and body temperature. Appl Anim Behav Sci 116:28–23

Schütz KE, Clark KV, Cox NR, Matthews LR, Tucker CB (2010a) Responses to short term exposure to rain and wind by dairy cattle: time budgets, shelter use, body temperature and feed intake. Anim Welf 19:375–383

Schütz KE, Rogers AR, Poulouin YA et al (2010b) The amount of shade influences the behavior and physiology of dairy cattle. J Dairy Sci 93:125–133

Schütz KE, Rogers AR, Cox NR et al (2011) Dairy cattle prefer shade over sprinklers: effects on behavior and physiology. J Dairy Sci 94:273–228

Schütz KE, Cox NR, Macdonald KA et al (2013) Behavioral and physiological effects of a short-term feed restriction in lactating dairy cattle with different body condition scores at calving. J Dairy Sci 96:4465–4476. https://doi.org/10.3168/jds.2012-6507

Schütz KE, Cox NR, Tucker CB (2014) A field study of the behavioral and physiological effects of varying amounts of shade for lactating cows at pasture. J Dairy Sci 97:3599–3605

Scibilia LS, Muller LD, Kensinger RS et al (1987) Effect of environmental temperature and dietary fat on growth and physiological responses of newborn calves. J Dairy Sci 70:1426–1433

Seigel JM (2005) Clues to the functions of mammalian sleep. Nature 437(7063):1264–1271. https://doi.org/10.1038/nature04285

Shum LWC, McConnel CS, Gunn AA, House JK (2009) Environmental mastitis in intensive high-producing dairy herds in New South Wales. Aust Vet J 87:469–475

Solano L, Barkema HW, Mason S et al (2016) Prevalence and distribution of foot lesions in dairy cattle in Alberta, Canada. J Dairy Sci 99:6828–6841

Somers J, O'Grady L (2015) Foot lesions in lame cows on 10 dairy farms in Ireland. Ir Vet J 68:10

Storry JE, Rook JAF (1962) The magnesium nutrition of the dairy cow in relation to the development of hypomagnesaemia in the grazing animal. J Sci Food Agric 13:621–627
Stull CL, Messam LLMV, Collar CA (2008) Precipitation and temperature effects on mortality and lactation parameters of dairy cattle in California. J Dairy Sci 91:4579–4459
Sutherland MA, Stewart M, Schütz KE (2013) Effects of two substrate types on the behaviour, cleanliness and thermoregulation of dairy calves. Appl Anim Behav Sci 147:19–27
Sutherland MA, Worth GM, Schütz KE, Stewart M (2014) Rearing substrate and space allowance influences locomotor play behaviour of dairy calves in an arena test. Appl Anim Behav Sci 154:8–14
Teagasc (2016) Lameness. In: Teagasc Dairy Manual, Section 8 – Animal Health, Chapter 47. Teagsc, Oak Park, Carlow, Ireland, pp 19–22
Thompson WR, Meiske JC, Goodrich RD et al (1983) Influence of body composition on energy requirements of beef cows during winter. J Anim Sci 56:1241–1252
Towers NR, Smith BL (1978) The protective effect of zinc sulphate in experimental sporidesmin intoxication of lactating dairy cows. N Z Vet J 26:199–202
Tucker CB, Rogers AR, Verkerk GA et al (2007) Effects of shelter and body condition on the behaviour and physiology of dairy cattle in winter. Appl Anim Behav Sci 105:1–13. https://doi.org/10.1016/j.applanim.2006.06.009
Tucker CB, Rogers AR, Schütz KE (2008) Effect of solar radiation on dairy cattle behaviour, use of shade and body temperature in a pasture-based system. Appl Anim Behav Sci 109:141–154
Turnpenny J, Wathes C, Clark J, McArthur A (2000) Thermal balance of livestock. 1. A parsimonious model. Agric For Meteorol 101:15–27
Vermunt JJ, Stafford KJ, Thompson KG (1995) Observations on colostral intake in newborn dairy calves. N Z Vet J 43:205–206
Vogels Z, Chuck GM, Morton JM (2013) Failure of transfer of passive immunity and agammaglobuinaemia in calves in south-west Victorian dairy herds: prevalence and risk factors. Aust Vet J 91:150–158
Waghorn GC, Collier K, Bryant M, Dalley DE (2018) Feeding fodder beet (Beta vulgaris L.) with either barley straw or pasture silage to non-lactating dairy cows. N Z Vet J 66:178–185
Webster JR, Stewart M, Rogers AR, Verkerk GA (2008) Assessment of welfare from physiological and behavioural responses of New Zealand dairy cows exposed to cold and wet conditions. Anim Welf 17:19–26
Wesselink R, Stafford KJ, Mellor DJ, Todd S, Gregory NG (1999) Colostrum intake by dairy calves. N Z Vet J 4:31–34
Wilson SC, Mitlöhner FM, Morrow-Tesch J et al (2002) An assessment of several potential enrichment devices for feedlot cattle. Appl Anim Behav Sci 76:259–265
Yang DA, Johnson WO, Müller KR et al (2019) Estimating the herd and cow level prevalence of bovine digital dermatitis on New Zealand dairy farms: a Bayesian superpopulation approach. Prev Vet Med 165:76–84
Young BA (1981) Cold stress as it affects animal production. J Anim Sci 52:154–163

# Welfare of Beef Cattle in Extensive Systems

# 6

Simon P. Turner, Maria Eugênia Andrighetto Canozzi, and Drewe Ferguson

## Contents



**Abstract**

Extensive beef cattle production is characterised by the management of animals on pasture, often in remote areas, and with infrequent contact with humans. Great diversity exists in environmental and management conditions, even within

S. P. Turner (✉)
Animal and Veterinary Sciences Department, Scotland's Rural College (SRUC), Edinburgh, UK
e-mail: Simon.turner@sruc.ac.uk

M. E. A. Canozzi
Programa Nacional de Investigación en Producción de Carne y Lana, INIA La Estanzuela, Colonia, Uruguay

D. Ferguson
Livestock and Aquaculture Program, CSIRO Agriculture and Food, FD McMaster Laboratory, Armidale, NSW, Australia



M. Haskell (ed.), *Cattle Welfare in Dairy and Beef Systems*, Animal Welfare 23,
https://doi.org/10.1007/978-3-031-21020-4_6

the confines of a single farm, as a result of topography and climate. The welfare challenges within the extensive beef sector are therefore highly variable in location and time. In this chapter, we explore the commonalities and diversity present in extensive beef systems and consider what farmers and the public regard as good welfare in the context of beef cattle. It is generally considered, particularly by the public, that access to an outdoor environment that allows animals to behave naturally is central to the concept of good welfare. We consider whether the scientific evidence supports the view that cattle value access to an outdoor space and to pasture specifically. Thereafter, the chapter discusses the primary welfare challenges faced by extensive beef cattle, which centre around weather or climate stressors, nutritional stressors from inadequate forage of a suitable quality, health stressors, and painful management stressors. Three case studies illustrate how these challenges are manifested, and how they are managed, in the contrasting extensive beef industries of Australia, Brazil, and the UK. Lastly, future industry developments in response to changing societal and environmental pressures are considered, and suggestions are made of the areas of research that should be prioritised to maximise animal welfare.

**Keywords**

Beef cattle · Welfare · Extensive · Pasture · Behaviour

## 6.1 Introduction

The term 'extensive' encapsulates a wide array of different systems united by the feature that the animals are pasture-based for all, or a majority, of the time (Matthews 1996). The term 'extensive' is nebulous and its interpretation differs between countries. Thus, a herd of beef cattle confined in a 10 hectare European field of improved pasture at a stocking rate of around 0.2 hectares per animal may be regarded as one end of a continuum of extensiveness (Fig. 6.1). At the opposite end of the continuum, farms in Australia's Northern Territory (Fig. 6.2) have a median herd size of 13,700 animals on an area of 304,000 hectares such that cattle are stocked at 22 hectares per animal (Bortolussi et al. 2005). In some countries a significant proportion of the national herd consists of herds of less than 10 cows and where the main source of income of the farmers is derived elsewhere (e.g. in England, 39% of beef holdings have fewer than 10 cattle; AHDB 2019). In practice, cattle used in beef production systems may not have had access to pasture for their entire lives. Cattle that enter intensive (e.g. feedlot) finishing systems may have only been bred and reared as young calves on extensive pasture (Petherick 2005).

Extensive beef production systems have existed since domestication of cattle around 10,000 years ago (O'Neill et al. 2010). Currently, the major beef producing countries by head of cattle slaughtered are the United States of America, Brazil, Russia, and China, each producing a similar or greater amount of beef than the European Union or Africa and more than double the quantity of other major beef

**Fig. 6.1** Beef cattle in lowland Scotland at the opposite end of the continuum of 'extensiveness' as compared to systems in northern Australia (see Fig. 6.2). (Photo courtesy of Marianne Farish)

**Fig. 6.2** Cattle on extensive pasture in northern Australia. Systems in this region are characterised by large herds roaming over vast areas of pasture (Image credit: Cooperative Research Centre for Beef Genetic Technologies (http://www.beefcrc.com))

producing countries such as Australia (Ritchie and Roser 2017). Beef cattle can be reared under a range of conditions. It is the aim of this chapter to discuss the welfare issues in extensive systems.

In the following sections, we begin by exploring the defining common characteristics of extensive beef systems, but also their diversity. The chapter will then consider how welfare is conceptualised in the context of an extensive environment and whether there is evidence that cattle value the opportunity to access pasture. It will build on this to consider the welfare challenges, illustrating these and how they are managed using case studies from contrasting countries. Finally, we will conclude with a discussion of the future priorities where management change, underpinned by research, could benefit animal welfare.

## 6.2 Similarities and Differences in Extensive Beef Systems

Generalisations about extensive beef production systems are difficult to make but some features are common to most extensive beef farms in most countries. The key defining characteristic of these systems is their pasture-based nature. Extensive systems can exploit improved grassland, but quite often they are based on low-input rangeland that is unsuitable for other agricultural systems such as crop production. This point is often missed in the debate surrounding land use and environmental concerns of livestock production. It could be argued that the capacity to convert plant biomass of variable quality, which cannot be utilised by humans, into nutritious food for human consumption is a key virtue of extensive beef production (see also Chap. 13). In some environments, meat from extensively managed ruminants is the only form of human food that can be derived from the land. Pasture-based production leads to extensive beef farms sharing other characteristics. In particular, these farms are characterised by reduced contact between humans and animals and infrequent handling. Farms are typically remote from centres of human population and hence the market for beef. Although more visible to the public than intensively managed housed livestock, the remoteness of extensive beef production reduces the opportunity for the public to observe animals and likely constrains their knowledge and understanding of welfare issues in extensive systems (Cornish et al. 2016). Remoteness also creates challenges regarding the availability of specialist advice, training, and skilled labour and necessitates the long-distance transport of animals to market and slaughter (FAWC 2019).

Extensive beef production usually has low or negative profit margins (Hocquette et al. 2018) which constrains welfare improvement, and the industry is supported by financial subsidies in some countries. In some regions investment is increasing and driving greater financial security. The lack of reliable data collection is common on these farms, and coupled with lack of farm business management training, probably perpetuates this situation (McLean and Holmes 2015). However, opportunities for improved technical efficiency may be genuinely few due to constraints imposed by cost:benefit considerations. However, beef carcasses are valuable and herd sizes are often lower than for other livestock systems. Therefore, each individual animal may be an important financial asset within the system, especially in small herds, which ought to increase the incentive to protect their health and welfare to ensure their survival and maximise their productivity.

Despite these defining characteristics, diversity exists in major attributes of extensive beef systems, such as breed and herd size. The wide diversity of extensive beef production systems is largely the result of matching the system to the environment and the domestic and export beef market requirements. For example, the large variation in stocking density referred to earlier occurs in response to the great heterogeneity in topography, pasture quality, rainfall, temperature, and disease risk in the regions where extensive beef farming is practised. Environmental heterogeneity may also be substantial across the area of a single farm and across time. A key aspect of the diversity of beef systems is the variety of breeds developed to meet specific production needs. The introduction and development of *Bos indicus* breeds in sub-tropical and tropical climates is a seminal example here. In contrast to *Bos taurus* breeds, indicine breeds display greater adaptation to heat stress and diseases of tropical environments.

Large integrated businesses exist in the beef industry. Within the same organisation, many thousands of cattle can be produced, transported, processed, and marketed. These organisations can be contrasted with fragmented industries in areas such as Europe (Hocquette et al. 2018) which often place significant value on established traditions of cattle husbandry (Waterhouse 1996) and whereby different businesses are involved in genetic selection, animal production, feed production, carcass processing, and retail. Where the industry is fragmented, communication and uptake of innovation is slow. Communication infrastructure may also be lacking which impedes greater flow of knowledge. Record keeping of production, health, and welfare metrics can be poor and there is a lack of opportunities or effort to perform within- and between-farm benchmarking which constrains health and welfare improvement (FAWC 2019).

Appreciating the shared characteristics of, but also diversity within, extensive beef systems is necessary to understand the welfare challenges faced in this sector. Before discussing the welfare challenges in detail, the next section will first consider how welfare is conceptualised and how this leads to a general lack of concern about the welfare of extensive beef cattle.

## 6.3 Conceptualising Welfare in the Context of Extensive Beef Production

Three well-recognised concepts of good animal welfare have been described. Good animal welfare is achieved (1) when animals lead natural lives through the development and use of their natural adaptations and capabilities; (2) when they feel good; and (3) when they function well as biological agents (Fraser et al. 2013; and see Chap. 1 in this volume). These can be summarised as the 'naturalness', 'feeling good', and 'biological functioning' welfare perspectives, respectively. A substantial body of literature suggests that the public regard the opportunity to have a natural existence as the principal determinant of good animal welfare (e.g. Blokhuis et al. 2003; Clark et al. 2016), a major component of which is the ability to behave naturally in an unconstrained environment. Recent work suggests that this view is labile and animal health concerns (part of the 'biological functioning' perspective) can, in certain contexts, lead the public to conclude that closer confinement is justified (Vigors et al. 2021a).

Despite debate about the relative importance given to health and naturalness, it is nonetheless clear that naturalness is a central component of how the public conceptualises animal welfare. Daily access to pasture is a key attribute valued by beef consumers, with a portion of consumers willing to pay more for beef derived from pasture. While studies in this area have mostly focussed on consumers in European countries and the USA (Stampa et al. 2020; Kühl et al. 2021), recent work in Brazil has also confirmed that the public regard pasture-based production as natural and beneficial for animal mental state (Yunes et al. 2017; Cardoso et al. 2018). As such, the welfare of extensive beef cattle has been of lower concern to the public than the welfare of intensively managed animals (Matthews 1996; Temple and Manteca 2020).

The weight given to naturalness by the public when appraising welfare is typically not shared by farmers who regard animal health and productivity as the central determinants of welfare (Spooner et al. 2016 (Canada); Cardoso et al. 2018 (Brazil); Vigors et al. 2021b (UK and Republic of Ireland); Bundle et al. 2021 (Australia)). However, not all farmers subscribe to this view (Vanhonacker et al. 2008). In pasture-based systems, the animals' opportunity for expression of autonomy and natural behaviour may be assumed by farmers to be a guaranteed outcome of the extensive farming system. Farmers are known to adopt systems of production that meet their own values, including naturalness (Dockès and Kling-Eveillard 2006). Therefore, farmers may simply assume that naturalness requires no further action, and hence their concept of good welfare may not represent the discordance with that of the public's perspective as it first appears. However, some work suggests that beef farmers regard it as acceptable that animals face challenges that they deem to be 'natural' but that could compromise their welfare (e.g. Spooner et al. 2016).

Extensive environments, when these are complex and the animals are well matched to them, provide opportunities for expression of choice and for animals to engage meaningfully and richly with their surroundings. The concept of positive affective engagement has been described in Chap. 7. Briefly, this encompasses all positive emotional experiences resulting from active engagement in motivated and rewarded behaviours (Mellor 2015; Lawrence et al. 2019). Such positive affective engagement may be integral to the achievement of positive welfare states (Lawrence et al. 2019). Complex extensive environments that allow beef cattle to exercise agency potentially allow greater opportunity for positive welfare outcomes than simple, monotonous, and predictable environments (Špinka and Wemelsfelder 2011).

However, extensive environments can pose severe challenges that cannot be readily mitigated by management solutions, such as thirst, starvation, and exposure to extremes of temperature. When the animals' adaptive mechanisms are unable to meet such challenges, the potential exists for prolonged and severe suffering. Given the reduced level of contact between farmers/managers and their animals in extensive beef systems, this risk is further exacerbated. We may predict that the emotional landscape that can be experienced by extensively managed beef cattle may be broader than for most other livestock types, assuming our livestock species each have similar capacity to experience emotions. Appleby (1996) has argued similarly in the context of all extensively managed livestock by stating '*Life in the wild can be tough, but it can also be good*'.

## 6.4 Do Cattle Want to Be in an Extensive Environment?

Our expectation that a natural environment benefits cattle welfare is predicated on the assumption that cattle would prefer to be in such an environment over a more intensively managed one. But is there any evidence that this assumption is correct? Most of the work on cattle environmental preferences has been performed with dairy animals. In temperate climates, dairy cattle exhibit a preference to be outdoors at pasture, especially for lying (Krohn et al. 1992; Ketelaar-de Lauwere et al. 2000). These animals have a preference for pasture that exceeds their preference to be outdoors on other soft surfaces, suggesting that the ability to graze is valued as well as the outdoor access *per se* (Smid et al. 2020). Haskell et al. (2013) have shown that dairy cattle do not change their use of an outdoor concrete 'loafing' area when this has an unrestricted view of pasture, indicating that access to pasture itself rather than the naturalistic view of pasture is important to the animals. Again, in dairy cattle, the preference for pasture is elastic and some studies find that they will choose to be indoors during hot, cold, or wet weather (Legrand et al. 2009; Charlton et al. 2011). When given a choice between feedlot or pasture environments, beef cattle perform the large majority of their lying on pasture (81%; Lee et al. 2013). Although we do not know how hard beef cattle will work to access pasture and the elasticity of this demand, the preference studies in both dairy and beef animals suggest that pasture access is likely to be important to them. It is noteworthy that, in the case of dairy cattle, their preference for pasture is partial and dependent upon the ambient conditions, suggesting an aversion to some of the conditions that we may regard as 'natural' such as rainfall and hot or cold weather. Lee et al. (2014) have also shown that beef cattle will choose to access a feedlot where they received concentrate feed instead of pasture where concentrate was not provided, indicating that pasture access may have a lower priority compared to feed of high nutrient density, although there may be individual variation around this. The extent to which beef cattle will tolerate hot, cold, or wet weather in order to remain at pasture is unknown, although they do seek shelter when at risk of heat or cold stress (Van Iaer et al. 2014). It is likely that their preference for pasture will, like dairy cattle, be partial and sensitive to ambient conditions and pasture quality, although beef animals may have a different threshold to dairy cattle when choosing to sacrifice pasture access for an alternative environment. Thus, the body of evidence suggests that cattle value access to pasture and will choose to use it when not under threat of heat or cold stress or nutritional challenge. Therefore, it would seem a reasonable working assumption that pasture access benefits welfare in the absence of thermoregulatory or nutritional challenges.

## 6.5 The Concept of Environmental Fit

Cattle were domesticated in the Fertile Crescent, Saharan Africa and the Indian subcontinent (O'Neill et al. 2010; Bollongino et al. 2012) but are now raised in environments quite different to these ancestral ones (McManus et al. 2009; Villalba et al. 2016). The assumption that naturalness contributes to good animal welfare, and belief in the capacity of existing systems to provide a natural existence, raises important questions. In regard to the concept of environmental fit, we can ask firstly whether the environments in which we farm beef cattle are sufficiently representative of those in which the species evolved, and secondly whether artificial selection has changed the animals themselves such that they have different requirements from the environment compared to their ancestors (Provenza 2008; Ferguson 2014). Both are likely to be critical in determining welfare outcomes. Examples exist in which beef breeds were introduced into regions for which they were poorly adapted; notable examples being the introduction of temperate-adapted *Bos taurus* breeds into sub-tropical northern Australia and the introduction of taurine breeds into Central and South America during the European migration in the sixteenth century. The correct choice of breed for the environment is probably the simplest way of avoiding welfare compromises in extensive beef production due to disease, parasites, and climatic challenges (heat, cold, and drought) (Petherick 2005; see also Chap. 7). Adaptation of genotypes to a new environment can be rapid, measured in a decade or less, such as the acquisition of resistance to cattle ticks in *Bos taurus* cattle introduced into tropical environments (O'Neill et al. 2010). Whilst subsequent generations of cattle likely benefit from this adaptation, welfare may suffer during the adaptation process if it is not carefully managed. Lastly, it should be remembered that adaptation is no guarantee that an animal will cope. Wild animals living in the environment in which they have evolved can still face environmental challenges that are beyond their ability to respond (Villalba and Manteca 2016). Extreme environmental challenges, such as prolonged drought, will harm the welfare of the most adapted animals unless mitigated by human intervention.

## 6.6 Challenges of Assessing Cattle Welfare in an Extensive Environment

Challenges to animal welfare are not static but differ in type and severity over time. Extensive environments exemplify this instability and require that we develop a welfare assessment framework that works in an inherently variable system. Major system variables such as the ambient temperature, pathogen load, and amount and quality of food and water can vary dramatically on both short and longer timescales in the same location (Villalba et al. 2016). The toolkit used to assess animal welfare needs to be capable of sensitively appraising the outcomes of a wide range of potential threats that can vary considerably in severity. It is not enough to assess whether the biological systems of the animal are able to cope with a challenge: it is also necessary to assess whether the animal suffers irrespective of whether it functionally copes or not (Broom 1991; Colditz et al. 2014). The inherent temporal instability and geographical variability of extensive systems coupled with the difficulty of observing freely behaving animals has resulted in less progress in the development of welfare indicators for extensive livestock than for intensively raised animals. The perception that welfare is under less threat in extensive environments has likely exacerbated this slower progress, coupled with a view that humans have diminished responsibility to animals living, or perceived to be living, in a natural environment (Appleby 1996; Petherick 2005). Throughout this chapter, we give examples of innovations that are overcoming the barriers presented by an extensive environment to develop new ways to assess and improve cattle welfare.

The infrequency of close-proximity inspection, sometimes only twice per year in specific regions of some countries such as Australia (Petherick 2005), renders the application of many welfare indicators developed for intensive systems unusable under extensive conditions. Indeed, in the most extensive systems, welfare assessment of beef cattle may share more parallels with welfare assessment of wild animals than intensively managed livestock. Indicators of a chronic welfare challenge, such as poor body condition, poor growth, and parasitism are easier to monitor due to their persistence, but these can be insensitive or non-specific indicators of the underlying challenge. For example, growth rate can be depressed for a multitude of reasons. Such indicators may also reflect a challenge that may have caused suffering for a prolonged period before detection (Petherick 2005). In addition to the environmental constraints on welfare assessment, cattle are a predated species and behaviourally stoical, tending to hide signs of pain or illness (Hudson et al. 2008). They are also often fearful of human contact and observing their undisturbed behaviour at close range is difficult.

Turner and Dwyer (2007) proposed a suite of welfare indicators suitable for application in extensive environments. Rather than the detailed resource-based assessments of environmental inputs commonly performed in intensive environments (e.g. water quality), this focussed on assessment of contingencies that the farmer should provide for periods of environmental challenge (e.g. contingencies to deal with water scarcity). Greater contingency planning has also been advocated by Petherick (2005) as a key approach to enhance welfare in extensive beef systems. This should be complemented by maximising use of animal-based welfare indicators during the infrequent handling episodes. Another approach proposed by Colditz et al. (2014) recognises the need for tools to aid self-assessment and continuous self-improvement by farmers for welfare on their own farm, instead of welfare assessment approaches designed for external market assurance or legal compliance. It combines farmer-led risk assessment and risk management with industry benchmarking and external auditing to ensure quality of on-farm recording. The farmer-led activity involves risk assessment, identification of corrective actions, and selection of monitoring metrics and uses data from the animals, resource inputs, and management that either contribute to welfare or are reflective of it.

It has been recognised for several decades that welfare assessment in the most extensive environments is deeply challenging (Matthews 1996). There remains a profound need for methods to accurately and sensitively assess welfare in livestock at pasture with enough specificity to identify the underlying challenge to inform management decisions (Temple and Manteca 2020). Technological developments are expected to revolutionise the quantity and quality of data available for welfare assessment in intensive environments (Berckmans 2014). Large potential exists for the use of technological solutions to aid welfare assessment and welfare management in extensive environments (Rutter 2014; Manning et al. 2017; Chap. 11) but much work is needed to exploit these opportunities. Under the case studies below, we consider examples of developments in this area.

## 6.7 Welfare Challenges

As noted by Villalba et al. (2016) and Temple and Manteca (2020), extensive environments provide the opportunity to perform natural behaviours, to rest on a soft surface, to exercise, avoid social stress, and to minimise exposure to pathogens that proliferate under more intensive conditions. The converse is that these same environments can present a plethora of welfare challenges that are diverse in form, hard to predict, difficult to detect due to infrequent inspection, and often difficult to mitigate. The difficulty of mitigation results from practical barriers to enacting change in a large and unpredictable landscape and exacerbated by the low profit margins that are common in these extensive beef systems (McLean and Holmes 2015; Hocquette et al. 2018). Several of the welfare challenges are not unique to extensive environments. For example, painful husbandry events, social challenges (weaning, regrouping), heat stress, poor diets, and transport all occur in intensive as well as extensive production systems (see Chap. 7 for a discussion of this topic). Our aim in the rest of this chapter is not to provide an exhaustive description of all welfare challenges in extensive beef systems. For comprehensive reviews and summaries, we refer the reader to Petherick (2005), Villalba and Manteca (2016), FAWC (2019), and Temple and Manteca (2020). Instead, we focus on the following four core issues since they can have a profound effect on the welfare of individual animals and because of the threat they pose to extensive beef cattle across a diverse range of production systems. They are: **weather or climate stressors; nutritional stressors from inadequate forage of a suitable quality; health stressors; and painful management stressors.** Table 6.1 lists the welfare impacts and outcomes associated with each of these core challenges. In the case studies below, we consider how the welfare impacts of these challenges are assessed and managed in regions that contrast in their production systems. In selecting these core challenges, we recognise that other challenges can also be significant and locally more important than those that we have selected (e.g. long-distance transport from extensive breeding farms to feedlot finishing units) (Petherick 2005; covered in Chap. 7); predation

**Table 6.1** Core welfare challenges and welfare impacts in extensive beef production. Welfare impacts are expressed in the terminology of the Affective Experience Domain of the Five Domains Model (Mellor and Beausoleil 2015). The welfare impacts below are not exhaustive but represent the primary impacts associated with these core stressors

| Core challenge | Welfare impacts | Key references |
|---|---|---|
| Weather or climate stressors | Over heating<br>Chilling<br>Thirst/dehydration<br>Hunger<br>Death | Silanikove (2000)<br>Morgan et al. (2009)<br>Salama et al. (2016) |
| Inadequate forage or feed of sufficient quality leading to nutritional stressors | Thirst<br>Hunger<br>Nutrition malaise<br>Death | Manteca and Smith (1994)<br>Hogan and Phillips (2016)<br>Villalba et al. (2016) |
| Health stressors | Breathlessness<br>Pain<br>Debility, weakness<br>Sickness, malaise<br>Nausea<br>Dizziness<br>Exhaustion<br>Death | Petherick (2005)<br>Goddard (2016) |
| Painful management stressors | Pain<br>Anxiety, fearfulness, panic<br>Neophobia<br>Depression | Lay et al. (1992)<br>Stafford et al. (2006)<br>Canozzi et al. (2017, 2018) |

(e.g. Ramler et al. 2014; Laundré 2016); and exposure to toxic plants (e.g. Pfister et al. 2016; Temple and Manteca 2020). However, the selected core challenges map closely onto those previously identified by Temple and Manteca (2020). They also impinge on four of the five domains in the Five Domains model of animal welfare, namely the environmental, nutritional, health, and affective (mental state) domains (Mellor 2016). Additionally, these challenges speak to four of the Five Freedoms (FAWC 1993).

## 6.8 Case Studies

In this section, we describe the production systems and welfare challenges for three different beef production systems that represent the diversity in this sector across the world.

### 6.8.1 Australian Case Study

With a national herd size of 25 million head, Australia is a relatively small beef producer compared to some other countries. However, Australia exports approximately 60% of its production, making it the second largest beef exporter by volume. Australia supplies a large number and diversity of global and domestic beef markets and this is mirrored in the product specifications. Importantly with respect to animal welfare, these specifications extend beyond the standard attributes of beef quality and safety to include other aspects specific to how the animals were raised and produced. To address current and future consumer concerns about animal production, it is inevitable that market specifications pertinent to animal welfare and environmental stewardship will increase in the future.

A key feature of extensive beef production systems in Australia is its diversity. This large heterogeneity is manifest across several different physical dimensions such as geography, topography, vegetation, climate, size, and scale of operation. These dimensions are also linked, for example, smaller farms (100–400 head) tend to be concentrated in the higher rainfall coastal regions of south-eastern Australia whereas the larger operations (1600–5400 head) are located predominantly across northern Australia. Significantly, these larger operations only account for 9% of all beef farms but 54% of Australia's beef herd (Thompson and Litchfield 2020). This geographic diversity is also reflected in the breed composition. Taurine breeds of British and European origin are more prominent in south-eastern Australia whereas tropically adapted indicine breeds (e.g. Brahman), crossbreds, and composites predominate in northern Australia. Against this brief background, effectively managing animal welfare within extensive beef production systems is incredibly challenging and complex.

#### 6.8.1.1 Managing Climatic Stressors: Weather or Climate Stressors and Nutritional Stressors

With the exception of Antarctica, Australia is the driest continent on the planet with many parts of the country where cattle are produced receiving ≤500 mm of rain annually. Moreover, rainfall is seasonal and at times, highly variable. The extreme manifestation of variable rainfall coupled with high temperature, is of course, droughts, which are not uncommon in Australia due to its geospatial location and geography. Severe long-term droughts such as the recent 2017–2019 drought over most of south-eastern Australia produced devastating economic, environmental, and social impacts. Droughts represent the greatest welfare challenge to extensive beef producers due to the impacts of reduced feed/forage and water availability and the increased risks of natural disasters such as bushfires.

When nutrient intake fails to meet metabolic requirements in the animal, an adaptive response is initiated, resulting in catabolism of fat and muscle tissue and subsequent reductions in liveweight and body condition. Cattle have evolved to adapt to some fluctuation in body condition, especially females during each pregnancy/lactation cycle. From an animal welfare perspective, these normal fluctuations are probably of minimal concern. However, during moderate and sustained undernutrition through drought where chronic hunger and significant weight and condition loss occurs, there is an increased risk of reduced animal welfare. Unfortunately, our understanding about the nature and magnitude of the welfare impacts of moderate and sustained under-nutrition in beef cattle is somewhat limited. However, some insight can be gained from dairy cattle research focused on the perennial issue of maintaining body condition during lactation. In their review of the association between body condition and dairy cow productivity, health, and welfare, Roche et al. (2010) concluded that low body condition was associated with productivity losses (e.g. milk production and impaired reproduction), reduced immune function, and an increased risk of discomfort in cold environments and metabolic disorders. A notable gap in our understanding is the impact of chronic undernutrition on the affective state of the animal and this clearly warrants more attention.

Drought is an inevitability in Australia and given the projected impacts of climate change, including increasing periods of drought (State of Climate 2020 Report; https://www.csiro.au/en/research/environmental-impacts/climate-change/state-of-the-climate), extensive beef producers must develop and implement drought management plans to ensure animal welfare is not compromised. Preparedness and contingency planning are paramount (Petherick 2005), and central here is the ability to make prescient decisions about reducing stocking rates and/or destocking to match declining pasture reserves. The emergence of sensor technologies and decision support tools/software will, in time, enhance and enable improved grazing management decisions (see Tedeschi et al. 2021). Increasing storage capacity for feed supplements (grain, silage, etc.) and upgrading water infrastructure are other critical elements of any drought management plan.

An additional risk posed by the Australian climate is heat stress. Adaptive behavioural responses to heat stress, such as increased drinking and shade seeking, may be constrained by the environment, particularly in drought conditions. Schutz et al. (2008) have shown that dairy cattle prefer to stand in the shade than lie in the sun during hot days, even if deprived of the opportunity to lie down for 12 h beforehand. Beef cattle may have a different tolerance for heat compared to dairy, and tropically adapted breeds cope better with prolonged high temperatures than temperate breeds (Beatty et al. 2006). However, when heat and humidity are extreme, it seems reasonable to expect that tropically adapted beef breeds will also show a strong aversion to conditions that cause heat stress when given the opportunity to express adaptive responses. Heat stress has significant biological consequences such as increased respiration rate and reduced feed intake and milk yield. However, the affective experience of heat stress is less well known. It is likely that the willingness of cattle to sacrifice performance of motivated behaviours in order to avoid heat stress (as discussed above for dairy cattle in the Schutz et al. 2008 study) is underpinned by an aversive affective experience, but this requires further investigation.

#### 6.8.1.2 Optimising Health and Welfare Outcomes

In a recent survey and economic analysis of endemic disease in the Australian red meat industries, Lane et al. (2015) reported that the top five health priorities in beef cattle based on their economic impact were: cattle tick ($156 million), bovine pestivirus ($117 million), buffalo fly ($98.1 million), dystocia ($97.8 million), and neonatal calf mortality ($96.2 million).

Cattle tick and buffalo fly are external parasites in sub-tropical and tropical northern production systems. In addition to the direct impacts that infection with these parasitic organisms causes, both are vectors for other significant diseases such as ephemeral fever and tick fever (e.g. Babesiosis). Whilst there are commercial vaccines available for these diseases, the rate of adoption has generally been low and variable (Bortolussi et al. 2005). A commercial vaccine against ticks was released in the 1990s (TickGARD), but it was subsequently withdrawn due to lack of uptake. The requirement for regular booster vaccinations throughout the year was deemed impractical in northern beef systems where cattle are not frequently handled. Notwithstanding this, the quest to develop an effective tick vaccine continues (Lew-Tabor and Rodriguez Valle 2016).

Reliance on *Bos indicus* genotypes coupled with use of acaricides and insecticides has been the prevalent strategy for managing ticks and buffalo fly. However, chemical control for ticks is proving problematic due to increasing acaracide resistance (Cutulle et al. 2009; see also Brazilian case study). Given this, and in lieu of an effective vaccine, future emphasis should be directed at breeding for tick resistance. Whilst tick resistance in cattle is moderately to highly heritable, a key constraint has been developing a simple and cost-effective method to phenotype the trait at scale (Burrow et al. 2019). In this context, selection for immunocompetence (Hine et al. 2016 also see Chap. 7) may also offer potential but the relationship between general immunocompetence and parasite resistance in beef cattle has not been evaluated to date.

Of the remaining endemic health priorities above, all are directly relevant to the critical issue of reproductive wastage (i.e. cow infertility and calf mortality). Reducing reproductive loss, or to put it in more positive light, improving reproductive efficiency in extensive beef systems requires an integrated approach incorporating genetic, nutritional, breeder management, and reproductive disease control strategies and this has been reviewed in detail by Burns et al. (2010).

As noted earlier, the lower frequency of inspection and handling may constrain the capacity to optimise health and welfare outcomes on extensive beef farms. However, inspection frequency can vary considerably depending on the region, production period, and size of the property. For example, breeding females on smaller farms may be monitored twice daily during the critical period of calving. In stark contrast, cattle on large properties in northern Australia may only be handled once or twice annually (Petherick 2005). The latter situation is a cause for concern given the lack of alignment with the agreed animal welfare principles for livestock production systems (see Fraser et al. 2013).

On a more positive note, the application of digital technologies to remotely monitor and manage livestock has the potential to transform extensive beef farming. The advances in sensor technologies and precision livestock farming were recently reviewed by Tedeschi et al. (2021) (see also Chap. 11). In addition to monitoring animals remotely, digital sensor technologies also offer benefits associated with

**Fig. 6.3** Cattle at pasture wearing the eShepherd virtual herding collars. Photo credit: Gallagher Animal Management (https://am.gallagher.com/)

optimising grazing and natural resource management. Whilst Tedeschi et al. (2021) noted a slower rate of penetration of digital technologies within extensive grazing systems, this is changing in Australia with improved digital connectivity and the commercial release of technologies such as Ceres Tag (www.cerestag.com) and eShepherd (https://am.gallagher.com/en-au/new-products/eShepherd; Campell et al. 2021) (Fig. 6.3). The eShepherd is a GPS-based virtual fencing and animal management system. The technology is based on associative learning principles where animals learn to recognise audio cues with the presence of a virtual boundary.

#### 6.8.1.3 Reducing the Impact of Painful Husbandry Procedures

There are several husbandry procedures that cause pain to cattle (e.g. castration, dehorning, branding, ear notching, and spaying (surgically rendering females infertile to prevent pregnancy in pasture rearing systems where males cannot be separated from females)). The justification for these procedures varies but is generally predicated on improving animal management and welfare, and operator safety. For example, castration reduces sexual activity and aggression, thus making cattle safer to manage and handle and obviates the risk of uncontrolled mating and unwanted pregnancies. Other procedures such as hot iron branding enable the identification of cattle and are a legal requirement in some state and territory jurisdictions in Australia.

The welfare impacts of some painful husbandry procedures such as castration, dehorning/disbudding, and spaying in cattle have been well established (e.g.

Stafford and Mellor 2005; Petherick et al. 2013; Canozzi et al. 2018). Given this and the associated concerns of consumers, the Australian beef industry has been proactive in implementing a strategy to reduce the impact of these procedures. This strategy is based on a 3R model—replace (develop alternatives to the procedure), relieve (provide pain relief), and refine (modify the procedure to reduce impact). According to benchmarking surveys conducted by the Australian Beef Sustainability Framework (https://www.sustainableaustralianbeef.com.au/) reasonable progress has been made to date. For example, 30% of beef producers are now applying pain relief when undertaking these procedures. This figure has been steadily rising (approximately 8% per year) from 2017 onwards. Beef producers have access to three commercial products; Tri-Solfen (topical anaesthetic—lignocaine + Bupivacaine) and two meloxicam-based analgesics (Buccalgesic and Metacam). Importantly, Buccalgesic is administered orally which has significant practical advantages.

Breeding for polledness is viewed as the best long-term strategy to obviate the need for dehorning/disbudding. The rate this can be achieved has been enhanced through the commercial availability of a gene marker test for polledness in 2010. This DNA test has subsequently been optimised to improve its accuracy to predict different horned phenotypes across multiple breeds (Randhawa et al. 2020).

The development of alternatives to castration and spaying has largely focused on immunisation against GnRH or more specific targets such as ovarian zona pellucida glycoprotein. Immunocontraceptive vaccines are available (e.g. BOPRIVA, Zoetis) but their adoption has been limited due to the relatively short period of immunocontraception and the requirement for frequent revaccination (D'Occhio 2013). There is ongoing research in Australia aimed at developing a more practical, longer acting immunocontraception.

#### 6.8.1.4 Summary

In summary, the greatest challenges for extensive beef cattle welfare in Australia result from drought, and the associated risk of starvation, thirst, and heat stress. Furthermore, endemic diseases that are difficult to treat in animals that are rarely handled, and the reliance on painful management procedures, are routine challenges to welfare.

### 6.8.2 Brazilian Case Study

In 2019, the Brazilian national herd (beef and dairy) was around 215 million animals kept on 157 million hectares of land (cultivated and natural pastures) (IBGE 2019). Brazil is the second largest global beef producer after the USA (FAOSTAT 2021; www.fao.org/faostat/en/). In 2020, 76.3% of the beef produced in Brazil went to the domestic market, while 23.6% was exported (ABIEC 2020).

Beef cattle production in Brazil is characterised by technological and management heterogeneity, although most beef is produced on specialised larger farms. Ninety percent of beef is produced in grass pasture-based systems (ANUALPEC

**Fig. 6.4** Brazilian pasture-based system with *Bos indicus* cattle. Image credit: Mateus Paranhos

2018), with a small percentage from native and cultivated low-input pastures of low productivity (Cesar et al. 2005). Animals with the genetic potential to grow quickly are typically raised at pasture until 18–21 months of age and then fed concentrate in feedlots for 3–4 months until slaughter to achieve the desired fat deposition to meet the market demands (Ferraz and Felício 2010). Thus, cattle spend most of their life at pasture.

Cattle are farmed on land with soils of low and intermediate fertility (Euclides et al. 2010). The breeds used are adapted to tropical climates (Fig. 6.4) and to the associated ectoparasites (ticks and flies). Since 1970, the population of the triploid (three homologous sets of chromosomes) indigenous breed Nellore has increased substantially (Euclides Filho 2000) and exploits *Brachiaria* grass pastures. Synthetic breeds, like Brangus, Braford, Canchim, and Santa Gertrudis, are also produced in all regions. In the southern region, which has a temperate climate, purebred *Bos taurus*, mainly Angus and Hereford, are reared (Ferraz and Felício 2010).

#### 6.8.2.1 Managing Climatic Stressors

In Brazil, the weather varies considerably from tropical in the north, to temperate climates south of the Tropic of Capricorn. Most of the country has moderate rainfall (1000 to 1500 mm a year), with the majority falling in the summer south of the Equator (December to March). It tends to be hot and dry in the interior of the country, changing to humid in the tropical rainforests of the Amazon (2000 to 3000 mm rainfall annually). Climate change is expected to increase temperatures and reduce precipitation across Brazil.

Heat stress is a risk in tropical regions with adverse consequences for production (lower growth rate), health (greater parasitism), reproduction (longer oestrous cycles and lower sperm motility), and behaviour (reduced feeding time, decreased lying time, crowding around water sources, and open-mouthed and laboured breathing under extreme situations) (van Iaer et al. 2014; Menegassi et al. 2015). Measuring respiration rate and the occurrence of panting appear to be the most accessible indicators of heat stress in extensive conditions (Silanikove 2000), although it may not be feasible to approach the animals closely enough to observe it. Conversely, cold stress occasionally occurs in Brazil, mainly in the mid-west region. A sudden drop in temperature combined with cold winds can lead to hypothermia and death (https://www.canalrural.com.br/noticias/pecuaria/tristeza-frio-surpreende-pecuaristas-e-causa-morte-de-gado-em-mato-grosso/) (see UK case study also).

Indigenous *Bos indicus* cattle are well adapted to hot, humid tropical environments. In contrast, European *Bos taurus* breeds, introduced to the tropics because of their better meat quality (tenderness and marbling), have experienced higher mortality rates and poorer reproductive performance due to their poorer adaptability to tropical climates, forages, and parasites (Ferraz and Felício 2010; Fraser et al. 2013). Breeding programmes have been used to combine *Bos indicus* and *Bos taurus* genotypes to improve environmental fit, such as the Santa Gertrudis (Zebu and Shorthorn), Ibagé (3/8 Nellore and 5/8 Aberdeen Angus), and Canchin/Charbray (*Bos indicus* breeds and Charolais). Some Brazilian beef cattle breed associations (e.g. Senepol Cattle Breeders Association) now provide information about expected progeny differences (a predictor of the genetic merit of an animal's progeny) in the slick hair coat characteristic. This trait is controlled by a single dominant gene expressed in tropically adapted animals that confers a short and sleek hair coat allowing superior thermoregulation and productivity (Flori et al. 2012; Huson et al. 2014).

The most common solution used by producers to counteract thermal stress in Brazil is the use of shelters to reduce wind speeds during cold weather or to provide shade during hot weather. This can be achieved by the introduction of trees or shrubs around the farm and adopting integrated systems, such as the Integrated Crop–Livestock–Forest System (ICLFS). When the ICLFS is not possible, an alternative is to graze paddocks that allow access to temporary shade or trees during the heat of the day, mainly from 10 am to 5 pm. In hot weather, shelter belts can generate an additional cooling effect through moisture evaporation from the vegetation; and in cold weather they reduce wind speed and create a small 'rain shadow' (Van Iaer et al. 2014). Providing an adequate source of clean and cool drinking water is also essential for promoting thermoregulation.

Technological change is important for the future of Brazilian pastoral systems (Euclides Filho 2004). A free natural disaster risk warning system, developed by the Brazilian government, works via SMS to a mobile phone and does not require Internet access. This system can be used by beef farmers to take proactive action to minimise the effects of adverse weather events on their cattle. In neighbouring

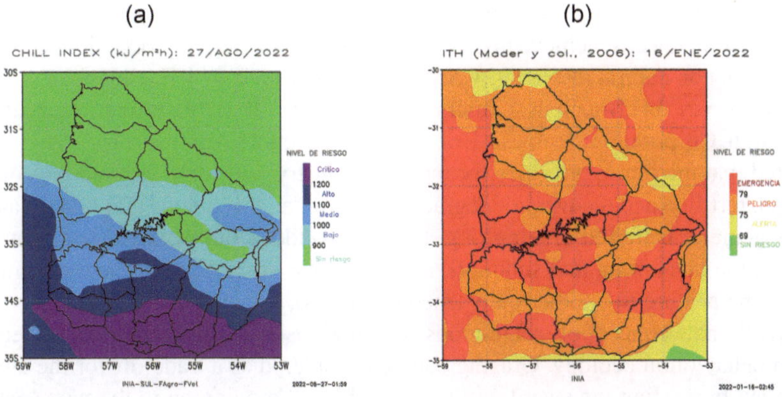


**Fig. 6.5** Examples of a map of Uruguay generated by Lambs forecast (**a**), based on the chill index, and INIA TermoEstrés (**b**), based on the temperature humidity index (INIA, Uruguay). Images courtesy of: Instituto Nacional de Investigación Agropecuaria (INIA Uruguay). 2021. (**a**) Chill Index. [http://www.inia.uy/gras/Alertas-y-herramientas/Prevision%20Corderos]; (**b**) INIA TermoEstrés [http://www.inia.uy/gras/Alertas-y-herramientas/Prevision-ITH-Vacunos/INIA-Termoestres]

Uruguay, a smartphone application was developed by INIA (National Institute of Agricultural Research; Instituto Nacional de Investigación Agropecuaria) to provide beef and dairy producers with a 7-day advanced warning of heat stress based on the temperature and humidity index (Fig. 6.5; 'INIA ThermoStress', 'INIA TermoEstrés'; http://www.inia.uy/gras/). A complementary system for Uruguayan sheep producers focusses on protecting neonatal lambs by providing predictions of the chill index and could be adapted for beef producers ('Lambs forecast', 'Provisión para corderos'; http://www.inia.uy/gras/). These risk warning systems help farmers and technicians to predict and mitigate challenges in a more manageable way, allowing cattle to withstand extreme and rapid thermal changes.

#### 6.8.2.2 Managing Nutritional Stress

Brazilian citizens associate free-range systems with 'naturalness', freedom of movement, and positive affective states (Yunes et al. 2017; Cardoso et al. 2018). However, in all six Brazilian biomes (Amazon Forest, Cerrado, Atlantic Forest, Caatinga, Pampa, and Pantanal), periods of low or nil pasture growth occurs. This occurs during the dry period in the mid-west region (from May to December), or during the winter season in sub-tropical areas (June/July to September/October), due to insufficient precipitation for grass growth. Forage quality and quantity may be insufficient to meet cattle requirements during these times with regard to dry matter digestibility and protein and mineral content (Cesar et al. 2005). In addition, since 2006, there has been a reduction in the land area used for livestock because of the expansion of crop production and restrictions imposed for conservation of forests and degraded grassland (IBGE 2019). Provision of clean water and sufficient food to satisfy voluntary food intake is the most basic of management requirements.

The challenge for livestock producers is to create new pastoral environments that satisfy animal needs, achieve highly competitive efficiency of production, and have low impact on natural resources (Carvalho 2013). A review by Euclides et al. (2010) and a survey of Brazilian farmers by Embrapa (2021; Brazilian Agricultural Research Corporation; Empesa Brasileira de Pesquisa Agropecuária) have highlighted that the future priorities for the beef industry will centre around the conservation, fertility, and recovery of degraded pastures, and techniques for better pasture management. Deferred grazing (e.g. taking a paddock out of the grazing rotation during the late spring and summer), and rotational grazing (movement of animals from one pasture to another on a scheduled basis), are well-known strategies. The relatively new concept of 'rotatinuous stocking', uses the Marginal Value Theorem from behavioural ecology. Grazing can be considered as a trade-off for the animal between harvesting the sward within a 'patch' and moving on to the next 'patch'. Monitoring both the animals' grazing behaviour and sward height is therefore a way of optimising animal intake and avoiding reducing the sward and pasture to the point where it becomes degraded (Carvalho 2013). This grazing management system is considered to be a major management innovation. Ideal sward heights have been identified for different forage species to determine when animals should be removed from the pasture to maintain biodiversity and reduce pasture degradation (Carvalho 2013). The BRS South Ruler (Régua BRS Sul), a technology generated by Embrapa, allows for adjustment of the stocking density on a particular cultivated pasture, based on the pasture height at the entry and exit of animals (Embrapa 2021). In Uruguay, a similar ruler was developed for native pasture by INIA and is called The Green Ruler (La Regla Verde; Jaurena et al. 2018). A further tool developed by INIA (the 'Grass and body condition monitor', 'Monitor de pasto y condición corporal') generates information on the average pasture quality and body condition of cattle in a local region based on an annual survey sent to technicians and farmers.

A common error is to attempt to correct poor grazing management decisions through the provision of expensive feed supplementation. Basic management must be correct to begin with, as poorly distributed salt dispensers, inadequate troughs/ponds or feeding spaces, and spoiled feed are sometimes observed on farms. Effective supplementation programmes ought to augment good management and increase the utilisation of forage, improve grazing distribution, and increase animal performance. Supplementation during the dry season is necessary with conserved fodder or concentrate (Euclides Filho 2004). Critical decisions centre around selecting a delivery method (hand-fed vs. self-fed) and frequency (daily vs. infrequent) that creates maximum benefit from supplement, fuel, and labour costs, considering the size and topography of the pasture, and number of animals. Grain-based confinement fattening for the last months before slaughter has increased in recent decades (6.09 million head, 12.6% of total slaughter in 2019; ABIEC 2020). This can exploit by-product feed sources and reduce the stocking rate in the pasture land during the dry season, thereby limiting pasture over-utilisation. More technological solutions are becoming available, such as individual automatic monitoring of feed and water intake, and automatic scales to monitor body weight (Intergado Beef, developed by Intergado; BalPass scale, developed by Embrapa/UFMS/Coimma).

However, the adoption of even simple forms of technology is poor, and the reasons for this are unclear. Technology adoption must be accompanied by training of personnel. As highlighted by Euclides Filho (2004), effective dissemination of existing knowledge with an emphasis on animal welfare and pasture management is still scarce, even in the 2020s.

Unfortunately, to this day, drought has led farmers to abandon cattle to thirst and hunger in Brazil (for a wider discussion of the effect of drought on cattle welfare, please see the Australian case study).

#### 6.8.2.3 Optimising Health Outcomes: Tick Resistance

As in Australia (see Australian case study above), the hematophagous ectoparasite *Rhipicephalus microplus* (cattle tick) is a serious pest of livestock in tropical and sub-tropical regions (Klafke et al. 2016; Miraballes and Riet-Correa 2018; Rodrigues et al. 2018; Vilela et al. 2020). Its major economic impacts on the Brazilian beef industry are both direct (treatment costs, poor growth, leather damage, myiasis, and mortality) and indirect (tick-borne diseases associated with Babesia spp. and/or *Anaplasma marginale*).

The application of chemical products that quickly suppress the tick population has been the primary method of control. However, due to frequent and indiscriminate use, multiple acaricide resistance has been described in various Brazilian regions (Klafke et al. 2016; Valsoni et al. 2020; Vilela et al. 2020), exacerbated by climate change which has affected the tick lifecycle (Miraballes and Riet-Correa 2018; Andreotti et al. 2019). The increased use of crossbred animals with *Bos taurus* genotypes may have facilitated the development of multiple resistance (Ferraz and Felício 2010; Rodrigues et al. 2018). There is evidence, for example, that greater tick burdens are observed in Brangus compared to Nellore cattle (110 vs. 17.7 ticks/day; Rodrigues et al. 2018). In addition, high stocking densities, poor grazing management, and transportation of animals between regions are important causes of the dispersal and proliferation of acaricide resistant populations.

Embrapa, in partnership with two private Brazilian companies (Conexão Delta G and GensysConsultoresAssociados), is investing in genomic selection for tick resistance in Hereford and Braford bulls through the Service for the Genomic Prediction of Tick Resistance (Embrapa 2021). Both technologies will be available soon for Angus and Brangus breeds as well. Efficient control of ticks depends upon early detection of the problem (at the end of the dry season, September to December, when the ticks could be eliminated), rapid determination of resistance (by applying the acaricide test profile), and a rational choice of which acaricide to use (Klafke et al. 2016; Andreotti et al. 2019; Vilela et al. 2020). New cattle brought onto a farm should also be quarantined to verify the presence of parasites (Andreotti et al. 2019; Vilela et al. 2020). Moreover, deferred grazing, stimulation of the natural immunity of animals up to 9 months of age, and elimination of animals that are very tick sensitive are recommended (Embrapa 2021). Individual traceability of cattle and their movements between herds should improve tick surveillance and control, as has been adopted in Uruguay (Miraballes and Riet-Correa 2018).

Transmitting information to farmers and technicians, the development of government policies to control the use of acaricides, and encouraging use of susceptibility tests in certified laboratories are the key actions required to address the issue of tick resistance to acaricides (Klafke et al. 2016; Miraballes and Riet-Correa 2018). Alternative strategies to control ticks other than acaricide use and promoting animal resistance to ticks are constrained by the scale of Brazil and its variation in climate, soils, land use, infrastructure, personnel, and political and economic conditions. Lastly, but importantly, the control or eradication of this ectoparasite must be individualised for each farm since different farms reflect different situations.

#### 6.8.2.4 Reducing the Impact of Painful Husbandry Procedures

Castration, prevention of horn growth (disbudding), and removal of horns (dehorning) are still routine practices in the Brazilian beef industry. Non-veterinarians, as well as veterinarians, conduct these procedures due to practical and cost constraints (Hötzel and Sneddon 2013; Cardoso et al. 2016). As shown by Canozzi et al. (2020), the prevalence of pain mitigation use before castration and horn removal is higher for adult cattle (60.5 and 38.8%, respectively) than for suckled (44.2 and 31.3%, respectively) and newborn calves (32.0 and 28.6%, respectively). Leaving males intact is a real alternative, a practice which is increasing. However, some marketing schemes (e.g. 'NovilhoPrecoce MS', ''MS Early Maturing Steer') actively encourage use of castration. Genetic strategies, the use of polled breeds, or crossing with polled breeds (e.g. by using Zebu cattle and their crosses with *Bos taurus* breeds (mainly Angus)), are methods that decrease the need for horn removal that are currently being applied by Brazilian farmers.

Currently, hot- or freeze-branding is the most common procedure performed on cattle. Branding is used to permanently identify the animal's breed (for breed association membership), owner, or as an individual identification mark. Additionally, animals may be branded as a record of management practices. For example, it is a requirement that animals that have been vaccinated against brucellosis are branded on the left side of the face, a highly innervated region, using a hot iron or liquid nitrogen (Ministério da Agricultura, Pecuária e Abastecimento 2017). Pain mitigation is impractical due to the cost, practical logistics, and duration required between application of pain relief and branding. From an animal welfare point of view, branding is discouraged. However, as Brazilian legislation requires branding, it should be conducted by an experienced employee with well-maintained equipment (grills and irons) on dry days (Schmidek et al. 2009). Furthermore, there is evidence that acute pain is less severe with freeze branding than hot branding (Lay et al. 1992; Schwartzkopf-Genswein et al. 1997).

The public debate about pain control is still emerging in Brazil. The continued reliance on painful procedures is the result of a combination of economic and practical pressures coupled with tradition. A gap between scientific research and technological development of methods to minimise pain and their adoption on Brazilian farms was identified by Hötzel and Sneddon (2013). However, there is also limited evidence to define best practice pain management for castration and which method is least painful (Canozzi et al. 2017). Furthermore, local anaesthesia is not effective in reducing short-term pain following disbudding or dehorning (Canozzi et al. 2018). Consequently, there is a need for better pain management methods that can be applied in the diverse and challenging conditions of the Brazilian industry considering constraints on time, labour, and skills in extensive environments. There is also a need to develop immunocastration, the use of topical anaesthesia, and to identify pain indicators for use in the field in grazing animals (e.g. facial expressions). Moreover, the dissemination of guidelines and training on best practice by public and private institutions is needed.

#### 6.8.2.5 Summary

Many welfare concerns in the Brazilian beef industry result from lack of shelters and shade to protect from heat stress, poor pasture management and feed supplementation, resistance of ticks to acaricides, and inadequate use of methods to reduce pain from husbandry procedures. In the case of painful procedures, greater adoption is required of immunocastration and topical anaesthesia to minimise castration pain, genetic tools to select for polled cattle to avoid dehorning, and branding of less sensitive body areas with appropriate equipment. Routine anaesthetic and analgesic drug administration could be a solution, keeping in mind the inherent characteristics of extensive production systems. Finally, solutions will only be practically adopted if innovative research results are disseminated to farmers and technicians working in the field.

### 6.8.3 UK Case Study

The UK contributes only 1.6% to global beef production but meat from purebred beef cattle has a reputation for high meat-eating quality and animal welfare standards. The UK, along with France, Spain, and Ireland has the largest populations of beef cows in Europe (Hocquette and Chatellier 2011). Beef cattle in the UK typically calve in the spring and may be raised in a range of conditions from improved lowland pasture through to unimproved mountainous terrain. Calves born from dairy cows contribute 37% of UK registered beef calves, making the dairy-cross sector a significant part of the UK beef industry (Agriculture and Horticulture Development Board data from 2021). Cattle enclosed in lowland fields represent the least 'extensive' of pasture-based production systems due to the higher stocking density and constraints on environmental complexity and freedom to range. The UK does not experience dramatic regional variations in climate (see Australian case study) but, due to its northern latitude and associated climate, pasture quality deteriorates rapidly with altitude (Fig. 6.6). Thus, even within a small country, cattle face different challenges that are easier to mitigate in some environments than others. The UK has a large variety of breeds and crosses designed to thrive best in different environments, with greater use of Continental European breeds (e.g. Limousin, Simmental, Charolais) on improved pasture and native UK breeds elsewhere (e.g. Angus, Hereford, Shorthorn). Cattle may be fattened on grass or indoors on a concentrate diet. The majority of cattle are housed over the winter due to the

**Fig. 6.6** Cattle on a Scottish beef farm. Note the variation in topography and pasture quality from improved grassland in the field in the foreground through to rough grazing on the hill ground in the distance (Photo courtesy of Marianne Farish)

wet and windy climate in the UK that risks cold stress (Morgan et al. 2009) and poaching (i.e. damage) to the ground, although some are kept outdoors all year (known as 'outwintering').

### 6.8.3.1 Managing Climate: Cold Stress

The predominant form of thermal challenge for UK beef cattle is cold stress, although acute periods of heat stress can also occur. The effect of cold stress on cattle welfare has received much less attention than heat stress (Salama et al. 2016). This may have resulted from a perception that cattle are highly cold tolerant, but this view fails to give sufficient weight to the chilling effect of a combination of precipitation (rain and snow) and wind which frequently occurs in UK conditions.

Many UK beef farms are not profitable without financial support from government subsidy. Even with subsidy support, profit margins are low and farmers are, therefore, motivated to minimise production costs. Reductions in availability of low-cost bedding and high machinery costs have driven an increase in outwintering, especially for pregnant cows. The lower critical temperature of cattle depends upon many factors including their body condition, age, feed intake, and coat cover (Morgan et al. 2009), as well as behavioural means of thermoregulation (e.g. orientating themselves to maximise exposure to solar radiation and presenting a small area to the wind). In still, dry conditions, lower critical temperature can be −25 °C (−13°F) (NRC 1981). Winter temperatures in the UK average 2 °C (36°F) and few cattle ever experience temperatures below −15 °C (5°F) at any point in their life. Beef cattle in other countries (e.g. Canada, Russia) routinely experience temperatures substantially colder than this. However, the temperate maritime climate of the

UK combines cold with high levels of precipitation and wind speeds which can dramatically raise the lower critical temperature to as high as 13 °C (55°F) in a typical UK winter (Morgan et al. 2009). Therefore, for a large part of the winter cattle may be under cold stress unless able to find shelter. Conventional practice in the UK allows cows to lose body condition during the winter, thus imposing both nutritional restrictions and thermal stress on pregnant animals. Poaching of the ground by cattle feet and machinery can cause severe drainage problems in the pastures, which may force cattle to lie down in wet or muddy areas, compounding the effects of a cold ambient temperature by a wet and muddy coat (Figs. 6.7 and 6.8). Housing cattle in the winter to avoid cold stress can present its own welfare challenges. Slatted systems in which faeces and urine drain through the floor into a slurry pit, are common in parts of Scotland that are distant from sources of affordable bedding. Animals are reluctant to lie down on slatted concrete floors (Lowe et al. 2020). Compared to housing indoors on slats, outwintering on a pad of woodchips in Ireland led to improved lying behaviour and health outcomes despite the greater risk of chilling (Hickey et al. 2002).

Given how many cattle are produced under cold conditions, remarkably little is known about how cold stress affects their welfare. Cattle reduce their lying time on cold, wet surfaces and voluntarily make use of shelters (Olson and Wallander 2002; Tucker et al. 2007; Webster et al. 2008). Shelter use, lying time, and coat cleanliness may be useful welfare indicators, and it may be possible to approach cattle closely enough on lowland UK farms to see shivering. Excessive condition loss may occur although, if this is judged by eye from a distance, it is likely to be detected only when the animals have been in a state of negative energy balance for some time.

**Fig. 6.7** Outwintering in wet climates can rapidly lead to poaching (damage) to the ground, especially around feeders (Photo courtesy of Simon Turner)

**Fig. 6.8** Frozen poached ground risks lameness and may affect the ability to rest comfortably (Photo courtesy of Simon Turner)

These indicators should be developed further and implemented to allow best practice in reducing cold stress in beef cattle to be identified.

Outwintering of cattle should be restricted to areas where the ground is well drained, the breed used is well adapted, animals are in good condition, and shelter (artificial structures or trees) is available. Where feeders are provided, they should be relocated frequently to prevent poaching, or forage crops strip-grazed by regular movement of a temporary fence. Contingencies ought to be in place to relocate cattle during severe weather and to guarantee feed provision. Indeed, some assurance standards require evidence of contingency plans for outwintered cattle and provision of shelter (e.g. RSPCA Assured).

Given that cattle are either housed indoors or outdoors for prolonged periods that encompass days of widely varying weather, it is not clear which of these two environments has the largest net benefit for welfare (i.e. which maximises 'quality of life'; Lawrence et al. 2019). Whilst gross effects of cold stress on survival, physiology, and productivity have been studied, and behavioural choices indicate an aversion to cold conditions (Olson and Wallander 2002; Morgan et al. 2009), it is currently difficult to gauge how severely cold stress compromises welfare, as compared to other challenges. A more sensitive indicator of welfare impact could be obtained by examining how elastic their demand for shelter is when the costs of accessing it increases, for example by altering its distance from feeders. This work is yet to occur.

#### 6.8.3.2 Managing Nutritional Stress

In the UK, the effects of weather and climate on feed availability and quality are modest compared to other regions of the world due to the milder and more stable climatic conditions, such that grass growth and quality are typically not restricted as they are in harsher climates. Stocking densities are set to control grass height to close to the optimum. Beef cattle managed on improved pasture are kept at a higher stocking density than in other systems in the UK and elsewhere. Beef cattle are expected to gain body condition during the grazing season from spring to autumn when grass growth is the greatest. Nutritional stressors are greatest in neonatal calves, calves at weaning, and in cows. Our focus here will be on cows as their nutrition is often given a lower priority, likely because they are viewed as being robust to nutritional challenges and because seasonal weight loss may also be regarded as natural. Cows are the only animals in the system which are deliberately allowed to lose body condition over the winter. This is done to minimise feed costs and avoid dystocia associated with obesity. Even when wintered outdoors on fodder crops or deferred grazing, most beef cattle are reliant on being supplied with conserved forages to some extent. Body condition can fall excessively during the winter due to demands for heat production together with pregnancy which may not always be supported by the quality and quantity of conserved feed provided, or the remaining grass if outwintered on pasture. Farmers are recommended to body condition score cattle by palpation of the subcutaneous fat depth at key points in the year as a guide to making nutritional management decisions. However, a study of UK beef farmers (Rutherford 2013) showed that as few as 4% use this hands-on approach and around a third do not condition score at all. Furthermore, 32% of observed farms had ≥10% of cows classed as 'very lean' according to the Welfare Quality® recording approach. Additionally, cattle are sorted into winter housing groups according to body condition on only 30% of farms, meaning that overweight and underweight animals are managed in the same social group with the same quality of feed on most UK farms. Social science studies are needed to understand the constraints to adoption of condition scoring. The perception that obesity needs to be avoided at all costs to prevent dystocia may be a pervading issue that drives excessive condition loss and, if so, a more balanced message may be needed. There may also be a market for a low-cost technological solution that automatically condition scores cows during routine handling such as pregnancy testing and vaccination.

As discussed in the Australian case study, the extent to which falling body condition reflects the experience of hunger and impacts affective state is unknown. Even if modern cattle breeds inherited a legacy from their ancestors to tolerate seasonal weight loss, it is unclear whether more productive modern genotypes perceive nutrient restriction in the same way. In the UK, the recommended condition loss over winter is around 0.5 condition score points from a 5-point scale (Defra 2001). However, unpublished data from 35 commercial farms in 2017 indicates that 29% of cows lost an average of 1 condition score point which equates to 13% of their body weight (SAC 2011). A minority of animals lost more than one point over a period of only 2 months, indicating a rapid and significant drop in condition. It would seem reasonable to assume that, where weight loss is substantial and rapid, cattle may experience hunger or some other aversive state. Typically, cattle which lose the most condition are those that were overweight at the end of the preceding grazing season and may benefit from reduced risk of dystocia as a result of calving at a more modest body condition. However, this may indicate a conflict between welfare as defined by biological functioning versus that defined by feeling good. The effects of a falling plane of nutrition on calf foetal and post-natal development have been poorly studied, but the welfare consequences of poor management of body condition potentially extend to the progeny as well as the cow.

#### 6.8.3.3 Optimising Health Outcomes

A recent survey of UK farmers (AHDB 2021) showed that digital dermatitis, bovine viral diarrhoea (BVD), paratuberculosis (Johne's disease), and fasciolosis (liver fluke) are the diseases perceived to have the greatest impact on beef cattle welfare in extensive environments. Estimates of the economic impact of such endemic diseases have wide margins of error due to poor understanding of the effects of the disease, variation in reporting, and annual fluctuations in incidence (Bennett 2003). For example, BVD was estimated to cost the UK cattle industry between £9.6 and £59.5 million from lost output and increased expenditure (values from Bennett (2003), adjusted for inflation to represent costs in 2020).

Most endemic diseases in livestock attract little political debate or economic analysis in the UK but often persist at a high prevalence and are tolerated as a routine risk (Carslake et al. 2011). This situation substantially compromises animal welfare in the UK and elsewhere (FAWC 2012). Where concerted action is taken, the results can be positive. For example, the mandatory BVD eradication programme in Scotland reduced the number of cattle exposed to BVD from 40% to 12% in 7 years (Scottish Government 2016). However, membership of voluntary BVD schemes in other parts of the UK appears not to improve farmer knowledge of appropriate biosecurity measures (Azbel-Jackson et al. 2018); a situation noted for beef and sheep health schemes more generally in the UK (Heffernan et al. 2008). In addition, encouraging UK farmers to become members of voluntary health schemes in the first place can be challenging and there is significant scope to improve membership of the schemes targeting the most economically damaging diseases (e.g. the Cattle Health Certification Standards; FAWC 2019).

Effective biosecurity is particularly challenging in the UK extensive beef sector. Farms commonly have fields that are immediately adjacent to grazed fields belonging to other farms. Where coordinated action is lacking, individual local action by farmers can be thwarted by the prevalence of infected herds that reintroduce pathogens by fence-line contact or movement of stock between farms. Heffernan et al. (2008) have shown that UK beef and sheep farmers are dismissive of many forms of biosecurity and regard endemic disease as a problem only for 'bad' farmers. The UK climate provides ideal conditions for pathogen survival and spread and contributes to the challenge of prevention and treatment of fasciolosis and digital dermatitis, amongst other diseases. Endemic diseases are therefore difficult to manage through a combination of a wet climate, frequent animal movements between farms, close proximity of grazing stock in adjacent farms, and inadequate biosecurity practices. The spread of exotic insect borne diseases into, and then throughout, the UK is being aided by climate change (FAWC 2019) and will become increasingly problematic.

There is an evident need to improve uptake of existing biosecurity measures. This requires a neutral communication platform with credible messages delivered by trusted parties that maximises farmer input to the design of the strategy (Heffernan et al. 2008). Strategies must also effectively engage an increasing proportion of smallholder and hobby farmers who may not be reached through other channels (Scottish Government 2016).

#### 6.8.3.4 Reducing the Impact of Painful Husbandry Procedures

Hot iron branding is prevented by the Mutilations (Permitted Procedures) Regulations 2007 but freeze-branding is allowed in the UK. Freeze-branding induces fewer behavioural changes indicative of pain than hot iron branding, but more than sham branding (Schwartzkopf-Genswein et al. 1997).

Polled beef breeds are available but polled genotypes are less common within the dairy industry. This is relevant to the UK beef sector, as crossbred dairy animals represent a significant part of the UK beef industry and disbudding (removal of developing horn buds) remains a routine procedure on most dairy farms. Disbudding is predominantly performed by hot iron cautery or a disbudding scoop and it is a legal requirement to use anaesthesia (Protection of Animals (Anaesthetics) Act 1954). Caustic disbudding paste is allowed during the first week of life but strongly discouraged by government welfare codes (e.g. Defra 2003). Despite results of a survey showing that veterinarians regard disbudding and dehorning as highly painful, only around 30% of UK cattle veterinarians routinely prescribe use of non-steroidal anti-inflammatories (NSAIDs) to reduce post-procedure pain (Remnant et al. 2017).

Eighty-five percent of slaughtered male cattle in the UK are castrated by surgery, or by impeding blood flow using a Burdizzo clamp or rubber ring. Legislation limits use of rubber rings to the first week of life and requires castration of animals over 8 weeks of age to be performed by a veterinarian with the use of anaesthesia. Compliance with these restrictions needs to be improved and castration without pain relief below 8 weeks of age is legal despite evidence that it causes pain and licensed analgesics and anaesthetics are available for use (FAWC 2019). The European Union and UK currently lack products licensed for immunocastration of cattle (Monleon et al. 2020). The use of NSAIDs to reduce post-procedure pain is increasing (e.g. two thirds of UK cattle veterinarians report greater use of NSAIDs in the previous 5–10 years; Remnant et al. 2017), but is still inadequate. NSAIDs are prescribed for pain relief following castration by only 30% of UK cattle veterinarians despite the same veterinarians scoring the associated pain at 7 out of a maximum of 10 points (Remnant et al. 2017). The higher costs of labour and increases in stress caused to animals by repeated handling are likely to be barriers to post-procedure pain management in extensive conditions. Raising awareness of how pain management can support recovery and long-term productivity should be a focus to improve adoption within the industry (Laven et al. 2012).

#### 6.8.3.5 Summary

The welfare impacts of cold stress are under-studied but the assumption that cattle are adequately adapted to the cold, wet and windy conditions of countries such as the UK should be robustly tested by research. The challenges of nutritional management are less severe in the UK than countries with more variable or extreme climates. However, considerable scope exists to better manage body condition through routine condition scoring and use of the data in decisions regarding the need to provide supplementary feed. Endemic diseases are costly to animal welfare and economic and environmental sustainability in the UK beef sector. Evidence shows that mandatory eradication schemes can be successful but there is a tendency to accept endemic disease as an inescapable fact of farming and to adopt poor biosecurity precautions. This position needs to be challenged. Painful procedures are routine in the UK extensive beef sector without provision of adequate pain relief in most cases. Effort is required to demonstrate how pain relief can not only benefit animal welfare, but facilitate recovery and productivity.

## 6.9 Future Priorities and Conclusions

These case studies illustrate common threads that affect beef cattle welfare in extensive systems globally. Central to this is that innovative solutions are needed to improve animal welfare in unpredictable and geographically large and diverse physical environments. This must harness new developments afforded by digital technology. It is encouraging that new solutions are being developed, often in order to improve productivity but, in doing so, with potential to improve some aspects of animal welfare simultaneously. The role of existing solutions also must not be forgotten, and there is considerable scope to improve adoption of interventions in the beef sector (Hocquette et al. 2018). As a case in point, inadequate food availability was highlighted as a major threat to welfare in all case studies. Where animals are handled, which may be infrequently, body condition scoring is a simple, quick, and free method to assess animal welfare and to proactively manage food provision. However, adoption of condition scoring and its use in management decisions in the beef industry falls behind that in other sectors (e.g. dairy and sheep). To date, there are too few examples of robust social science methods to understand why beef farmers fail to adopt such management solutions to improve animal welfare. As we develop new solutions, social science ought to be embedded in the process. Specifically, the opportunity exists to make greater use of human behavioural change theories, such as the COM-B model or Behavioural Change Wheel (Michie et al. 2011; Carroll and Groarke 2019), and participatory research, to develop management interventions that have greater likelihood of adoption.

The acceptance of endemic disease and painful management procedures as a routine part of production needs to be continually challenged. Progress is being made in these areas, but they would benefit from greater policy focus. The lack of policy interest or closely defined legislative requirements likely reflects public perception that animal welfare is safeguarded by the ability to lead a natural life, and hence the smaller public focus placed on extensive than intensive systems. Low profitability in the extensive beef sector constrains management change. However, it also ought to drive innovation in reducing production inefficiencies caused by poor health and other environmental challenges. This will only be achieved when evidence is provided of the production and economic penalties of poor health and welfare outcomes. Even for challenges with well-documented biological impacts, such as a specific disease, we have a poor understanding of its economic impacts on a farming business. For other welfare challenges, the biological consequences themselves are also poorly known. For example, a cow may survive a nutritional challenge, but the effect on her calf through compromised foetal development and milk intake is poorly understood (Zago et al. 2019, 2020). To motivate management change, we must understand the totality of biological outcomes from welfare challenges and estimate the economic consequences. In the example of the nutritional challenge, if calf productivity or survival is impacted by poor maternal nutrition, the rational economic decision may be to provide extra feed to the cow; a decision that would not be made based purely on consideration of the outcome for the cow.

The greatest welfare harms in extensive systems result from unpredicted events for which there are no contingencies. The choice of contingencies may be severely constrained in an extensive environment, but greater effort is needed to predict extreme environmental events and to plan accordingly. Climate change will present more frequent and more extreme weather events, as well as more generally affecting pasture quality and pathogen exposure. Choice of breed and selection for local environmental conditions has played a major part in reducing welfare challenges, and continued work in this area is important. Ultimately, however, there is a limit to biological adaptation and contingencies are needed for even the most adapted animals.

As well as responding to climate change, the extensive beef sector will need to reduce its greenhouse gas emissions. The effect of GHG mitigation methods on animal welfare are not well studied (Llonch et al. 2017). Greater intensification of beef production systems to maximise output in kilograms of product per unit of $CO_2$ equivalent will exacerbate the welfare challenges associated with these intensive systems (Shields and Orme-Evans 2015). The effects of GHG reduction methods on extensive beef cattle welfare are poorly understood, although any change in breeding, feed quality and quantity, and farm location could impact animal welfare. However, improved health and reduction of stress could benefit welfare whilst making a valuable contribution to improving productivity for the same or smaller amount of GHGs produced. The industry will also have to adapt to changing markets, characterised by a projected reduction in beef consumption in developed countries and increased consumption in developing countries, with an associated relocation of production to meet demand. The nature and geographical location of welfare challenges facing extensive beef cattle are not static and the public's prioritisation of welfare amongst other societal concerns is also likely to change in ways that are hard to predict. Improving animal welfare is often seen as costly, but it is an important contributor to sustainability. Animal welfare needs to become more central to helping the extensive beef sector meet the environmental, economic, and social challenges it faces.

**Acknowledgements** The authors would like to thank Júlio Otávio Jardim Barcellos (Universidade Federal do Rio Grande do Sul, UFRGS, Porto Alegre) and Matheus José Rodrigues Paranhos da Costa (UniversidadeEstadualPaulista, UNESP Jaboticabal, São Paulo) for their valuable input in the Brazilian case studies discussion.

## References

ABIEC (2020) Beef Report – Brazilian Livestock Profile 2020. ABIEC, Rio de Janeiro, p 50

AHDB (2019) The UK Cattle Yearbook. Agriculture and Horticulture Development Board. Available at (accessed 02/02/2023): https://projectblue.blob.core.windows.net/media/Default/Imported%20Publication%20Docs/CattleYearbook2019_190718_WEB_FINAL.pdf

AHDB (2021) Ruminant Health and Welfare survey results. Presentation available at (accessed 02/02/2023): https://ahdb.org.uk/events/ruminant-health-and-welfare-survey-results

Andreotti R, Garcia M, Koller WW (2019) Controle estratégico dos carrapatos em bovinos. In: Andreotti R, Garcia M, Koller WW (eds) Carrapatos na cadeia produtiva de bovinos. Embrapa, Brasília, Brasil, p 240

ANUALPEC (2018) Anuário da Pecuária Brasileira. Informa Economics FNP, São Paulo, Xp

Appleby MC (1996) Can we extrapolate from intensive to extensive conditions? Appl Anim Behav Sci 49:23–27

Azbel-Jackson L, Heffernan C, Gunn G, Brownlie J (2018) Exploring the role of voluntary disease schemes on UK farmer bio-security behaviours: findings from the Norfolk-Suffolk Bovine Viral Diarrhoea control scheme. PLoS ONE 13(2):e0179877. https://doi.org/10.1371/journal.pone.0179877

Beatty DT, Barnes A, Taylor E, Pethick D, McCarthy M, Maoloney SK (2006) Physiological responses of Bos taurus and Bos indicus cattle to prolonged continuous heat and humidity. J Anim Sci 84:972–985

Bennett R (2003) The direct costs of livestock disease: the development of a system of models for the analysis of 30 endemic livestock diseases in Great Britain. J Agric Econ 54:55–71

Berckmans D (2014) Precision livestock farming technologies for welfare management in intensive livestock systems. OIE Sci Tech Rev 33:189–196. https://doi.org/10.20506/rst.33.1.2273

Blokhuis HJ, Jones RB, Geers R, Miele M, Viessier I (2003) Measuring and monitoring animal welfare: transparency in the food product quality chain. Anim Welf 12:445–455

Bollongino R, Burger J, Powell A, Mashkour M, Vigne J-D, Thomas MJ (2012) Modern Taurine cattle descended from small number of Near eastern founders. Mol Biol Evol Lett 29:2101–2104. https://doi.org/10.1093/molbev/mss092

Bortolussi G, McIvor JG, Hodgkinson JJ, Coffey SG, Holmes CR (2005) The northern Australian beef industry, a snapshot: 1. Regional enterprise activity and structure. Aust J Expt Agric 45:1057–1073. https://doi.org/10.1071/EA03096

Broom DM (1991) Animal welfare: concepts and measurement. J Anim Sci 69:4167–4175

Bundle EA, Bray HJ, Ankeny RA (2021) "Of course we care!": a qualitative exploration of Australian livestock producers' understandings of farm animal welfare issues. J Rural Stud 83:50–59

Burns BM, Fordyce G, Holroyd RG (2010) A review of factors that impact on the capacity of beef cattle females to conceive, maintain a pregnancy and wean a calf—implications for reproductive efficiency in northern Australia. Anim Reprod Sci 122:1–22

Burrow HM, Mans BJ, Cardoso FF, Birkett MA, Kotze AC, Hayes BJ, Mapholi N, Dzamal K, Marufu MC, Githaka NW, Djikeng A (2019) Towards a new phenotype for tick resistance in beef and dairy cattle: a review. Anim Prod Sci 59:1401–1427. https://doi.org/10.1071/AN18487

Campell DLM, Marini D, Lea JM, Keshavarzi H, Dyall TR, Lee C (2021) The application of virtual fencing technology effectively herds cattle and sheep. Anim Prod Sci. https://doi.org/10.1071/AN20525

Canozzi MEA, Mederos A, Manteca X, Turner S, McManus C, Zago D, Barcellos JOJ (2017) A meta-analysis of cortisol concentration, vocalization, and average daily gain associated with castration in beef cattle. Res Vet Sci 114:430–443. https://doi.org/10.1016/j.rvsc.2017.07.014

Canozzi MEA, Mederos A, Turner S, Manteca CX, McManus DC, Menegassi SRO, Barcellos JOJ (2018) Dehorning and welfare indicators in beef cattle – a meta-analysis. Anim Prod Sci 59:801–814. https://doi.org/10.1071/AN17752

Canozzi MEA, Borges JAR, Barcellos JOJ (2020) Attitudes of cattle veterinarians and animal scientists to pain and painful procedures in Brazil. Prevent Vet Med 177:104909. https://doi.org/10.1016/j.prevetmed.2020.104909

Cardoso CS, von Keyserlingk MA, Hötzel MJ (2016) Trading off animal welfare and production goals: Brazilian dairy farmers' perspective on calf dehorning. Livest Sci 187:102–108. https://doi.org/10.1016/j.livsci.2016.02.010

Cardoso CS, von Keyserlingk MA, Hötzel MJ (2018) Views of dairy farmers, agricultural advisors, and lay citizens on the ideal dairy farm. J Dairy Sci 201:1811–1821. https://doi.org/10.3168/jds.2018-14688

Carroll GA, Groarke JM (2019) The importance of social sciences in reducing tail biting prevalence in pigs. Animals 9:591. https://doi.org/10.3390/ani9090591

Carslake D, Grant W, Green LE, Caves J, Greaves J, Keeling M, McEldowney J, Weldegebriel H, Medley GF (2011) Endemic cattle diseases: comparative epidemiology and governance. Philos Trans R Soc B 366:1975–1986. https://doi.org/10.1098/rstb.2010.0396

Carvalho PFC (2013) Harry Stobbs Memorial Lecture: can grazing behavior support innovations in grassland management? Trop Grasslands – ForrajesTropicales 1:137–155

Cesar IM, Queiroz HP, Thiago LRLS, Cassales FLG, Costa FP (2005) Sistemas de produção de gado de corte no Brasil: uma descrição com ênfase no regime alimentar e no abate. Embrapa Gado de Corte, Campo Grande, p 40

Charlton GL, Rutter SM, East M, Sinclair LA (2011) Preference of dairy cows: indoor cubicle housing with access to a total mixed ration vs. access to pasture. Appl Anim Behav Sci 130:1–9

Clark B, Stewart GB, Panzone LA, Kyriazakis I, Frewer LJ (2016) A systematic review of public attitudes, perceptions and behaviours towards production diseases associated with farm animal welfare. J Agric Environ Ethics 29:455–478. https://doi.org/10.1007/s10806-016-9615-x

Colditz IG, Ferguson DM, Collins T, Matthews M, Hemsworth PH (2014) A prototype tool to enable farmers to measure and improve the welfare performance of the farm animal enterprise: the Unified Field Index. Animals 4:446–462. https://doi.org/10.3390/ani4030446

Cornish A, Raubenheimer D, McGreevy P (2016) What we know about the public's level of concern for farm animal welfare in food production in developed countries. Animals 6:74. https://doi.org/10.3390/ani6110074

Cutulle C, Jonsson NN, Seddon J (2009) Population structure of Australian isolates of the cattle tick *Rhipicephalus (Boophilus) microplus*. Vet Parasit 161:283–291. https://doi.org/10.1016/j.vetpar.2009.01.005

D'Occhio MJ (2013) GonaCon™ trial in heifers. MLA Final Report B.AWW.0194. Meat and Livestock Australia, Sydney

Defra (Department for Food, Environment and Rural Affairs) (2001) Condition scoring of beef suckler cows and heifers. PB6491. Available at (accessed 02/02/2023): https://assets.publishing.service.gov.uk/government/uploads/system/uploads/attachment_data/file/69370/pb6491-cattle-scoring-020130.pdf

Defra (Department for Food, Environment and Rural Affairs) (2003) Code of Recommendations for the Welfare of Livestock: Cattle. Department for Environment, Food and Rural Affairs. PB7949. Available at (accessed 02/02/2023): https://assets.publishing.service.gov.uk/government/uploads/system/uploads/attachment_data/file/69368/pb7949-cattle-code-030407.pdf

Dockès AC, Kling-Eveillard F (2006) Farmers' and advisors' representations of animals and animal welfare. Livest Sci 103:243–249. https://doi.org/10.1016/j.livsci.2006.05.012

EMBRAPA (2021) Revista da Embrapa Pecuária Sul. Embrapa, Bagé. Published by EMBRAPA, Bagé, Brasil, 40 p. Available at (accessed 04/02/2023): https://www.embrapa.br/documents/1355035/2751150/Revista+mar%C3%A7o+2021/6442d7ec-b31c-612e-eb64-590fc2bc254b

Euclides Filho K (2000) Produção de bovinos de corte e o trinômio genótipo-ambiente-mercado. Embrapa Gado de Corte, Campo Grande, p 61

Euclides Filho K (2004) Supply chain approach to sustainable beef production from a Brazilian perspective. Livest Prod Sci 90:53–61

Euclides VPB, Valle CB, Macedo MCM, Almeida RG, Montagner DB, Barbosa RA (2010) Brazilian scientific progress in pasture research during the first decade of XXI century. Rev Bras Zootec 39:151–168

FAWC (Farm Animal Welfare Council) (1993) Second Report on Priorities for Research and Development in Farm Animal Welfare. Available at (accessed 04/02/2023) https://edepot.wur.nl/134980

FAWC (Farm Animal Welfare Committee) (2012) Farm animal welfare: health and disease. Available at (accessed 04/02/2023): https://www.gov.uk/government/publications/fawc-report-on-farm-animal-welfare-health-and-disease

FAWC (Farm Animal Welfare Committee) (2019) Opinion on the welfare of cattle kept for beef production. Available at (accessed 02/02/2023): https://assets.publishing.service.gov.uk/government/uploads/system/uploads/attachment_data/file/902524/FAWC_Opinion_on_the_welfare_of_cattle_kept_for_beef_production.pdf

Ferguson DM (2014) Key features of 'environmental fit' that promote good animal welfare in different husbandry systems. Rev Sci Tech Off Int Epiz 33:161–169

Ferraz JBG, Felício PE (2010) Production systems – an example from Brazil. Meat Sci 84:238–243. https://doi.org/10.1016/j.meatsci.2009.06.006

Flori L, Gonzatti MI, Thevenon S, Chantal I, Pinto J, Berthier D, Aso PM, Gautier M (2012) A quasi-exclusive European ancestry in the Senepol tropical cattle breed highlights the importance of the slick locus in tropical adaptation. PLoS ONE 7(5):e36133

Fraser D, Duncan IJH, Edwards SA, Grandin D, Gregory NG, Guyonnet V, Hemsworth PH, Huertas SM, Huzzey JM, Mellor DJ, Mench JA, Spinka M, Whay HR (2013) General principles for the welfare of animals in production systems: the underlying science and its application. Vet J 198:19–27. https://doi.org/10.1016/j.tvjl.2013.06.028

Goddard P (2016) The health of livestock in extensive systems. In: Villalba JJ, Manteca X (eds) Animal welfare in extensive production systems. 5m Publishing, Sheffield. ISBN 978-1-910455-54-8

Haskell MJ, Masłowska K, Bell DJ, Roberts DJ, Langford FM (2013) The effect of a view to the surroundings and microclimate variables on use of a loafing area in housed dairy cattle. Appl Anim Behav Sci 147:28–33

Heffernan C, Nielsen L, Thomson K, Gunn G (2008) An exploration of the drivers to bio-security collective action among a sample of UK cattle and sheep farmers. Prevent Vet Med 87:358–372

Hickey MC, French P, Grant J (2002) Out-wintering pads for finishing beef cattle: animal production and welfare. Anim Sci 75:447–458

Hine B, Ingham A, Dominik S, Colditz I (2016) Assessment of immune competence in beef cattle. Final Report B.STU.0244. Meat and Livestock Australia, Sydney. https://www.mla.com.au/download/finalreports?itemId=3344

Hocquette J-F, Chatellier V (2011) Prospects for the European beef sector over the next 30 years. Anim Front 1:20–28

Hocquette J-F, Ellies-Oury M-P, Lherm M, Pineau C, Deblitz C, Farmer L (2018) Current situation and future prospects for beef production in Europe — a review. Asian-Australasian J Anim Sci 31:1017–1035. https://doi.org/10.5713/ajas.18.0196

Hogan JP, Phillips CJC (2016) Starvation in ruminant livestock. In: Phillips CJC (ed) Nutrition and welfare of farm animals. Springer International Publishing, Switzerland. ISBN 978-3-319-27354-9 ISBN 978-3-319-27356-3 (eBook). https://doi.org/10.1007/978-3-319-27356-3

Hötzel MJ, Sneddon JN (2013) The role of extensionists in Santa Catarina, Brazil, in the adoption and rejection of providing pain relief to calves for dehorning. J Dairy Sci 96:1535–1548. https://doi.org/10.3168/jds.2012-5780

Hudson C, Whay H, Huxley J (2008) Recognition and management of pain in cattle. In Pract 30:126–134

Huson HJ, Kim E-S, Godfrey RW, Olson TA, McClure MC, Chase CC, Rizzi R, O'Brien AMP, Van Tassell CP, Garcia JF, Sonstegard TS (2014) Genome-wide association study and ancestral origins of the slick-hair coat in tropically adapted cattle. Front Genet 5:101

IBGE (2019) Instituto Brasileiro de Geografia e Estatística. Censo Agropecuario 2017. IBGE, Rio de Janeiro, 109

Jaurena M, Porcille V, Baptista R, Carriquiry E, Días S (2018) La Regla Verde: una herramienta para el manejo del campo natural. Revista INIA 54:26–29

Ketelaar-de Lauwere CC, Ipemaa AH, Lokhorsta C, Metza JHM, Noordhuizen JPTM, Schouten WGP, Smits AC (2000) Effect of sward height and distance between pasture and barn on cows' visits to an automatic milking system and other behaviour. Livest Prod Sci 65:131–142

Klafke GM, Webster A, Dall'Agnol B, Pradel E, Silva J, La Canal LH, Becker M, Osório MF, Mansson M, Barreto R, Scheffer R, Souza UA, Corassini VB, Santos J, Reck J, Martins JR (2016) Multiple resistance to acaricides in field populations of *Rhipicephalus microplus* from Rio Grande do Sul state, Southern Brazil. Ticks Tick-Borne Dis 8:73–80. https://doi.org/10.1016/j.ttbdis.2016.09.019

Krohn CC, Munksgaard L, Jonasen B (1992) Behaviour of dairy cows kept in extensive (loose housing/pasture) or intensive (tie stall) environments. I. Experimental procedure, facilities, time budgets - diurnal and seasonal conditions. Appl Anim Behav Sci 34:37–47

Kühl S, Busch G, Gauly M (2021) How should beef be produced? Consumer expectations and views on local beef production in South Tyrol (Italy). Br Food J 123:1578–1595

Lane J, Jubb T, Shephard R, Webb-Ware J, Fordyce G (2015) Priority list of endemic diseases for the Australian red meat industries. Final Report B.AHE.0010, Meat and Livestock Australia, Sydney

Laundré JW (2016) Predation. In: Villalba JJ, Manteca X (eds) Animal welfare in extensive production systems. 5M Publishing Ltd, Sheffield

Laven R, Chambers P, Stafford K (2012) Using non-steroidal anti-inflammatory drugs around calving: maximizing comfort, productivity and fertility. Vet J 192:8–12

Lawrence AB, Vigors B, Sandøe P (2019) What is so positive about positive animal welfare? A critical review of the literature. Animals 9:783. https://doi.org/10.3390/ani9100783

Lay DC Jr, Friend TH, Randel RD, Bowers CL, Grissom KK, Jenkins OC (1992) Behavioral and physiological effects of freeze or hot-iron branding on crossbred cattle. J Anim Sci 70:330–336

Lee C, Fisher AD, Colditz IG, Lea JM, Ferguson DM (2013) Preference of beef cattle for feedlot of pasture environments. Appl Anim Behav Sci 145:53–59

Lee C, Lea J, Colditz I, Fisher A, Ferguson F (2014) Assessing cattle motivation for access to pasture or feedlot environments. Meat and Livestock Australia final report of project B.FLT.0149. Available at (accessed 04/02/2023): https://www.mla.com.au/contentassets/e61ffb7645594ab1be38b5170c678a22/b.flt.0149_final_report.pdf

Legrand AL, von Keyserlingk MAG, Weary DM (2009) Preference and usage of pasture versus free-stall housing by lactating dairy cattle. J. Dairy Sci 92:3651–3658. https://doi.org/10.3168/jds.2008-1733

Lew-Tabor AE, Rodriguez Valle M (2016) A review of reverse vaccinology approaches for the development of vaccines against ticks and tick-borne diseases. Tick Tick-Borne Dis 7:573–585. https://doi.org/10.1016/j.ttbdis.2015.12.012

Llonch P, Haskell MJ, Dewhurst RJ, Turner SP (2017) Current available strategies to mitigate greenhouse gas emissions in livestock systems: an animal welfare perspective. Animals 11:274–284

Lowe DE, Gordon AW, Lively FA (2020) Effect of overlaying rubber on fully slatted concrete floors on hoof health and lying postures in finishing dairy-origin bulls offered two contrasting diets. Animals 14:1043–1051

Manning J, Cronin G, González L, Hall E, Merchant A, Ingram L (2017) The behavioural responses of beef cattle (Bos taurus) to declining pasture availability and the use of GNSS technology to determine grazing preference. Agriculture 7:45. https://doi.org/10.3390/agriculture7050045

Manteca X, Smith AJ (1994) Effects of poor forage conditions on the behaviour of grazing ruminants. Trop Anim Health Prod 26:129–138. https://doi.org/10.1007/BF02241068

Matthews LR (1996) Animal welfare and sustainability of production under extensive conditions: a non-EU perspective. Appl Anim Behav Sci 49:41–46

McLean I, Holmes P (2015) Improving the performance of northern beef enterprises. Key findings for producers from the northern beef report, 2nd edn. Meat and Livestock Australia. ISBN 9781740363006. Available at (accessed 04/02/2023): https://futurebeef.com.au/wp-content/uploads/Improving-the-performance-of-northern-beef-enterprises.pdf

McManus C, Prescott E, Paludo GR, Bianchini E, Louvandini H, Mariante AS (2009) Heat tolerance in naturalized Brazilian cattle breeds. Livest Sci 120:256–264. https://doi.org/10.1016/j.livsci.2008.07.014

Mellor DJ (2015) Enhancing animal welfare by creating opportunities for positive affective engagement. NZ Vet J 63:3–8

Mellor DJ (2016) Updating animal welfare thinking: moving beyond the "Five Freedoms" to a "Life worth living". Animals 6:21. https://doi.org/10.3390/ani6030021

Mellor DJ, Beausoleil NJ (2015) Extending the 'Five Domains' model for animal welfare assessment to incorporate positive welfare states. Anim Welf 24:241–253. https://doi.org/10.7120/09627286.24.3.241

Menegassi SRO, Barcellos JOJ, Días EA, Koetz C Jr, Pereira GR, Peripolli V, McManus C, Canozzi MEA, Lopes FG (2015) Scrotal infrared digital thermography as a predictor of seasonal effects on sperm traits in Braford bulls. Int J Biomet 59:357–364. https://doi.org/10.1007/s00484-014-0847-z

Michie S, Van Stralen MM, West R (2011) The behaviour change wheel: a new method for characterising and designing behaviour change interventions. Implement Sci 6:42

Ministério da Agricultura, Pecuária e Abastecimento (2017) Instrução Normativa n. 10, de 3 de março de 2017. Diário Oficial da União, DF, Brasil

Miraballes C, Riet-Correa F (2018) A review of the history of research and control of Rhipicephalus (Boophilus) microplus, babesiosis and anaplasmosis in Uruguay. Exp Appl Acarol 75:383–398. https://doi.org/10.1007/s10493-018-0278-3

Monleon E, Noya A, Garza MC, Ripoll G, Sanz A (2020) Effects of an anti-gonadotrophin releasing hormone vaccine on the morphology, structure and function of bull testes. Theriogenology 141:211–218

Morgan CA, McIlvaney K, Dwyer CM, Lawrence AB (2009) Assessing the welfare challenges to out-wintered pregnant suckler cows. Animals 3:1167–1174

NRC (National Research Council) (1981) Effect of environment on nutrient requirements of domestic animals. National Academy Press, Washington, DC

O'Neill CJ, Swain DL, Kadamideen HN (2010) Evolutionary process of Bos taurus cattle in favourable versus unfavourable environments and its implications for genetic selection. Evol Applic 3:422–433

Olson BE, Wallander RT (2002) Influence of winter weather and shelter on activity patterns of beef cows. Can J Anim Sci 82:491–502

Petherick JC (2005) Animal welfare issues associated with extensive livestock production: the northern Australian beef cattle industry. Appl Anim Behav Sci 92:211–234. https://doi.org/10.1016/j.applanim.2005.05.009

Petherick JC, McCosker K, Mayer DG, Letchford P, McGowan M (2013) Evaluation of the impacts of spaying by either the dropped ovary technique or ovariectomy via flank laparotomy on the welfare of Bos indicus beef heifers and cows. J Anim Sci 91:382–394. https://doi.org/10.2527/jas.2012-5164

Pfister JA, Green BT, Welch KD, Provenza FD, Cook D (2016) Impacts of toxic plants on the welfare of grazing livestock. In: Villalba JJ, Manteca X (eds) Animal welfare in extensive production systems. 5M Publishing Ltd, Sheffield

Provenza FD (2008) What does it mean to be locally adapted and who cares anyway? J Anim Sci 86:271–284. https://doi.org/10.2527/jas.2007-0468

Ramler JP, Hebblewhite M, Kellenberg D, Sime C (2014) Crying wolf? A spatial analysis of wolf location and depredations on calf weight. Am J Agr Econ 96:631–656. https://doi.org/10.1093/ajae/aat100

Randhawa IAS, Burns BM, McGowan MR, Porto-Neto LR, Hayes BJ, Ferretti R, Schutt KM, Lyons RE (2020) Optimized genetic testing for polledness in multiple breeds of cattle. Genes Genomes Genetics 10:539–544. https://doi.org/10.1534/g3.119.400866

Remnant JG, Tremlett A, Huxley JN, Hudson CD (2017) Clinical attitudes to pain and use of analgesia in cattle: where are we 10 years on? Vet Rec 181(15):400. https://doi.org/10.1136/vr.104428

Ritchie H, Roser M (2017) Meat and dairy production. Published online at OurWorldInData.org. Retrieved from: 'https://ourworldindata.org/meat-production' [Online Resource] on 03/05/2021

Roche JR, Friggens NC, Kay JK, Fisher MW, Stafford KJ, Berry DP (2010) Body condition score and its association with dairy cow productivity, health and welfare. J Dairy Sci 92:5769–5801. https://doi.org/10.3168/jds.2009-2431

Rodrigues VS, Bonatte P Jr, Garcia MV, Higa LOS, Piña FTB, Zimmermann NP, Duarte PO, Barros JC, Andreotti R (2018) Efficacy profile of Cypermethrin and Chlorpyrifos based acaricides on *Rhipicephalus microplus* control on cattle in the rearing phase, naturally infested and exposed to tick fever agents in central Brazil. Vet Parasit: Regional Stud Rep 12:43–48. https://doi.org/10.1016/j.vprsr.2018.02.001

Rutherford KMD (2013) Early environment effects on animal welfare, health and productivity. Final report for Defra project AW0509. Available at (accessed 02/02/2023): http://sciencesearch.defra.gov.uk/Default.aspx?Menu=Menu&Module=More&Location=None&Completed=0&ProjectID=16389

Rutter SM (2014) Smart technologies for detecting animal welfare status and delivering health remedies for rangeland systems. OIE Sci Tech Rev 33:181–187. https://doi.org/10.20506/rst.33.1.2274

SAC (Scottish Agricultural College) (2011) Ensuring good welfare for out-wintered suckler cows. SAC Technical Note TN634. ISSN 0142 7695. ISBN 1 85482 901 7. Available at (accessed 02/02/2023): https://ww1.sruc.ac.uk/media/3s3hmckg/tn634-welfare-winter-suckler.pdf

Salama AAK, Caja G, Hamzaoui S, Such X, Albanell E, Badaoui B, Loor JJ (2016) Thermal stress in ruminants: responses and strategies for alleviation. In: Villalba JJ, Manteca X (eds) Animal welfare in extensive production systems. 5M Publishing Ltd., Sheffield. ISBN 978-1-910455-54-8

Schmidek A, Durán H, Paranhos da Costa MJ (2009) Boas práticas de manejo: identificação. FUNEP, Jaboticabal, p 39

Schutz KE, Cox NR, Matthews LR (2008) How important is shade to dairy cattle? Choice between shade or lying following different levels of lying deprivation. Appl Anim Behav Sci 114:307–318

Schwartzkopf-Genswein KS, Stookey JM, Welford R (1997) Behavior of cattle during hot-iron and freeze-branding and the effects on subsequent handling ease. J Anim Sci 75:2064–2072
Scottish Government (2016) Animal health and welfare in the livestock industry: strategy 2016-2021. Available at (accessed 02/02/2023): https://www.gov.scot/binaries/content/documents/govscot/publications/strategy-plan/2016/03/animal-health-welfare-livestock-industry-strategy-2016-2021/documents/00497126-pdf/00497126-pdf/govscot%3Adocument/00497126.pdf
Shields S, Orme-Evans G (2015) The impacts of climate change mitigation strategies on animal welfare. Animals 5:361–394. https://doi.org/10.3390/ani5020361
Silanikove S (2000) Effects of heat stress on the welfare of extensively managed domestic ruminants. Livest Prod Sci 67:1–18. https://doi.org/10.1016/S0301-6226(00)00162-7
Smid AMC, Weary DM, von Keyserlingk MAG (2020) The influence of different types of outdoor access on dairy cattle behavior. Front Vet Sci 7:257. https://doi.org/10.3389/fvets.2020.00257
Špinka M, Wemelsfelder F (2011) Environmental challenge and animal agency. CAB International, Anim Welf Wallingford, pp 27–44
Spooner JM, Schuppli CA, Fraser D (2016) Attitudes of Canadian beef producers towards animal welfare. Anim Welf 21:273–283
Stafford KJ, Mellor DJ (2005) The welfare significance of the castration of cattle: a review. NZ Vet J 53:271–278. https://doi.org/10.1080/00480169.2005.36560
Stafford KJ, Chambers JP, Mellor DJ (2006) The alleviation of pain in cattle: a review. CAB Rev Perspect Agric Vet Sci Nutr Nat Resour 1:7. https://doi.org/10.1079/PAVSNNR20061032
Stampa E, Chipmann-Schwarze C, Hamm U (2020) Consumer perceptions, preferences and behavior regarding pasture-raised livestock products: a review. Food Qual Pref 82:103872
Tedeschi LO, Greenwood PL, Halachmi I (2021) Advancements in sensor technology and decision support intelligent tools to assist smart livestock farming. J Anim Sci 99:1–11
Temple D, Manteca X (2020) Animal welfare in extensive production systems is still an area of concern. Front Sustain Food Syst 4:545902. https://doi.org/10.3389/fsufs.2020.545902
Thompson T, Litchfield F (2020) Australian Beef: financial performance of beef farms 2017–18 to 2019–20. ABARES, Canberra. https://doi.org/10.25814/naed-7641
Tucker CB, Rogers AR, Ververk GA, Kemball PE, Webster JR, Matthews LR (2007) Effects of shelter and body condition on the behaviour and physiology of dairy cattle in winter. Appl Anim Behav Sci 105:1–13
Turner SP, Dwyer CM (2007) Welfare assessment of extensive animals: challenges and opportunities. Anim Welf 16:189–192
Valsoni LM, Freitas MG, Echeverria JT, Borges DGL, Tutija J, Borges FA (2020) Resistance to all chemical groups of acaricides in a single isolate of Rhipicephalus microplus in Mato Grosso do Sul, Brazil. Int J Acarol 46:276–280. https://doi.org/10.1080/01647954.2020.1765867
Van Iaer E, Moons CPH, Sonck B, Tuyttens FAM (2014) Importance of outdoor shelter for cattle in temperate climates. Livest Sci 159:87–101. https://doi.org/10.1016/j.livsci.2013.11.003
Vanhonacker F, Verbeke W, Van Poucke E, Tuyttens FAM (2008) Do citizens and farmers interpret the concept of farm animal welfare differently? Livest Sci 116:126–136
Vigors B, Ewing DA, Lawrence AB (2021a) Happy or healthy? How members of the public prioritise farm animal health and natural behaviours. PLoS ONE 16(3):e0247788. https://doi.org/10.1371/journal.pone.0247788
Vigors B, Ewing DA, Lawrence AB (2021b) The importance of farm animal health and natural behaviors to livestock farmers: findings from a factorial survey using vignettes. Front Anim Sci 2:638782. https://doi.org/10.3389/fanim.2021.6387
Vilela VLR, Feitosa TF, Bezerra RA, Klafke GM, Riet-Correa F (2020) Multiple acaricide-resistant Rhipicephalus microplus in the semi-arid region of Paraíba State, Brazil. Ticks Tick-Borne Dis 11:101413. https://doi.org/10.1016/j.ttbdis.2020.101413
Villalba JJ, Manteca X (2016) Animal welfare in extensive production systems. 5M Publishing Ltd., Sheffield

Villalba JJ, Manteca X, Vercoe PE, Maloney SK, Blache D (2016) Integrating nutrition and animal welfare in extensive systems. In: Phillips CJC (ed) Nutrition and the welfare of farm animals. Springer, Berlin, Heidelberg, pp 135–163. https://doi.org/10.1007/978-3-319-27356-3_7

Waterhouse A (1996) Animal welfare and sustainability of production under extensive systems – a European perspective. Appl Anim Behav Sci 49:29–40

Webster JR, Stewart M, Rogers AR, Verkerk GA (2008) Assessment of welfare from physiological and behavioural responses of New Zealand dairy cows exposed to cold and wet conditions. Anim Welf 17:19–26

Yunes MC, von Keyserlingk MAG, Hötzel MJ (2017) Brazilian citizens' opinions and attitudes about farm animal production systems. Animals 7:75. https://doi.org/10.3390/ani7100075

Zago D, Canozzi MEA, Barcellos JOJ (2019) Pregnant cow nutrition and its effects on foetal weight – a meta-analysis. J Agric Sci 157:83–95

Zago D, Canozzi MEA, Barcellos JOJ (2020) Pregnant beef cow's nutrition and its effects on post-natal weight and carcass quality of their progeny. PLoS ONE 15(8):e0237941

# Welfare of Beef Cattle in Intensive Systems

# 7

Hannah Salvin, Karen Schwartzkopf-Genswein, Caroline Lee, and Ian Colditz

## Contents



H. Salvin (✉)
NSW Department of Primary Industries, Livestock Industries Centre, Armidale, NSW, Australia
e-mail: Hannah.salvin@dpi.nsw.gov.au

K. Schwartzkopf-Genswein
Agriculture and Agri-Food Canada, Science and Technology Branch, Lethbridge Research Centre, Lethbridge, AB, Canada
e-mail: Karen.Genswein@Canada.ca

C. Lee · I. Colditz
CSIRO Agriculture and Food, Animal Behaviour and Welfare, FD McMaster Laboratory, Armidale, NSW, Australia
e-mail: Caroline.lee@csiro.au; Ian.colditz@csiro.au


M. Haskell (ed.), *Cattle Welfare in Dairy and Beef Systems*, Animal Welfare 23,
https://doi.org/10.1007/978-3-031-21020-4_7

**Abstract**

Intensive management of cattle in open-air and enclosed housing systems known as feedlots is a widespread and growing practice for production of beef. Feedlots can expose cattle to harms with the potential to compromise behavioural, physiological and mental activities. In addition, there is a growing realisation that livestock production systems need to enable positive welfare outcomes through provision of opportunities for positive affective engagement with the environment. This chapter describes aspects of the physical environment, management practices and animal genotype that influence welfare outcomes in feedlots. Topics considered include housing, food and water, thermal comfort, health, pregnancy management, environmental enrichment, weaning practices, transport, painful husbandry procedures, backgrounding, preconditioning, induction practices, mixing, vaccination, non-specific immune stimulation, antibiotic use, metaphylaxis, stockperson attitudes, breed, temperament, immune competence and resilience. Attention is drawn to the potential for genotype, prior experience and the physical and psychological structures of the environment to enhance competence of animals to adapt to and engage positively with their environment and to be resilient to daily challenges posed by the feedlot environment. It is concluded that adaptability, resilience and positive affective engagement are complements to good animal management, providing important additional pathways towards positive welfare outcomes during finishing in well-designed feedlots.

**Keywords**

Animal welfare · Beef cattle · Feedlot · Intensive beef · Positive welfare

## 7.1 Introduction

Beef cattle are kept in a variety of intensive production systems globally. Historically, cattle in cold climates have spent at least part of the year indoors; however, intensive feeding of beef cattle in large scale commercial feedlots did not develop in North America (the United States and Canada) until the 1920s and in Australia until the 1960s (Gaughan and Sullivan 2014). Commercial feedlots can also be found in other parts of the world including South America; South Africa; New Zealand; parts of Asia including Japan, South Korea, China, the Philippines and Indonesia; and the European Union, predominantly Spain. The scale of feedlots and the proportion of the beef market they occupy vary considerably between countries. For example, South Africa has the largest feedlot in the southern hemisphere with a capacity of 150,000 heads, with feedlot beef making up 80% of South Africa's formal meat market (Deblitz 2012).

This chapter focuses on the Australian and North American commercial feedlot systems, which are intensive indoor and outdoor production systems that utilise high-energy concentrate diets to finish cattle for various domestic and export markets. An increasing shift towards intensive environments and greater public concern for the welfare of livestock (Eurobarometer 2007; Spain et al. 2018) has brought into focus the importance of maintaining good welfare in feedlots. There is also a growing acceptance that avoidance of harms such as pain and suffering is not sufficient to provide good welfare, but that animals should be provided with a 'life worth living' through positive affective engagement with physical and psychological dimensions of their environment (Lawrence et al. 2019; Mellor 2016).

While generally similar, some key differences in feedlot design, management and climate between these two regions may have implications for welfare as identified below. A glossary of terms has been provided (see Box 7.1), which provides definitions for the types of feedlot and confinement systems used, and the facilities incorporated as well as definitions of specific terms used for management practices commonly used in feedlots. As of October 2022, there were 11.4 million cattle in feedlots with a 1000+ head capacity in the United States and 1.5 million in Canada, representing approximately 13% and 10% of the total beef cattle population within each country, respectively (USDA 2019, 2022; Statistics Canada 2022). Two confinement systems predominate, with approximately three quarters of cattle finished in open-air lots, with or without shelter (Fig. 7.1), and one quarter in bedded or deep-pit indoor confinement systems (Schulz 2014). Confinement systems are more common in the eastern United States and Canada where precipitation and humidity are greater than in the west, requiring that the cattle have shelter. More than half of the cattle placed into feedlots are recently weaned calves (150–200 kg; 5–8 months of age) with the remainder being yearlings (Schulz 2014; Endres and Schwartzkopf-Genswein 2018). In Australia, intensive systems have a capacity of 1.5 million cattle at any one time, constituting approximately 4% of the country's total cattle population and contributing 30%–40% of the meat market (Australian Lot Feeders Association 2015). In contrast to North America however, Australian cattle generally experience a longer backgrounding period on pasture and don't enter the feedlot environment until they reach weights of 280–500 kg (1.5–2 years) (Andrews 2015). Indoor confinement systems are also less common.

**Box 7.1 Glossary of Terms**

*Feedlot:* An area of open ground (either indoors or outdoors) surrounded by a fence or wall (pen) and containing feed and water troughs. In outdoor settings, the dirt floor of the pens is sloped to facilitate drainage following rain events, with runoff collected into holding ponds to avoid environmental contamination. Bedding (straw, woodchips, etc.) is added to the pen floor for animal comfort and to mitigate muddy conditions. Structures that provide shade over a portion of the pen may be provided in climatic zones with high summer temperatures.

*Bedded confinement*: Cattle are kept indoors on solid floors covered with bedding, usually straw or another crop residue.

*Deep-pit confinement*: Cattle are kept indoors on slatted floors in which the manure falls through into a pit below.

*Backgrounding*: A period of weeks to months prior to feedlot entry where young, weaned cattle are grazed at pasture until they reach feedlot entry weight. Cattle may be introduced to a concentrated feed ration to familiarise them with this feed type. Vaccination programs to protect cattle against bovine respiratory disease may be commenced during backgrounding.

*Pre-conditioning*: Familiarisation of cattle with a concentrated feed ration and vaccination against bovine respiratory disease during backgrounding. In addition to these practices, calves in North America must be weaned for a minimum of 30 days before they are transported to a feedlot.

*Induction*: The process of entering an animal into a feedlot, including completion of paperwork, first weighing and any required husbandry or health procedures.

*Mounds*: Areas of ground within the pen approximately 1.2–2.5 m (4–8 ft) higher than the rest of the pen to allow cattle a dry place to rest.

*Bunk*: The concrete trough used for holding feed.

*Pen rider*: Stockperson inspecting cattle for signs of illness and ill-thrift.

*Fed cattle*: Cattle finished to slaughter weight through a feedlot.

*Short-fed*: Cattle fed in a feedlot for between 70 and 150 days, usually for the Australian domestic market.

*Medium-fed*: Cattle fed in feedlots for 150–200 days, usually for the Japanese and Korean markets.

*Long-fed*: Cattle fed in feedlots for more than 200 days (up to 550 days), usually for the top Japanese markets.

This chapter aims to identify aspects of the physical environment, management practices and the cattle's genotype which may infringe on the behavioural, physiological and mental requirements of cattle in feedlots. Some future research directions that may help ensure the sustainability of feedlot production from animal welfare and social licence perspectives are also identified.

**Fig. 7.1** A typical open-air feedlot pen in which shelter and shade is not provided. In climatic zones exposed to high summer temperatures, shading may be provided over a portion of each feedlot pen. (Photo courtesy of Dominic Niemeyer, CSIRO, Armidale, Australia)

## 7.2 Physical Environment

### 7.2.1 Housing

The environment in which cattle are housed under intensive conditions has a strong influence on their health and comfort and therefore their welfare. Ability to manage climatic challenges through behaviour may also be limited depending on the system and is covered in detail under thermal comfort in Sect. 7.2.3.

Minimum space recommendations for feedlot cattle in Australia are 9 m²/head (National Guidelines for Beef Cattle Feedlots in Australia 2012). A survey of feedlot producers in Australia found space per head allocations ranged between 7 and 29 m², although this work was presented in an industry report and was not peer-reviewed (Perkins 2013). A similar survey in the high plains region of the United States found that space per head ranged between 9 and 23 m²/head (Simroth et al. 2017). Pen space recommendations for Canadian feedlots can vary significantly by region and type of animal. For example, cattle in the east are recommended to have between 28 and 56 m²/head (calves) and between 37 and 75 m²/head (finishing), while recommendations for feedlots in the west are between 16 and 23 m²/head (calves) and between 18 and 28 m²/head (finishing) (OMAFRA 2020; Alberta Feedlot Management Guide 1996). Pen space requirements are generally lower in geographic locations where annual precipitation is low.

**Fig. 7.2** Mud poses a significant welfare risk to animal comfort and health, potentially reducing lying time and increasing the incidence of lameness, injury and infectious claw diseases. (Photo courtesy of Dr. Karen Schwartzkopf-Genswein, Agriculture and Agri-Food Canada, Alberta, Canada)

Research regarding the impact of stocking density is generally concentrated around pen surface management and air quality (McGinn et al. 2003). However, antagonistic interactions are likely to occur at the stocking densities generally provided by feedlots and may be a substantial cause of social stress (Bouissou 1980). Kondo et al. (1989) found that antagonistic interactions in adult cattle housed on pasture increased as space per head decreased below 20 $m^2$, with the least number of antagonistic interactions occurring when space allowances were over 360 $m^2$/head. Adequate space allowance per animal at feeders and water troughs is also an important consideration in minimising antagonistic interactions and will be discussed in Sect. 7.2.2.

Pen surface conditions need to be managed carefully to prevent detrimental effects on production and animal comfort. Muddy pen surfaces in outdoor feedlots (Fig. 7.2) have been identified as a significant risk to animal welfare (Grandin 2016) and may impact their patterns of feeding and lying behaviour (Wilson et al. 2005) as well as increase the incidence of lameness and injury because mud creates slippery conditions and facilitates spread of infectious claw-related disease such as foot rot and digital dermatitis (Stokka and Goldsmith 2015; Davis-Unger et al. 2019). Management of pen surface to improve animal comfort may involve the use of mounds (Mader 2003), seasonal adjustments in stocking density (Grandin 2016), bedding (Mader 2011; Watts et al. 2015), pen cleaning frequency and ensuring pen drainage by sloping pens 2–6° away from the feed bunks (National Guidelines for Beef Cattle Feedlots in Australia 2012; Alberta Feedlot Management Guide 1996). Success of these management actions may be assessed through evaluation of coat cleanliness (Grandin 2016).

### 7.2.2 Feed and Water

Two aspects of feeding management have the potential to impact welfare in the feedlot environment: space allocation at the feed bunk and ration composition. In Australia, the Model Code of Practice for Cattle recommends 25–46 cm/head for feed troughs (PISC 2004). This is similar to the 25 cm/head recommended for Canadian feedlots where cattle are fed ad libitum, while 45–66 cm/head is recommended for restrictively fed cattle. The increase in space allows all cattle to feed at once (OMAFRA 2020; Alberta Feedlot Management Guide 1996). The average bunk space in the high plains region of the United States was reported to be 22–30 cm/head (Simroth et al. 2017).

As indicated by the space per head recommendations, the effect of trough space on cattle welfare is dependent on feed delivery schedules. Cattle fed under a restricted feeding regime are likely to experience increased competition at feeding, and adequate space for all animals to feed simultaneously becomes of greater importance than in ad libitum feeding regimes, where feed is always available. Research in dairy cattle suggests that when competition at the feed bunk is increased, cattle maintain dry matter intake by having fewer meals per day but increasing speed of consumption (Collings et al. 2011; Hosseinkhani et al. 2008). In feedlot cattle, increasing social pressure from 7.5 to 15 heifers per feeding station and offering ad libitum feed also resulted in a reduction of time spent feeding per day, indicating that heifers compensated for shorter daily feeding times by eating at a faster rate (González et al. 2011). However, in other feedlot cattle studies, no effect of reducing bunk space per head on feeding behaviour was identified between 60 and 80 cm/head, when fed ad libitum (Gottardo et al. 2004) and between 15 and 60 cm/head when restricted feeding (Zinn 1989).

Feedlot rations are usually designed to maximise growth and performance and minimise risk of metabolic disorders. However, the rapid transition (typically over a 21–28 day period) from roughage-based diets to a grain-based diet can be problematic if not managed carefully (Galyean and Rivera 2003; Schwartzkopf-Genswein et al. 2003). This transition would typically commence upon feedlot entry in Australia or approximately 90 days after feedlot entry in North American systems. Preference testing has demonstrated that cattle given access to both pasture and a feedlot ration will choose to obtain most of their daily nutritional requirements from the feedlot ration (Lee et al. 2013). However, other research suggests that this preference may shift if animals are experiencing ruminal acidosis (DeVries et al. 2014). It has been estimated that digestive disorders account for 30%–42% of mortalities in feedlots (Smith 1998) with the most prevalent disorders being acidosis, liver abscesses and bloat (Galyean and Rivera 2003). Individual variation in susceptibility to disorders such as acidosis have been shown (Galyean and Rivera 2003; Bevans et al. 2005; Schwartzkopf-Genswein et al. 2003); however, to avoid mortalities, it is still necessary to adapt any management protocols to the most susceptible individuals.

Adequate provision of water is a key consideration in the design and running of feedlots, with both the quality of the water provided (Wright 2007), and the water temperature (Schütz et al. 2018) potentially impacting water consumption and therefore welfare. Water storage, emergency provisions and flow rates are also important as individual animals can drink between 26 and 66 L/day (Olkowski 2009), with some industry sources suggesting up to 75 L/day in hot weather (Watts et al. 2016). Sowell et al. (1999) found cattle spent approximately 4 min a day drinking on entry to the feedlot, increasing to 7–8 min further into the period spent in the feedlot. Cattle may also drink 87% more water in summer than in winter (Arias and Mader 2011), highlighting the importance of considering seasonal variation in water supply requirements. Water is also used for many other aspects of feedlot operation that may impact cattle welfare including dust abatement, cattle washing and cooling (Watts et al. 2016). Consideration of additional usage must therefore be made when determining overall water requirements.

### 7.2.3 Thermal Comfort

All cattle have an ideal climatic temperature range, the thermoneutral zone, within which they do not need to expend energy to warm or cool themselves. Temperatures outside this zone have the potential to cause heat or cold stress leading to compromised welfare and possibly death (Busby and Loy 1997). The upper limit of the thermoneutral zone for *Bos taurus* cattle has been suggested to be around 24°C, although this may be influenced by several factors. Heat load can also build up over several days if the overnight temperatures are not sufficiently cool to allow

**Fig. 7.3** Windbreak fencing is commonly used in North America and Canada to reduce the effect of windchill on maintaining body temperature. Windbreaks are not used in Australia due to the milder climate, where heat rather than cold is likely to be the biggest source of thermal stress. (Photo courtesy of Dr. Karen Schwartzkopf-Genswein, Agriculture and Agri-Food Canada, Alberta, Canada)

accumulated heat to dissipate back into the environment (Gaughan et al. 2013). While all cattle may be exposed to environmental extremes, the effects on outdoor feedlot cattle can be greater as they have limited ability to seek out microclimates, depending on the confinement system used. For example, a heat wave in 1995 in Iowa, United States, resulted in over 3500 cattle deaths, with mortality rates of 0.2% in feedlots with shade compared to a rate of 4.8% in lots without shade (Busby and Loy 1997). Feedlot cattle are also fed high-energy diets which can generate larger amounts of metabolic heat (Mader and Griffin 2015).

There are several management techniques that can be used in feedlots to manage the thermal comfort of cattle (Mader 2003). Indoor housing systems provide protection against both cold and heat stress; however, they are more costly to construct and maintain (Lawrence et al. 2001; Euken et al. 2015), and there may be trade-offs in airflow and air quality (Euken et al. 2015). Insulation under metal roofs may also be needed to prevent radiant heat from the roof structure exacerbating heat stress in hot climates (Sparke et al. 2001). Microclimates may also be generated in outdoor feedlots using windbreaks (Mader et al. 1997a), bedding (Mader 2003), shade structures (Mader et al. 1997b, 1999) and water sprinklers or misters (Mader et al. 2007; Davis et al. 2003; Parola et al. 2012) with varying success. The biggest concern in cold climates is the effect of high wind chill which can increase energy demand in cattle (Ames and Insley 1975; Ames 1988). This is one reason why windbreak fencing is common in North America, particularly in western Canada (Fig. 7.3). Allowing cattle time to adjust to the prevailing weather conditions before entering the feedlot can be helpful. Cattle which have been previously exposed to temperatures at the limits of their thermoneutral zone may be less stressed by sudden temperature increases or decreases outside of their thermoneutral zone than those that have not had time to adjust (Young 1985). A web-based tool using weather forecasts is available to enable feedlot managers in Australia to implement strategies to reduce heat load before periods of increased risk of heat stress (Gaughan et al. 2008).

In addition to managing the climatic sources of thermal stress, it is also possible to manage the contribution of metabolic heat to the total heat load the animal experiences. Restricted feeding has been shown to reduce the body temperature of cattle during hot weather; however, daily weight gain was also impacted (Mader et al. 2002). Conversely, low-fibre, high-starch diets in cold conditions provide greater benefit by increasing metabolizable energy compared with the benefits of increasing metabolic heat production through increasing fibre consumption (Mader 2003). Other strategies to reduce cold weather effects on cattle include increasing feed deliveries, feeding in the evening which increases heat production during the cooler part of the day (Bergen et al. 2007) and providing ample dry bedding (Schwartzkopf-Genswein et al. 2017). It is recommended that these strategies are only employed during hot and or cold periods, respectively, to ensure that production and other aspects of welfare are not adversely impacted (i.e. cold stress may be exacerbated by restricting feed during cold periods and heat stress exacerbated by keeping the energy level of the diet high during hot periods).

Finally, careful selection of cattle genotype to suit the climate is important. Poor temperament is associated with increased susceptibility to heat stress, with

Brown-Brandl et al. (2006) reporting that excitable heifers had 3.2% higher heat stress levels compared with calm heifers. Dark-coated cattle are particularly susceptible to heat stress, reaching peak body temperature up to 2 h earlier than light-coated cattle during a heat load event (Mader et al. 2002). *Bos indicus* cattle are also more heat tolerant than *Bos taurus* cattle due to differences in coat characteristics, sweating and metabolic rate (Blackshaw and Blackshaw 1994). With the increasing risk of temperature extremes in a changing global climate, thermal comfort is one of the greatest challenges to cattle welfare in the feedlot.

### 7.2.4 Health

A survey of Australian feedlot producers found that bovine respiratory disease (BRD) accounted for 84% of all removals to the hospital pens, followed by muscular conditions including traumatic injuries, foot conditions and septic arthritis. A similar trend was seen in mortalities (Perkins 2013). Similar findings were reported in a Canadian study where 46% of all morbidity was due to BRD, 32% to lameness and 22% to other diagnoses (Davis-Unger et al. 2019). The highest number of mortalities occur in short-fed cattle (spending less than 85 days on feed), with a peak at between 4 and 5 weeks after entry into the feedlot (Perkins 2013). Some illnesses may also have longer-term impacts on welfare even if occurring prior to feedlot entry. Brown-Brandl et al. (2006) found that animals who had been treated for pneumonia at any time during their life had greater heat stress responses while in the feedlot than those who had never been diagnosed or treated.

The identification of morbidities in the feedlot environment can be difficult. Recognition of morbid animals is generally reliant on the subjective evaluation of behavioural indicators and is therefore highly dependent on the skill and training of the stockperson (Weary et al. 2008a). Although the value of the stockperson to animal welfare is undisputed, attracting and retaining skilled workers to the role can be a challenge (Daigle and Ridge 2018). One development that will increase the accuracy of morbidity monitoring is the emerging use of technology to automatically record behaviours indicative of illness. For example, reduced intake of feed and increased intake of water around 4–5 days after feedlot entry has been found to be predictive of subsequent BRD morbidity (Sowell et al. 1998, 1999; Buhman et al. 2000; Moya et al. 2015; Wolfger et al. 2015). As yet, however, there has not been widespread uptake of this technology by the industry.

### 7.2.5 Environmental Enrichment

Environmental enrichment has been defined as 'improvement in the biological functioning of captive animals resulting from modifications to their environment'. Enrichments should result in measurable improvements to welfare by increasing the performance of natural behaviours, reducing abnormal behaviours, reducing negative emotional states, improving health and improving the use of environmental resources (Newberry 1995). Environmental enrichment has been more comprehensively studied in other intensively housed livestock, such as pigs and chickens, while feedlot cattle enrichment studies are few. Despite this, existing studies suggest that environmental enrichment may provide improved welfare to cattle housed in confined conditions.

The most commonly studied enrichment in feedlot cattle is a brush that is provided within the pen to enable scratching/rubbing and self-grooming. The use of brushes for environmental enrichment in feedlot pens has been reported to reduce stereotypic (Park et al. 2019a) and aggressive behaviours (Park et al. 2019b). The brush was shown to be a more preferred enrichment than a scented device and was used frequently by cattle within the feedlot (Wilson et al. 2002). In addition, the frequency of brush usage was maintained for more than 6 months in Japanese Black steers, perhaps due to the importance of self-grooming, a natural behaviour in cattle (Ninomiya 2019).

Giving cattle the opportunity to access environmental enrichments at adequate levels must take into account the number of animals present and is an important factor in preventing aggression and antagonistic interactions between cattle. When a straw bale was used as an enrichment in the feedlot, there was an increase in aggressive behaviours due to the limited quantity available (Pelley et al. 1995). The opportunity for feedlot cattle to access enrichments can be manipulated through adjusting the stocking density to reduce the chances of antagonistic behaviours (Meneses et al. 2019). A study investigated the effects of providing three enrichments to calves reported that providing wooden partitions during feeding reduced antagonistic behaviour in Japanese Shorthorn calves and increased affiliative behaviour in Japanese Black calves (Ninomiya and Sato 2009). The same study found that the enrichment of providing clean straw bedding improved rest quality with increased sternal lying and lying with the head touching the flank in calves (Ninomiya and Sato 2009). These postures may reflect better sleep quality as they are positions that support the head and enable rapid eye movement sleep (Ternman et al. 2014). At this time, however, few feedlots in Australian and North America use environmental enrichment strategies.

## 7.3 Management Practices

### 7.3.1 Preinduction Management

#### 7.3.1.1 Weaning Practices

Calves are typically weaned between 5 and 8 months of age (USDA 2008). Weaning can occur on the breeding farm (as in Australia), or at the time calves are separated from their mothers for sale. In North American production systems, many calves are transported to a feedlot environment on the day that weaning occurs. Weaning exposes calves to a combination of nutritional, physical and psychological stressors (Weary et al. 2008b). For example, calves may be exposed to a change of diet, new social interactions with conspecifics, a new physical environment, frequent interactions with humans and husbandry procedures such as castration, vaccination and tagging for individual identification (Lynch et al. 2019). Consequently, newly weaned calves are at increased risk of greater morbidity once they enter the feedlot, particularly when additional stressors such as transportation and commingling with unfamiliar calves are added (Wilson et al. 2017).

The weaning experience can influence subsequent social and physical performance of the calf (Lynch et al. 2019). During weaning in Australia, calves are usually separated from their mothers, confined to yards and fed a processed feed ration from troughs together with good-quality roughage for around 7–10 days before release to pasture. This method has been shown to lead to better weight gain and reduced incidence of BRD during the feedlot period, compared with alternative methods (Walker et al. 2007; Hay et al. 2016a). Alternative methods of weaning include fence-line weaning where auditory and visual contact is retained between the cow and calf or two-stage weaning where a nose-flap first stops the calf from suckling and is removed when the cow and calf are separated (Haley et al. 2005). These methods reduce the number of stressors calves must face at any one time. On a more positive note, however, the period of confinement during yard weaning may facilitate the establishment of social skills that enhance later adaptation to the feedlot environment.

#### 7.3.1.2 Husbandry Procedures

Routine management procedures such as castration, spaying, dehorning, branding, tagging and vaccination are typically conducted at early ages, while calves are on pasture with their dams, well before they are marketed to a feedlot. The most common time to conduct these procedures in North America is between 1 week and 3 months of age (Moggy et al. 2017a, b); however, this can vary greatly by individual ranch and geographic location. Veterinarians and animal scientists advocate that these procedures be done as early in life as possible, as removal of larger and more developed testes and horns may increase tissue trauma and risk of infection, blood loss or death (Schwartzkopf-Genswein et al. 2012).

One Canadian study indicated that 53%, 51% and 52% of farms surveyed castrated, dehorned and branded at <1 week of age, respectively (Moggy et al. 2017a). Unfortunately, a portion of calves entering the feedlot remain intact (testicles, ovaries and horns) due to poor animal management on the ranch, or in the case of castration, the use of incorrect technique where one or both of the testicles remain. The procedures indicated above are considered painful at all ages and for all methods (Coetzee 2011; Duffield et al. 2010; Heinrich et al. 2009). However, castration (Meléndez et al. 2017; Marti et al. 2017a) and dehorning (Duffield et al. 2010; Heinrich et al. 2009) are more invasive when conducted between 5 and 8 months old, the age calves typically enter feedlots in North America. Consequently, the use of pain mitigation strategies by cow-calf producers and feedlot managers has increased substantially in Canada over the past 5 years, but to a lesser extent in the United States, mostly due to the lack of availability of registered long-acting non-steroidal anti-inflammatory drugs for cattle. The increased use of pain control drugs can also be attributed to the requirements of Canadian codes of practice (NFACC

2013) that may be used as regulatory control. Similarly, in Australia, the Australian Animal Welfare Standards and Guidelines for Cattle recommend the use of pain relief for painful husbandry procedures, and this practice is steadily increasing (Animal Health Australia 2014).

In Canada, more feedlot managers are refraining from dehorning or tipping (removal of the insensitive extremity of the horn) and branding at induction due to a combination of greater awareness and concern for animal welfare, requirements for pain control, increased labour costs and negative effects on growth performance in the case of dehorning (Goonewardene and Hand 1991). In contrast, branding is still commonly conducted on ranch calves by commercial feedlots that feed cattle for multiple owners, and on cattle purchased using money borrowed from lending institutions (i.e. bank or feeder association), which are required to be branded as proof of ownership (Endres and Schwartzkopf-Genswein 2018).

### 7.3.2 Transport

The marketing of calves, yearlings and fed cattle ultimately means they must be transported off the ranch or feedlot. Numerous transport- and animal-related factors (alone or in combination) can reduce an animal's ability to cope with transport, leading to increased incidence of lameness, becoming non-ambulatory or death. Transport-related factors include loading density, transport duration, trailer design and ventilation, driving and handling quality, road and environmental conditions and fitness of the animals. Animal-related factors include pre-transport management and cattle age, breed and condition (Schwartzkopf-Genswein et al. 2016; Tucker et al. 2015; Goldhawk et al. 2015).

Negative welfare outcomes have been associated with long (>30 h) transport durations, ambient temperatures >20°C and < −15°C, low (1.5 $m^2$) and high (0.5 $m^2$) space allowance and excessive weight loss (shrink >10% of body weight) (González et al. 2012a, b, c). Poor handling, driving quality and/or facilities can increase slipping and falling and the potential for injury and lameness (Schwartzkopf-Genswein and Grandin 2019). Cattle transported by drivers having more than 5 years of experience hauling livestock had fewer poor welfare outcomes than drivers with fewer than 5 years of experience. This was attributed to calm handling (i.e. minimal use of prods yelling and running), good driving technique (i.e. smooth cornering and gradual stopping) and the ability to manage risk (i.e. selecting optimal transport times or routes to reduce the effects of extreme weather and or poor road conditions) (González et al. 2012c).

The likelihood of an animal experiencing negative welfare during and after transportation is highly dependent upon their fitness for transport (age and health condition at the time of loading) in combination with the factors previously listed (Schwartzkopf-Genswein and Grandin 2019; Tucker et al. 2015). For example, cull cows and calves were shown to be more susceptible to transport stress as they demonstrated higher incidences of lameness, non-ambulation and death compared to feeder and fat cattle. The increased susceptibility was attributed to reduced energy reserves, ability to thermoregulate and overall health (González et al. 2012c).

Reducing the duration of transport to the feedlot reduces the incidence of BRD (Cusack and Mahony 2019). One study reported transport durations >6 h either on the day before induction or on the day of induction increased the risk of BRD (Hay et al. 2014). Loading and unloading are the most stressful stages of transport (Pettiford et al. 2008) and require careful management to minimise their impact. The few existing studies assessing the effects of rest stops on cattle welfare have contradictory findings (Meléndez et al. 2020; Marti et al. 2017b; Cooke et al. 2013), indicating more research is needed in this area. In addition, research assessing the association between rest stop duration and quality and health and welfare outcomes in cattle once they enter the feedlot is lacking.

### 7.3.3 Backgrounding and Preconditioning

Backgrounding describes a period of weeks to months from weaning until feedlot entry where calves are fed at pasture (as shown also in the Glossary). This grow-out phase is often undertaken by specialist producers who source calves from multiple breeding enterprises in order to make up uniform lines of cattle that meet feedlot entry specifications such as body weight, breed and vaccination history. Management practices used during backgrounding that are aimed at improving feedlot health and performance of cattle are termed preconditioning.

The majority of animals entering North American feedlots are calves between 5 and 8 months of age. These calves may be at high risk of disease on entry to the feedlot due to prior exposure to numerous stressors conducted at the same time, within a few days before leaving the ranch. Potential stressors include weaning, castration, dehorning, spaying, branding, vaccination, handling, transport and commingling with other calves sold through auctions. Preconditioning has been advocated as a way to mitigate stress by conducting husbandry procedures over a longer period of time, well before the calves are transported and while the calves are still with their dams. Although there is currently no standard protocol for preconditioning, guidelines usually include some criteria for the minimum time prior to leaving the ranch at which certain procedures should be conducted. These are 35–45 days for weaning and 21 days for feeding from a bunk, castration and dehorning. Vaccination is also recommended prior to leaving the ranch, at around 4 months of age (Radostits 2000). Feeding from a trough and vaccination with BRD vaccines during backgrounding tend to reduce the risk of BRD during feedlot finishing (Hay et al. 2016b). The welfare and production benefits of preconditioning are well known and include improved rate of gain (Karren et al. 1987) as well as the reduced incidence of BRD in the first 28 days in the feedlot (Macartney et al. 2003). A study assessing the effect of conditioning, combined with long and short haul transport on calf welfare, reported that conditioning calves prior to transport reduced transport and handling stresses (Schwartzkopf-Genswein et al. 2006). Other studies have found the benefits of BRD vaccination during backgrounding to be equivocal

(Cusack and Mahony 2019). An epidemiological study found that mixing cattle from different sources during backgrounding at least 4 weeks before feedlot entry reduced the risk of BRD, in comparison with cattle mixed closer to the time of feedlot entry (Hay et al. 2014). Many studies on the benefits of preconditioning have provided equivocal results, perhaps due to variability in animal genetics, previous history, weaning practices, climatic conditions and methods used in the studies (Wilson et al. 2017). The practice of preconditioning has not been consistently adopted in North America. This is due to the fact that premiums are not typically paid for preconditioned calves marketed through an auction because buyers for feedlots cannot confirm if they have been preconditioned or not (Endres and Schwartzkopf-Genswein 2018).

### 7.3.4 Induction

The procedures that cattle are exposed to at feedlot induction are usually tailored to their previous management history. Typical practices include animal identification, vaccination, treatment for internal and external parasites, cutting the tips off horns and combining animals from multiple sources into pen groups. The stress of induction is a major contributor to BRD risk (Cusack and Mahony 2019), and the strong focus within weaning and backgrounding practices is to pre-adapt cattle to the feedlot environment in order to improve their resilience to the stress of induction.

#### 7.3.4.1 Social Interactions

Allocation of cattle to pens at feedlot induction can involve regrouping. Mixing unfamiliar cattle is a known stressor associated with increased agonistic interactions during the establishment of a new social hierarchy (Mench et al. 1990). To reduce the negative impacts of regrouping, mixing cattle well before feedlot entry, reducing the number of groups mixed per pen and avoiding purchasing of cattle out of saleyards for direct transfer to the feedlot are recommended. The stress of regrouping may also influence the incidence of BRD with increased risk if regrouping occurs close to feedlot entry (Barnes et al. 2014). Providing cattle with a longer period to recover from the stress of mixing before entering the feedlot is proposed to provide immunological benefits and thereby contribute to reducing the incidence of BRD (Cusack and Mahony 2019). Regrouping once in the feedlot is not recommended; however, there may be instances where this is necessary, for example, when removing sick animals or if animals are held back to meet market specification. Regrouping within 2 weeks of slaughter may impact meat quality (Colditz et al. 2007; Warren et al. 2010).

A behavioural problem involving social interactions is termed "buller syndrome" and is characterised by repeated mountings of one steer by other steers (Brower and Kiracofe 1978; Edwards 1995). Buller syndrome can cause injuries such as swelling and trauma on the rump and tail head and increased incidence of health issues in the 'buller' or recipient steer and penile injuries in the initiating steer. Several factors have been suggested to induce buller syndrome, including mud, dusty pens, group size and the use of anabolic agents (Tucker et al. 2015). At present, there is no clear solution to the problem except to remove the recipient animals from the pen.

#### 7.3.4.2 Vaccination

Vaccines are available against several clostridial diseases of cattle and against several viral and bacterial agents involved in BRD. Despite efficacy in clinical trials, the benefits of BRD vaccines administered prior to or at feedlot entry in reducing disease incidence in commercial feedlots are equivocal (Cusack and Mahony 2019; Theurer et al. 2015). This observation does not provide an argument against vaccination but suggests that further refinements to vaccination protocols may be needed to capture the benefits vaccination can confer (Stokka and Goldsmith 2015).

All calves and yearlings that enter the backgrounding feedlot undergo initial processing which may include vaccination and/or revaccination, growth implants, identification with an ear tag or tags and performing castration or dehorning on those calves that were not done on the ranch. Calves can be further handled during arrival processing, when they are allotted to home pens to achieve the desired number of head per pen or more uniform groups.

#### 7.3.4.3 Non-specific Immune Stimulation

An alternative to the development of specific immunity against disease pathogens provided by vaccination is the potential to augment non-specific immune defence by administering immune stimulants as a prophylactic treatment at induction or as a therapeutic treatment for clinical disease. The strategy aims to harness non-specific activities of the innate immune system which can remain enhanced for extended periods following activation and exhibit heightened activity upon subsequent reactivation. This phenomenon is known as trained immunity (Netea et al. 2016). The strategy shows some promise but requires further research to better establish efficacy and appropriate protocols for use (Nickell et al. 2016; Nosky et al. 2018; Woolums et al. 2019).

#### 7.3.4.4 Antibiotic Usage

The use of antibiotics during feedlot finishing is strongly influenced by the regulations in place within the different jurisdictions. Three major uses are as growth promotants, to reduce disease risk and to treat clinical disease (Badger et al. 2020). While prevention and treatment of disease are strongly supported on welfare grounds, the important and complex issues of antimicrobial stewardship from a One Health perspective are beyond the scope of this chapter but are an important consideration. Direct welfare benefits from the use of growth promotants include a reduction of the incidence of acidosis (Callaway et al. 2003). Growth promotants may also allow animals to meet market specifications sooner, thereby reducing the number of days the animal is exposed to the feedlot environment.

#### 7.3.4.5 Metaphylaxis

Metaphylaxis refers to the mass medication of a group of animals to eliminate or minimize an expected outbreak of disease. Bovine respiratory disease continues to be the leading cause of morbidity and mortality in feedlot cattle (Wilson et al. 2017). Antibiotic metaphylaxis is a common feedlot practice used to mitigate potential, costly outbreaks of BRD in newly received, high-risk calves (recently weaned, lightweight, non-preconditioned and commingled calves transported long distances) (Griffin 1997). This practice has been shown to reduce calf mortality and morbidity, days on feed and occasionally medication costs, while improving carcass and offal quality (Schumann et al. 1990; Van Donkersgoed 1992; Cernicchiaro et al. 2012, 2013; Tennant et al. 2014). Metaphylaxis is the preferred method of managing high-risk cattle because it is more cost-effective than the 'pull and treat' method which is characterized by administering antibiotic treatments to only those individual animals exhibiting clinical symptoms (Dennis et al. 2018).

The use of mass medication with antibiotics by feedlots is coming under more scrutiny due to public concerns regarding the transfer of antimicrobial resistance to the human population, although little evidence for this currently exists. Consequently, a significant amount of research is being conducted with the goal of reducing antibiotic use, altering management to reduce stress, boosting animal immunity and finding alternatives to antibiotics such as prebiotics and probiotics which are also known as direct fed microbials (Chaucheyras-Durand and Durand 2010; Mingmongkolchai and Panbangred 2018). The most significant challenge facing the beef industry in the coming decade will be finding a balance between metaphylaxis with antibiotics and limited or loss of use. Loss of antibiotic use would represent a significant risk in reducing animal health and welfare. Current strategies to reduce the use of antibiotics in food animals in North America include increased drug residue testing, specific mandatory drug withdrawal periods, greater veterinary oversight (required prescriptions for all types of antibiotics) and producer education regarding prudent use.

### 7.3.5 Pregnancy Management

Pregnancy and calving in the feedlot environment are situations that feedlot managers try to avoid as they reduce productivity and generate welfare risks for both the heifer and calf. In addition to the normal risks and stressors associated with calving, heifers may experience additional stress if they are not able to isolate themselves during calving as they usually would (Herring 2014). Ideally, to remove the welfare risk of having pregnant females in feedlots, managers should ensure all incoming females are pregnancy tested as open (not pregnant) or have been spayed. In Australia, the average proportion of spayed females entering the feedlot ranged from 13% to 39% depending on feedlot capacity (Perkins 2013). Although spaying is conducted less frequently in North America than Australia, it is more popular in western Canada and the United States where more ranches exist and where management goals are focused on increasing production and performance of feedlot heifers. Both the Willis dropped ovary and surgical flank approaches to spaying have been shown to impact the health and welfare of heifers with evidence of acute pain following the procedure and reduced weight gain for up to 6 weeks (McCosker et al.

2010). However, studies have shown that spaying-related pain can be mitigated using a combination of anaesthetics and analgesics (Lauder et al. 2020). Alternatively, abortifacients may be used on entry either in combination with pregnancy testing, or for all females, particularly at high-risk times of year. However, there are several welfare risks associated with the administration of abortifacients which increase with increasing gestational stage and include retained foetus or membranes, metritis, increased disease susceptibility and death (Buhman et al. 2003; Bergman 2019). Finally, for females who do calve in the feedlot, the emphasis must then be placed on early detection of parturition and segregation into designated calving pens or onto pasture (Bergman 2019).

### 7.3.6 Human-Animal Interaction

Cattle housed in feedlots will encounter humans on a more frequent basis than cattle in extensive conditions. The nature of the human-animal interaction in feedlots is important as cattle are dependent on humans to provide adequate nutrition, maintain their health and handle them in a way that minimises fear and stress. Stockmanship is used to describe aspects of personality and behaviour of the animal carer in relation to attitude, attentiveness and handling of animals. Stockpersons' behaviour and attitudes have been widely reported to impact animal productivity, health and welfare (Hemsworth 2003). Aspects of human personality such as self-esteem and job satisfaction can influence behaviour towards animals and subsequently animal welfare (Waiblinger et al. 2002; Boivin et al. 2003). A study in beef cattle found that formal training in animal handling resulted in improved attitudes of the handler, improved handling consistency and reduced incidence of undesirable animal behaviours (Ceballos et al. 2018). Ensuring stockpersons are adequately trained in animal care and handling will lead to welfare improvements in cattle feedlots.

## 7.4 Genotype

Genetic factors can influence how readily individual animals adapt to the feedlot environment and how susceptible they are to the environmental harms that can compromise welfare. The capacity of an organism to flourish within an environment is termed environmental fit (Ferguson 2014). Physiological and psychological processes that maintain homeostasis are effortful and consume resources. When homeostatic processes are overtaxed, the animal's welfare can be compromised (Moberg 2000). Selecting genotypes of cattle with improved environmental fit should therefore be able to influence welfare in feedlots. This section describes several genetic factors that influence how well cattle cope with feedlot environments.

### 7.4.1 Breed

The breeds of cattle used in feedlots differ in many phenotypic characteristics that influence health and welfare outcomes in the feedlot (Cusack and Mahony 2019). In addition, feedlots are located across a range of environments suggesting that a single breed will not suit all feedlot production systems. Nonetheless, a systematic influence of breed on health outcomes was observed in an epidemiological study of over 35,000 cattle finished in 107 cohorts through 14 commercial open-air feedlots in a broad geographic spread across subtropical and temperate Australia (Hay et al. 2016b). Of 8285 cases of suspected illness, 73.6% were attributed to BRD. Cumulative incidence of BRD in the first 50 days following feedlot entry was 17.6% (Hay et al. 2014). Breed had a strong effect on incidence of BRD. Using Angus as the reference population for estimating odds ratios (OR), Herefords had an increased risk (OR: 2.0, 95% credible interval: 1.5–2.6) and tropically adapted breeds, and their crosses had a reduced risk (OR: 0.5, 95% credible interval: 0.3–0.7) of developing BRD. Associations of breeds with other health outcomes in the feedlot were not reported.

### 7.4.2 Temperament

Pioneering work in the 1960s and 1980s recognised that behavioural responses of beef cattle to standardised handling procedures could be quantified to reveal differences between individuals that are expressed in a relatively stable manner over time and that are heritable (Burrow et al. 1988; Fordyce et al. 1982; Tulloh 1961). These repeatable responses are described as manifesting an aspect of the animal's temperament (Burrow 1997; Finkemeier et al. 2018; Haskell et al. 2014; MacKay and Haskell 2015). The behavioural manifestations of temperament are associated with physiological (Cafe et al. 2011) and affective differences (Lee et al. 2018) between individuals that are of direct importance to the welfare (Fell et al. 1999) and commercial performance of feedlot cattle.

Two commonly used methods for assessing temperament are a subjective score of behaviour while contained within a crush or chute, and an objective measure of the time taken when released from the crush/chute to traverse a defined distance, typically 1.7–1.8 m, although other distances can also be used reliably (e.g. see Sebastian et al., 2011). Crush score (CS) is typically based on a five-point categorical scale, while escape from the crush is usually expressed as flight time (FT) in seconds or fight speed (FS) as velocity in metres per second. Heritability of the traits is moderate to high (reviewed by Haskell et al., 2014), and genetic correlations between the traits tend to be moderate (Kadel et al., 2006).

Animals with poor temperament associated with high FS or CS values have higher basal plasma cortisol, glucose, lactate, nonesterified fatty acids (Cafe et al. 2011; Curley et al. 2006) and higher basal rectal temperatures (Lees et al. 2020) than calm animals. A number of studies have found poor temperament to have unfavourable associations with growth rate and health outcomes during feedlot finishing and with tenderness of meat from feedlot finished animals (see Ferguson et al., 2006; Haskell et al., 2014). These associations indicate the potential to improve welfare of feedlot cattle by breeding for improved temperament or by classing animals on temperament to determine their suitability for the feedlot environment before entry. Genetic evaluations for temperament traits have been available from some breeds for over two decades. The use of seedstock with superior temperament genetics by Limousin breeders in Australia has resulted in an annual improvement in docility score of +1.9% per year over 20 years (Walkom et al. 2018). This result illustrates the potential for a breeding program to modify temperament of national cattle population. Classing cattle for temperament before feedlot entry to determine suitability for the feedlot environment does not appear to have been widely adopted.

### 7.4.3 Immune Competence

The strength of innate and adaptive immune responses to challenges such as vaccination and infection differ between individual cattle (Rossi et al. 1978) and a portion of the interindividual difference is heritable (Kelm et al. 1997; Lie 1979). Selection for the strength of the immune response to vaccination, termed immune competence, has been applied in commercial breeding programs in dairy cattle and pigs where high immune competence is associated with decreased incidence and severity of several infectious diseases and increased productivity (reviewed by Larmer and Mallard 2016; Mallard et al. 1992). The adaptive immune system includes antibody- and cell-mediated components. Hine et al. (2019) observed heritabilities (represented by $h^2$ values) for antibody-mediated immune competence of $h^2 = 0.32 \pm 0.09$ and for cell-mediated immune competence of $0.27 \pm 0.08$ in Angus cattle. Antibody- and cell-mediated immune competence were favourably genetically correlated with flight time ($r = 0.63 \pm 0.31$ and $0.60 \pm 0.29$, respectively), meaning calmer cattle had stronger immune responses. This observation is in accord with studies in other species that show an association between temperament (personality) and immune function (Hessing et al. 1995; Koolhaas 2008). Preliminary studies during feedlot finishing indicate that steers with high immune competence have lower health treatment costs and lower death rates (Hine et al. 2016). These preliminary studies suggest that selection for immune competence may confer health and welfare benefits during feedlot finishing. An additional benefit may be the potential for selection for immune competence to contribute to a reduced reliance on antimicrobial drugs for treatment of disease.

### 7.4.4 Resilience

Studies on temperament and immune competence indicate that individuals differ in their capacity to cope with the psychological and physical aspects of management by humans within breeder and feedlot production environments. A capacity to be minimally affected by environmental stressors or to return rapidly to the performance trajectory that existed prior to disturbance is known as resilience (Berghof et al. 2018; Colditz and Hine 2016). Resilience of an individual is influenced both by prior developmental experience and by genotype (Russo et al. 2012). In the study of Hine et al. (2019), immune competence was measured while animals were undergoing the stress of weaning. In accord with typical beef cattle production practices in Australia, calves were born and raised at pasture then weaned at 5–9 months of age using confinement and hand feeding for at least 7 days in the open-air yards used for routine animal husbandry on commercial farms. Strong immune responses were favourably genetically correlated with growth rate over the yard weaning period and with calm temperament. The genetic associations are in accord with phenotypic studies in a number of species which suggest that a 'resilience syndrome' confers adaptive fitness to animals exposed to short-term environmental fluctuations (Colditz and Hine 2016; Koolhaas and van Reenen 2016). Implications of resilience for welfare are discussed in Sect. 7.5.

## 7.5 Future Directions for Improved Welfare

### 7.5.1 Design of Facilities

As noted above, design features to minimise harms such as bedding gradients to improve drainage, providing adequate space allocation at the feed trough to reduce competition and avoidance of sharing water troughs between pens to reduce disease transmission are well recognized (Hay et al. 2016b; Tucker et al. 2015). Approaches to generate positive affective experiences through enhancing environmental complexity include fixed and automated grooming brushes, scents, novel objects and tasks such as extracting straw from a feeder (Pelley et al. 1995; Wilson et al. 2002). Behavioural demand studies indicate cattle will work for access to a mechanical grooming device (McConnachie et al. 2018) and make use of the feature when it is provided in a feedlot pen (Ninomiya 2019). Further research on welfare benefits of these features is desirable. Studies on sows show that providing animals with the means to control social interactions at the feeder has positive welfare benefits (Manteuffel 2015; Zebunke et al. 2011). Potential benefits of systems for feedlot cattle that provide animals with control over their exposure to potential threats and their access to resources deserve attention.

### 7.5.2 Management Practices

Early life experiences can confer the psychological and behavioural competence to cope with challenges later in life. In pre-weaned lambs, exposure to a feedlot ration in the company of their dams improves acceptance of the feed and reduces prevalence of shy feeders at subsequent feedlot entry (Savage et al. 2008). Early exposure of beef calves destined for feedlot finishing to feed rations in the presence of their mothers may confer similar benefits. In production systems where calves are grazed at pasture for several months after weaning, the weaning period provides calves with an opportunity to develop competence for coping with the feedlot environment. As described in Sect. 7.2, calves confined and fed in open-air yards on their property of birth for 7 or more days at weaning have better growth rates and lower disease rates during feedlot finishing than fence-weaned calves (Walker et al. 2007). Further research on animal handling procedures that develop the competence of cattle for adapting to handling by humans and novel environments is warranted. Studies on timely intervention points for euthanasia of moribund feedlot animals are needed as well as refinements in methods of euthanasia (AVMA 2020; Schwartzkopf-Genswein et al. 2017).

### 7.5.3 Genetic Selection

Several traits associated with adaptability, resilience and environmental fit provide opportunities for genetic selection to improve welfare outcomes in feedlot cattle. The heritability of temperament traits measured as FS and CS provide a practical method to select animals for improved health and productivity during feedlot finishing (reviewed by Haskell et al. 2014). Additionally, learning ability differs between individual cattle (Campbell et al. 2019; Webb et al. 2015) and is likely to be a component of adaptation to new handling procedures and to the physical infrastructure of new environments (Wechsler and Lea 2007); however, there appears to be no studies to date on heritability of learning ability in beef cattle. The rate of adaptation to handling in yards during weaning may provide an opportunity to phenotype cattle for adaptive learning skills relevant to feedlot finishing (Monk et al. 2018). Together these genetic strategies to improve environmental fit and capacity to cope with the feedlot environment have strong potential to improve welfare.

A large volume of research is addressing the potential for selection of beef cattle for heat tolerance (Bradford et al. 2016; Carabaño et al. 2019; Mackinnon et al. 1991); however, implementation within industry is currently limited. A capacity for animals to perform uniformly across a diversity of macroenvironments such as tropical, subtropical, temperate and Mediterranean climatic zones or across production systems such as pasture versus feedlot is termed robustness (Knap 2005). In contrast, as discussed above, the capacity to maintain uniformity of performance across the day-to-day microenvironmental variations that occur within a given macroenvironment is termed resilience (Berghof et al. 2018; Colditz and Hine 2016). Mulder and colleagues have demonstrated the potential for analysing production data such as daily milk yield and body weight to estimate genetic parameters for resilience in dairy cows (Elgersma et al. 2018; Poppe et al. 2020) and layer hens (Berghof et al. 2019). These methods depend on individual animal data recorded at a high frequency (Lung et al. 2020). The results suggest that high-frequency recording of performance data on feedlot cattle such as body weight or feed intake (Putz et al. 2018) could be used to estimate breeding values for resilience for feedlot finishing. The health and production benefits for lot-fed steers of high immune competence indicate the potential for this genetic strategy to improve resilience for feedlot production (Hine et al. 2019). However, it is very important to note that high productivity per se does not necessarily equate to high resilience. Rather, for output measures of animal function like growth rate, resilience to microenvironmental fluctuations is evidenced by a capacity to maintain a constant performance trajectory in accord with the individual's genetic potential. For intermediary measures of function such as metabolism, body temperature and behavioural repertoire, it is a capacity to maintain a normal circadian dynamic range and comprehensive repertoire that is indicative of resilience. The decrease in heart rate variability during stress provides an illustration of an intermediate measure of low resilience. Further work is needed to refine measures of resilience in feedlot cattle and to determine whether there needs to be a short-term trade-off for production when selecting for high resilience. From the animal welfare perspective, it is considered that resilience traits confer better welfare during periods of (micro)environmental variability. Short-term costs may confer longer-term economic benefits through a capacity to maintain production in the presence of increasing climatic variability.

### 7.5.4 Prediction of Environmental Fit

Despite the implementation of breeding programmes to improve the suitability of progeny for feedlot finishing, there remains phenotypic variation between progeny in their capacity to cope and thrive within the feedlot environment. This variation provides an opportunity to select individuals before feedlot entry in order to further minimise the risk of adverse health and welfare outcomes. A coarse level of prediction is provided by selecting breeds suited to the climatic zone a feedlot is located in. Minimising exposure to a number of risk factors associated with the history of animals is employed by some feedlot managers in order to improve health outcomes. These criteria include sourcing animals from suppliers with a track record of providing animals with good health outcomes during feedlot finishing, sourcing animals that have been yard weaned and vaccinated before feedlot entry and excluding animals bought through saleyards from feedlot entry. Additional refinement of prediction of environmental fit could be achieved by phenotypic classing of animals, for instance, on temperament. At present, this practice does not appear to have been adopted.

Haematological parameters, especially the prevalence and activity of leukocyte populations, have shown some promise as predictors of health and welfare outcomes in the feedlot (Colditz et al. 2007; Colditz and Hennessy 2001; Fell et al. 1999). Further research on phenotypic predictors of environmental fit seems warranted. An additional approach attracting considerable interest is genomic prediction. A proof of principle study has demonstrated the potential for genomic prediction to identify animals with superior potential for growth and production of high-value carcases when finished in a feedlot (Angus Select 2020). This strategy depends on genomic analysis of individual animals prior to feedlot entry in order to identify individuals with superior potential for feedlot finishing. In view of the heritability of immune competence and resilience, genomic analyses should also have the potential to enable prediction of health and welfare outcomes and could provide an additional pathway towards improving the welfare of feedlot cattle.

### 7.5.5 The Need to Address Positive Welfare Outcomes

Improving welfare is widely understood to include both minimising harms and enabling the animal to achieve positive welfare outcomes (Mellor 2015; Yeates and Main 2008). Despite its well-recognised importance, what constitutes positive welfare and how it can be measured are topics that are under development (reviewed by Lawrence et al. 2019; Rault et al. 2020). Lawrence et al. (2019) suggest that positive welfare includes (1) positive affective states, (2) engagement with the environment in a manner that provides positive affective experiences, (3) quality of life arising from the balance of positive and negative states and (4) happiness across the whole of the animal's life. This formulation valorises agency and competence realized as hedonic positive affective experiences through psychological and behavioural engagement of the animal with its environment (Spinka and Wemelsfelder 2011; Špinka 2019; White 1959).

It is noteworthy that agency and competence realised as environmental fitness and resilience through physiological and immunological competence also deserve consideration as dimensions of positive welfare (Colditz 2018; Colditz and Hine 2016; Nordenfelt 2011). This view draws support from studies of wellbeing and the consequences of adversity on immune function and health outcomes in humans (Cole et al. 2015). Chronic stressors such as social isolation are associated with a shift in gene expression in the innate and adaptive immune systems away from a pattern associated with strong antiviral and antibody defence towards a pattern associated with inflammatory defence. The inflammatory gene expression profile, termed a conserved transcriptional response to adversity, is associated with increased susceptibility to a range of inflammatory and infectious diseases and has also been observed in rhesus monkeys and rodents following exposure to chronic social stressors (Cole 2019). Studies of wellbeing in humans identify two aspects: hedonic wellbeing (that can be characterised with the epigram 'I feel well; therefore, I am well') and eudaemonic wellbeing ('I function well; therefore, I flourish') (Fredrickson 2016). Across a number of independent studies, eudaemonic wellbeing has been found to align with the pattern of immune gene expression that favours strong antiviral defence and better health outcomes, whereas hedonic wellbeing does not (Fredrickson 2016; Fredrickson et al. 2015). For the concept of positive welfare, these findings on wellbeing in humans and on immune competence and resilience in cattle suggest that adaptive agency as a capacity for the animal to engage positively with its environment in order to flourish arises from a conjunction of immunological, physiological and behavioural competences as well as psychological competence and positive affect (Fig. 7.4). Good functioning can furnish more than good health; it provides a foundation for eudaemonic wellbeing and for the capacity to flourish (Nordenfelt 2011).

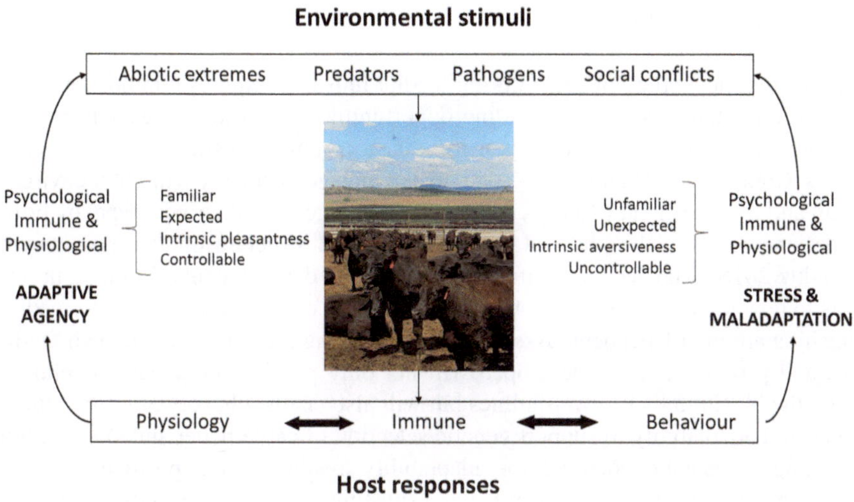


**Fig. 7.4** Feedlot cattle engage with the fluctuating conditions in their environment through physiological, behavioural and immunological host actions. Positive engagement through familiarity, prediction, control and intrinsic pleasantness confers adaptive agency to the animal and generates positive welfare outcomes. When environmental conditions are unpredictable, uncontrollable, unfamiliar and intrinsically unpleasant, the animal can become psychologically, physiologically and immunologically stressed and maladapted to the adverse conditions and experiences poor welfare. (Photo courtesy of Dominic Niemeyer, CSIRO, Armidale, Australia)

This model builds a bridge grounded in enactivism (i.e. the concept that organisms gain 'meaning' of the world through actively interacting with it; Colditz 2019) between *good* welfare based on feelings, biological functions and naturalness (Fraser et al. 1997; Fraser 2008) and *positive* welfare realised through an enhanced capacity for positive eudaemonic function based on physiological, immunological and behavioural engagement with environmental challenges, as well as through hedonic positive affective engagement. As described above, methods are being developed for measuring immune competence and physiological resilience. Similarly, an ability to quantify the psychological dimensions of positive welfare is in its infancy, but is being supported by advances in recent years in measuring positive hedonic affective states in a range of farm animal species including cattle (Ede et al. 2019; Mendl and Paul 2020). Of importance for feedlots (and other beef production environments) is the need for methods capable of the assessment of positive affective states of animals during routine handling through yards or while they are undertaking voluntary activities within their home pen (Mattiello et al. 2019). Promising candidates requiring further research for quantitative application in feedlots include vocalizations (Tucker et al. 2015), lateralization (Mendl and Paul 2020), attention bias (Lee et al. 2018) and qualitative behavioural assessment of activity and demeanour (Wemelsfelder and Lawrence 2001).

## 7.6 Conclusions

Optimising the welfare of cattle during feedlot finishing requires minimising exposure to potential harms and providing opportunities for positive engagement with the physical and social environment in order to promote positive welfare experiences. Competence of animals to adapt to and engage positively with their environment and to be resilient to daily challenges is influenced by their genotype, by prior experience and by the physical and psychological structure of the environment (Colditz 2018). The design of infrastructure and the development of training protocols to develop competence and to provide positive experiences relevant to feedlot finishing are interdependent tasks within an emerging area of research. Such facilities and procedures once developed will not only provide immediate benefits to individuals within their own lifetimes but will also enable phenotyping for competence and adaptability to support genetic selection of cattle better suited to feedlot finishing. A genetic potential for adaptability, resilience and positive affective engagement with the environment is not a substitute for but rather provides a complement to good animal management from birth until finishing in well-designed feedlots.

## References


Alberta Feedlot Management Guide (1996) Alberta Agriculture Food and Rural Development, Edmonton

American Veterinary Medical Association (AVMA) (2020) AVMA guidelines for the euthanasia of animals, 2020 edn, p 121

Ames D (1988) Adjusting rations for climate. Vet Clin N Am Food Anim Pract 4:543–550

Ames DR, Insley LW (1975) Wind-chill effect for sheep and cattle. J Anim Sci 40:161–165

Andrews T (2015) Primefact 621 second edition – market specifications for cattle. NSW Department of Primary Industries. NSW. https://www.dpi.nsw.gov.au/animals-and-livestock/beef-cattle/management/market-information/beef-market-specifications. Accessed 5 Dec 2019

Angus Select (2020) Prediction of breeding values for commercial animals now a reality. Angus Bulletin Winter 2020. https://www.angusaustralia.com.au/news-2/publications/angus-bulletin/. Accessed 16 July 2020

Animal Health Australia (2014) Australian Animal Welfare Standards and Guidelines for Cattle. www.animalwelfarestandards.net.au. Accessed 1 Mar 2020

Arias RA, Mader TL (2011) Environmental factors affecting daily water intake on cattle finished in feedlots. J Anim Sci 89(1):245–251

Australian Lot Feeders Association (2015). http://www.feedlots.com.au/home. Accessed 2 Nov 2022

Badger SM, Sullivan KF, Jordan D, Caraguel CGB, Page SW, Cusack PMV, Frith D, Trott DJ (2020) Antimicrobial use and stewardship practices on Australian beef feedlots. Aust Vet J 98:37–47

Barnes T, Hay K, Morton J, Schibrowski M, Ambrose R, Fowler E, Mahony T (2014) Epidemiology and management of bovine respiratory disease in feedlot cattle - final report. Meat and Livestock Australia

Bergen RD, Kennedy AD, McAllister TA, Schwartzkopf-Genswein KS (2007) Effect of morning vs evening feeding on thermoregulation and performance of beef heifers in winter. Can J Anim Sci 87:307–313

Berghof TV, Poppe M, Mulder HA (2018) Opportunities to improve resilience in animal breeding programs. Front Genet 9

Berghof T, Bovenhuis H, Mulder H (2019) Body weight deviations as indicator for resilience in layer chickens. Front Genet 10:1216

Bergman E (2019) Feedlot best practice management: pregnant heifers. Meat & Livestock Australia, Sydney

Bevans DW, Beauchemin KA, Schwartzkopf-Genswein KS, McKinnon JJ, McAllister TA (2005) Effect of rapid or gradual grain adaptation on subacute acidosis and feed intake by feedlot cattle. J Anim Sci 83(5):1116–1132

Blackshaw JK, Blackshaw AW (1994) Heat stress in cattle and the effect of shade on production and behaviour: a review. Aust J Exp Agric 34:285–295

Boivin X, Lensink J, Tallet C, Veissier I (2003) Stockmanship and farm animal welfare. Anim Welf 12(4):479–492

Bouissou MF (1980) Social relationships in domestic cattle under modern management techniques. Italian J Zool 47(3–4):343–353

Bradford HL, Fragomeni BO, Bertrand JK, Lourenco DAL, Misztal I (2016) Genetic evaluations for growth heat tolerance in Angus cattle. J Anim Sci 94:4143–4150

Brower GR, Kiracofe GH (1978) Factors associated with the Buller-Steer syndrome. J Anim Sci 46(1):26–31

Brown-Brandl TM, Eigenberg RA, Nienaber JA (2006) Heat stress risk factors of feedlot heifers. Livest Sci 105(1–3):57–68

Buhman MJ, Perino LJ, Galyean ML, Wittum TE, Montgomery TH, Swingle RS (2000) Association between changes in eating and drinking behaviors and respiratory tract disease in newly arrived calves at the feedlot. Am J Vet Res 61:1163–1168

Buhman MJ, Hungerford LL, Smith DR (2003) An economic risk assessment of the management of pregnant feedlot heifers in the USA. Prev Vet Med 59(4):207–222

Burrow HM (1997) Measurements of temperament and their relationship with performance traits of beef cattle. In: Burrow HM (ed) Measurements of temperament and their relationship with performance traits of beef cattle. Animal breeding abstracts 65, pp 477–495

Burrow HM, Seifert GW, Corbet NJ (1988) A new technique for measuring temperament in cattle. Proc Aust Soc Anim Prod 17:154–157

Busby D, Loy D (1997) Heat stress in feedlot cattle: producer survey results. Beef Research Report, 1996. 26. http://lib.dr.iastate.edu/beefreports_1996/26. Accessed 11 Feb 2020

Cafe LM, Robinson DL, Ferguson DM, Geesink GH, Greenwood PL (2011) Temperament and hypothalamic-pituitary-adrenal axis function are related and combine to affect growth, efficiency, carcass, and meat quality traits in Brahman steers. Domest Anim Endocrinol 40:230–240

Callaway TR, Edrington TS, Rychlik JL, Genovese KJ, Poole TL, Jung YS, Bischoff KM, Anderson RC, Nisbet DJ (2003) Ionophores: their use as ruminant growth promotants and impact on food safety. Curr Issues Intest Microbiol 4:43–51

Campbell DL, Lea JM, Keshavarzi H, Lee C (2019) Virtual fencing is comparable to electric tape fencing for cattle behavior and welfare. Front Vet Sci 6:445

Carabaño MJ, Ramón M, Menéndez-Buxadera A, Molina A, Díaz C (2019) Selecting for heat tolerance. Anim Front 9:62–68

Ceballos MC, Sant'Anna AC, Boivin X, Costa FDO, Carvalhal MVDL, Paranhos da Costa MJR (2018) Impact of good practices of handling training on beef cattle welfare and stockpeople attitudes and behaviors. Livest Sci 216:24–31

Cernicchiaro N, White BJ, Renter DG, Babcock AH, Kelly L, Slattery R (2012) Associations between the distance traveled from sale barns to commercial feedlots in the United States and overall performance, risk of respiratory disease, and cumulative mortality in feeder cattle during 1997 to 2009. J Anim Sci 90:1929–1939

Cernicchiaro N, White BJ, Renter DG, Babcock AH (2013) Effects and economic implications of metaphylactic treatment of feeder cattle with two different dosages of tilmicosin on the incidence of bovine respiratory disease (BRD) – a summary of two studies. Am J Vet Res 74:300–309

Chaucheyras-Durand F, Durand H (2010) Probiotics in animal nutrition and health. Benefic Microbes 1(1):3–9
Coetzee JF (2011) A review of pain assessment techniques and pharmacological approaches to pain relief after bovine castration: practical implications for cattle production within the United States. Appl Anim Behav Sci 135:192–213
Colditz IG (2018) Objecthood, agency and mutualism in valenced farm animal environments. Animals 8:50
Colditz IG (2019) A consideration of physiological regulation from the perspective of Bayesian enactivism. Physiol Behav 214:112758
Colditz IG, Hennessy DW (2001) Associations between immune system, growth and carcass variables in cattle. Aust J Exp Agric 41:1051–1056
Colditz IG, Hine BC (2016) Resilience in farm animals: biology, management, breeding and implications for animal welfare. Anim Prod Sci 56:1961–1983
Colditz, IG, Ferguson, DM, Greenwood, PL, Doogan, VJ, Petherick, JC, Kilgour, RJ. 2007. Regrouping unfamiliar animals in the weeks prior to slaughter has few effects on physiology and meat quality in Bos taurus feedlot steers. Aust J Exp Agric 47, 763–40]769
Cole SW (2019) The conserved transcriptional response to adversity. Curr Opin Behav Sci 28:31–37
Cole SW, Levine ME, Arevalo JM, Ma J, Weir DR, Crimmins EM (2015) Loneliness, eudaimonia, and the human conserved transcriptional response to adversity. Psychoneuroendocrinology 62:11–17
Collings LKM, Weary DM, Chapinal N, Von Keyserlingk MAG (2011) Temporal feed restriction and overstocking increase competition for feed by dairy cattle. J Dairy Sci 94(11):5480–5486
Cooke RF, Guarnieri Filho TA, Cappellozza BI, Boohnert DW (2013) Rest stops during road transport: impacts on performance and acute-phase protein responses of feeder cattle. J Anim Sci 91:5448–5454
Curley KOJ, Pfeiffer CE, King DA, Savell JW, Vann RC, Welsh TH Jr, Randel RD (2006) Relationships of cattle temperament and physiologic responses to handling during typical management situations. J Anim Sci 84(Suppl. 2):32
Cusack PMV, Mahony TJ (2019) Evaluation of practices used to reduce the incidence of bovine respiratory disease in Australian feedlots. Meat & Livestock Australia, Sydney. https://www.mla.com.au/globalassets/mla-corporate/research-and-development/program-areas/feeding-finishing-and-nutrition/cic_104383_mla-brd-manual_06-web.pdf. Accessed 29 Jan 2020
Daigle CL, Ridge EE (2018) Investing in stockpeople is an investment in animal welfare and agricultural sustainability. Anim Front 8:53–59
Davis MS, Mader TL, Holt SM, Parkhurst AM (2003) Strategies to reduce feedlot cattle heat stress: effects on tympanic temperature. J Anim Sci 81(3):649–661
Davis-Unger JK, Schwartzkopf-Genswein EA, Pajor S, Hendrick S, Marti S, Dorin C, Orsel K (2019) Prevalence and lameness-associated risk factors in Alberta feedlot cattle. Translational Anim Sci 3:595–606
Deblitz C (2012) Feed lots: a new tendency in global beef production. Agri Benchmark. https://literatur.thuenen.de/digbib_extern/dn054620.pdf. Accessed 9 July 2018
Dennis EJ, Schroeder TC, Renter DG, Pendell DL (2018) Value of arrival metaphylaxis in U.S. cattle industry. J Agric Resour Econ 43(2):233–250
DeVries TJ, Schwaiger T, Beauchemin KA, Penner GB (2014) The duration of time that beef cattle are fed a high-grain diet affects feed sorting behavior, both prior to and after acute ruminal acidosis. J Anim Sci 92:1728–1737
Duffield TF, Heinrich A, Millman ST, DeHaan A, James S, Lissemore K (2010) Reduction in pain response by combined use of local lidocaine anesthesia and systemic ketoprofen in dairy calves dehorned by heat cauterization. Can Vet J 51:283–288
Ede T, Lecorps B, von Keyserlingk MA, Weary DM (2019) Symposium review: scientific assessment of affective states in dairy cattle. J Dairy Sci 102:10677–10694
Edwards TA (1995) Buller syndrome: what's behind this abnormal sexual behavior? Large Anim Vet 50(40):6–8

Elgersma GG, De Jong G, Van der Linde R, Mulder HA (2018) Fluctuations in milk yield are heritable and can be used as a resilience indicator to breed healthy cows. J Dairy Sci 101:1240–1250
Endres M, Schwartzkopf-Genswein K (2018) Overview of cattle production systems. In: Tucker CB (ed) Advances in cattle welfare. Woodhead Publishing, pp 1–25
Euken R, Doran BE, Clark CA, Shouse SC, Loy D, Schulz LL (2015) Beef feedlot systems manual. Extension and outreach publications. 95. http://lib.dr.iastate.edu/extension_pubs/95. Accessed 10 Feb 2020
Eurobarometer (2007) Attitudes of EU citizens towards animal welfare. European Commission. http://ec.europa.eu/commfrontoffice/publicopinion/archives/ebs/ebs_270_en.pdf. Accessed 5 Nov 2018
Fell LR, Colditz IG, Walker KH, Watson DL (1999) Associations between temperament, performance and immune function in cattle entering a commercial feedlot. Aust J Exp Agric 39:795–802
Ferguson DM (2014) Key features of 'environmental fit' that promote good animal welfare in different husbandry systems. Revue scientifique et technique (International Office of Epizootics) 33:161–169
Ferguson DM, Johnston D, Burrow HM, Reverter A (2006) Relationships between temperament, feedlot performance and beef quality. In: Beef CRC, Australian beef - the leader conference, Armidale, pp 161–165
Finkemeier MA, Langbein J, Puppe B (2018) Personality research in mammalian farm animals: concepts, measures, and relationship to welfare. Front Vet Sci 5:131
Fordyce G, Goddard ME, Seifert GW (1982) The measurement of temperament in cattle and the effect of experience and genotype. Proc Aust Soc Anim Prod J 4:329–332
Fraser D (2008) Understanding animal welfare. Acta Vet Scand 50:S1
Fraser D, Weary DM, Pajor EA, Milligan BN (1997) A scientific conception of animal welfare that reflects ethical concerns. Anim Welf 6:187–205
Fredrickson BL (2016) The eudaimonics of positive emotions. In: Handbook of eudaimonic well-being. Springer, pp 183–190
Fredrickson BL, Grewen KM, Algoe SB, Firestine AM, Arevalo JM, Ma J, Cole SW (2015) Psychological well-being and the human conserved transcriptional response to adversity. PLoS One 10:e0121839
Galyean ML, Rivera JD (2003) Nutritionally related disorders affecting feedlot cattle. Can J Anim Sci 83(1):13–20
Gaughan JB, Sullivan ML (2014) Australian feedlot industry. In: Kahn L, Cottle D (eds) Beef cattle production and trade. CSIRO Publishing, pp 205–233
Gaughan J, Mader T, Holt S, Lisle A (2008) A new heat load index for feedlot cattle. J Anim Sci 86:226–234
Gaughan JB, Bonner SL, Loxton I, Mader TL (2013) Effects of chronic heat stress on plasma concentration of secreted heat shock protein 70 in growing feedlot cattle. J Anim Sci 91(1):120–129
Goldhawk C, Janzen E, Gonzalez KA, Crowe T, Kastelic J et al (2015) Trailer temperature and humidity during winter transport of cattle in Canada and evaluation of indicators used to assess the welfare of cull beef cows before and after transport. J Anim Sci 93:3639–3653
González LA, Manteca X, Calsamiglia S, Schwartzkopf-Genswein KS, Ferret A (2011) Ruminal acidosis in feedlot cattle: review from a feeding behaviour perspective. Anim Feed Sci Technol 172:66–79
González LA, Schwartzkopf-Genswein KS, Bryan M, Silasi R, Brown F (2012a) Space allowance during commercial long distance transport of cattle in North America. J Anim Sci 90:3618–3629
González LA, Schwartzkopf-Genswein KS, Bryan M, Silasi R, Brown F (2012b) Factors affecting body weight loss during commercial long haul transport of cattle in North America. J Anim Sci 90:3630–3639
González LA, Schwartzkopf-Genswein KS, Bryan M, Silasi R, Brown F (2012c) Relationships between transport conditions and welfare outcomes during commercial long haul transport of cattle in North America. J Anim Sci 90:3640–3651

Goonewardene LA, Hand RK (1991) Studies on dehorning steers in Alberta feedlots. Can J Anim Sci 71:1249–1252

Gottardo F, Ricci R, Preciso S, Ravarotto L, Cozzi G (2004) Effect of the manger space on welfare and meat quality of beef cattle. Livest Prod Sci 89(2):277–285

Grandin T (2016) Evaluation of the welfare of cattle housed in outdoor feedlot pens. Vet Anim Sci 1(2):23–28

Griffin D (1997) Economic impact associated with respiratory disease in beef cattle. Vet Clin N Am Food Anim Pract 13:367–377

Haley DB, Bailey DW, Stookey JM (2005) The effects of weaning beef calves in two stages on their behavior and growth rate. J Anim Sci 83:2205–2214

Haskell MJ, Simm G, Turner SP (2014) Genetic selection for temperament traits in dairy and beef cattle. Front Genet 5. https://doi.org/10.3389/fgene.2014.00368

Hay KE, Barnes TS, Morton JM, Clements ACA, Mahony TJ (2014) Risk factors for bovine respiratory disease in Australian feedlot cattle: use of a causal diagram-informed approach to estimate effects of animal mixing and movements before feedlot entry. Prev Vet Med 117:160–169

Hay KE, Morton JM, Schibrowski ML, Clements ACA, Mahony TJ, Barnes TS (2016a) Associations between prior management of cattle and risk of bovine respiratory disease in feedlot cattle. Prev Vet Med 127:37–43

Hay KE, Morton JM, Clements ACA, Mahony TJ, Barnes TS (2016b) Associations between feedlot management practices and bovine respiratory disease in Australian feedlot cattle. Prev Vet Med 128:23–32

Heinrich A, Duffield TF, Lissemore KD, Squires EJ, Millman ST (2009) The impact of meloxicam on postsurgical stress associated with cautery dehorning. J Dairy Sci 92:540–547

Hemsworth PH (2003) Human–animal interactions in livestock production. Appl Anim Behav Sci 81:185–198

Herring AD (2014) Beef cattle production systems. CABI, Texas

Hessing MJC, Coenen GJ, Vaiman M, Renard C (1995) Individual differences in cell-mediated and humoral immunity in pigs. Vet Immunol Immunopathol 45:97–113

Hine B, Ingham A, Dominik S, Colditz I (2016) Assessment of immune competence in beef cattle. https://www.mla.com.au/download/finalreports?itemId=3344. Accessed 29 Sept 2019

Hine BC, Bell AM, Niemeyer DDO, Duff CJ, Butcher NM, Dominik S, Ingham AB, Colditz IG (2019) Immune competence traits assessed during the stress of weaning are heritable and favourably genetically correlated with temperament traits in Angus cattle. J Anim Sci 97:4053–4065

Hosseinkhani A, DeVries TJ, Proudfoot KL, Valizadeh R, Veira DM, Von Keyserlingk MAG (2008) The effects of feed bunk competition on the feed sorting behavior of close-up dry cows. J Dairy Sci 91(3):1115–1121

Kadel MJ, Johnston DJ, Burrow HM, Graser HU, Ferguson DM (2006) Genetics of flight time and other measures of temperament and their value as selection criteria for improving meat quality traits in tropically adapted breeds of beef cattle. Crop Pasture Sci 57:1029–1035

Karren DB, Basarab JA, Church TL (1987) The growth and economic performance of preconditioned calves and their dams on the farms and of calves in the feedlot. Can J Anim Sci 67:327–336

Kelm SC, Dettilleux JC, Freeman AE, Kehrli ME Jr, Dietz AB, Fox LK, Butler JE, Kasckovics I, Kelley DH (1997) Genetic association between parameters of innate immunity and measures of mastitis in periparturient Holstein cattle. J Dairy Sci 80:1767–1775

Knap PW (2005) Breeding robust pigs. Aust J Exp Agric 45:763–773

Kondo S, Sekine J, Okubo M, Asahida Y (1989) The effect of group size and space allowance on the agonistic and spacing behaviour of cattle. Appl Anim Behav Sci 24:127

Koolhaas JM (2008) Coping style and immunity in animals: making sense of individual variation. Brain Behav Immun 22:662–667

Koolhaas JM, van Reenen CG (2016) Interaction between coping style/personality, stress, and welfare: relevance for domestic farm animals. J Anim Sci 94:2284–2296

Larmer SG, Mallard BA (2016) High immune response sires reduce disease incidence in North American large commercial dairy populations. Cattle Pract 24:80

Lauder JK, Marti S, Schwartzkopf-Genswein KS, Jelinski MD, Janzen ED (2020) Measuring behavioral and physiological responses to pain mitigation for ovariectomy in Bos taurus yearling beef heifers. J Anim Sci 98(1):skz386

Lawrence J, Harmon J, Lorimor J, Edwards W, Loy D, Klein W, Miller W (eds) (2001) Beef feedlot systems manual. PM 1867. Iowa State University, Ames

Lawrence AB, Vigors B, Sandöe P (2019) What is so positive about positive animal welfare? Critical review of the literature. Animals 9:783

Lee C, Fisher AD, Colditz IG, Lea JM, Ferguson DM (2013) Preference of beef cattle for feedlot or pasture environments. Appl Anim Behav Sci 145(3):53–59

Lee C, Cafe LM, Robinson SL, Doyle RE, Lea JM, Small AH, Colditz IG (2018) Anxiety influences attention bias but not flight speed and crush score in beef cattle. Appl Anim Behav Sci 205:210–215

Lees AM, Salvin HE, Colditz IG, Lee C (2020) The influence of temperament on body temperature responses to handling in Angus cattle. Animals 10:172

Lie Ø (1979) Genetic analysis of some immunological traits in young bulls. Acta Vet Scand 20:372

Lung LHDS, Carvalheiro R, Neves HHDR, Mulder HA (2020) Genetics and genomics of uniformity and resilience in livestock and aquaculture species: a review. J Anim Breed Genet 137(3):263–280

Lynch E, McGee M, Earley B (2019) Weaning management of beef calves with implications for animal health and welfare. J Appl Anim Res 47:167–175

Macartney JE, Bateman KG, Ribble CS (2003) Health performance of feeder calves sold at conventional auctions versus special auctions of vaccinated or conditioned calves in Ontario. J Am Vet Med Assoc 223:677–683

MacKay JR, Haskell MJ (2015) Consistent individual behavioral variation: the difference between temperament, personality and behavioral syndromes. Animals 5:455–478

Mackinnon MJ, Meyer K, Hetzel DJS (1991) Genetic variation and covariation for growth, parasite resistance and heat tolerance in tropical cattle. Livest Prod Sci 27:105–122

Mader TL (2003) Environmental stress in confined beef cattle. J Anim Sci 81(14 Suppl 2):E110–E119

Mader TL (2011) Mud effects on feedlot cattle. Nebraska Beef Cattle Reports. 613. https://digitalcommons.unl.edu/animalscinbcr/613/. Accessed 23 Apr 2018

Mader TL, Griffin D (2015) Management of cattle exposed to adverse environmental conditions. Vet Clin Food Anim Pract 31(2):247–258

Mader TL, Dahlquist JM, Gaughan JB (1997a) Wind protection effects and airflow patterns in outside feedlots. J Anim Sci 75:26–36

Mader TL, Fell LR, McPhee MJ (1997b) Behavior response of non-Brahman cattle to shade in commercial feedlots. In: Proceedings of the 5th international livestock environmental symposium, American Society of Agricultural Engineering, St. Joseph, pp 795–802

Mader TL, Dahlquist JM, Hahn GL, Gaughan JB (1999) Shade and wind barrier effects on summertime feedlot cattle performance. J Anim Sci 77:2065–2072

Mader TL, Holt SM, Hahn GL, Davis MS, Spiers DE (2002) Feeding strategies for managing heat load in feedlot cattle. J Anim Sci 80(9):2373–2382

Mader TL, Davis MS, Gaughan JB (2007) Effect of sprinkling on feedlot microclimate and cattle behavior. Int J Biometeorol 51(6):541–551

Mallard BA, Wilkie BN, Kennedy BW, Quinton M (1992) Use of estimated breeding values in a selection index to breed Yorkshire pigs for high and low immune and innate resistance factors. Anim Biotechnol 3:257–280

Manteuffel C (2015) The technical manipulation of the behaviour of sows exemplified by call feeding and active crushing prevention. PhD thesis, University of Kiel, pp 1–117

Marti S, Meléndez DM, Pajor EA, Moya D, Heuston CEM, Gellatly D, Janzen ED, Schwartzkopf-Genswein KS (2017a) Effect of band and knife castration of beef calves on welfare indicators of pain at three relevant industry ages: II. Chronic Pain J Anim Sci 95:4367–4380

Marti S, Wilde RE, Moya D, Heuston CE, Brown F, Schwartzkopf-Genswein KS (2017b) Effect of rest stop duration during long distance transport on welfare indicators of recently weaned beef calves. J Anim Sci 95:1–9

Mattiello S, Battini M, De Rosa G, Napolitano F, Dwyer C (2019) How can we assess positive welfare in ruminants? Animals 9:758

McConnachie E, Smid AM, Thompson AJ, Weary DM, Gaworski MA, von Keyserlingk MA (2018) Cows are highly motivated to access a grooming substrate. Biol Lett 14:20180303

McCosker K, Letchford P, Petherick JC, Meyer D, McGowan M (2010) Morbidity, mortality and body weight gain of surgically spayed, yearling Brahman heifers. Aust Vet J 88(12):497–503

McGinn SM, Janzen HH, Coates T (2003) Atmospheric ammonia, volatile fatty acids, and other odorants near beef feedlots. J Environ Qual 32(4):1173–1182

Meléndez DM, Marti S, Pajor EA, Moya D, Heuston CEM, Gellatly D, Janzen ED, Schwartzkopf-Genswein KS (2017) Effect of band and knife castration of beef calves on welfare indicators of pain at three relevant industry ages: I. Acute Pain J Anim Sci 95:4352–4366

Meléndez DM, Marti S, Haley DB, Schwinghamer TD, Schwartzkopf-Genswein KS (2020) Effect of transport and rest stop duration on the welfare of conditioned cattle transported by road. PLoS One. https://doi.org/10.1371/journal.pone.0228492

Mellor DJ (2015) Positive animal welfare states and encouraging environment-focused and animal-to-animal interactive behaviours. N Z Vet J 63:9–16

Mellor DJ (2016) Moving beyond the "five freedoms" by updating the "five provisions" and introducing aligned "animal welfare aims". Animals 6(10):59

Mench JA, Swanson JC, Stricklin WR (1990) Social stress and dominance among group members after mixing beef cows. Can J Anim Sci 70:345–354

Mendl M, Paul ES (2020) Assessing affective states in animals. In: Mcmillan FD (ed) Mental health and well-being in animals. CABI, Wallingford, pp 328–344

Meneses XCA, Park RM, Daigle CL (2019) Impact of sociometric status on brush utilization in environmentally-enriched, feedlot. J Anim Sci 97:249–249

Mingmongkolchai S, Panbangred W (2018) Bacillus probiotics: an alternative to antibiotics for livestock production. J Appl Microbiol 124(6):1334–1346

Moberg GP (2000) Biological response to stress: implications for animal welfare. In: Moberg GP, Mench JA (eds) The biology of animal stress. CABI, Wallingford, pp 1–22

Moggy MA, Pajor EA, Thurston WE, Parker S, Greter AM, Schwartzkopf-Genswein KS, Campbell JR, Windeyer MC (2017a) Management practices associated with pain in cattle on western Canadian cow–calf operations: a mixed methods study. J Anim Sci 95:958–969

Moggy MA, Pajor EA, Thurston WE, Parker S, Greter AM, Schwartzkopf-Genswein KS, Campbell JR, Windeyer MC (2017b) Management practices associated with stress in cattle on western Canadian cow–calf operations: a mixed methods study. J Anim Sci 95:1836–1844

Monk JE, Hine BC, Colditz IG, Lee C (2018) A novel protocol to assess acclimation rate in Bos taurus heifers during yard weaning. Animals 8:51

Moya D, Silasi R, McAllister TA, Genswein B, Crowe T, Marti S, Schwartzkopf-Genswein KS (2015) Use of pattern recognition techniques for early detection of morbidity in receiving feedlot cattle. J Anim Sci 93:3623–3638

National Farm Animal Care Council (NFACC) (2013) Code of practice for the care and handling of beef cattle. http://www.nfacc.ca/pdfs/codes/beef_code_of_practice.pdf. Accessed 12 Mar 2020

National Guidelines for Beef Cattle Feedlots in Australia (2012) Meat and Livestock Australia, North Sydney

Netea MG, Joosten LA, Latz E, Mills KH, Natoli G, Stunnenberg HG, O'Neill LA, Xavier RJ (2016) Trained immunity: a program of innate immune memory in health and disease. Science 352:aaf1098

Newberry RC (1995) Environmental enrichment: increasing the biological relevance of captive environments. Appl Anim Behav Sci 44:229–243

Nickell JS, Keil DJ, Settje TL, Lechtenberg KF, Singu V, Woolums AR (2016) Efficacy and safety of a novel DNA immunostimulant in cattle. Bov Pract 50:9–20

Ninomiya S (2019) Grooming device effects on behaviour and welfare of Japanese black fattening cattle. Animals 9:186

Ninomiya S, Sato S (2009) Effects of 'Five freedoms' environmental enrichment on the welfare of calves reared indoors. Anim Sci J 80:347–351

Nordenfelt L (2011) Health and welfare in animals and humans. Acta Biotheor 59:139–152

Nosky B, Biwer J, Alkemade S, Prunic B, Milovanovic A, Maletic M (2018) Effect of a non-specific immune stimulant (AmplimuneTM) on the health and production of light feedlot calves. J Dairy Vet Anim Res 6:10–15406

Olkowski AA (2009) Livestock water quality – a field guide for cattle, horses, poultry and swine. Agriculture and Agri-Food Canada. www.agriculture.gov.sk.ca/Livestock_Water_Quailty_Guide. Accessed 9 Mar 2020

Ontario Ministry of Agriculture, Food and Rural Affairs (OMAFRA) (2020) Housing requirements for backgrounding beef cattle. http://www.omafra.gov.on.ca/english/livestock/beef/facts/info_housreq.htm. Assessed 9 March 2020

Park RM, Bova R, Jennings JS, Daigle CL (2019a) Environment enrichment reduces aggression and stereotypic behaviors in feedlot steers. J Anim Sci 97:67–67

Park RM, Jennings JS, Daigle CL (2019b) Impact of environmental enrichment on feedlot steer productivity and aggression. J Anim Sci 97:226–226

Parola F, Hillmann E, Schütz KE, Tucker CB (2012) Preferences for overhead sprinklers by naïve beef steers: test of two nozzle types. Appl Anim Behav Sci 137(1):13–22

Pelley MC, Lirette A, Tennessen T (1995) Observations on the responses of feedlot cattle to attempted environmental enrichment. Can J Anim Sci 75:631–632

Perkins N (2013) Animal health survey of the Australian feedlot industry 2010. MLA final report P.PSH.0547. Meat and Livestock Australia, Sydney

Pettiford SG, Ferguson DM, Lea JM, Lee C, Paull DR, Reed MT, Hinch GN, Fisher AD (2008) Effect of loading practices and 6-hour road transport on the physiological responses of yearling cattle. Aust J Exp Agric 48:1028–1033

PISC (ed) (2004) Model code of practice for the welfare of animals – cattle, 2nd edn PISC Report 85. CSIRO Publishing, Collingwood

Poppe M, Veerkamp RF, Van Pelt ML, Mulder HA (2020) Exploration of variance, autocorrelation, and skewness of deviations from lactation curves as resilience indicators for breeding. J Dairy Sci 103:1667–1684

Putz AM, Harding JC, Dyck MK, Fortin F, Plastow GS, Dekkers JC, Canada P (2018) Novel resilience phenotypes using feed intake data from a natural disease challenge model in wean-to-finish pigs. Front Genet 9. https://doi.org/10.3389/fgene.2018.00660

Radostits O (2000) Control and prevention of diseases in feedlot cattle. In: Marx T (ed) Alberta feedlot management Guide. Alberta Agriculture Food and Rural Development, Edmonton

Rault JL, Hintze S, Camerlink I, Yee J (2020) Positive welfare and the like: distinct views and a proposed framework. Front Vet Sci 7:370

Rossi CR, Kiesel GK, Kramer TT, Hudson RS (1978) Cell-mediated and humoral immune responses of cattle to Brucella abortus, Mycobacterium bovis, and tetanus toxoid: immunization of the fetus. Am J Vet Res 39:1742

Russo SJ, Murrough JW, Han MH, Charney DS, Nestler EJ (2012) Neurobiology of resilience. Nat Neurosci 15:1475–1484

Savage DB, Ferguson DM, Fisher AD, Hinch GN, Mayer DG, Duflou E, Lea JM, Baillie ND, Raue M (2008) Preweaning feed exposure and different feed delivery systems to enhance feed acceptance of sheep. Aust J Exp Agric 48:1040–1043

Schulz KE (2014) Iowa Beef Centre 2014 Feedlot Operator Survey. Iowa Beef Centre, Ames, Iowa. https://store.extension.iastate.edu/product/Iowa-Beef-Center-2014-Feedlot-Operator-Survey. Accessed 5 Dec 2019

Schumann FJ, Janzen ED, McKinnon JJ (1990) Prophylactic tilmicosin medication of feedlot calves at arrival. Can Vet J 31:285–288

Schütz KE, Lee C, DeVries TJ (2018) Cattle priorities. Feed and water selection, ability to move freely and to access pasture. In: Tucker CB (ed) Advances in cattle welfare. Woodhead Publishing, pp 93–122
Schwartzkopf-Genswein KS, Grandin T (2019) Cattle transport in North America. In: Grandin T (ed) 5th edition of livestock handling and transport. CABI International, pp 153–183
Schwartzkopf-Genswein KS, McAllister TA, Beauchemin KA, Gibb DJ, Crews DH Jr, Hickman HH, Streeter M (2003) Effect of bunk management on feeding behavior, ruminal acidosis and performance of feedlot cattle: a review. J Anim Sci 81(E. Suppl. 2):E149–E158
Schwartzkopf-Genswein KS, Booth ME, McAllister TA, Mears GJ, Schaefer AL, Cook NJ, Church JS (2006) Effects of pre-haul management and transport distance on beef cattle performance and welfare. Appl Anim Behav Sci 108:12–30
Schwartzkopf-Genswein K, Stookey J, Berg J Campbell J Haley D, Pajor E, McKillop I (2012) Code of practice for the care and handling of beef cattle: review of scientific research on priority issues. National Farm Animal Care Council. https://wwwnfaccca/resources/codes-of-practice/beef-cattle/Beef_Cattle_Review_of_Priority_Welfare_Issues_Nov_2012pdf. Accessed 12 Mar 2020
Schwartzkopf-Genswein KS, Ahola J, Edwards-Callaway L, Hale D, Paterson J (2016) Symposia paper: transportation issues impacting cattle well-being and considerations for the future. Prof Anim Sci 32:707–716
Schwartzkopf-Genswein KS, Gellatly D, Mélendez D, Janzen E, Marti S (2017) Welfare issues in feedlot cattle. In: Engle TE, Klingborg D, Rollin B (eds) 1st edition of cattle welfare. CRC Press, p 386
Sebastian T, Watts J, Stookey J, Buchanan F, Waldner C (2011) Temperament in beef cattle: methods of measurement and their relationship to production. Can J Anim Sci 91:557–565
Simroth JC, Thomson DU, Schwandt EF, Bartle SJ, Larson CK, Reinhardt CD (2017) A survey to describe current cattle feedlot facilities in the High Plains region of the United States. Prof Anim Sci 33(1):37–53
Smith RA (1998) Impact of disease on feedlot performance. A review. J Anim Sci 76:272–274
Sowell BF, Bowman JGP, Branine ME, Hubbert ME (1998) Radio frequency technology to measure feeding behavior and health of feedlot steers. Appl Anim Behav Sci 59:277–284
Sowell BF, Branine ME, Bowman JGP, Hubbert ME, Sherwood HE, Quimby W (1999) Feeding and watering behaviour of healthy and morbid steers in a commercial feedlot. J Anim Sci 77:1105–1112
Spain C, Freund D, Mohan-Gibbons H, Meadow R, Beacham L (2018) Are they buying it? United States consumers' changing attitudes toward more humanely raised meat, eggs, and dairy. Animals 8:128
Sparke EJ, Young BA, Gaughan JB, Holt M, Goodwin PJ (2001) Heat load in feedlot cattle. MLA final report FLOT307-309. Meat and Livestock Australia, North Sydney
Špinka M (2019) Animal agency, animal awareness and animal welfare. Anim Welf 28:11–20
Spinka MARE, Wemelsfelder F (2011) In: CABI International (ed) Environmental challenge and animal agency. Animal welfare. Wallingford, pp 27–43
Statistics Canada (2022) Number of cattle, by class and farm type. https://www150.statcan.gc.ca/t1/tbl1/en/tv.action?pid=3210013001. Accessed 2 Nov 2022
Stokka G, Goldsmith TJ (2015) Feedlot vaccination: does it really matter? Vet Clin Food Anim Pract 31:185–196
Tennant TC, Ives SE, Harper LB, Renter DG, Lawrence TE (2014) Comparison of tulathromycin and tilmicosin on the prevalence and severity of bovine respiratory disease in feedlot cattle in association with feedlot performance, carcass characteristics, and economic factors. J Anim Sci 92:5203–5213
Ternman E, Pastell M, Agenas S, Strasser C, Winckler C, Nielsen PP, Hanninen L (2014) Agreement between different sleep states and behaviour indicators in dairy cows. Appl Anim Behav Sci 160:12–18
Theurer ME, Larson RL, White BJ (2015) Systematic review and meta-analysis of the effectiveness of commercially available vaccines against bovine herpesvirus, bovine viral diarrhea

virus, bovine respiratory syncytial virus, and parainfluenza type 3 virus for mitigation of bovine respiratory disease complex in cattle. J Am Vet Med Assoc 246:126–142

Tucker CB, Coetzee JF, Stookey JM, Thomson DU, Grandin T, Schwartzkopf-Genswein KS (2015) Beef cattle welfare in the USA: identification of priorities for future research. Anim Health Res Rev 16:107–124

Tulloh NM (1961) Behaviour of cattle in yards. II. A study of temperament. Anim Behav 9:25–30

USDA (2008) Part I: reference of beef cow-calf management practices in the United States, 2007–08. National Animal Health Monitoring System. USDA, Fort Collins. http://www.aphis.usda.gov/animal_health/nahms/beefcowcalf/downloads/beef0708/Beef0708_dr_PartI_rev.pdf. Accessed 11 Mar 2020

USDA (2019) 2017 Census of agriculture, United States summary and state data: volume 1, geographical area series, part 51. National Agricultural Statistics service. www.nass.usda.gov. Accessed 22 June 2018

USDA (2022) Cattle on feed report, National Agricultural Statistics service. www.nass.usda.gov. Accessed 2 Nov 2022

Van Donkersgoed J (1992) Meta-analysis of field trials of antimicrobial mass medication for prophylaxis of bovine respiratory disease in feedlot cattle. Can Vet J 33:786–795

Waiblinger S, Menke C, Coleman G (2002) The relationship between attitudes, personal characteristics and behavior of stockpeople and subsequent behavior and production of dairy cows. Appl Anim Behav Sci 79(3):195–219

Walker KH, Fell LR, Reddacliff LA, Kilgour RJ, House JR, Wilson SC, Nicholls PJ (2007) Effects of yard weaning and training on the behavioural adaptation of cattle to a feedlot. Livest Sci 106:210–217

Walkom SF, Jeyaruban MG, Tier B, Johnston DJ (2018) Genetic analysis of docility score of Australian Angus and Limousin cattle. Anim Prod Sci 58:213–223

Warren LA, Mandell IB, Bateman KG (2010) Road transport conditions of slaughter cattle: effects on the prevalence of dark, firm and dry beef. Can J Anim Sci 90:471–482

Watts PJ, Yan MJ, Sullivan TJ, Luttrell MM, Davis RJ, Keane OB (2015) Feedlot bedding study. MLA final report B.FLT.0237, Meat and Livestock Australia, Sydney

Watts PJ, Davis RJ, Keane OB, Luttrell MM, Tucker RW, Stafford R, Janke S (2016) Beef cattle feedlots: design and construction. Meat and Livestock Australia, Sydney

Weary D, Huzzey JM, Keyserlingk M (2008a) BOARD-INVITED REVIEW: using behavior to predict and identify ill health in animals. J Anim Sci 87:770–777

Weary DM, Jasper J, Hötzel MJ (2008b) Understanding weaning distress. Appl Anim Behav Sci 110:24–41

Webb LE, Van Reenen CG, Jensen MB, Schmitt OA, Bokkers EAM (2015) Does temperament affect learning in calves? Appl Anim Behav Sci 165:33–39

Wechsler B, Lea SE (2007) Adaptation by learning: its significance for farm animal husbandry. Appl Anim Behav Sci 108:197–214

Wemelsfelder F, Lawrence AB (2001) Qualitative assessment of animal behaviour as an on-farm welfare-monitoring tool. Acta Agric Scand Sect A Anim Sci 51:21–25

White RW (1959) Motivation reconsidered: the concept of competence. Psychol Rev 66:297

Wilson SC, Mitlöhner FM, Morrow-Tesch J, Dailey JW, McGlone JJ (2002) An assessment of several potential enrichment devices for feedlot cattle. Appl Anim Behav Sci 76:259–265

Wilson SC, Dobos RC, Fell LR (2005) Spectral analysis of feeding and lying behavior of cattle kept under different feedlot conditions. J Appl Anim Welf Sci 8:13–24

Wilson BK, Richards CJ, Step DL, Krehbiel CR (2017) Best management practices for newly weaned calves for improved health and well-being. J Anim Sci 95:2170–2182

Wolfger BKS, Schwartzkopf-Genswein HW, Barkema EA, Pajor ML, Orsel K (2015) Feeding behavior as an early predictor of bovine respiratory disease in North American feedlots. J Anim Sci 93:377–385

Woolums AR, Karisch BB, Parish JA, Park J, Seo KS, Badial P, Olsen SC (2019) Effect of a DNA-based immunostimulant on growth, performance, and expression of inflammatory and

immune mediators in beef calves abruptly weaned and introduced to a complete ration. J Anim Sci 97:111–121
Wright CL (2007) Management of water quality for beef cattle. Vet Clin N Am Food Anim Pract 23(1):91–103
Yeates JW, Main DCJ (2008) Assessment of positive welfare: a review. Vet J 175:293–300
Young BA (1985) Physiological responses and adaptations of cattle. In: Yousef MK (ed) Stress physiology in livestock, Ungulates, vol 2. CRC Press, Boca Raton, pp 101–109
Zebunke M, Langbein J, Manteuffel G, Puppe B (2011) Autonomic reactions indicating positive affect during acoustic reward learning in domestic pigs. Anim Behav 81:481–489
Zinn RA (1989) Manger space requirements for limit-fed feedlot steers. J Anim Sci 67:853–857

# Part III
# Cattle Welfare in Different Contexts

# The Welfare of Cattle at Slaughter

# 8

Temple Grandin

## Contents



T. Grandin (✉)
Colorado State University, Fort Collins, CO, USA
e-mail: Cheryl.miller@ColoState.edu



M. Haskell (ed.), *Cattle Welfare in Dairy and Beef Systems*, Animal Welfare 23,
https://doi.org/10.1007/978-3-031-21020-4_8

**Abstract**

Maintaining high animal welfare standards at a slaughter plant requires well-maintained stunning equipment, trained people who use behavioral principles of cattle handling, and managers who have a strong commitment to welfare. This chapter discusses the key issues and principles affecting welfare of cattle at this important point in their lives, where the potential for a very poor experience is high. The first of these issues is the condition of cattle on arrival at the slaughter plant. Increasingly, cattle are arriving with pre-existing lameness, lack of experience of being handled, or other conditions that should have been dealt with by the producers. Once at the abattoir, a combination of good design of the facilities, good animal handling through knowledge of the principles of animal behavior, and adherence to good slaughter practices are required to ensure high welfare standards are achieved. Welfare can be monitored throughout the process by using indicators such as lameness scoring, number of vocalizations, and use of electric prods. Above all, good management and commitment to animal welfare is key. Constant monitoring and animal-based scoring will help prevent deterioration of the quality of handling and stunning practices.

**Keywords**

Animal welfare · Slaughter · Stunning · Handling · Cattle · Abattoir

## 8.1 Introduction

Across the world, many millions of cattle are slaughtered every day. The number of cattle slaughtered in the United States in 2019 totaled 2.94 million (NASS 2019) and 183,000 in the United Kingdom (Defra 2020). Although the time spent at the abattoir is relatively short, the potential for the animal to experience very poor welfare is high when the operators do not have the necessary skills or motivation, and the appropriate facilities and equipment are not present or of a good standard (e.g., EFSA AHAW Panel (EFSA Panel on Animal Health and Welfare) et al. 2020). Animal welfare should be a top priority for livestock producers, consumers, and businesses (Edwards-Callaway and Calvo-Lorenzo 2020). The key to achieving good standards of welfare during the slaughter process is to understand the principles of animal behavior, welfare measurement, and their application in practice.

In this chapter you will learn how to assess cattle welfare shortly before slaughter. It will cover both animal-based measures of welfare and provide recommendations on methods to improve cattle handling. Signs of poor management on the farm that are detrimental to welfare can also be assessed at the abattoir. Some of them are lameness, poor body condition, injuries, or dirty cattle (Grandin 2017).

Many times I get asked, "Do cattle know if they will get slaughtered?" This is a question I had to answer for myself when my career first started. I discovered that the behavior of cattle walking into the slaughter plant was the same as their behavior going into a chute to be vaccinated. If they knew they were going to get slaughtered, they should have been much more agitated while walking into the abattoir. A review of the literature indicated that the level of cortisol a stress hormone was similar in both places (Grandin 1997, 2014; Mitchell et al. 1988). The cortisol levels ranged from high to low, but they were in the same range at both the farm and the abattoir.

## 8.2 Sources of Stress in the Slaughterhouse

There are many ways that cattle can become stressed at the abattoir shortly before slaughter. This section will describe sources of stress, scientific evidence of their detrimental effects on welfare, and animal-based methods for measuring them.

During handling at a slaughter plant, it is really important to avoid leaving a single bovine or sheep alone. Since direct information on cattle is not available, we can infer from studies with other farm animals that isolation is stressful. In sheep, isolation is highly stressful (Apple et al. 1993). Schaeperkoetter et al. (2021) found that a single sheep left alone for 3 min had higher lactate levels compared to groups of sheep stunned as a group on the floor. Lactate is an easy-to-use measurement for assessing stress shortly before stunning (Edwards et al. 2010a, b; Burfiend and Heuwiesser 2012). In pigs, higher blood lactate levels are associated with aversive handling events such as being shocked with an electric prod or getting jammed in the single-file race (chute) (Edwards et al. 2010a, b). The use of electric prods shortly before stunning will toughen beef (Warner et al. 2007) and increase physiological measures of stress in cattle (Hemsworth et al. 2011). An electric prod should never be the primary driving tool to move cattle. Abattoirs with good handling practices can almost eliminate electric prods. Their use should be restricted to a few animals at the stun box entrance. Another factor that may contribute to stress at the abattoir is the novelty of the strange environment. Suddenly being brought into a novel environment can be stressful (Grandin 1997). Bourquet et al. (2010) reported that the cattle that became the most agitated when confronted with a sudden novel event on the farm were the same ones that had the highest cortisol levels at the abattoir. They were reacting to the novelty of the new environment. When intact bulls are being processed, they must be kept in their original groups. If bulls are mixed, they will fight and have more dark-cutting meat, due to the stress of interacting with unfamiliar animals causing depletion of glycogen (Price and Tennessen 1981).

## 8.3 Vocalizations

### 8.3.1 Scoring Vocalizations to Detect Handling and Equipment Problems

A measure that can be easily used to identify severe welfare problems at an abattoir is counting vocalizations. Ninety-nine percent of the cattle vocalizing (moo or bellow) in the stunning area had experienced an obvious stressful or painful event such as electric prods, a gate slammed on them, excessive pressure from a restraint device, or being pinched by a sharp edge (Grandin 1998a, b, 2001). Simple improvements in equipment and a reduction of electric prod use greatly reduce the percentage of cattle that vocalize. In one study, reducing the pressure applied to a bovine's neck by a head restraint reduced the percentage of cattle that vocalized from 23% to 0% (Grandin 2001). Installation of a light over a dark restrainer entrance made it possible to reduce the percentage of cattle that vocalized during entry from 8% to 0% (Grandin 2001). Vocalization was reduced because the use of the electric prod was greatly reduced. When the light was installed, fewer cattle balked and refused to enter. These examples suggest that noting the points in the abattoir where vocalizations are common can tell us where the animals are experiencing stress or fear, which can help identify where changes must be made. The last example also serves to illustrate that lighting is really important in a livestock handling system. Animals often refuse to enter a dark place. Changes in lighting can be used to enhance animal movement (Grandin 2001; Van Putten and Elshof 1978; Grandin 1982).

Data collected in slaughter plants indicates that vocalization in cattle during handling and restraint is associated with elevated physiological measures of stress (Dunn 1990; Hemsworth et al. 2011; Warriss et al. 1994). When vocalization scoring is used to assess cattle welfare at an abattoir, each animal is scored as either silent or as a vocalizer (Grandin 1998a, b, 2001). All cattle that vocalize in the stun box or restrainer are counted. In the single-file race that leads up to the stun box, only vocalizations that occur when people are moving animals are counted. Active handling is defined as a person moving the cattle. Since vocalization is a handling measurement, it is not scored when cattle are waiting in the lairage. Bulls held in the lairage will often bellow and another bull will respond and bellow.

### 8.3.2 Vocalization Behavior Is Different in Cattle and Sheep

There is a difference between sheep and cattle or pigs. Pigs and cattle will vocalize in direct response to an aversive event such as electric prods, excessive pressure from a restraint device, or becoming jammed in a race (Grandin 1998a, 2001; Edwards et al. 2010a; Hemsworth et al. 2011). Sheep are stoic because they are a defenseless prey species animal. They will vocalize in response to isolation stress but remain silent when poked with an electric prod. In conclusion of this section, scoring the percentage of vocalizing cattle can be used to detect handling and restraint problems. This scoring can be used to identify individual animals and areas within the facility that should be examined.

## 8.4 Poor Condition of Incoming Cattle Is a Major Welfare Problem

When I visit a large slaughter plant, some of the worst animal welfare problems I observe are due to cattle being brought to the abattoir in poor condition. Many of these cows should have been euthanized on the farm. A common reason for culling a dairy cow from the herd is lameness. Cows should be brought in when they are still fully mobile. Audits conducted at abattoirs that process cull dairy cows indicate that failure to market the animal in a timely manner is a major problem (Harris et al. 2017; Edwards-Callaway et al. 2018). Another serious problem is neglected health problems such as severe cancer of the eye where the eyeball has ruptured or a necrotic rotten prolapse. The Canadian Code of Practice prohibits marketing a cow if the eyeball has ruptured. Another serious problem is dairy cows being marketed without having their milk dried up. The US National Beef Quality Audit indicated that 8% of old dairy cows arrived at abattoirs with a full udder (Harris et al. 2017). They should have been dried off before leaving the farm.

### 8.4.1 Feedlot Beef Cattle Problems

In the 1990s, problems with lame finished cattle arriving from feedlots was seldom a problem. Today cattle originating from certain feedlots arrive stiff and lame (Davis-Unger et al. 2019). This is due to a combination of factors that I call "biological system overload" (Grandin and Whiting 2018). The factors that may contribute to either lameness or deaths close to the end of the fattening period are as follows:

- Overselection for carcass traits which may contribute to poor leg conformation or cardiac problems
- Heavier at a younger age
- A finishing diet with 95% grain
- High doses of beta-agonist growth promotors (such as ractopamine or zilpaterol)

The author has observed that it is especially a problem during hot weather. For further information in both cattle, sheep, and pigs, refer to Poletto et al. (2009, 2010), Marcias-Cruz et al. (2010), Ritter et al. (2017), Longeragen et al. (2014), and Peterson et al. (2015).

## 8.5 Measurement of Animal Welfare Problems

People can manage the things that they measure. Measurement of welfare indicators provides the information that can be used to assess whether conditions in abattoirs and farms are improving or getting worse. The measurements also allow comparisons between abattoirs to be made (Grandin 2015). Benchmarking is a good way to provide feedback to producers that they should change some of their practices. This enables a producer to see how they compare to other producers (Von Keyserlingk et al. 2012). Below is a list of measures that should be made on incoming cattle (Grandin 2017).

- *Lameness scoring*—The simplest scoring system measures normal, lame, and downed (Welfare Quality Network 2009). Edwards-Callaway et al. (2017) developed a four-point scoring system of (1) normal, (2) lame but keeps up with the walking group, (3) lame but does not keep up, and (4) almost a downer. Care must be taken when comparing lameness data from different sources. In some scoring systems, a normal animal is rated a zero, while in another system, a normal bovine is given a score of one (Welfare Quality Network 2009; Edwards-Callaway et al. 2017). Lameness has many different causes. A Canadian survey showed that there were big differences between the best and the worst feedlots. The best feedlot had only 1.3% lame cattle, and the worst one had 46% (Davis-Unger et al. 2019).
- *Body condition scoring*—Severely emaciated cattle should have been euthanized on the farm.
- *Sores, lesion, and swollen hocks*—A good scoring tool for accessing swollen hocks on dairy cattle is given by Fulwider et al. (2007). Skin lesions are scored on a 0–3 scale where 0 represents no hair loss or lesions, through to score 3 for severe swelling. Poor design or lax management of freestalls (cubicles) is a major cause of swollen hocks (Fulwider et al. 2007).
- *Old injuries from abusive handing*—Some examples are broken tails or damaged hides due to being poked with sticks with nails in them.
- *Liver abscesses*—Some cattle originating from feedlots may have high levels of liver abscesses (Amachawadi and Nagaraja 2016). When they are severe, they may have adhesions to the body wall. One cause of liver abscesses is feeding high-grain diets. Adding roughage to the diet will reduce liver abscesses (Reinhardt and Hubbert 2015).
- *Bruises*—Scoring of bruises is an excellent method to detect problems with poor handling. Fresh bruises will be red (Hamdy et al. 1957). Old bruises that are several days old may have yellowish mucous on them (McCausland and Dougherty 1978). If bruises are occurring in the plant, the bruises will be on cattle from many different origins. Older yellowish bruises are common in cattle that have been moved through auctions. Common causes of bruises that occur at the slaughterhouse are as follows:
  - Striking the animal's back with stun box gate (Strappini et al. 2013; Meischke and Horder 1976)

- Two cattle stuck in the truck door during unloading
- Sharp edges on equipment
- Poor handling in the yards or at the farm of origin (Grandin 1981; Mendonça et al. 2018; Benthancourt et al. 2019)

Common causes of bruises occurring during transport:

- Truck compartment ceiling height is too low (Lee et al. 2017).
- Overloaded vehicles (Eldridge et al. 1988; Tarrant et al. 1988).
- Poor driving throws cattle off-balance (Tarrant et al. 1988).
- A fallen animal gets stepped on by other cattle.

### 8.5.1 Other considerations for animals at the abattoir

*Cattle that are difficult to handle*—Cattle that have been exclusively handed by people on horses can be dangerous to handle at the abattoir. The man on the horse is perceived as safe and familiar, and the person walking on the ground is new and frightening (Grandin 2014; Grandin and Deesing 2008).

*Bobby calves*—The author has worked in slaughter plants that process day-old bobby bull dairy calves. It was awful to attempt to move baby animals that had difficulty walking. The best solution to bobby calf slaughter is to find alternative markets and grow the calves into bigger cattle for beef. In the United States, the majority of Holstein bull calves are fed in feedlots for beef.

## 8.6 Behavioral Principles of Cattle Handling

Understanding the behavioral principles of cattle handling will greatly improve the movement of the animals through the abattoir. This will improve the efficiency of the plant and also improve animal welfare. There are a number of key principles of cattle behavior to be aware of and some key ways of responding for cattle handlers and managers.

### 8.6.1 Remove Distractions

Cattle are extremely sensitive to small visual distractions that people often do not notice. Removal of visual distractions in a handling facility will improve movement of cattle through the race. The first step to improve ease of movement into the stun box is to check air flow through the stun box door. Cattle may refuse to enter, if air is blowing toward them (Grandin 1996). Changing air flow may greatly improve cattle movement. The following visual distractions can cause cattle to stop and refuse to move. If you are a good observer, the cattle will show you where the distractions are. They will stop and look at them.

**Fig. 8.1** This crowd pen has shadows caused by the overhead structure. To eliminate these shadows may require building a roof over it. You need to be observant and determine if the shadows are making the cattle stop (Photo courtesy of Temple Grandin)

- Paper towels hanging down and moving
- Shiny metal that jiggles and glare (Klingimair et al. 2011)
- Seeing people or moving machinery in front of them
- Hose on the floor
- Drains or changes in flooring such as concrete transitioning to metal (Grandin 1996)
- Shadows or sharp contrasts of dark and light (Fig. 8.1) (Grandin 1980a, 1996; Willson et al. 2021)

### 8.6.2 Effects of Lighting

Cattle and other animals will often refuse to move into a dark place. Adding a light to illuminate a dark race (chute) entrance will improve movement (Grandin 1982, 2001; Van Putten and Elshof 1978). Moving of overhead lamps sideways will often remove reflections. One person needs to move the lights, and the other person should stand in the chute and determine when the reflection goes away.

### 8.6.3 Flight Zone Principles and Point of Balance

Employees who work with cattle need to understand the basic principles of cattle behavior. Some of the principles are (1) flight zone, (2) point of balance at the shoulder, (3) cattle want to return to where they came from, and (4) natural following behavior. Cattle that are not completely tame will have a large flight zone. They will move away when people enter the edge of the flight zone. A common problem is cattle rearing in either the single-file chute (race) or stun box. Figure 8.2 shows the single-file chute leading to the stun box. This occurs when a person is standing inside the flight zone and the animal is not able to move away. When the bovine rears up, the person should back away and get out of its flight zone. A common mistake made by handlers when they are moving cattle through a single file race is to stand in front of the animal's head and poke its rear. Figure 8.2 shows a steer at the stun box door. To move this steer forward, the handler must be positioned behind the shoulder. The person *must* be behind the point of balance at the shoulder to make the cow move forward. Further information for handlers on the flight zone and point of balance can be found at Grandin and Deesing (2008), Grandin (2014), and www.grandin.com.

### 8.6.4 Moving Small Groups of Cattle

Another major mistake is moving large groups of cattle from the lairage (stockyard) into the crowd pen that leads to the single-file race. Handlers need to be taught how to use natural following behavior. To take advantage of following behavior, the next group should be brought up when the single-file race is partially empty. The cattle should pass through the crowd pen that leads to the single-file race without stopping and follow the leader into the single-file race. If the cattle wait in the crowd pen, they will turn around and attempt to return to the lairage. The manager at each abattoir must determine the correct number of cattle that works quietly and efficiently. In a small slaughter plant, it may be four or five large cattle. In a larger facility, it may be 10 or 12.

### 8.6.5 Single Lone Cattle Get Highly Stressed

A single lone cow, bull, or steer can become highly stressed. It may also be dangerous for people to handle. To calm it down, some other cattle should be put in with it.

## 8.7 Tips on Facility Design

A well-designed slaughter facility will mean that animal movement through the plant is as smooth as possible, and stunning and slaughter are efficient and humane. There are some key aspects to consider.

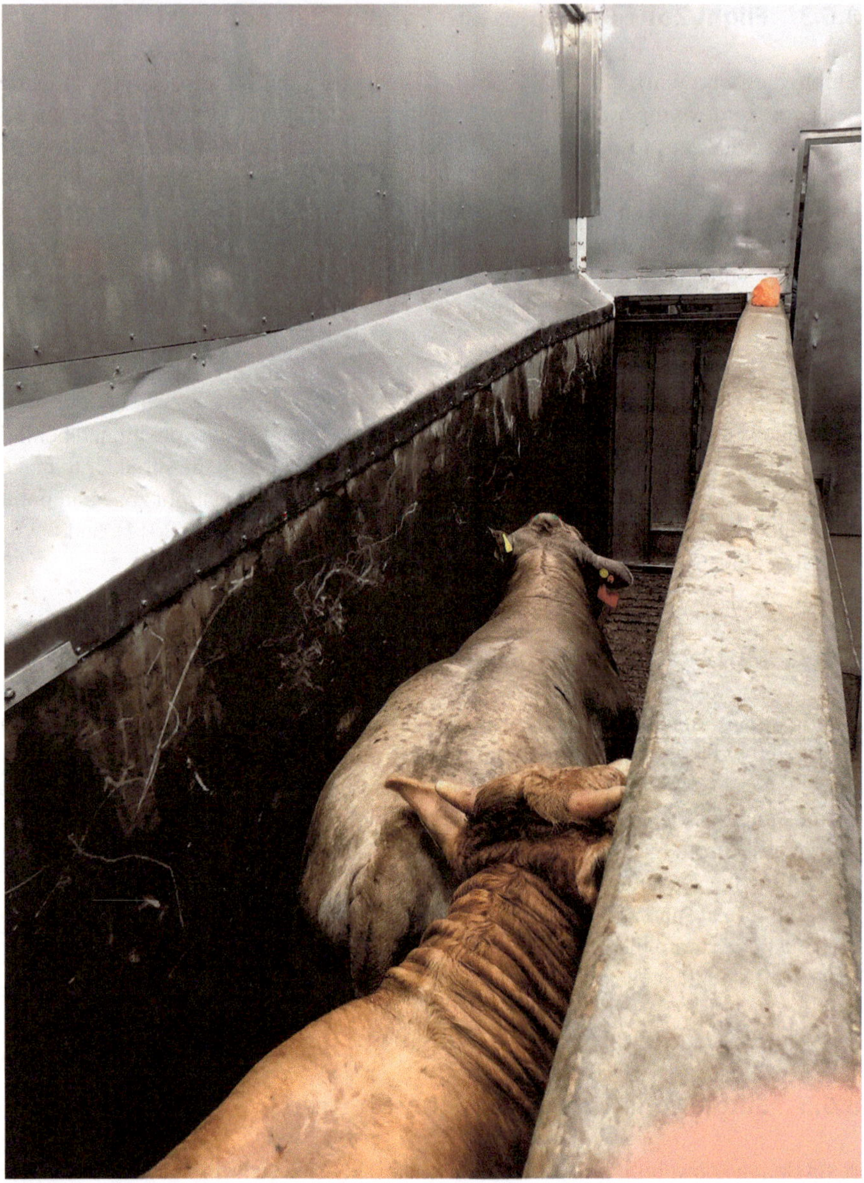

**Fig. 8.2** Steer standing at the stun box door. After the door is opened, the steer will easily move into the box if the handler quickly walks past it in the opposite direction of desired movement. The handler must never stand at the steer's head and poke it in the rear (Photo courtesy of Temple Grandin)

**Fig. 8.3** Well-designed nonslip flooring for a cattle slaughter facility (Photo courtesy of Temple Grandin)

### 8.7.1 Nonslip Flooring

Problems with slipping and falling during handling are often caused by flooring that is too slick. Figure 8.3 shows a good nonslip floor for cattle. It has grooves with an 8 inch (20 cm) square or diamond pattern. The grooves should be a minimum of 1 inch (2.5 cm) deep. This flooring is recommended for high-traffic areas such as unloading ramps, main drive alleys, and stun box floors. One of the problems with flooring is that it wears out gradually, and people do not realize that slipping and falling may be increasing. This is why scoring of slips and falls is strongly recommended (Grandin 1998a, b; OIE 2019a, b; Welfare Quality Network 2009). When slips and falls are quantified, it makes it easier for managers to determine that the problem is increasing.

### 8.7.2 Stun Box Design

One of the biggest problems that the author has observed in stun boxes is animals becoming agitated due to slipping on the floor. When an animal repeatedly makes small slips, it will not stand still. In many abattoirs, stunning accuracy was improved when an old worn out stun box floor was replaced with a new nonslip floor.

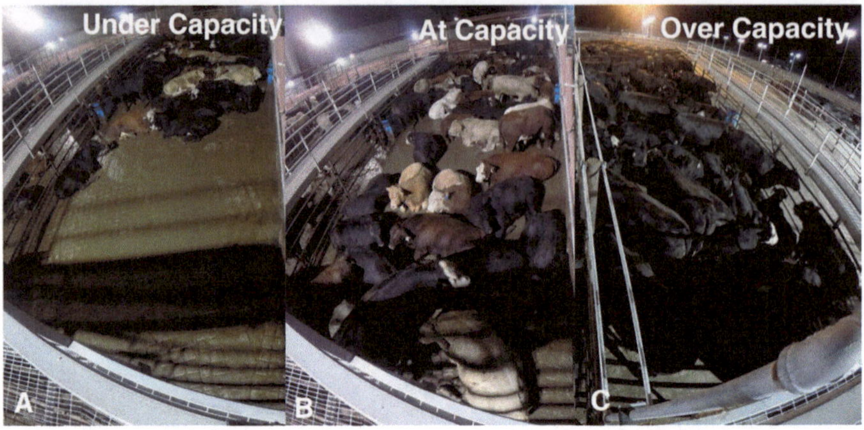


**Fig. 8.4** Stocking densities in pens provided for overnight lairage at a commercial slaughter facility. In photograph (**a**) the pen stocking density is under capacity provided, in (**b**) it is at capacity, and in (**c**) the pen is over capacity with too many animals housed in it (Source: Kline et al. (2019a). Image reproduced with permission of Elsevier)

Two other common problems in stun boxes are (1) cattle refusing to enter because they see people or equipment through the head holder, or (2) the rear entry gate slams on the animal's back and causes bruises. To prevent cattle from seeing people or equipment through the headholder, a solid barrier should be installed 1 m (3 ft.) in front of the headholder. Controls for the rear stun box door must be designed so that the operator can easily stop downward movement of the door. If the controls are poorly designed, the door may slam on the animal after the operator has pushed the control to make the door go up.

### 8.7.3 Design of Stockyards and Lairages

The best designs have one-way traffic through the lairage pens. Cattle enter through one end of the pen and exit to move through the other end, with curved raceways to promote movement. Layouts are shown in at Grandin and Deesing (2008), Grandin (1984, 2014), and www.grandin.com. When cattle are held overnight, there must be sufficient space for all the cattle to lie down (Kline et al. 2019a). Figure 8.4 shows the correct lairage stocking density. Cows and large feedlot cattle need 20 to 22 square feet (1.85–2 $m^2$) per animal (NAMI 2019).

## 8.8 Stunning Practices

### 8.8.1 Stunning Practices Using a Captive Bolt Stunner

There are two types of captive bolt stunners. They are penetrating with a retractable rod that penetrates the brain and non-penetrating. On adult cattle, penetrating

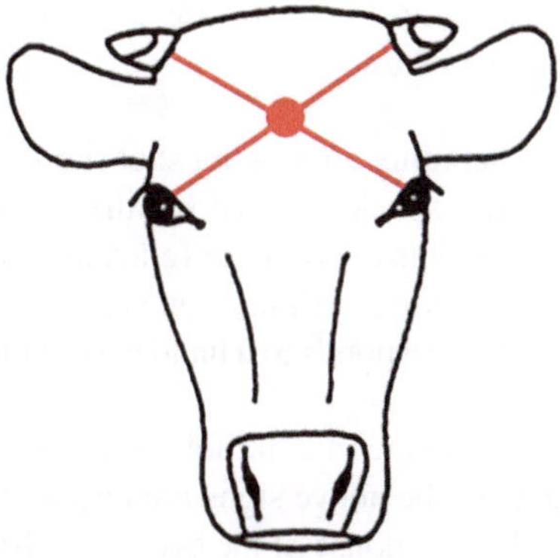

**Fig. 8.5** Diagram showing the ideal position for use of a captive bolt or firearm (Image copyright Humane Slaughter Association, 2013, reproduced with permission; www.hsa.org.uk)

captive bolts are more effective than non-penetrating (Gibson et al. 2019; Oliveira et al. 2018). The use of non-penetrating captive bolt is not recommended for bulls or mature cattle (AVMA 2020). When a powerful pneumatic-penetrating captive bolt is used on *Bos taurus* English/Continental steers and heifers, they can be instantly rendered unconscious without visible damage to the brain stem (Kline et al. 2019b). Good sources of open-access information for the use of either captive bolt or firearms is in OIE (2019a, b), AVMA (2020), and Humane Slaughter Association (2005).

The best position for shooting cattle is the center of the forehead and never between the eyes. Figure 8.5 shows the correct position on the head of the animal, and further information is available (AVMA 2013; Humane Slaughter Association 2005; Humane Slaughter Association, United Kingdom; NAMI 2019). Shooting in the hollow behind the bovine's poll should not be used as the primary shooting position. This position should only be used if the frontal position is not accessible.

### 8.8.2 Determining Unconsciousness

People who are working in an abattoir must be trained to determine if an animal has been both rendered insensible to pain and is rendered unconscious. If there is any question that an animal is returning to consciousness, it must be immediately reshot. There are three stages:

1. Definitely fully conscious
2. Transition zone between consciousness and brain death
3. Brain death

These three stages relate to the presence of brain functioning or dysfunction following effective stunning (a fully scientific explanation of this can be found in Terlouw et al., 2016). Indicators of consciousness are shown below (and shown in a clear simple easy-to-use chart in NAMI (2019)) which includes the indicators shown below.

1. Standing after being shot (no loss of posture)
2. Has species typical vocalizations (moo, bellow)
3. Retains the righting reflex and is able to lift up its head. If the animal is hung on the rail, it will have an arched back and may lift up its head.
4. Eye responds to a hand waved in front of it without touch (menace/threat reflex).

A bovine is definitely brain-dead when the corneal reflex, rhythmic breathing, and all the above signs are absent. When captive bolt is done correctly, the corneal reflex (response of the eye to touch) can be easily abolished after a single shot from a captive bolt (Grandin 2001). In commercial practice, the AVMA (2013) and Gregory (2007) state that the corneal reflex must be absent. If the bovine has a corneal reflex, it must be immediately reshot.

### 8.8.3 Ignore Limb Kicking

Kicking limbs must be ignored. A leg may kick even after the head is removed or the spinal cord has been severed (Terlouw et al. 2015). When properly stunned cattle are hung on the rail, the head should hang straight down, and the neck should be limp and floppy (Grandin 2002). If a soft flaccid tongue is extended, the bovine will be unconscious (Grandin 2002). There will be some properly stunned cattle where the tongue may not become extended because it is trapped inside the jaw.

### 8.8.4 Good Maintenance Is Essential

One of the biggest problems observed by the author with captive bolt sunning is poor maintenance or damp cartridges (Grandin 1998a, b, 2002). For a captive bolt to be effective and cause instantaneous unconsciousness, it must be well maintained. If the cartridges for a powder-activated captive bolt becomes damp, they may lose power.

## 8.9 Animal Welfare and Religious Slaughter Without Stunning

From an animal welfare standpoint, slaughter without stunning is controversial. In the United States and many other countries, slaughter without stunning is allowed to enable people in the Jewish and Muslim faiths to practice their religion (*Humane Slaughter Act*, 1958; OIE 2019a, b).

There are three main welfare issues when slaughter without stunning is performed. They are as follows:

- The restraint method used to hold the conscious animal in position.
- Time for the animal to lose consciousness after the throat cut.
- Does the throat cut cause pain?

The author has worked extensively on improving the methods to restrain cattle. In the United States, it is legal to hang a fully conscious adult cow or steer upside down by a chain wrapped around a single back leg. In a large plant, the author observed that a high percentage of the cattle were bellowing (Grandin 1980b). In this situation, it was almost impossible to separate the variables of a highly stressful restraint method form the animal's reaction to the throat cut. The OIE (2019a, b) has strong recommendations against suspension of mammals by one back leg. Recommendations for improvements in restraining devices can be found in the following publications (Grandin 1988, 1991, 1992, 2003; Giger et al. 1977; Deroin 2003; Grandin 2014; Westervelt et al. 1976). Some of the most important improvements in restrainers are (1) elimination of sharp edges and pinch points, (2) prevent excessive pressure from being applied by a restraint device that causes cattle to vocalize, (3) the best restraint devices hold the cattle in a comfortable upright position, and (4) either stun the animal or perform the throat cut within 10 s after the head is fully restrained. A major cause of vocalization is excessive pressure applied by a restraint device (Bourquet et al. 2012). When low-stress handling and restraint is used, the percentage of cattle vocalizing before the throat is cut will be 5% less (Grandin 2012). Excessive pressure from a restraint device or the use of electric prods on almost 100% of the cattle can cause the percentage of cattle vocalizing to rise. A number of studies have all shown increases in vocalizations to various degrees (23%: Grandin, 2001; 25%: Bourquet et al., 2012; 47%: Hayes et al., 2016; and 32%: Grandin, 1998a, b).

### 8.9.1 Time to Lose Consciousness

Welfare concerns during slaughter without stunning are a much bigger issue for cattle compared to sheep or goats. There are two reasons for this. Cattle take longer to lose consciousness compared to sheep (Blackmore 1984; Blackmore et al. 1983, 1986). This is due to differences in the anatomy in sheep and cattle (Baldwin and Bell 1963a, b). Since cattle are larger and heavy, they are much more difficult to restrain compared to sheep. To restrain a bovine in a low-stress manner requires expensive equipment. Sheep can be easily restrained in an upright position by a person straddling them. When good technique is used, sheep will lose consciousness in 2–14 s and cattle in 17–85 s (Blackmore 1984; Daly et al. 1989; Blackmore and Newhook 1982). When poor procedures are used, cattle may remain conscious for several minutes. Both Grandin (2012) and Gregory et al. (2010) report that when good technique is used, 90% of the cattle will collapse within 30 s. Cutting position will make a difference. Cutting at the C1 (cervical 1) position will induce more rapid collapses (Gibson et al. 2015; Gregory et al. 2012).

### 8.9.2 Painfulness of the Cut

The results of the research are mixed on assessing painfulness of the cut. Grandin (1994) and Grandin and Regenstein (1994) reported that when the special long kosher knife was used, the cattle did not appear to feel the cut. Gibson et al. (2009a, b, c) reported that the cut was definitely painful. The difference in the results may be due to differences in the knives that were used. Another problem is aspiration of blood into the respiratory tract (Gregory et al. 2009). It is likely that aspiring large amounts of blood into the respiratory tract would be highly stressful.

### 8.9.3 Recommendations for Improvements

Many Muslim religious authorities will accept pre-slaughter stunning (Nakyinsige et al. 2013; Fuseini et al. 2018). They will often accept stunning if they can be assured that the animals actually die due to the throat cut (Fuseini et al. 2018). Death is usually defined when the beating heart stops. Muslims will often accept captive bolt stunning because the heart will continue to beat for several minutes (Vimini et al. 1983). Fully reversible electrical stunning of cattle, sheep, or poultry is also acceptable in many Halal markets. There are good reviews by Sabow et al. (2016, 2017, 2019).

Religious authorities in the Jewish faith have been more reluctant to accept pre-slaughter stunning. The author's own opinion is that kosher slaughter using the special long supersharp knife can be acceptable from an animal welfare standpoint. It requires much greater attention to the details of the procedure compared to slaughter where the animal is pre-stunned. When slaughter without stunning is done poorly with lax supervision by management, the animal's welfare will be severely compromised. To maintain an acceptable level of welfare requires an abattoir manager who strictly supervises handling, restraint, and the cutting method. The author has observed severe welfare problems when slaughter without stunning is done sloppily.

## 8.10 The Importance of Management Commitment to Cattle Welfare

A common mistake is to think that a new piece of state-of-the-art equipment will replace good management. New equipment does not provide automatic management. Managers and quality assurance managers must constantly monitor and supervise handling, restraint, stunning, and other procedures. The author strongly recommends the use of animal-based numerical scoring. This will help prevent deterioration of handling and stunning practices. The five critical control points should be measured on a regular basis: (1) efficacy of stunning, (2) induction of unconsciousness, (3) vocalization during handling and restraint, (4) electric prod use, and (5) cattle falling down. There must also be no acts of abuse such as dragging conscious cattle, deliberately slamming gates on cattle, poking sensitive areas such as the eyes or rectum, or beating.

## 8.11 Conclusions

The welfare of the animal at the slaughterhouse is affected by its condition when it leaves the farm and the quality of the transportation experience. Animals must arrive at the slaughterhouse in good condition and in good health. Ensuring high standards of welfare at the slaughter facility requires well-trained staff that understand the principles of animal behavior and well-designed, well-maintained equipment. Ongoing monitoring and measuring of welfare outcomes is essential to identify problems, and scoring systems are available that are easy to implement. The management personnel are also important to the welfare of cattle as they can provide the resources to ensure that equipment is updated and maintained, that staff are trained, and that monitoring of welfare outcomes is carried out and acted on where necessary.

## References

Amachawadi RG, Nagaraja TG (2016) Liver abscesses in cattle: a review of incidence in Holsteins and of bacteriology and vaccine approaches to control in feedlot cattle. J Anim Sci 94:1620–1632

Apple JK, Minton JE, Parsons KM, Unruh JA (1993) Influence of repeated restraint and isolation stress and electrolyte administration on pituitary-adrenal secretions, electrolytes and other blood constituents of sheep. J Anim Sci 71:71–77

AVMA (2013) Guidelines for euthanasia of animals, 2013th edn. American Veterinary Medical Association, Schaumberg

AVMA (2020) Guidelines for depopulation of animals, 2019th edn. American Veterinary Medical Association, Schaumberg

Baldwin BA, Bell FR (1963a) The effect of temporary reduction in cephalic blood flow on the EEG of sheep and calf. Electroencephalogr Clin Neurophysiol 15:465–473

Baldwin BA, Bell FR (1963b) The anatomy of the cerebral circulation of the sheep and ox: the dynamic distribution of the blood supplied by the carotid and vertebral arteries to cranial regions. J Anat 97:203–215

Benthancourt AJ et al (2019) Pre-slaughter factors affecting the incidence of severe bruising in cattle carcasses. Livest Sci 222:41–48

Blackmore DK (1984) Differences in behavior between sheep and cattle during slaughter. Res Vet Sci 37:223–236

Blackmore DK, Newhook JC (1982) Electroencephalographic studies of stunning and slaughter in sheep and calves – part 3. The duration of insensibility induced by electrical stunning in sheep and calves. Meat Sci 7:19–28

Blackmore DK, Newhook JC, Grandin T (1983) Time of onset of insensibility in four- to six-week-old calves during slaughter. Meat Sci 9:145–149

Blackmore NL, Cheetham K, Blackmore DK (1986) Differences in blood supply to the cerebral cortex between sheep and calves during slaughter. Res Vet Sci 40:252–254

Bourquet C, Deiss V, Gobert M et al (2010) Characterizing emotional reactivity of cows to understand and predict their stress reactions to slaughter procedures. Appl Anim Behav Sci 125:9–21

Bourquet C, Deiss V, Tannugi CC, Terlouw EMC (2012) Behavioral and physiological reactions of cattle in a commercial abattoir, relationships with organizational aspects of the abattoir and animal characteristics. Meat Sci 68:158–168

Burfiend O, Heuwiesser W (2012) Validation of handheld meters to measure blood L-lactate concentration in dairy cows and calves. J Dairy Sci 95:6449–6456

Daly CC, Kallweit E, Ellendorf F (1989) Cortical function in cattle during conventional captive bolt stunning followed by exsanguination compared to shechita slaughter. Vet Rec 122:325–329

Davis-Unger J, Schwartzkoph-Genswein KG, Pajor EA et al (2019) Prevalence of lameness-associated risk factors in Alberta feedlot cattle. Transl Anim Sci 3:595–606

Defra (2020) United Kingdom slaughter statistics – March 2020. National Statistics, Department for Environment, Food and Rural Affairs

Deroin KL (2003) Headgate assembly for restraining livestock. U.S. Patent 6,537, 145B1 U.S. Patent Office, Washington, D.C.

Dunn CS (1990) Stress reactions of cattle undergoing ritual slaughter using two methods of restraint. Vet Rec 126:522–525

Edwards LN, Engle TE, Correa JA et al (2010a) The relationship between exsanguination blood lactate concentration and carcass quality in slaughter pigs. Meat Sci 85:435–440

Edwards LN, Grandin T, Engle TE et al (2010b) Use of exsanguination blood lactate to assess the quality of pre-slaughter handling. Meat Sci 86:384–390

Edwards-Callaway LN, Calvo-Lorenzo MS (2020) Animal welfare in the U.S. slaughter industry—a focus on fed cattle. J Anim Sci 98:1–21. https://doi.org/10.1093/jas/skaa040

Edwards-Callaway LN, Calvo-Lorenzo MS, Scanga JA et al (2017) Mobility scoring of finished cattle. Vet Clin North Am Food Anim Pract 33:235–250

Edwards-Callaway LN, Walker J, Tucker C (2018) Perspective: culling decisions and dairy cattle welfare during transport and slaughter in the United States. Front Vet Sci. https://doi.org/10.3380/fveets.2018.00343

EFSA AHAW Panel (EFSA Panel on Animal Health and Welfare), Nielsen SS, Alvarez J, Bicout DJ et al (2020) Scientific opinion on the welfare of cattle at slaughter. EFSA J 18(11):6275. https://doi.org/10.2903/j.efsa.2020.6275

Eldridge GA, Winfield GG, Cahill DJ (1988) Responses of cattle to different space allowances, pen sizes, and road conditions during transport. Aust J Exp Agric 28:155–159

Fulwider WK, Grandin T, Garrick DJ, Engle TE et al (2007) Influence of free-stall base on tarsal joint lesions and hygiene in dairy cows. J Dairy Sci 90:3559–3566

Fuseini A, Teye M, Wotton SB et al (2018) Electrical water bath stunning of Halal poultry meat production: animal welfare issues and compatibility with Halal rules. CABI Rev 13(016). https://doi.org/10.10179/PAVSNNR2018/3016

Gibson TJ, Johnson CB, Murrell JC et al (2009a) Electroencephalographic responses of halothane-anesthetized calves to slaughter by ventral-neck incision without stunning. N Z Vet J 57:77–85

Gibson TJ, Johnson CB, Murrell JC et al (2009b) Components of electroencephalographic responses to slaughter in halothane-anaesthetized calves: effects of cutting neck tissues compared to major blood vessels. N Z Vet J 57:84–89

Gibson TJ, Johnson CB, Murrell JC et al (2009c) Amelioration of electroencephalographic responses to slaughter by nonpenetrative captive bolt stunning after ventral neck incision in halothane anesthetized calves. N Z Vet J 57:96–101

Gibson TJ, Dadios N, Gregory NG (2015) Effect of neck cut position on time to collapse in Halal slaughtered cattle without stunning. Meat Sci 110:310–314

Gibson TJ, Oliveira S, Dalla-Costa FA et al (2019) Electroencephalographic assessment of pneumatically powered captive bolt stunning of bulls. Meat Sci 151:54–59

Giger W, Prince RP, Westervelt RG et al (1977) Equipment for low stress animal slaughter. Trans Am Soc Agric Eng 20:571–578

Grandin T (1980a) Observations of cattle behavior applied to the design of cattle handling facilities. Appl Anim Ethol 6:10–31

Grandin T (1980b) Problems with kosher slaughter. Int J Study Anim Probl 1(6):375–390

Grandin T (1981) Bruises on southwestern feedlot cattle. J Anim Sci 53(Sul. 1):213. (Abstract)

Grandin T (1982) Pig behavior studies applied to slaughter plant design. Appl Anim Ethol 9:141–151
Grandin T (1984) Race system for slaughter plants with 1.5 m radius curves. Appl Anim Behav Sci 13:295–299
Grandin T (1988) Double rail restrainer conveyor for livestock handling. J Agric Eng Res 41:327–338
Grandin T (1991) Double rail restrainer for handling beef cattle. Paper, no. 91-5004, Am Soc Agric Engs. St. Joseph, Michigan
Grandin T (1992) Observations of cattle restraint devices for stunning and slaughtering. Anim Welf 1:85–91
Grandin T (1994) Euthanasia and slaughter of livestock. J Am Vet Med Assoc 204:1354–1360
Grandin T (1996) Factors that impede animal movement at slaughter plants. J Am Vet Med Assoc 209:757–759
Grandin T (1997) Assessment of stress during handling and transport. J Anim Sci 75:249–257
Grandin T (1998a) Objective scoring of animal handling and stunning practices in slaughter plants. J Am Vet Med Assoc 212:36–93
Grandin T (1998b) The feasibility of using vocalization scoring as an indicator of poor welfare during slaughter. Appl Anim Behav Sci 56:121–128
Grandin T (2001) Cattle vocalizations are associated with handling and equipment problems in slaughter plants. Appl Anim Behav Sci 71:191–201
Grandin T (2002) Return to sensibility problems after penetrating captive bolt stunning of cattle in commercial beef slaughter plants. J Am Vet Med Assoc 221:1258–1261
Grandin T (2003) Transferring results from behavioral research to industry to improve animal welfare on the farm, ranch and slaughter plants. Appl Anim Behav Sci 81:216–228
Grandin T (2012) Developing measures to audit welfare of cattle and pigs at slaughter. Anim Welf 21:351–356
Grandin T (2014) Livestock handling and transport, 4th edn. CABI, Wallingford
Grandin T (2015) The importance of measurement to improve welfare of livestock, poultry and fish. In: Improving animal welfare – a practical guide, 2nd edn. CAB International, Oxfordshire
Grandin T (2017) On-farm conditions that compromise animal welfare that can be monitored at the slaughter plant. Meat Sci 132:52–58
Grandin T, Deesing M (2008) Humane livestock handling. Storey Publishing, North Adams
Grandin T, Regenstein JM (1994) Religious slaughter and animal welfare: a discussion for meat scientists. In: Meat focus international. CAB International, Wallingford, pp 115–123
Grandin T, Whiting M (2018) Are we pushing animals to their biological limits? CABI, Wallingford
Gregory NG (2007) Animal welfare and meat production. CABI, Wallingford
Gregory NG, von Wenzlawowicz M, von Hollenben KV (2009) Blood in the respiratory track during slaughter with and without stunning in cattle. Meat Sci 82:13–16
Gregory NG, Fielding HR, von Wenzlawowicz M et al (2010) Time to collapse following slaughter without scanning of cattle. Meat Sci 85:66–69
Gregory NG, Schuster P, Mirabito I et al (2012) Arrested blood flow during false aneurysm formation in the carotid arteries of cattle slaughtered with and without stunning. Meat Sci 90:368–372
Hamdy M, Kunkle L, Rheins M et al (1957) Bruised tissue III some factors affecting experimental bruises. J Anim Sci 16(2):496–501
Harris KA, Eastwood LC, Boykin CA et al (2017) National Beef Quality Audit – 2016: transportation, mobility, live cattle and carcass assessments of targeted producer-related characteristics that affect value of market cows and bulls their carcasses and by-products. Transl Anim Sci 1:510–574
Hayes NS, Schwartz CA, Phelps KJ et al (2016) The relationship between preharvest stress and carcass characteristics of beef heifers that qualified for kosher designation. Meat Sci 100:134–138
Hemsworth PH, Ric M, Karlen MG et al (2011) Human-animal interactions at the abattoir, relationships between handling and animal stress in sheep and cattle. Appl Anim Behav Sci 135:24–33
Humane Slaughter Association (2005) Humane killing of livestock using firearms, 2nd edn. Humane Slaughter Association, United Kingdom. ISBN 1 871561 37 X

Humane Slaughter Association, United Kingdom. www.hsa.org.uk. Accessed 6 Sept 2019
Kline HC, Edwards-Calloway LN, Grandin T (2019a) Field observation: pen stocking capacities for overnight lairage of finished steers and heifers at a commercial slaughter facility. Appl Anim Sci 35:130–132
Kline HC, Wagner DR, Edwards-Calloway LN et al (2019b) Effect of captive bolt lengths on brain trauma and post stunning hind limb activity. Meat Sci 155:69–73
Klingimair K, Stevens KB, Gregory NG (2011) Luminaire and glare in indoor handling facilities. Anim Welf 20:263–269
Lee TL, Reinhardt CD, Bartle SJ et al (2017) Assessment of risk factors contributing to carcass bruising in fed cattle at commercial slaughter facilities. Transl Anim Sci 1(4):489–497
Longeragen GH, Thomson DU, Scott HM (2014) Increased mortality in groups and cattle administered the β-adrenergic agonists ractopamine hydrochloride and zilpaterol hydrochloride. PLoS One 9(3):e91177. https://doi.org/10.1371/journal.pone.0091177
Marcias-Cruz U, Alvarez-Valenzuela FD, Torrentera-Olivera NG et al (2010) Effect of zilpaterol hydrochloride on feedlot performance and carcass characteristics of ewe lambs during heat-stress conditions. Anim Prod Sci 50:983–989
McCausland I, Dougherty R (1978) Histological aging of bruises in lambs and calves. Aust Vet J 54(11):525–527
Meischke HRC, Horder JC (1976) A knocking box effect on bruising in cattle. Food Technol Aust 28:369–371
Mendonça FS, Vaz RZ, Vaz FN et al (2018) Causes of bruises in carcasses of beef cattle during farm, transport, and slaughter-house handling in Brazil. Anim Sci J 90(2)
Mitchell G, Hattingh J, Ganhao M (1988) Stress in cattle assessed after handling, transport, and slaughter. Vet Rec 123:201–205
Nakyinsige K, Che Man YB, Aghwan AZ et al (2013) Stunning and animal welfare from an Islamic and scientific perspective. Meat Sci 95:352–361
NAMI (2019) Recommended animal handling guidelines and audit guide. North American Meat Institute, Washington, D.C.
NASS (2019) Livestock slaughter (2019). National Agricultural Statistics Service, Agricultural Statistics Board, United States Department of Agriculture. ISSN: 0499-0544
OIE (2019a) Terrestrial animal health code: guidelines for the slaughter of animals for human consumption. Organization mundial de la Sante Animals. World Organization and Animal Health, Paris. http://www.oie.int/eng/en_index.htm
OIE (2019b) Terrestrial animal health code: guidelines for the land transport of animals. Organization Mundial de la Santa animals. World Organization for Animal Health, Paris
Oliveira SEO, Gregory NG, Dalla-Costa OA et al (2018) Effectiveness of pneumatically powered penetrating and non-penetrating captive bolts in stunning cattle. Meat Sci 140:9–13
Peterson CM, Pilcher CM, Rothe HM et al (2015) Effect of feeding ractopamine hydrochloride on the growth performance and responses to handling and transport in heavy-weight pigs. J Anim Sci 93:1239–1249
Poletto R, Rostagno MH, Richert BT et al (2009) Effects of a 'step up' ractopamine feeding program, sex and social rank on growth performance, foot lesions and enterobacteriaceae shedding in finishing pigs. J Anim Sci 87:304–313
Poletto R, Cheng HW, Neisel RL et al (2010) Aggressiveness and brain amine concentrations in dominant and subordinate finishing pigs fed the β-adrenoreceptor agonist ractopamine. J Anim Sci 88:310–3120
Price MA, Tennessen T (1981) Preslaughter management and dark cutting in carcasses of young bulls. Can J Anim Sci 61:205–208
Reinhardt CD, Hubbert ME (2015) Control of liver abscesses in feedlot cattle: a review. Prof Anim Sci 31:101–108
Ritter MJ, Johnson AK, Benjamin ME et al (2017) Review: effects of ractopamine hydrochloride (Paylean) on welfare indicators for market weight pigs. Transl Anim Sci 1:533–558
Sabow AB, Colt YM, Zulkiffi I et al (2016) Blood parameters and electroencephalographic responses of goats to slaughter without stunning. Meat Sci 121:148–156

Sabow AB, Nakyisige K, Adeyemi KD, Sazilli AD, Johnson CB, Webster J, Farouk MM (2017) High frequency stunning of ruminants: a review. Meat Sci 202:124–134
Sabow AB, Goh YM, Zolkifli I et al (2019) Electroencephalographic and blood parameter changes in anaesthetized goats subjected to slaughter without stunning and slaughter following different electrical stunning methods. Anim Prod Sci 59:849–860
Schaeperkoetter M, Weller Z, Kness D et al (2021) Impacts of group stunning on the behavioral and physiological parameters of pigs and sheep in a small abattoir. Meat Sci 170:108538
Strappini AC, Metz JH, Gallo C et al (2013) Bruises on culled cows, when, where, and how they are inflicted. Animal 7:485–491
Tarrant V, Kenny FJ, Harrington D (1988) The effect of stocking density during 4 hour transportation to slaughter on behavior, blood constituents and carcass bruising in Friesian steers. Meat Sci 24:209–222
Terlouw EMC, Bourquet C, Deiss V et al (2015) Origins of movements following stunning and during bleeding in cattle. Meat Sci 110:135–144
Terlouw C, Bourquet C, Deiss V (2016) Consciousness, unconsciousness and death in the context of slaughter, part II, evaluation methods. Meat Sci 118:147–156
Van Putten G, Elshof WJ (1978) Observations of the effects of transport on the well-being and lean quality of slaughter pigs. Anim Regul Stud 1:247–271
Vimini RJ, Field A, Riley ML et al (1983) Effect of delayed bleeding after captive bolt stunning and heart activity and blood removal in cattle. J Anim Sci 57:628–631
Von Keyserlingk MG, Barrientos A, Ito K et al (2012) Benchmarking cow comfort on North American freestall dairies, lameness, leg injuries, lying time, facility design, and management for high producing Holsteins. J Dairy Sci 95:7399–7408
Warner RD, Ferguson DM, Cottrell JJ et al (2007) Acute stress induced by preslaughter use of electric prodders causes tougher meat. Aust J Exp Agric 47:782–788
Warriss PD, Brown S, Adams SJM (1994) Relationship between subjective and objective assessment of stress at slaughter and meat quality in pigs. Meat Sci 38:329–340
Welfare Quality Network (2009) Assessment protocols. www.welfarequality.net. Accessed 26 Oct 2019
Westervelt RG, Kinsman D, Prince RP et al (1976) Physiological stress measurements during slaughter of calves and lambs. J Anim Sci 42:831–834
Willson D, Baier FS, Grandin T (2021) An observational field study on the effects of changes in shadow contrasts and noise on cattle movement in a small abattoir. Meat Sci 179:108539

# The Human-Animal Relationship and Cattle Welfare

# 9

Susanne Waiblinger and Stephanie Lürzel

## Contents



**Abstract**

The human-animal relationship, defined as the perception humans and animals have of each other, is crucial for the welfare of cattle. However, there is large variation between farms: the quality of relationship ranges from very poor to

S. Waiblinger (✉) · S. Lürzel
University of Veterinary Medicine, Vienna, Austria
e-mail: susanne.waiblinger@vetmeduni.ac.at



M. Haskell (ed.), *Cattle Welfare in Dairy and Beef Systems*, Animal Welfare 23,
https://doi.org/10.1007/978-3-031-21020-4_9

very good, which is reflected in a continuum from poor handling behaviour and fearful animals to calm, competent handling using positive interactions, resulting in cattle trusting their handlers. After a brief outline of the history of the human-cattle relationship and its importance in modern agriculture, we define the human-animal relationship in detail and describe its variability on farms. We discuss different influential factors, including personality and attitudes on the human side and the quality and quantity of interactions with humans on the animal's side. Predictability, control over the situation and experiential and genetic makeup are further influential factors. We then describe the direct effects of the quality of the human-animal relationship on cattle welfare, in terms of the potential to elicit stress reactions but also antistress effects, which have an impact on productivity, health and risk of accidents. The farmers' perception of the animals also affects their decision-making and job satisfaction. Finally, we provide recommendations for establishing and maintaining a good human-animal relationship.

**Keywords**
Human-animal relationship · Cattle welfare · Cattle handling · Stockpersonship · Cattle behaviour · Occupational safety

## 9.1 Cattle and Humans: An Introduction

Cattle were an important part of the human environment long before their domestication, as reflected in the famous paintings of aurochs (*Bos primigenius*) in the caves of Lascaux and Chauvet aging back 15,000 and 35,000 years, respectively. Besides being prey animals, wild cattle were venerated in many societies (Marom and Bar-Oz 2013; Clutton-Brock 1999), and this was a potential driver for domestication: evidence of sacrificed cattle/ritual treatment of cattle has been found at the sites of first domestication (Clutton-Brock 1999). Later, cattle had a crucial role in the development of modern civilisation due to their usefulness as draught animals in the context of ploughing and transportation, besides directly providing food (meat and milk), hides for clothing and housing and manure to be used as fertilizer and fuel. The aurochs, the ancestor of our taurine and indicine cattle, was first domesticated about 9000 BC in the Middle East, and signs of sedentism at this time are found only in the same region, indicating an important role of domesticated cattle (Clutton-Brock 1999; Bollongino et al. 2012). The sedentary cattle-based societies gained ascendancy over the nomadic goat- or sheep-based societies (Schwabe 1978, cited in Albright and Arave 1997).

Nowadays, cattle are spread around the globe. They are the farm animal species with the highest number of individuals (1.5 billion) after chickens and with by far the largest body mass globally (Waiblinger 2019, based on FAOSTAT 2017). Cattle can convert plants that are inedible for humans into high-quality proteins and energy for human consumption and can thus add substantially to food security of the growing human population (Weltagrarbericht 2009). However, in intensive or industrialized production systems and global trade, this advantage is often lost by feeding high amounts of human-edible feed, leading to a loss in net food supply (Ertl et al. 2015). Similarly, the ecological impacts of cattle depend strongly on the farming system (Knaus 2016). Therefore, cattle play an important role for humankind and will do so in the future, and the development of more sustainable worldwide cattle systems is crucial (see Chap. 13 for a discussion on sustainability in cattle systems).

Cattle production systems vary greatly around the world, and they differ in the inherent level of contact[1] between humans and animals, but the humans responsible for the cattle always have a crucial role in establishing and maintaining good cattle welfare and sustainability of the system. Production systems include extensive or intensive systems with very limited contact between humans and animals, e.g. free-range cattle or feedlot cattle, as well as systems with close contact several times a day, such as in dairy production (Waiblinger 2019). These systems will be familiar to many; however, the range of cattle production systems extends to systems that can be defined as a two-species social system, demonstrated by some nomadic cattle cultures in Africa (Kratli 2008; Lott and Hart 1979), where there is constant and often close contact between the animals and the human. However, even within production systems, the amount of contact with the animals, the kind of interactions between humans and cattle and the resulting relationships can vary and can have a considerable impact not only on cattle welfare but also on productivity and human welfare. After defining the human-animal relationship (HAR), this chapter deals with the variation in human-animal interactions (HAI) and HARs, their underlying causes, their association with the quality of stockpersonship and their effects on animals and humans. We will also derive recommendations for practical handling.

## 9.2 What Is the Human-Animal Relationship?

In the strictest sense, a relationship develops between two individuals that know each other from former interactions and thus have formed expectations regarding their respective behaviour (Estep and Hetts 1992). In this sense, the HAR is the mutual perception of two individuals, a human and an animal, which develops based on the mutual behaviour and also expresses itself therein (Waiblinger et al. 2006a). The relationship can range from poor, where the interaction partner is perceived negatively (e.g. as frightening, and negative emotions are involved during interactions), to good, where the partner is perceived as a social partner and/or a source of pleasurable emotions (Waiblinger et al. 2006a; Waiblinger 2017). The HAR is not

[1] The term 'contact' is used as in Waiblinger (2019): 'In contrast with "interaction" the term "contact" is used when there is the possibility for human-animal interaction (HAI), but it does not necessarily happen to all animals in a herd or flock (e.g. the time a person is in the barn, in the same room as the animals, is the time of human contact; however as long as the animals do not look at the person, they do not interact visually), or to point out that the human is the initiator or active part (e.g. the human provides physical contact with an animal)'.

static but can change with new experiences during HAI (Waiblinger et al. 2006a). This implies that there is a need to keep continually providing positive contact to maintain a good-quality relationship but also that there is the possibility of improving the HAR even in adult animals.

Cattle can discriminate between different individual people (e.g. Rybarczyk et al. 2001; Taylor and Davis 1998), but when there are no other learnt associations that influence the response, will react to unfamiliar humans in a similar way as they would to their familiar caretaker (Rousing and Waiblinger 2004, Hemsworth et al. 1995, cited in Breuer et al. 2000). This means that they generalise their experience with their caretaker to other humans, allowing their relationship with their caretaker(s) to be assessed by recording their behaviour towards an unknown experimenter (in cattle: Waiblinger et al. 2003b; Windschnurer et al. 2008, 2009a, b; Destrez et al. 2018b). Similarly, it has been shown in sheep that the presence of an unfamiliar human can reduce stress in the animals, given that a good relationship with a familiar caretaker is present (Boivin et al. 1997). However, this familiar caretaker is more effective in reducing stress, thus suggesting that generalised and individual relationships can exist in parallel. Individual cognitive abilities, location of interactions (de Passillé et al. 1996; Rushen et al. 1998, 1999) and possibilities that promote or hamper the learning of individual characteristics of handlers (which depends on herd size, the number and consistency of caretakers, duration of contact and distance during exposure) can influence the processes of generalisation and discrimination between humans. If individual recognition is difficult, the animals probably develop generalised expectations about humans (de Passillé et al. 1996) depending on the interactions they have experienced. Thus, even without an individual relationship, an animal's perception of humans can range from frightening to pleasant. Humans may also show generalised attitudes and behaviour towards animals of one species or even animals in general (Hemsworth and Coleman 2011). In accordance with Waiblinger (2019), we thus use the term HAR in a wider sense, to encompass the perception of the other species in general, not only the individualised relationships. In the following discussion, when we want to explicitly refer to the animal's perspective—the relationship of the animal to humans—we use the term animal-human relationship (AHR). When referring to the perspective of the human, we will use a term such as farmer-cattle relationship.

## 9.3 Variation in Human-Cattle Interactions and Relationships

As mentioned above, production systems differ inherently in the frequency and type of interactions, and this may lead to differences in the HAR. For example, free-ranging beef suckler cattle and fattening cattle in large feedlots generally have less frequent and less close contact with humans compared with dairy animals that are milked twice daily. Accordingly, the relationship between cattle and humans can vary. Eight-month-old heifers that had been raised in a beef suckler herd with minimal human contact from birth onwards showed both more fear of and more aggression towards humans in a 'docility test' than animals that were kept separately from their dams for the first 3 months of life but were handled twice daily to let them suckle their dam (Boivin et al. 1992b, 1994). For a more general discussion on differences between production systems with respect to the type of interactions, see Waiblinger (2019).

However, the actual interactions (type, quantity, quality) and relationships can also vary greatly within a specific production system and even within a husbandry system depending on the exact farm situation and particularly the behaviour of the humans responsible for the animals. In a sample of 150 dairy farms in Austria, Switzerland and Germany, milkers used 0–11 positive vocal and tactile behaviours per milked cow, while the number of shouting or forceful hitting events—clearly negative interactions—ranged from 0 to 1 per cow (Rouha-Mülleder et al. 2009; Waiblinger and Menke 1999; Waiblinger et al. 2002), with some milkers never and some milkers exclusively showing positive behaviours. Similar variation was also found in other studies in Europe and Australia, which sometimes included much larger herd sizes (Breuer et al. 2000; Hemsworth et al. 2000; Ivemeyer et al. 2018; Waiblinger et al. 2007). Accordingly, the behaviour of cows towards humans varies remarkably. For example, the median avoidance distance, i.e. the distance maintained from an approaching test person, ranged from 0 to 1.5 m in the 150 farms mentioned above, and there were herds where up to 80% of the cows accepted human touch and others where not a single cow could be touched (Rouha-Mülleder et al. 2010; Waiblinger et al. 2002, 2003b). In these on-farm studies, animals' reactions showed clear associations with human behaviour, supporting the concept of HARs described above. In addition to the quality of interactions, the type of interactions and their frequency and duration also vary. French farmers keeping beef suckler cattle reported monitoring their animals once daily or less (for cows: 0.8 ± 0.4 times per day, for heifers 0.5 ± 0.4), with only 55% of them monitoring them on foot (Destrez et al. 2018b). Consequently, the avoidance distance of the heifers also showed high variation ranging from 0 to 20 m. In a survey on handling practices in Austria, 27% of the 146 responding farmers[2] answered that they did not check on their cows outside the times of milking and feeding. Forty-six percent did additionally observe their animals from outside the pen, and 27% of farmers checked on their animals by walking through the herd (Waiblinger et al. 2007). On the same farms, the time spent in contact with the animals per day ranged from 0.9 to 22.5 min/dairy cow and from 1 to 25 min/calf. Similarly, daily contact time on foot ranged from 2 to 33 min/cow (mean ± SD; 7.6 ± 7.24) on 32 dairy farms in Germany (Ebinghaus et al. 2018).

[2] Farmers generally are the ones making decisions about details concerning the housing and management as well as caring for the animals, while other stockpersons may often be limited in their decision-making power and may also carry out the farmer's decisions.

## 9.4 Why Do Human-Animal Interactions and Human-Animal Relationships Differ Between Farms?

Across production and husbandry systems, it is the human who mainly determines the possibilities for, and characteristics of, the HAI: whether interactions take place at all, their type and timing and the environment in which the interactions can take place. In addition, farmers influence the genetic background of the herd by selecting breeds or individual animals. Thus, the farmer has the largest influence on the development of the human-cattle relationship on a given farm. Before we review what we know about effects of different characteristics of HAI on the animals' relationship to humans and briefly touch the role of genetics, we consider the question of why humans differ in their interactions and relationships with their animals.

### 9.4.1 Factors Influencing Humans' Behaviour and Relationship to Their Cattle

Personality and attitudes are personal variables that have been widely used in psychology to explain human behaviour, and their importance has also been confirmed for understanding stockperson behaviour towards farm animals, including cattle. Other personal variables, such as empathy, knowledge and experience, and external variables, such as the actual situation (e.g. time pressure, workload) and peer pressure, can be influential as well (Spoolder and Waiblinger 2009; Waiblinger and Spoolder 2007; for more detailed reviews, see Ajzen 1988; Hemsworth and Coleman 2011).

Personality traits may directly affect the behaviour of stockpeople towards cattle, as suggested by the correlations found on dairy farms, which showed that patient and agreeable dairy farmers interacted with the cows more positively and less negatively during milking (Waiblinger et al. 2002; Waiblinger 1996). Other personality characteristics act indirectly by influencing attitude formation (Waiblinger et al. 2002), in line with the theory of Ajzen and Fishbein (1980): The more pessimistic stockpeople were, the less positive beliefs about cattle they had and the more negative behavioural attitudes. In detail, more pessimistic stockpeople rated regular positive contact with the animals as less important, had a lower intention to use patient behaviour during milking and felt less comfortable managing cattle (Waiblinger et al. 2002).

Attitudes are often considered the most important causal factor of a person's behaviour towards social objects, and this seems to hold true also for HAI (Hemsworth and Coleman 2011). Differences in attitudes towards animals and towards HAI can explain differences in humans' behaviour towards animals, in interaction with the other influencing variables mentioned above, such as personality. Variability in attitudes has also been found in stockpersons working with cattle, and the influence of these attitudes on their interactions with the cattle, and on the animals' subsequent behaviour and productivity, has been confirmed (Hemsworth et al. 2002; Lensink et al. 2000b; Waiblinger et al. 2002). Stockpeople who reported greater enjoyment derived from the contact with the animals and agreed more strongly with statements indicating the importance of regular positive contact and the intention to use patient and calming behaviour during handling of animals showed more positive and fewer negative interactions with their cows during milking (Waiblinger et al. 2002). Accordingly, stockperson attitudes are associated with cattle behaviour, with a lower avoidance shown by dairy cows when stockpeople reported stronger intentions to use patient, calm behaviours during milking and moving, enjoyed contact with their cows more and had more positive beliefs about cows (Ebinghaus et al. 2018; Waiblinger et al. 2007). But attitudes are not just factors influencing HAI and as a consequence, the HAR. Cognitive and affective attitudes are strongly linked to the way in which animals and interactions with them are perceived during encounters, and thus, they form an integral part of a human's relationship with animals.

Attitudes are learnt. Their formation starts in childhood, and they can be modified later by new information or experiences (Ajzen 1988; Paul and Serpell 1993). This makes attitudes an important target when it comes to attempts to improve HAI and HAR. Indeed, training programs based on cognitive-behavioural intervention (Hemsworth and Coleman 2011) were successful in changing human attitudes and behaviour, and subsequently, animal behaviour and production on dairy and beef suckler farms (Hemsworth et al. 2002; Ruis et al. 2010; Windschnurer et al. 2010). Such training programs can interrupt the vicious cycle in which negative handling leads to difficulties in handling, which then reinforce negative beliefs about cattle. Stockpeople that had received formal training regarding best practices for cattle handling scored more positively and less negatively on attitude scales and used more positive and fewer negative behaviours during a 1-day vaccination event on farms with extensive cattle production compared to non-trained persons (Ceballos et al. 2018b). On these farms, there were also stockpersons that had not been trained formally but could have been informally trained because they were part of a team in which at least one person was trained. Interestingly, their values were recorded as being in between the trained and non-trained stockpersons and showed that they used more positive behaviours than the latter. In an earlier publication, Seabrook (2001) also showed that handling styles can be adopted from colleagues.

Herd size and facility design can also influence HAI and HAR. On dairy farms, milking parlour design affects the possibilities for interactions. In tandem milking parlours, there is constant visual contact, and the milker has access to the 'whole cow', including the head and neck regions, which allows for tactile interactions in these preferred body regions, whereas in side-by-side milking parlours, the milker only sees the hind view of the cow, and the cow hardly has any visual contact with the human. The herringbone parlour is in-between the other two parlour types. In a comparison of these three parlour types, milkers in tandem milking parlours indeed used the most positive behaviours (Waiblinger 1996), including stroking cows in the head-neck area, and the lowest numbers of moderately negative behaviours (Waiblinger and Rouha-Mülleder 2015), while milkers in side-by-side parlours used the lowest number of positive interactions. Moderately negative behaviours are mainly used for moving the cows in and out of the milking parlour (Waiblinger et al.

2003a), which explains the lowest use in tandem and highest use in herringbone parlours (Waiblinger and Rouha-Mülleder 2015). Nevertheless, personality and attitude explained most of the variation in milker behaviour, while milking parlour design was of relatively low explanatory value (Waiblinger and Menke 1999; Waiblinger 1996). Additionally, it cannot be excluded that farmers with a better relationship with their cows are more likely to choose a tandem milking parlour.

Automatic milking systems (AMS) can lead to an even lower duration of interactions between cow and milker (Wildridge et al. 2020), but research on their effect on the HAR is still scarce. The few studies that are available suggest that the effect of the transition to AMS depends on the basal quality of the HAR. On farms that transitioned from conventional milking to an AMS and had a poor HAR to begin with, cows showed a somewhat reduced level of fear of humans after the transition, probably due to the lack of exposure to the negative behaviours used by handlers moving the cows to and out of the conventional parlour (Wildridge et al. 2020). In contrast, in a cross-sectional study, herds milked in an AMS had a comparable relationship with humans as herds milked in herringbone parlours, while they showed less fear of humans on farms with tandem milking parlours (Ebinghaus et al. 2018). A study in France found a slightly higher number of animals that could be touched on farms with AMS in the univariate analyses; however, the milking system was no predictor in multivariate analysis, and the level of fear of humans was again relatively high on all farms (de Boyer des Roches et al. 2016). These results add further evidence for a crucial role of stockperson behaviour.

The intensity of contact, measured as the frequency of positive interactions per animal and the recognition of individual animals by the farmer, decreased with herd size on dairy and veal farms (Waiblinger and Menke 1999; Lensink et al. 2000b), and correspondingly, the proportion of cows with a very good AHR also declined (Waiblinger and Menke 1999). Similarly, the probability of calves allowing themselves to be touched was higher when stockpersons were responsible for a lower number of calves (Leruste et al. 2012). However, the quality of HAI assessed during handling on the dairy farms and the level of negative interactions in veal farms was independent of herd size (Lensink et al. 2000b), confirming results on pig farms (Hemsworth and Coleman 2011). Furthermore, attitudes and personality explained most of the variation in contact intensity, explaining why high-intensity positive contact and a reasonable quality of AHR can be found in larger herds as well as in smaller herds (Waiblinger and Menke 1999). Despite the high influence of human personal variables on the quality of the HAR on a given farm, there is nevertheless a higher risk of an animal being fearful of humans in larger herds (Ebinghaus et al. 2018). Lower intensity of contact, a higher number of stockpeople and more frequent changes of stockpeople in larger herds may contribute to this risk (Waiblinger and Menke 1999; Knierim and Waran 1993; Schlichting 1974).

A high workload and time pressure can have a negative influence on HAI. Stockpersons reported time pressure to be a reason for negative behaviour (Seabrook 2001). The longer the total daily working time of dairy farmers, the more negative and the lower proportion of positive interactions were used (Waiblinger et al. 2003b).

### 9.4.2 Effects of Different Human-Animal Interactions on the Animal-Human Relationship

#### 9.4.2.1 Quality of the Interaction

An animal may perceive an interaction as negative (aversive, unpleasant emotions involved), neutral (neither pleasant nor unpleasant, no change in the affective state) or positive (pleasant emotions elicited). While some interactions are perceived as negative by their very nature, especially when being painful (e.g. dehorning without anaesthesia, forceful hitting with a hand or stick), the perception of others depends on the animals' existing relationship with humans (Waiblinger et al. 2006a). Based on previous experience, touch by a human can elicit fear reactions, be perceived indifferently or elicit pleasant emotions (Lange et al. 2020b). Nevertheless, interactions can generally be categorized according to their inherent quality based on their potential to activate specific receptors (e.g. a strong strike activating pressure-sensitive nociceptors or touch activating touch-sensitive receptors (Zimmermann et al. 2014)) and thus the potential to elicit pleasant or unpleasant emotions. The evidence for the precise nature of an interaction is found by testing for changes in animals' reactions towards humans and thus their relationship with humans after having experienced a specific type of interaction. While negative interactions worsen the relationship by increasing fear, and thus increase avoidance and reduce approach behaviour, neutral interactions lead to reduced fear by habituation to humans and thus reduced avoidance. The latter is also true for positive interactions, but to classify an interaction as positive, signs of positive emotions associated with the interaction need to be shown as well (Rault et al. 2020). These could be immediate signs of relaxation or appetent behaviour. All kinds of external sensory perception can be involved during HAI (i.e. acoustic, visual, tactile, olfactory and gustatory). However, we know most about visual and tactile interactions, while the other senses have barely been investigated at all.

##### 9.4.2.1.1 Negative Interactions

Interactions that cause pain are among the most obviously negative interactions. Some painful interactions occur during routine management, for example, disbudding (Stafford and Mellor 2005), branding (Tucker et al. 2014) or castration (Bergamasco et al. 2021), if they are carried out without the effective use of anaesthetics and analgesics. There are some examples that illustrate the impact of painful interactions on the AHR. One day after disbudding, the avoidance distance of calves was increased (Lürzel et al. 2015a). Moreover, calves that had been stroked during the first 14 days of life had a lower avoidance distance before disbudding than non-stroked calves, but this difference disappeared after disbudding (Lürzel et al. 2015a). Taschke (1995) found poorer 'tameness scores' (a combined measure of approach and avoidance) in calves up to 6 days after disbudding compared to before, similarly indicating a negative effect on the AHR. Additionally, some interactions commonly used while moving or restraining cattle are perceived as aversive and can even cause pain, such as hitting the animals or using an electric cattle prod (Pajor et al. 2000, 2003).

It is not only interactions that cause pain that are perceived as negative but also shouting and loud noise. Moving cattle towards a place where they expect to be shouted at required more force than moving them towards a control treatment (a person moving but not interacting with the animal) (Pajor et al. 2000), and when they had the choice between being shouted at and a control treatment, they chose the control treatment significantly more often (Pajor et al. 2003). Interestingly, there was no clear difference between shouting and either hitting or using an electric prod. It may be that shouting has a stronger impact than previously assumed and may potentially elicit a very unpleasant or even painful acoustic sensation. Similarly, the noise created by milking machines can be perceived as aversive (Arnold et al. 2008), which indicates that other types of noise (e.g. knocking a stick against milking parlour equipment) might also be unpleasant for cattle. Cattle hear higher frequencies than humans (up to 35 kHz, optimal hearing around 8 kHz) and thus are more sensitive to high-frequency sounds, such as metallic noises. In addition to this, events that occur suddenly or without warning can elicit fear and startle reactions (Greiveldinger et al. 2007). Accordingly, sudden, loud sounds and quick movements (Lanier et al. 2000), especially if suddenly occurring in the visual field of the animal, are often perceived negatively. Cows that had experienced milkers displaying waving motions with their arms when moving them were less likely to approach an experimenter in an unfamiliar test arena (Breuer et al. 2000).

#### 9.4.2.1.2 Neutral Interactions

As a stimulus without a consequence for the animal, neutral interactions facilitate habituation of cattle to humans. Examples of neutral interactions include the routine observation of the animals from a distance, walking through the herd or doing work such as cleaning in the barn. Low-intensity interactions used in the context of moving cows, such as tactile interactions using the hand or a stick with little force or talking to the cows in a dominant, determined way, have been categorized as neutral in some observational studies (Waiblinger et al. 2002; Ivemeyer et al. 2011, 2018). However, based on the correlations with cow behaviour, Waiblinger et al. (2002) argue that these interactions might be perceived as negative.

Neutral interactions seem to play a role in improving a previously poor AHR and maintaining a neutral AHR once it has been established, so that the human presence elicits neither fear nor pleasant expectations. There are indications that a neutral human presence also leads to a habituation effect in chickens, as the concept would explain tier effects on the AHR in multitier battery cage systems in which animals closest to human traffic are less fearful of humans (Barnett et al. 1994). However, there has been no systematic research on the effectiveness of this type of interaction in improving the AHR in cattle. In some studies, there is a control treatment including the presence of a person, but no control without it, which would allow the effect of human presence itself to be assessed. For example, the latency to approach a person to within 1 meter in a test arena decreased over 3 weeks when the cows experienced close presence of the person for 5 min/day for 5 days/week, but latency to contact stayed the same, in contrast to stroking treatments (Schmied et al. 2008a). With a 'no person' treatment, it would have been possible to find out whether the effect on approach latency was due to habituation to the person or to another factor, such as repeated testing. In beef suckler herds, the avoidance distance was lower in herds with higher monitoring frequency (Destrez et al. 2018b); however, the occurrence of positive interactions in addition to neutral ones during these control visits cannot be ruled out. While neutral interactions can likely improve the AHR to a certain extent, they have a limited effect, especially when it comes to neutralizing or overcoming previous aversive experiences (de Passillé et al. 1996).

#### 9.4.2.1.3 Positive Interactions

##### 9.4.2.1.3.1 Gentle Tactile Interactions

Tactile stimulation such as stroking or brushing has been shown to improve the AHR in numerous experiments, both in terms of reduced avoidance distances (beef calves: Probst et al. 2012; dairy calves: Lürzel et al. 2015a; dairy heifers: Lürzel et al. 2016; dairy cows: Lürzel et al. 2018, Windschnurer et al. 2009a) and increased approach behaviour (Lange et al. 2020b, dairy cows: Schmied et al. 2008a; veal calves: Lensink et al. 2000a). The idea that the stroking or brushing of cattle might be perceived as positive by them is based on their own natural affiliative behaviour, in which social licking plays an important role (Reinhardt et al. 1986), and stroking and brushing are thought to mimic social licking. Cattle often solicit social licking from other animals (Sato et al. 1991; Laister et al. 2011) but also brushing (Bertenshaw and Rowlinson 2008) or stroking from a human (SW, personal observation; anecdotally in Murphey et al. 1981), indicating a positively valenced experience for the animals. During the interaction, the licked or stroked animal often stretches its neck and may let its ears hang down, which—due to the positive valence of the interaction—can be interpreted as an indicator of enjoyment and/or relaxation, a low-arousal positive state (Reinhardt et al. 1986; Schmied et al. 2008b, 2005; Lange et al. 2020a; Lürzel et al. 2015a).

The area of the body that is stroked plays an important role in determining how stroking is perceived by the animals. Stroking dairy cows in the ventral neck area or at the withers elicited a longer duration of neck stretching and ear hanging than stroking in the lateral chest area (Schmied et al. 2008b). Accordingly, the AHR was improved especially after stroking the ventral neck area once a day for 3 weeks (Schmied et al. 2008a), with the effect lasting for about 8 weeks, while the effect was less pronounced and shorter-lasting after stroking the withers and nonexistent after stroking the chest. Stroking cows on either the withers or the ventral part of the neck also improved their reactions to rectal palpation, compared with cows that had experienced only the close presence of the experimenters previously (Schmied et al. 2010). This underlines the importance of tactile interactions in the context of reducing the aversiveness of necessary unpleasant procedures. In some studies reporting only weak, if any effects of stroking, calves were stroked in non-preferred areas (Boivin et al. 1998), or the stroking procedure including the body areas stroked was not described (Jago et al. 1999), which indicates that the body region involved might be a crucial factor for the perception of tactile interactions by the animals. Different body regions are also preferred in intraspecific allogrooming in cattle (Schmied et al. 2005) and may differ in the density of touch receptors (in humans: Taylor-Clarke et al. 2004).

#### 9.4.2.1.3.2 Gentle Vocal Interactions

Although cattle are not a very vocal farm animal species (Watts and Stookey 2000) and most vocalizations occur in the context of high-arousal, negative situations (Bouissou et al. 2001), vocalizations also play a role in positive contexts, such as during cow-calf bonding shortly after parturition, with the dam directing low-frequency calls at her calf (Padilla de la Torre et al. 2015). The features of acoustic stimuli seem to carry intrinsic meaning: in dogs, a change in fundamental frequency and signal duration had an effect on motor activity in the context of learning: puppies learned more easily to come with a command of four short-rising notes as compared to one long note (McConnell 1990). This phenomenon is probably species-independent as suggested by studies on the use of different sounds by animal trainers (McConnell 1991) as well as humans' ability to perceive the emotional valence of animal vocalisations (Pongrácz et al. 2005, 2006; Tallet et al. 2010). Thus, quiet talking with long vowels and a decreasing pitch at the end of speech elements (described as calming or soothing) has been categorized as positive milker behaviour in several observational studies (e.g. Waiblinger et al. 2002; Ivemeyer et al. 2018). In a number of experiments, talking in a gentle voice has been combined with other interactions (e.g. Schütz et al. 2012; Lürzel et al. 2015a; Rushen et al. 1999), but it has not been thoroughly investigated on its own. However, cattle are sensitive to human vocal stimuli: calves and cows can learn to follow specific calls to go to the milking parlour (Murphey and Duarte 1983; Albright et al. 1966). In a choice experiment, cows chose the maze arm containing a person talking to them in a gentle voice numerically more often than the one with a person not interacting with them, especially during the later trials, when learning had progressed, but the difference was not significant (Pajor et al. 2003).

#### 9.4.2.1.3.3 Feeding

Feeding has been shown to be perceived as positive by cattle, as expected due to its proximate effect, the satiation of hunger and its ultimate importance for survival. In a Y-maze, cows chose hand-feeding of hay and concentrates, and calves chose pail-feeding of calf starter more frequently than a noninteracting handler (Pajor et al. 2003). Feeding has also successfully been used as a reward in the context of training dairy heifers to accept an injection (Lomb et al. 2021).

Feeding manually or at least in the presence of a person could be a method for improving the AHR by creating an association between humans and feed. Many studies have found a potential positive effect of interactions involving hand-feeding or feeding in the presence of humans on the AHR (Boivin et al. 1992b; Lensink et al. 2000a; Munksgaard et al. 1997; Waiblinger et al. 2004). However, in these studies, feeding was used alongside gentle tactile and vocal interactions, and the effects cannot be separated. It is probable that the improvement of the AHR can be largely attributed to the tactile component because later similar studies using stroking without feeding were also effective (Schmied et al. 2008a, 2010). Some studies investigated the effect of feeding itself in calves. Calves that had received their milk rations from a person in the pen had a lower latency to approach an unfamiliar person in the home pen than calves that had been fed without seeing a person (Jago et al. 1999). If calves are provided with milk by a person, it may thus lead to an improved AHR, as hormones involved in filial bonding (oxytocin, cholecystokinin) are released not only during suckling (Nowak and Boivin 2015; Lupoli et al. 2001) but also during bucket feeding (Lupoli et al. 2001), allowing a positive association with human contact (Waiblinger et al. 2020). Accordingly, more veal calves on farms with bucket feeding could be touched and fewer calves had high levels of fear of humans compared to calves fed using trough feeding systems (Leruste et al. 2012). Regarding effects of feed provision in adult cattle, manual feed provision was a predictor of less fearful responses of cows to human interactions at the feeding place (touch, release from feeding rack) but not for the avoidance distance in the feeding rack in a cross-sectional study on dairy farms (Ebinghaus et al. 2018), suggesting only limited effects.

#### 9.4.2.2 Quantity of Contact: Total and Relative Amount of Human-Animal Interactions of Different Qualities

Few studies have investigated the effects of the quantity of different HAI explicitly, but some studies allow inferences to be drawn or report the change in animal reactions throughout the course of the experiment. In a study comprising two handling periods separated by 4 weeks without handling, the avoidance distance of dairy cows at the feeding place (ADF) was lower than that of non-handled controls after just the first period of ten 3-min sessions of stroking in the milking parlour (5 days, two times daily, in total 30 min); however, the effect vanished within 4 weeks (Windschnurer et al. 2009a). The ADF was lower again after a second treatment period of seven stroking sessions. The avoidance distance in the barn differed only after the second treatment period, i.e. after a total of 51 min. In unrestrained cows, a duration of 45 min of HAI was effective for almost all individuals in the group to accept some physical contact, and additional 15 min were required for them to enjoy stroking or at least not avoid it (Lange et al. 2020b). However, we cannot exclude the fact that the additional visual contact might have increased the effect, as the animals were not prevented from watching the experimenters' interactions with other individuals (Munksgaard et al. 2001). Experimental studies finding effects of gentle contact rarely used less than 50 min of total contact time (for review see Windschnurer et al. 2009a). One hour of gentle handling of heifers during the first hour after calving reduced fear of humans (Hemsworth et al. 1989), and it would be interesting to find out whether a shorter duration during this potentially sensitive period would be as effective.

Although a single-farm study showed a greater level of avoidance of humans in cows in their first lactation than in subsequent lactations (Haskell et al. 2012), a general age effect within dairy herds does not exist on commercial farms—both lower and higher avoidance distances in younger than in older animals were found in a study on 35 farms, with a lack of an age effect on most farms (Waiblinger et al. 2003b). This indicates again the importance of the quality of contact and experiences throughout life: if the regular contact during milking is perceived as negative, the AHR will deteriorate with increasing age of the cows, but if it is perceived as positive, older cows may have a better AHR than heifers. However, if a good AHR was established already in heifers, further improvements may not be detectable.

A positive AHR on farms does not necessarily mean that stockpeople never use (moderately) negative interactions, but the use of positive interactions outweighs negative interactions by far (Waiblinger et al. 2002). Neutral interactions alone do not seem to be fully effective in overcoming negative experiences (de Passillé et al. 1996). There is no study in cattle that has investigated the effect of the relative amount of positive and negative interactions. However, the reaction to humans of pigs that had experienced positive vs. negative interactions in a ratio of 5:1 was more comparable to the reaction of pigs that had been treated only negatively than to the reaction of pigs that had been treated positively or minimally (Hemsworth et al. 1987). However, besides the amount, the level of predictability of interactions of different qualities may be important, as negative and positive interactions were provided unpredictably in this experiment.

#### 9.4.2.3 Predictability, Animal Agency and Control Over the Situation

Lack of predictability and/or controllability can aggravate a stressful situation (Koolhaas et al. 2011). Human behaviour that increases the unpredictability or decreases controllability of a situation for the animals is likely to be perceived negatively. This includes inconsistency, for instance, sometimes positive, sometimes negative behaviour in comparable situations and unclear signalling, for example, when slapping the animal to initiate movement is not stopped as soon as the animal starts moving. Human behaviour may be perceived as less predictable by cows on farms with a higher number of stockpeople and changes in staff and thus contribute to a poorer AHR and/or stress (Knierim and Waran 1993; Schlichting 1974; Waiblinger and Menke 1999).

Predictability and control over the environment or the situation are closely connected (Bassett and Buchanan-Smith 2007). Being in control is an important aspect of agency, and it is well-known that agency is an important factor contributing to human wellbeing (Franks and Higgins 2012). In recent years, the concept has gained traction also in animal welfare research (Špinka 2019). It has long been suspected that the degree of control cows have over their interactions with humans is important for the improvement of their relationship with humans (Windschnurer et al. 2009a; Lürzel et al. 2016; Le Neindre et al. 1993). In a recent study, human interactions with unrestrained dairy cows in their home barn improved an initially poor AHR more than interacting with the cows when they were restrained in the feeding rack, in comparison with a control group that only experienced routine handling (Lange et al. 2020b). For the unrestrained group, the handler progressed from talking to the cows from a distance to establishing physical contact and stroking them across the study period. The restrained group experienced the stroking and talking right from the start, and thus received 'imposed' stroking and more stroking overall. After 20 days of treatment, all but one unrestrained animal accepted physical contact, and thereafter the stroking duration increased quickly, indicating that (1) the AHR improved by only visual and vocal interactions first and (2) a larger improvement occurs when animals can actually be stroked and accept it voluntarily. The AHR of the restrained cows improved numerically over the course of the experiment, but not as strongly as the AHR of the unrestrained cows, which had control over their interactions with the handler. A study in dairy heifers with a good relationship with humans (Lange et al. 2021) found more neck stretching if the animals were stroked while they were free to move than during restraint, although signs of enjoyment were also shown during restraint. Gentle vocal and tactile interactions thus seem to be effective in both situations (imposed or voluntary), but if possible, cattle-human interactions designed to improve the AHR should take place while the animals are unrestrained.

#### 9.4.2.4 Early Experiences and Sensitive Periods

Socialisation is the process by which an animal develops appropriate social behaviour towards conspecifics and forms social relationships and attachments (Dietz et al. 2018; Scott 1962). Many species have sensitive periods for socialisation. One of these sensitive periods is directly after birth, when filial imprinting or attachment takes place; however the weaning and adolescent periods are also important for socialisation (Sachser et al. 2013), as they represent periods of social change. In dogs and cats, it is well-known that there are also sensitive periods for the socialisation with humans, i.e. where contact with humans promotes the formation of a dog-human relationship (for review, see Waiblinger 2017). In farm animals, the timing of human contact can also be important, although the early socialisation period seems less crucial.

The periods soon after birth, weaning and parturition seem to constitute sensitive periods for improving the relationship with humans in cattle, in that the same amount and type of contact has a stronger impact than at other times (Boivin et al. 2003; Waiblinger et al. 2006a). For example, calves that were stroked and talked to during milk feeding during their first 4 days of life showed significantly more approach behaviour at the age of 20 and 40 days than calves that had not experienced this treatment, whereas calves that had experienced the same treatment later (days 6–9 or days 11–14) did not differ significantly from controls at the age of 40 days (Krohn et al. 2001). Also, heifers handled gently in the first 2 weeks after weaning accepted stroking and hand-feeding by a human to a much higher degree than heifers handled in the same manner 6 weeks after weaning (Boivin et al. 1992a). The latter, however, were still less fearful of humans than animals not handled at all. The time around calving has also been suggested as a sensitive period for cows: if a handler with foetal fluid from the calf on their hands allowed the cows to sniff them during the first hour after their first calving, they showed fewer flinch-step-kick responses during milking, especially in the first 2 weeks of lactation, and more approach behaviour 6 weeks after calving (Hemsworth et al. 1987, 1989).

However, a prolonged period of regular handling seems to be necessary to maintain an established relationship. In heifers, 30 days of gentle handling reduced the fear of humans most strongly if they were spread out over the first 9 months of the heifers' lives, in contrast to a treatment period covering months 0–3 or months 6–9 of life (Boissy and Bouissou 1988). At the age of approximately 1 year, there was no longer any difference between heifers that had been stroked or not during their first 14 days of life and had differed in their responses to humans after this treatment (Lürzel et al. 2015a, 2016). However, the animals had experienced negative handling (e.g. disbudding) in the year after the treatment. Nevertheless, a 10-day treatment using gentle tactile and vocal interactions was highly effective at reducing these heifers' avoidance distance at that later age, independently of the treatment given at the earlier age. Even if they were not stroked during the early life period, all of the heifers had experienced close human contact (tube-feeding of colostrum, bucket feeding for up to 10 days, moving from single box to group housing, being taught to drink from an automatic feeder) in the first days of life. This may be sufficient for a primary socialisation effect with humans, which allows for relatively quick improvements later in life.

Nevertheless, long-lasting effects have been found for early positive contact. In beef suckler calves, stroking in the first 4 weeks of life had a positive effect on the AHR that lasted for about 9 months (Probst et al. 2012). In dairy calves, the amount of human contact during the first 5 days of life resulted in differences in their behaviour towards humans at the age of 4 weeks (Waiblinger et al. 2020). One year later, these differences had declined partially but were still visible, but by the time of the first calving, they had disappeared. Even though these two studies did not test specifically for a sensitive phase for socialisation, as there was no treatment in which animals were handled at a different age, they point towards the great potential for long-lasting effects of contact during early life.

Thus, it is most beneficial for the cow's relationship with humans that regular positive contact with humans takes place during sensitive periods early in life and also later on. This is in agreement with research on intraspecific social relationships of cattle, where the strongest bonds are formed earlier in life and are regularly reinforced later by positive social interactions (Reinhardt and Reinhardt 1981; Bouissou et al. 2001; Reinhardt 1980). It must be emphasised, however, that adequate socialisation with conspecifics is also important for the welfare of cattle both in terms of their intraspecific social behaviour (Wagner et al. 2012) as well as interactions with humans. Bulls reared in individual pens with physical isolation were more aggressive when compared to group-housed bulls (Price and Wallach 1990). In extreme cases, hand-reared animals without sufficient contact with conspecifics may display courtship and sexual behaviour preferably or exclusively towards humans, depending on the duration of isolation from conspecifics (Sambraus and Sambraus 1975).

#### 9.4.2.5 Social Environment and Social Learning

An animal's behaviour towards humans can be affected by the behaviour of its conspecifics, especially its mother, and the effects might be greater than those created by direct contact with humans (reviewed by Waiblinger 2017). For example, foals whose dams were brushed and hand-fed by a human in the foals' presence in the 5-day postpartum were less fearful of humans even 1 year after the experience, compared with foals whose mares did not experience this additional contact (Henry et al. 2005). Furthermore, foals whose dams showed protective behaviour during brushing and feeding spent less time close to the experimenter than foals whose dams did not show protective behaviour (Henry et al. 2005). Thus, it was probably the dams' behaviour towards the human that led to the decrease in fearfulness in the foals.

This mechanism might also reconcile the seemingly contrasting results of the studies by Krohn et al. (2003), who found that the presence of the dam reduced the effect of positive handling of calves on their AHR, and Probst et al. (2012), who achieved a long-lasting improvement of the AHR by providing gentle tactile contact to calves while their dams were present. An explanation might be that the dams' behaviour differed. In the study by Probst et al. (2012), the dams and calves were not separated, and the cows behaved calmly while their calves were in contact with humans (Probst, oral communication), whereas in Krohn et al.'s study, dams may have shown signs of unease, being close to the calf but unable to prevent the contact because they were physically separated.

One study shows evidence for cows adapting their behaviour towards a handler depending on their observations of the handler's interactions with other cows (Munksgaard et al. 2001). Another study found a correlation between scores of calves handled in a 'docility test' and those of their dams (Boivin et al. 2009), which also points towards a maternal influence on the AHR; however, the genetic inheritance of docility traits affecting this result cannot be ruled out. To date, the area of social transmission of the AHR is still under researched.

### 9.4.3 Animal Characteristics

The AHR of an animal is in part influenced by its genetic makeup. Dairy cattle showed lower avoidance towards humans than beef cattle (Murphey et al. 1980, 1981), and this difference was mainly caused by breed differences and only marginally affected by the rearing system (Murphey et al. 1980). However, dairy breed calves were less approachable by humans compared to dual purpose breed and crossbred calves in more recent studies (Calderon-Amor et al. 2020; Leruste et al. 2012). A genetic influence on the AHR has also been found within the same breed. The reactions of heifers to humans in a docility test were in part explained by their sire (Grignard et al. 2001). Heritability of behaviour during milking ('milking temperament', which is assessed by the farmer) has been estimated to be approximately 0.5 in Holstein-Friesian cows (Dickson et al. 1970), while the heritability of behaviour of beef heifers towards humans in the docility test was 0.28 (Le Neindre et al. 1993). On the other hand, a recent study in Holstein cows (Stephansen et al. 2018) found low heritabilities of handling temperament (0.13) and farmer-assessed temperament (0.10). This large variability can be explained by the different measures, procedures of assessment and ages of the animals. Studies in various species indicate that the influence of an animal's genetic background on its AHR is the result of underlying heritable differences in various personality traits, mainly fearfulness/boldness, aggressiveness, curiosity and sociability. These can influence the perception of human-animal interactions (Waiblinger et al. 2006a; Waiblinger 2017) and thus ease or complicate the establishment of a neutral or good AHR.

Although the genetic disposition of an animal plays a role in the development of its AHR, the AHR is mainly affected by experience, as described above. However, the AHR should be included as a criterion in breeding decisions, as the desired disposition will facilitate the establishment of a good AHR. Accordingly, some farmers include behaviour as a criterion for selection, and a statistical tendency was found for this practice to be associated with a lower avoidance distance on pasture in a study on 20 French beef cattle farms (Destrez et al. 2018b), confirming an earlier finding on dairy farms (Waiblinger 1996).

## 9.5 Direct and Indirect Effects of the Human-Animal Relationship on Cattle Welfare and Production

### 9.5.1 Stockpersonship, the Human-Animal Relationship and Animal Welfare

The welfare of an animal can range from poor to good according to the ease or difficulty of coping with the environment and the negative and positive emotions elicited. For good welfare, not only should negative emotions be avoided as far as possible but animals should also experience positive emotions (Broom 2007; Boissy et al. 2007; Waiblinger 2012). HAI and the AHR are inseparably intertwined with animal emotions and therefore have a direct impact on animal welfare. A good relationship can not only minimise or prevent negative emotions and stress during interactions but can also provide opportunities for pleasant emotions, contributing to good welfare.

The stockpersons' handling skills, their attitudes and interactions with the animals thus directly affect animal welfare. In addition, farmers and stockpersons have a large impact on all other dimensions of cattle welfare via their decisions on housing and management; the quality of daily care (e.g. cleaning, feeding, milking), their ability to recognize problems, be it disease, social stress or dysfunctional equipment; and their willingness and ability to solve them (e.g. Waiblinger 1996; English et al. 1992; Seabrook and Bartle 1992). All these aspects of stockpersonship are at least partly linked to the relationship with their animals, constituting the indirect effects of HAR on welfare (see Sect. 9.5.3).

### 9.5.2 Direct Effects of the Human-Animal Relationship and of Human-Animal Interactions on Cattle Welfare

The direct effects of HAI and HAR on welfare are based on the emotions involved and on the associated behavioural and physiological stress reactions or antistress effects. These may subsequently affect the productivity and health of the animals as well as their long-term affective states and behaviour (Fig. 9.1).

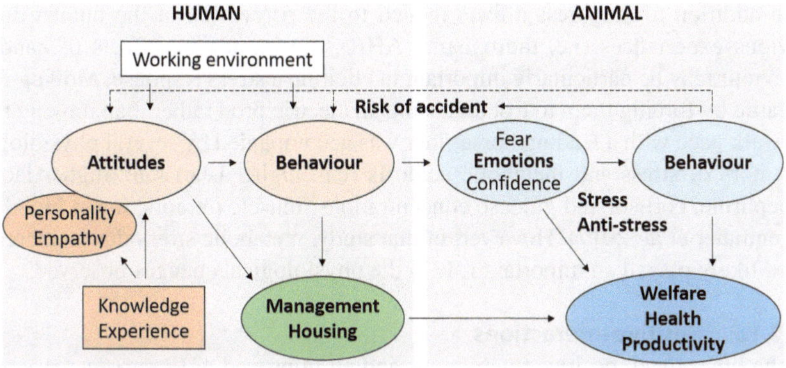


**Fig. 9.1** Simplified model of the human-cattle relationship illustrating its direct and indirect effects on welfare and the main influencing factors. The lightly coloured ovals represent the core components of the HAR, the darker elements other features that can influence the HAR or be influenced by it. The boxes represent key features that can be targeted by interventions for improving the HAR. The dashed line indicates the feedback of cattle behaviour on human attitudes and behaviour

#### 9.5.2.1 Physiological Responses

Previous interactions, and thus the AHR, not only influence animals' behavioural reactions towards humans but are also linked to physiological reactions during human-animal encounters in the short as well as in the long term.

##### 9.5.2.1.1 Negative Interactions and Fear

Cattle that experience fear in the presence of humans as a result of negative interactions, a lack of habituation to humans or a combination of both, can exhibit marked physiological stress responses. The mere presence of a person that has previously handled the animals negatively elicits acute stress responses in dairy heifers (increased plasma cortisol concentration, Breuer et al. 2003) and cows (increased heart rate, impaired milk ejection indicating inhibition of oxytocin, Rushen et al. 1999) and in beef cattle (increased heart rate, Grignard et al. 2001). Veal calves raised on farms with poor handling behaviour practised by the stockpeople did not only show behavioural signs of a higher level of fear of humans in the rearing period but had also higher heart rates during loading for transport to the slaughterhouse as compared to calves from farms with stockpeople providing positive interactions (Lensink et al. 2001b). Besides acute stress responses, negatively handled heifers also showed signs of chronic stress, i.e. higher basal cortisol concentrations as compared to positively handled heifers (Breuer et al. 2003). Fear of humans may not necessarily lead to physiological stress responses if the animal does not encounter humans or is able to keep a sufficient distance to humans (Munksgaard et al. 2001), thus coping with the stressor behaviourally (Moberg 2000). Extensive conditions with low levels of human contact will therefore not create the problem of chronic stress caused by HAI. However, non-habituated animals from extensive conditions will experience stress during necessary handling procedures (Ceballos et al. 2018a).

In addition to the stress effects related to the perception of the human due to previous experiences (i.e. the existing AHR), the immediate effects of handling behaviour may be particularly important in eliciting a stress response. Moving feedlot cattle by forcing them to trot and using an electric prod rather than moving them at a walk pace with a leading horse rider was accompanied by several physiological indicators of stress and metabolic acidosis (e.g. higher heart rate; higher lactate, epinephrine, cortisol and glucose concentrations; muscle tremors; lower blood pH; Hagenmaier et al. 2017). However, in that study, metabolic stress due to increased speed likely played an important role in the physiological changes observed.

#### 9.5.2.1.2 Positive Interactions

On the other hand, positive interactions and an improved AHR can decrease acute stress in cattle during HAI. Beef cattle receiving gentle tactile stimulation in the first weeks of life had an improved AHR at the age of 10 months when they were slaughtered, and experienced less stress during slaughter, reflected by less avoidance behaviour in the stunning box, a lower concentration of cortisol in exsanguination blood and increased tenderness of the meat (Probst et al. 2012). Gentle interactions with heifers in the first hour after their first calving improved their AHR and decreased the mean cortisol concentration in milk in the first 5 months of lactation (Hemsworth et al. 1987, 1989). Dairy cows that had experienced stroking for a period of 3 weeks had lower heart rate increases and showed fewer behavioural signs of stress during a rectal palpation compared with control cows that either had not been handled in addition to routine management practices (Waiblinger et al. 2004) or experienced close human presence (Schmied et al. 2010). Positive tactile interactions can further reduce the perceived aversiveness of (necessary) aversive events, such as veterinary procedures, restraint or isolation. In the study by Waiblinger et al. (2004), cows had a lower increase in heart rate and showed less restlessness when they were provided with gentle contact by the handler or a familiar stockperson during rectal palpation than provided with gentle contact by an unknown person or without any person present. Similarly, cows had a lower heart rate during isolation from other cows when brushed by a 'positive handler' when compared to being isolation but without brushing (Rushen et al. 2001).

An improved AHR and thus lower levels of fear and stress during close contact with humans may account for the effects of previous handling described above. However, immediate and longer-term physiological antistress effects (e.g., lowered blood pressure and cortisol concentrations) may result from tactile stimulation triggering oxytocin release (Grahn et al. 2021; Uvnäs-Moberg 1997) given a certain quality of AHR (Lürzel et al. 2020; Schmied et al. 2008b), although results are inconclusive in cattle. Stroking cows with a good AHR (especially stroking that resembles intraspecific social grooming with respect to preferred body regions) elicits behavioural signs of relaxation and enjoyment, which are associated with a reduction in heart rate (Schmied et al. 2008b) and with an increase of the concentration of salivary oxytocin (Lürzel et al. 2020). However, an increase in salivary oxytocin after stroking has not been found in two studies in heifers with a good AHR (Lürzel et al. 2020; Lürzel, unpublished data), which might be due to not all heifers enjoying the interactions in every trial. Also, only peripheral oxytocin concentrations were measured in these studies, which does not necessarily allow conclusions to be drawn about central oxytocin release (Rault 2016). Regarding longer-term effects, calves that were stroked during the first 2 weeks of life had lower basal cortisol concentrations than controls that did not receive such tactile stimulation, although calves of both treatments had a similarly good relationship with humans (Lürzel et al. 2015b). This finding suggests specific beneficial physiological effects of stroking and may be due to the release of oxytocin in the first days of life, which was shown to have long-lasting organisational effects on physiology in other species (Holst et al. 2002).

#### 9.5.2.2 Productivity

Early studies in the field of human-farm animal relationship focused on effects on productivity. Both experimental and on-farm studies show the positive association between the quality of HAI, HAR and animal productivity. Seabrook (1972, 1984, 1986) pioneered studies on the human-cattle relationship. He compared 20 dairy farms run by single stockmen but all belonging to a larger property (and thus having otherwise comparable housing and management conditions) and found that the milk yield differed up to 12% between farms depending on the personality of the stockman. Differences in productivity corresponded to differences in HAR: on high-yielding farms, stockmen talked more to the cows and touched them more often, while cows showed lower avoidance and more approach towards them and entered the milking parlour more quickly compared to cows on low-yielding farms (Seabrook 1984, 1986). The stockmen not only talked more to their cows but also in a different way. Seabrook (1984) differentiated between talking *to* the cows, which implies that the stockperson was merely issuing 'commands' or 'comments', and talking *with* the cows, where the stockperson was almost 'expecting a response' (e.g. 'How are you old girl this afternoon?'). Stockpeople on lower-yielding farms not only talked much less *to* the cows, but never talked *with* the cows, while stockpeople of the higher-yielding herds talked *with* the cows much more than *to* the cows. These differences in the form of communication with the cows might reflect different attitudes of the stockpeople to their cows and the working situation. Similarly, milk yield was higher on farms where cows had names rather than numbers (Bertenshaw and Rowlinson 2009), probably reflecting different attitudes and thus differences in HAR.

Later on-farm studies have confirmed the association of HAR and milk yield but also found relationships with reproduction in cows and productivity in veal calves. Negative relationships were found between fear of humans or milkers' negative behaviour and milk yield (Breuer et al. 2000; Hemsworth et al. 2000; Waiblinger et al. 2002). The conception rate after artificial insemination was negatively correlated with the use of negative interactions and positively with the use of positive interactions (correlation coefficients 0.31–0.37) in both an Austrian (indoor housing) and an Australian study (pasture-based system), consequently with varying degrees of human-animal contact (Hemsworth et al. 2000; Waiblinger et al. 2006b). In veal calf production, weight gain and feed conversion rate as well as overall productivity (a composite measure of weight gain, feed conversion efficiency and mortality) were higher on farms with more positive contacts and stockpeople using slow and careful movements (Lensink et al. 2000b, 2001a).

Acute and chronic stress responses can both contribute to reduced (re-)production where there are negative HAI and a poor HAR. Stress can reduce fertility by inducing different reproductive disorders depending on the period of the cycle when stress occurs and on its intensity (Dobson and Smith 2000). Release of catecholamins or glucocorticoids due to the activation of the sympatho-adrenal axis or hypothalamic-pituitary-adrenal axis, respectively, can inhibit both milk let-down and uterus motility after insemination by antagonising oxytocin or reducing oxytocin release (Dobson and Smith 2000; Unshelm 1988; van Reenen et al. 2002; Bruckmaier and Wellnitz 2008). Chronic stress induces catabolic metabolism, reducing milk yield and growth.

Experimental studies support the on-farm associations of HAR and productivity, where improved management and housing in the case of more positive attitudes may sometimes confound results (see Sect. 9.5.3). Brushing and talking to heifers in the weeks before calving not only decreased avoidance behaviour and increased approach towards a human handler (Bertenshaw and Rowlinson 2008) but also led to 19% faster milk let-down than in control heifers during the first 4 weeks after calving (Bertenshaw et al. 2008). Positively handled heifers kicked less in human presence and kicks in human presence accounted for 7.1% of the variation in milk yield. Dairy calves that had experienced stroking during the first 2 weeks of life had an up to 7% higher average daily gain from birth until about 12 weeks of age (Lürzel et al. 2015b). Meat quality was better in veal calves and beef suckler cattle after gentle handling and improvement of the HAR (Lensink et al. 2000b; Probst et al. 2012), although the growth rate was unaffected in the veal calves and not investigated in the beef cattle study. In the veal calf study, control animals had a poorer AHR but might have been able to maintain a distance from the human, thus coping behaviourally and not developing chronic physiological stress. Maintaining a consistent set of people caring for and handling the animals seems important as well, likely due to a reduced predictability of the mutual behaviour if human and animals do not know each other well. Frequent changes of milkers as well as a higher number of milkers negatively affected milk yield (up to 14%) in a study by Schlichting (1974); increased duration of care by the same stockperson was accompanied by an increase in milk yield. Effects were stronger in tie-stall systems where contact between stockperson and cow is closer. Also Seabrook and Bartle (1992) found a reduction of milk yield after every change of the milker, even when the milk yield eventually reached a higher level than before the change. Similarly, cows have a lower milk yield and a higher heart rate increase when milked by relief milkers compared to their regular milker (Knierim and Waran 1993).

#### 9.5.2.3 Immune Function and Health

Relatively few studies have dealt with the association between the HAR and cattle health. Improved immune responses in positively handled animals have been shown in other farm animal species (lambs: Caroprese et al. 2006, laying hens: Barnett et al. 1994, chickens: Gross and Siegel 1979, Gross and Siegel 1980, Gross and Siegel 1982). The underlying mechanism is the physiological stress or antistress response that affects immune responses (Moberg 2000). In addition, acute behavioural stress responses increase the risk of injury: fear of humans and/or negative handling will increase the number of escape attempts or quick withdrawal reactions, leading to hasty foot placement and consequently slipping, stumbling or falling. Accordingly, high fear of humans was associated with more traumatic incidents during unloading in veal calves, and with more animal injuries and deaths (Fordyce et al. 1985; Lensink et al. 2001a). Impatient handling during moving is associated with higher prevalence of lameness in dairy cows (reviewed by Hemsworth et al. 1995). For instance, in a New Zealand case-controlled study, the patience of the stockperson was the most important predictor of lameness, accounting for more than 20% of the variance (Chesterton et al. 1989). Similarly, in indoor cubicle housing, farmer attitudes and behaviour were predictors for lameness prevalence, although housing design was most important (Rouha-Mülleder et al. 2009). More recent studies confirm the high importance of handling behaviour on claw disorders or lameness (Moreira et al. 2019). In addition, forcing cows to move more quickly may hinder directed foot placement. Besides increasing the risk for traumatic claw disorders, a poorer HAR can contribute to the risk for infectious claw disorders and impaired healing via negative effects on the immune system.

Udder health shows associations with the HAR as well. In a study on 80 dairy farms with cubicle loose housing in Austria (Waiblinger et al. 2006b), the milk somatic cell count (SCC) of the cows was explained largely by milker attitude and behaviour. A general positive attitude of the milkers towards the cows was related to lower SCC and a general negative attitude to higher SCC. Later studies on Swiss, German and Danish dairy farms confirmed the association of improved udder health on farms with a better HAR. The results suggest that both a decreased susceptibility to infections (based on the connection between physiological responses and HAI) and a lower risk of pathogen distribution (due to calmer cows kicking off milking cups less frequently) may contribute to better udder health on farms with a better HAR (Ivemeyer et al. 2011, 2018). In these studies, a better HAR was characterised by milkers using more positive interactions and having more positive attitudes (e.g. enjoying contact with cows more and reporting a higher intention to use patient behaviour during handling of cows), and by cows having lower avoidance distances and behaving more calmly. No routine change of milkers, more contact time during routine working and additional barn and herd monitoring were associated with better HAR and also with better udder health (Ivemeyer et al. 2018). It is worth noting that these findings cannot be explained by differences in milking hygiene or other management factors, as these were included in the statistical models.

Effects on health have also been shown in studies in calves. Positively handled veal calves developed fewer abomasal ulcers (Lensink et al. 2000c), a pathology associated with stress. Disease incidence and mortality was lower on farms where stockpeople used a higher number of positive interactions, slower and more careful movements and fewer negative contacts (Lensink et al. 2000b, 2001a).

#### 9.5.2.4 Behaviour

The level of fear and stress, or conversely of calmness, can also affect behaviours apart from those directed towards humans and even influence the development of personality traits. In heifers, prolonged and regular positive handling early in life reduced general fearfulness (Boissy and Bouissou 1988). Negative and unpredictable human behaviour might enhance aggression, while positive interactions might reduce it (Waiblinger et al. 2001). In a small study in 1 dairy herd of about 80 cows, agonistic interactions increased after 'stress milkings' where 10 cows were milked in a rough manner, including shouting, light hitting and using cold water for cleaning the udders (Menke 1986). Similarly, in a more recent experiment including two dairy herds, levels of agonistic interactions between cows were increased after negative milkings (the milker displayed negative behaviours) compared to after positive milkings (Reiter 2017). Such effects can be caused by frustration, pain, received aggression and stress, which all increase aggression (Scott 1958; Fox 1968; Neumann and Steinbeck 1971; Reinhardt 1980), and the release of oxytocin after gentle physical contact, which is generally associated with reduced aggression and higher levels of socio-positive behaviours (Uvnäs-Moberg 2004).

### 9.5.3 Indirect Effects of the Human-Animal Relationship on Cattle Welfare

Farmers' and stockpersons' attitudes towards their animals and handling them are also connected with their behaviour when it comes to decisions on management and housing. The farmers who had the stronger intentions to use patient, calm behaviours during milking, who rated regular positive interactions as more important, and who enjoyed working with, and having contact to the cows more strongly, were those who considered the needs of cows with regard to housing design and management to a greater extent (Waiblinger et al. 2006c). The fundamental perception of the animals as individuals with needs that have to be respected may be the basis for such positive cognitive and affective attitudes as well as for the readiness to adapt the environment to the animals (Waiblinger et al. 2006c). Consequently, all attitudes related to animals are consistent with each other (Hemsworth and Coleman 2011). In veal calf production, more positive attitudes towards the calves and the work also seem to be linked not only to more positive behaviour but also to better surveillance and care for the animals (Lensink et al. 2001a).

However, farmer decision-making and stockpersonship are influenced by the HAR not only via underlying attitudes. A better farmer-cow relationship, characterised by a higher intensity and quality of contact and the use of more positive interactions during milking, was associated strongly with better management and, in turn, improved welfare (less agonistic behaviour, fewer injuries) in a study on 35 farms with horned dairy cow herds in loose housing (Waiblinger 1996; Waiblinger et al. 2001). In addition to the probable common basis of an attitude that acknowledges the importance of respecting cows' needs, this association is likely based on a better understanding of the individual cow and the herd through close contact with the animals, which in turn facilitates problem recognition and solving, as suggested already by Seabrook (1984). These results, together with former work on management styles (Ploeg van der 1994), indicate a broader influence of the farmers' and stockpersons' relationship with their animals on several aspects of stockpersonship, such as attention to detail, readiness to solve problems and decision-making. Figure 9.1 depicts the direct and indirect effects of the HAR and some external influences on cattle welfare. A good HAR is thus an important component of good stockpersonship, interacting with further aspects such as technical knowledge, experience and skills in safeguarding animal welfare.

## 9.6 Effects of the Human-Animal Relationship on the Risk of Accidents and on Human Welfare

In cattle with a better AHR, the ease of handling is higher, and risk of accidents for humans and also animals is lower, as they show less defensive behaviour such as kicking, less aggression and less flight behaviour (Bertenshaw et al. 2008; Boivin et al. 1992b; Hemsworth et al. 1989; Waiblinger et al. 2004). Fear of humans or of the handling situation and poor handling behaviour, on the other hand, can induce strong flight reactions, which may lead to the animal falling or slipping, or knocking over or even running over the human handler. Similarly, fear can result in defensive behaviour targeted directly at the handler (e.g. attacks or kicking; Ceballos et al. 2018c, Lindahl et al. 2016). All of these behaviours clearly increase the risk of injuries to the human handler as well as to the animals (Lindahl et al. 2016).

Breeding bulls with a history of attacking humans differed in their resistance to handling and in their attack frequency depending on the caretaker. Resistance and attacks were highest when they were handled by 'excitable' caretakers, while such behaviours were rarely seen when they were handled by caretakers assessed as 'calm, well-balanced' (Renger 1975). Easier handling (leading, moving or loading) in animals with a better AHR and when the handler used more positive HAI was confirmed in further studies with breeding bulls (Wenz and Laube 1989), beef suckler heifers (Boivin et al. 1992a, b) and dairy heifers (Boissy and Bouissou 1988). Easier handling, reflecting more relaxed animals, is clearly associated with a lower risk of accidents (Lindahl et al. 2016) but can also save time. For example, weighing of beef calves took less time when better handling practices were applied (Destrez et al. 2018a).

Besides the physical welfare of humans, mental wellbeing is improved by a better HAR. Farmers reported higher job satisfaction after completing a handling course that changed their attitudes and behaviours and, subsequently, dairy cows' behaviour and milk yield (Hemsworth et al. 2002; Hemsworth and Coleman 2011). On dairy farms with a more positive HAR, farmers consistently enjoyed contact with their animals more (Ebinghaus et al. 2018; Waiblinger et al. 2002), thus experiencing positive emotions during their work.

## 9.7 Cattle Handling Recommendations

### 9.7.1 Establishment and Maintenance of a Good Human-Animal Relationship

Having covered the theoretical background to the establishment and maintenance of a good HAR in previous sections, we will now discuss how this works in practice on farms. A good AHR can be established by using mainly positive interactions, i.e. interactions the animals perceive as pleasant, and by avoiding negative interactions, which are perceived as aversive, as far as possible. Therefore, it is of utmost importance to adhere to good handling practices, as outlined below. Some negative interactions cannot be avoided completely, as they are part of necessary management procedures (e.g. veterinary treatment, artificial insemination). Nevertheless, even the negative emotions caused by these procedures can be mitigated by a good AHR if positive interactions also occur during the procedure (Waiblinger et al. 2004) or by training the animals to the procedure using positive reinforcement, which might also positively influence the AHR (Lomb et al. 2021).

The first opportunity to establish a good AHR is when feeding colostrum and, later, milk to calves. The calves then have the opportunity to associate the handler with the positive stimulus of feeding, allowing for classical conditioning (Toates 2001; de Passillé et al. 1996) and thus improvement of the relationship with humans (Jago et al. 1999). For this effect to occur, the animals have to perceive the person feeding them, and there must not be any simultaneous negative experiences. For best results, feeding should be accompanied by additional positive tactile interactions, as described in a study on veal farms (Lensink et al. 2001b). In systems where calves suckle their dams or where feeding is automated, some form of positive HAI will be beneficial to compensate for the lost possibility of interacting during feeding. Provision of colostrum in a bottle helps to ensure calf health and contact during suckling allows monitoring of calves as well as improving the AHR. Contact in association with milk intake in the first week of life may be especially effective (Nowak and Boivin 2015; Waiblinger et al. 2020). It should, however, be kept in mind that feeding alone is not sufficient to establish and maintain a good AHR if handling at other times does not adhere to the principles outlined in this section.

As discussed above, there are reports of long-term effects of positive contact with humans (Waiblinger et al. 2020; Probst et al. 2012), but other studies have shown that positive interactions at a young age do not necessarily have lasting effects (Boissy and Bouissou 1988; Lürzel et al. 2015a). Thus, while a positive or at least neutral AHR in calves is a good basis, it is necessary to maintain regular positive contact with cattle over their lifetime. Such contact can be provided during routine work with or close to the animals (e.g. milking, herd check-ups in the barn or on pasture, inspection of equipment, daily cleaning and maintenance) and thus does not necessarily require additional time—although in case of a good HAR, farmers often enjoy stroking or brushing the animals and thus invest some additional time. In dairy cows, a good opportunity for positive contact is during milking. In addition to good milking practices, such as addressing a cow verbally while approaching and touching her on the leg before cleaning the udder and attaching the cluster, which serve to increase predictability, the milker can talk to the cows and stroke them. During work in the animal area of the barn, the person should behave calmly, and the practice of talking soothingly to the animals and touching or stroking the ones approaching or that are standing close by, can easily be integrated when moving through the herd. It is not necessary to touch every animal during every checkup, but to act in a calm, positive manner in general. If every animal experiences positive interactions regularly from time to time, that should be sufficient for the maintenance of a good AHR (Boissy and Bouissou 1988). It is important to maintain this type of contact also with youngstock or other cattle that do not require daily handling, such as beef cattle on pasture, because a longer period of reduced contact might cause a decline in the AHR due to its dynamic nature.

It is important to avoid eliciting fear during the interactions with cattle, which is most easily achieved in unrestrained animals by allowing them sufficient space to maintain their preferred distance from the person. In addition, this approach facilitates a high level of animal agency. Cattle are often very curious and might approach a person to sniff or lick their clothes, which offers the stockperson an opportunity to attract them. On the other hand, cattle will usually tolerate the approach of a person if the person approaches only when their presence has not elicited any avoidance for some time. However, the AHR can also be improved by interacting with the animals during restraint, if interactions are not feasible when the animals are free to move.

One important point is to clearly distinguish between fear and respect when speaking about the AHR and cattle handling. The HAR needs to be characterized by mutual respect. While animals clearly should have no fear of humans, they have to respect them and follow their commands. This is in line with the principles of cattle social behaviour where affiliative and dominance relationships regulate mutual interactions and the threat of a dominant animal is followed by the withdrawal or submissive gesture of the subordinate (Bouissou et al. 2001). Likewise, humans need to send clear signals (e.g. with their body posture and by retaining the power of decision about whether an animal is allowed to establish physical contact with them) to the animals as well as providing positive interactions (see, for example, the Fulani herdsmen; Lott and Hart 1979).

### 9.7.2 Stress-Free Handling of Cattle

Three main aspects facilitate stress-free handling of cattle during moving, catching, examining or other handling procedures. These are a good AHR, good handling skills avoiding negative and using positive interactions and the use of suitable facilities and procedures. The suitability of facilities, such as the characteristics of a well-designed trail or ramps for transport vehicles, are reviewed in detail elsewhere (e.g. Grandin 2021, Grandin 2014b; see also Chap. 8). Here, we focus on the second aspect—how human handlers should behave and why.

To ensure animal and human safety and to avoid stress, some general principles should be followed. It is extremely important to consider the animals' sensory capacities (Grandin 2014a). Cattle have a wide field of vision of 300°; however, they cannot see an area of about 60° behind them and thus may startle if a person approaches them too closely from behind. Also, their field of binocular vision is relatively small, which means that if a cow needs to assess the height of a step, for example, she will lower her head and hesitate a moment before taking the step. The same behaviour is shown if there are shadow contrasts on the floor (Willson et al. 2021). Furthermore, cattle are very sensitive to movement and visual acuity is relatively low; thus quick movements can easily frighten them. Alleyways should be well-lit, and cattle should not have to move from light to darkness as their eyes only slowly adapt to different lighting conditions. Moving them towards very bright light should also be avoided. The range of hearing in cattle is much wider than that of humans. They can hear higher frequency sounds than humans and are very sensitive to high frequencies (often included in metallic noises). Loud noise is aversive to them, including shouting (Pajor et al. 2000, 2003; Arnold et al. 2008). The sense of smell has not been thoroughly investigated (Rørvang et al. 2017), but it has been shown that cattle react with increased signs of fear to pheromones of stressed conspecifics, as it took them longer to pass an alley if it contained urine of a stressed conspecific (Boissy et al. 1998). This makes handling of the following animals more difficult if a previous one was stressed.

Another crucial aspect is that the handler needs to know about the natural behaviour of cattle (Grandin 2014a) to be able to both foresee and interpret an animal's behavioural responses and adapt their own behaviour. Cattle are social animals, which means that they have the capabilities of dealing with social encounters and follow their species-specific social rules. If humans behave in a way that follows these rules, cattle handling will become easier and safer. Cattle are gregarious animals and isolated animals are easily stressed (e.g. Boissy and LeNeindre 1997), so handling will be facilitated if several animals are handled at once. If an animal has to be separated from the herd, it should still have visual contact with its herd mates for as long as possible. It is also helpful to handle nervous, inexperienced animals together with a calm, trained animal, allowing for social facilitation and a reduction of fearfulness in the nervous animals. Cattle can perceive and interpret social signals emitted by their conspecifics, and some of these signals can be mimicked by a human handler (e.g. direction of walking, body posture, tactile interactions). In addition, it is important for handlers to be able to interpret these social signals as well as the behavioural expression of emotional state to allow them to anticipate a potential attack or panic response and distinguish it from a fearful, submissive or friendly and relaxed animal.

Handlers should also be aware of the response of cattle to frightening stimuli. Sudden movements or noises can cause panic and flight, and while an animal may resort to aggressive behaviour if flight is not possible, it may also inadvertently injure a handler who is standing in its flight path by knocking them over.

Another aspect concerns the flight distance or avoidance distance, which is related to the individual distance. The individual distance describes the minimum distance between two animals that does not elicit agonistic behaviour, i.e. aggression by the dominant or avoidance by the subordinate animal (Hediger 1940, cited in Aschwanden et al. 2008), whereas flight or avoidance distance describes the distance at which an animal retreats from an approaching person (Waiblinger and Menke 1999; Hemsworth et al. 2002). We prefer the term 'avoidance distance' as we consider it semantically more appropriate because this term includes reactions of animals that are not fear responses towards humans, which would be expressed as flight behaviour, but includes the behaviour of animals that simply prefer to keep their distance. The flight/avoidance zone is the perimeter around an animal defined by its avoidance distance (Grandin 2014a). If used appropriately during moving animals, it increases the efficiency of the procedure and reduces stress during moving. The balance point at the shoulder should be used to determine the direction in which an animal will move away (Fig. 9.2). If a handler enters the avoidance zone of the animal behind the balance point at the shoulder, the animal will go forward, while the animal will either turn or, if this is not possible, walk backwards, if the handler enters the zone ahead of the shoulder. The handler should create the lowest amount of pressure necessary to elicit a reaction in the animal, for example, by approaching it slowly until the avoidance zone is entered. When it shows the desired reaction, the animal should be rewarded immediately by removing the pressure, e.g. by letting the animal increase the distance to the handler. To keep the animal moving, the handler should move at the edge of the avoidance zone—moving too quickly and thus entering the avoidance zone too far may cause the animal to run away, with the handler losing control. If the reward of decreasing pressure is timed precisely and applied consistently, the animals learn to understand the signals given by the handler and to show the appropriate reaction, making handling easier in the long run. Flighty, nervous animals become calmer because they learn to not overreact to every movement of the handler, as the handler's behaviour becomes predictable for

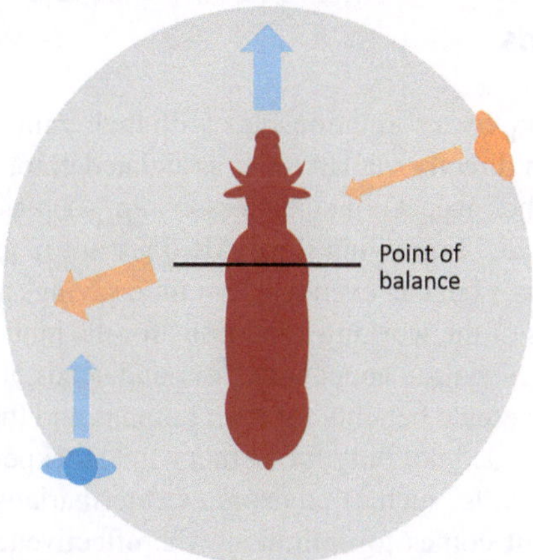


**Fig. 9.2** Diagram showing the avoidance zone (grey), the point of balance (black) and the effect of handler position on animal movement direction, comparing animal movement when the handler approaches from behind (blue handler position, animal moving to the front) or from the front and side (orange handler position for turning the animal to the side). The thin arrows represent handler movement, the thick arrows animal movement (Diagram by S Lürzel)

them. On the other hand, animals with a very low or nonexistent avoidance distance may be difficult to move. Here, clear, consistent signalling using intraspecific rules is important to induce movement. The animals learn to move if requested to do so by increasing the pressure consistently in a fixed sequence of actions until they show a reaction; the increase in pressure will become predictable to these animals, and they will likely move at a lower pressure during future interactions.

This way of influencing an animal's behaviour (often termed 'low-stress stockmanship') allows for calm handling of cattle especially during moving, making use of their natural behaviour. While moving cattle, it is crucial to avoid putting pressure on animals that are already moving or are not able to move, e.g. by moving too far into their avoidance zone or even shouting or slapping them, because then they cannot learn which behaviour the handler wants to elicit. One example is the practice of putting intense pressure (e.g. by shouting or hitting) on individuals at the back of the herd where they cannot avoid the pressure by moving forward because other, often higher-ranking animals are in their way. In these situations, the animals cannot learn how to react appropriately, have very little agency and experience a high level of stress.

Patience is crucial, particularly in the context of moving animals. Cattle should never be rushed, as the risk of damage to their claws will increase and animals can easily become stressed if they have to move faster than they would normally do. If the environment is unfamiliar, even more patience is needed, as the animals probably experience some level of fear and they need time for exploration. Some features of the environment prompt cattle to investigate them, such as differences in floor height or sharp contrasts. Optimally, the environment should be constructed in a way that eliminates these features (Willson et al. 2021; Grandin 2014a).

## 9.8 Conclusions

The farmers' and stockpersons' relationships with their animals have a huge impact on cattle welfare, both directly via HAI and associated emotions and indirectly via the connection to further aspects of stockpersonship, such as decision-making and problem-solving abilities. Improving the HAR does not require additional time to be spent with the animals but can even result in time savings and improve the physical and mental aspects of the working conditions for the humans involved. Training in cattle handling and education about behaviour and needs of cattle are thus of high importance and would create benefits for both humans and their cattle. This type of training should be provided not only for farmers and stockpersons but also to other groups working with cattle, such as advisors or veterinarians. While there are still open questions when it comes to optimizing the effectiveness of human contact, implementing existing knowledge would already make a large difference. The education of people working with cattle should include this important aspect of stockpersonship to increase the sustainability of cattle production.

## References


Ajzen I (1988) Attitudes, personality, and behavior. Open University Press, Milton Keynes

Ajzen I, Fishbein M (1980) Understanding attitudes and predicting social behaviour. Prentice-Hall, Inc., Upper Saddle River

Albright JL, Arave CW (1997) The behaviour of cattle. CAB International, Wallingford

Albright JL, Gordon WP, Black WC et al (1966) Behavioral responses of cows to auditory training. J Dairy Sci 49:104–106

Arnold NA, Ng KT, Jongman EC et al (2008) Avoidance of tape-recorded milking facility noise by dairy heifers in a Y maze choice task. Appl Anim Behav Sci 109:201–210

Aschwanden J, Gygax L, Wechsler B et al (2008) Social distances of goats at the feeding rack: influence of the quality of social bonds, rank differences, grouping age and presence of horns. Appl Anim Behav Sci 114:116–131

Barnett JL, Hemsworth PH, Hennessy DP et al (1994) The effects of modifying the amount of human contact on behavioural, physiological and production responses of laying hens. Appl Anim Behav Sci 41:87–100

Bassett L, Buchanan-Smith HM (2007) Effects of predictability on the welfare of captive animals. Appl Anim Behav Sci 102:223–245

Bergamasco L, Edwards-Callaway LN, Bello NM et al (2021) Unmitigated surgical castration in calves of different ages: cortisol concentrations, heart rate variability, and infrared thermography findings. Animals 11:2719

Bertenshaw C, Rowlinson P (2008) Exploring heifers' perception of 'positive' treatment through their motivation to pursue a retreated human. Anim Welf 17:313–319

Bertenshaw C, Rowlinson P (2009) Exploring stock managers' perceptions of the human-animal relationship on dairy farms and an association with milk production. Anthrozoös 22:59–69

Bertenshaw C, Rowlinson P, Edge H et al (2008) The effect of different degrees of 'positive' human-animal interaction during rearing on the welfare and subsequent production of commercial dairy heifers. Appl Anim Behav Sci 114:65–75

Boissy A, Bouissou M-F (1988) Effects of early handling on heifers' subsequent reactivity to humans and to unfamiliar situations. Appl Anim Behav Sci 20:259–273

Boissy A, LeNeindre P (1997) Behavioral, cardiac and cortisol responses to brief peer separation and reunion in cattle. Physiol Behav 61:693–699

Boissy A, Terlouw C, Le Neindre P (1998) Presence of cues from stressed conspecifics increases reactivity to aversive events in cattle: evidence for the existence of alarm substances in urine. Physiol Behav 63:489–495

Boissy A, Manteuffel G, Jensen MB et al (2007) Assessment of positive emotions in animals to improve their welfare. Physiol Behav 92:375–397

Boivin X, Le Neindre P, Chupin JM (1992a) Establishment of cattle-human relationships. Appl Anim Behav Sci 32:325–335

Boivin X, LeNeindre P, Chupin JM et al (1992b) Influence of breed and early management on ease of handling and open-field behaviour of cattle. Appl Anim Behav Sci 32:313–323

Boivin X, LeNeindre P, Garel JP et al (1994) Influence of breed and rearing management on cattle reactions during human handling. Appl Anim Behav Sci 39:115–122

Boivin X, Nowak R, Desprès G et al (1997) Discrimination between shepherds by lambs reared under artificial conditions. J Anim Sci 75:2892–2898

Boivin X, Garel JP, Durier C et al (1998) Is gentling by people rewarding for beef calves? Appl Anim Behav Sci 61:1–12

Boivin X, Lensink J, Tallet C et al (2003) Stockmanship and farm animal welfare. Anim Welf 12:479–492

Boivin X, Gilard F, Egal D (2009) The effect of early human contact and the separation method from the dam on responses of beef calves to humans. Appl Anim Behav Sci 120:132–139

Bollongino R, Burger J, Powell A et al (2012) Modern taurine cattle descended from small number of near-eastern founders. Mol Biol Evol 29:2101–2104

Bouissou M-F, Boissy A, Le Neindre P et al (2001) The social behaviour of cattle. In: Keelicheng LJ, Gonyou HW (eds) Social behaviour in farm animals. CAB International, Wallingford
Breuer K, Hemsworth PH, Barnett JL et al (2000) Behavioural response to humans and the productivity of commercial dairy cows. Appl Anim Behav Sci 66:273–288
Breuer K, Hemsworth PH, Coleman GJ (2003) The effect of positive or negative handling on the behavioural and physiological responses of nonlactating heifers. Appl Anim Behav Sci 84:3–22
Broom DM (2007) Quality of life means welfare: how is it related to other concepts and assessed? Anim Welf 16:45–53
Bruckmaier RM, Wellnitz O (2008) Induction of milk ejection and milk removal in different production systems. J Anim Sci 86:15–20
Calderon-Amor J, Beaver A, von Keyserlingk MAG et al (2020) Calf- and herd-level factors associated with dairy calf reactivity. J Dairy Sci 103:4606–4617
Caroprese M, Napolitano F, Albenzio M et al (2006) Influence of gentling on lamb immune response and human-lamb interactions. Appl Anim Behav Sci 99:118–131
Ceballos MC, Gois KCR, Sant'Anna AC et al (2018a) Frequent handling of grazing beef cattle maintained under the rotational stocking method improves temperament over time. Anim Prod Sci 58:307–313
Ceballos MC, Sant'Anna AC, Boivin X et al (2018b) Impact of good practices of handling training on beef cattle welfare and stockpeople attitudes and behaviors. Livest Sci 216:24–31
Ceballos MC, Sant'Anna AC, Gois KCR et al (2018c) Investigating the relationship between human-animal interactions, reactivity, stress response and reproductive performance in Nellore heifers. Livest Sci 217:65–75
Chesterton RN, Pfeiffer DU, Morris RS et al (1989) Environmental and behavioural factors affecting the prevalence of foot lameness in New Zealand dairy herds-a case control study. N Z Vet J 37:135–142
Clutton-Brock J (1999) A natural history of domesticated animals. Cambridge University Press, Cambridge
de Boyer des Roches A, Veissier I, Boivin X, Gilot-Fromont E, Mounier L (2016) A prospective exploration of farm, farmer, and animal characteristics in human-animal relationships: an epidemiological survey. J Dairy Sci 99(7):5573–5585
de Passillé AM, Rushen J, Ladewig J et al (1996) Dairy calves' discrimination of people based on previous handling. J Anim Sci 74:969–974
Destrez A, Haslin E, Boivin X (2018a) What stockperson behavior during weighing reveals about the relationship between humans and suckling beef cattle: a preliminary study. Appl Anim Behav Sci 209:8–13
Destrez A, Haslin E, Elluin G et al (2018b) Evaluation of beef herd responses to unfamiliar humans and potential influencing factors: an exploratory survey on French farms. Livest Sci 212:7–13
Dickson DP, Barr GR, Johnson LP et al (1970) Social dominance and temperament of Holstein cows. J Dairy Sci 53:904–907
Dietz L, Arnold AMK, Goerlich-Jansson VC et al (2018) The importance of early life experiences for the development of behavioural disorders in domestic dogs. Behaviour 155:83–114
Dobson H, Smith RF (2000) What is stress, and how does it affect reproduction? Anim Reprod Sci 60-61:743–752
Ebinghaus A, Ivemeyer S, Knierim U (2018) Human and farm influences on dairy cows' responsiveness towards humans - a cross-sectional study. PLoS One 13:e0209817
English P, Burgess G, Segundo R et al (1992) Stockmanship: improving the care of the pig and other livestock. Farming Press, Ipswich
Ertl P, Klocker H, Hörtenhuber S et al (2015) The net contribution of dairy production to human food supply: the case of Austrian dairy farms. Agric Syst 137:119–125
Estep DQ, Hetts S (1992) Interactions, relationships, and bonds: the conceptual basis for scientist-animal relations. In: Davis H, Balfour AD (eds) The inevitable bond - examining scientist-animal interactions. Cambridge University Press, Cambridge
Fordyce G, Goddard M, Tyler R et al (1985) Temperament and bruising of Bos indicus cross cattle. Aust J Exp Agric 25:283–285

Fox M (1968) Abnormal behaviour in animals. W.B. Saunders and Co.

Franks B, Higgins ET (2012) Chapter six - effectiveness in humans and other animals: a common basis for well-being and welfare. In: Olson JM, Zanna MP (eds) Advances in experimental social psychology. Academic Press

Grahn P, Ottosson J, Uvnäs-Moberg K (2021) The oxytocinergic system as a mediator of anti-stress and instorative effects induced by nature: the calm and connection theory. Front Psychol 12:617814

Grandin T (2014a) Behavioural principles of handling cattle and other grazing animals under extensive conditions. In: Grandin T (ed) Livestock handling and transport, 4th edn. CAB International, Wallingford

Grandin T (2014b) Livestock handling and transport. CAB International, Wallingford

Grandin T (2021) The visual, auditory, and physical environment of livestock handling facilities and its effect on ease of movement of cattle, pigs, and sheep. Front Anim Sci 2:744207

Greiveldinger L, Veissier I, Boissy A (2007) Emotional experience in sheep: predictability of a sudden event lowers subsequent emotional responses. Physiol Behav 92:675–683

Grignard L, Boivin X, Boissy A et al (2001) Do beef cattle react consistently to different handling situations? Appl Anim Behav Sci 71:263–276

Gross WB, Siegel PB (1979) Adaption of chickens to their handlers and experimental results. Avian Dis 23:708–714

Gross WB, Siegel PB (1980) Effects of early environmental stresses in chicken body weight, antibody response to RBC antigens, feed efficiency and response to fasting. Avian Dis 24:549–579

Gross WB, Siegel PB (1982) Socialization as a factor in resistance to infection, feed efficiency, and response to antigen in chickens. Am J Vet Res 43:2010–2012

Hagenmaier JA, Reinhardt CD, Bartle SJ et al (2017) Effect of handling intensity at the time of transport for slaughter on physiological response and carcass characteristics in beef cattle fed ractopamine hydrochloride. J Anim Sci 95:1963–1976

Haskell MJ, Bell DJ, Gibbons JM (2012) Is the response to humans consistent over productive life in dairy cows? Anim Welf 21:319–324

Hediger H (1940) Biologische Gesetzmäßigkeiten im Verhalten von Wirbeltieren. Mitteilungen der Naturforschenden Gesellschaft in Bern 1941:39–55

Hemsworth PH, Coleman GJ (2011) Human-livestock interactions: the stockperson and the productivity of intensively farmed animals. CAB International, Wallingford

Hemsworth PH, Hansen C, Barnett JL (1987) The effects of human presence at the time of calving of primiparous cows on their subsequent behavioural response to milking. Appl Anim Behav Sci 18:247–255

Hemsworth PH, Barnett JL, Tilbrook AJ et al (1989) The effects of handling by humans at calving and during milking on the behaviour and milk cortisol concentrations of primiparous dairy cows. Appl Anim Behav Sci 22:313–326

Hemsworth PH, Barnett JL, Breuer K, Coleman GC, Matthews LR (1995) An investigation of the relationships between handling and human contact and the milking behaviour, productivity and welfare of commercial dairy cows. Research Report on Dairy Research and Development Council Project, Attwood

Hemsworth PH, Coleman G, Barnett JL et al (2000) Relationships between human-animal interactions and productivity of commercial dairy cows. J Anim Sci 78:2821–2831

Hemsworth PH, Coleman G, Barnett JL et al (2002) The effects of cognitive behavioral intervention on the attitude and behavior of stockpersons and the behavior and productivity of commercial dairy cows. J Anim Sci 80:68–78

Henry S, Hemery D, Richard MA et al (2005) Human-mare relationships and behaviour of foals toward humans. Appl Anim Behav Sci 93:341–362

Holst S, Uvnäs-Moberg K, Petersson M (2002) Postnatal oxytocin treatment and postnatal stroking of rats reduce blood pressure in adulthood. Auton Neurosci 99:85–90

Ivemeyer S, Knierim U, Waiblinger S (2011) Effect of human-animal relationship and management on udder health in Swiss dairy herds. J Dairy Sci 94:5890–5902

Ivemeyer S, Simantke C, Ebinghaus A et al (2018) Herd-level associations between human-animal relationship, management, fecal cortisol metabolites, and udder health of organic dairy cows. J Dairy Sci 101:7361–7374

Jago JG, Krohn CC, Matthews LR (1999) The influence of feeding and handling on the development of the human-animal interactions in young cattle. Appl Anim Behav Sci 62:137–151

Knaus W (2016) Perspectives on pasture versus indoor feeding of dairy cows. J Sci Food Agric 96:9–17

Knierim U, Waran NK (1993) The influence of the human-animal interaction in the milking parlour on the behaviour, heart-rate and milk yield of dairy cows. In: Nichelmann M, Wierenga HK, Braun S (eds) Humboldt Universität, Institut für Verhaltensbiologie und Zoologie, pp 169–173

Koolhaas J, Bartolomucci A, Buwalda B et al (2011) Stress revisited: a critical evaluation of the stress concept. Neurosci Biobehav Rev 35:1291–1301

Kratli S (2008) Time to outbreed animal science? A cattle-breeding system exploiting structural unpredictability: the WoDaaBe herders in Niger. STEPS working paper 7. STEPS Centre, Brighton

Krohn CC, Jago JG, Boivin X (2001) The effect of early handling on the socialisation of young calves to humans. Appl Anim Behav Sci 74:121–133

Krohn CC, Boivin X, Jago JG (2003) The presence of the dam during handling prevents the socialization of young calves to humans. Appl Anim Behav Sci 80:263–275

Laister S, Stockinger B, Regner AM et al (2011) Social licking in dairy cattle - effects on heart rate in performers and receivers. Appl Anim Behav Sci 130:81–90

Lange A, Franzmayr S, Wisenöcker V et al (2020a) Effects of different stroking styles on behaviour and cardiac parameters in heifers. Animals 10:426

Lange A, Waiblinger S, Heinke A et al (2020b) Gentle interactions with restrained and free-moving cows: effects on the improvement of the animal-human relationship. PLoS One 15:e0242873

Lange A, Waiblinger S, van Hasselt R et al (2021) Effects of restraint on heifers during gentle human-animal interactions. Appl Anim Behav Sci 243:105445

Lanier JL, Grandin T, Green RD et al (2000) The relationship between reaction to sudden, intermittent movements and sounds and temperament. J Anim Sci 78:1467–1474

Le Neindre P, Trillat G, Chupin JM et al (1993) Genetic and epigenetic variation factors in the relationships between humans and animals. In: Nichelmann M, Wierenga HK, Braun S (eds) International congress on applied ethology, 1993 Berlin. Humboldt University, Institut für Verhaltensbiologie und Zoologie, Berlin, pp 161–168

Lensink BJ, Boivin X, Pradel P et al (2000a) Reducing veal calves' reactivity to people by providing additional human contact. J Anim Sci 78:1213–1218

Lensink J, Boissy A, Veissier I (2000b) The relationship between farmers' attitude and behaviour towards calves, and productivity of veal units. Ann Zootech 49:313–327

Lensink J, Fernandez X, Boivin X et al (2000c) The impact of gentle contacts on ease of handling, welfare, and growth of calves and on quality of veal meat. J Anim Sci 78:1219–1226

Lensink J, Fernandez X, Cozzi G et al (2001a) The farmers' influence on calves' behaviour, health and production of a veal unit. Anim Sci 72:105–116

Lensink J, Fernandez X, Cozzi G et al (2001b) The influence of farmers' behavior on calves' reactions to transport and quality of veal meat. J Anim Sci 79:642–652

Leruste H, Bokkers EAM, Heutinck LFM et al (2012) Evaluation of on-farm veal calves' responses to unfamiliar humans and potential influencing factors. Animal 6:2003–2010

Lindahl C, Pinzke S, Herlin A et al (2016) Human-animal interactions and safety during dairy cattle handling-comparing moving cows to milking and hoof trimming. J Dairy Sci 99:2131–2141

Lomb J, Mauger A, von Keyserlingk MAG et al (2021) Effects of positive reinforcement training for heifers on responses to a subcutaneous injection. J Dairy Sci 104:6146–6158

Lott DF, Hart BL (1979) Applied ethology in a nomadic cattle culture. Appl Anim Ethol 5:309–319

Lupoli B, Johansson B, Uvnäs-Moberg K, Svennersten-Sjaunja K (2001) Effect of suckling on the release of oxytocin, prolactin, cortisol, gastrin, cholecystokinin, somatostatin and insulin in dairy cows and their calves. J Dairy Res 68:175–187

Lürzel S, Münsch C, Windschnurer I et al (2015a) The influence of gentle interactions on avoidance distance towards humans, weight gain and physiological parameters in group-housed dairy calves. Appl Anim Behav Sci 172:9–16

Lürzel S, Windschnurer I, Futschik A et al (2015b) Effects of gentle interactions on the relationship to humans and on stress-related parameters in group-housed calves. Anim Welf 24:475–484

Lürzel S, Windschnurer I, Futschik A et al (2016) Gentle interactions decrease the fear of humans in dairy heifers independently of early experience of stroking. Appl Anim Behav Sci 178:16–22

Lürzel S, Barth K, Windschnurer I et al (2018) The influence of gentle interactions with an experimenter during milking on dairy cows' avoidance distance and milk yield, flow and composition. Animal 12:340–349

Lürzel S, Bückendorf L, Waiblinger S et al (2020) Salivary oxytocin in pigs, cattle, and goats during positive human-animal interactions. Psychoneuroendocrinology 115:104636

Marom N, Bar-Oz G (2013) The prey pathway: a regional history of cattle (*Bos taurus*) and pig (*Sus scrofa*) domestication in the Northern Jordan Valley, Israel. PLoS ONE 8(2):e55958

McConnell PB (1990) Acoustic structure and receiver response in domestic dogs, *Canis familiaris*. Anim Behav 39:897–904

McConnell PB (1991) Lessons from animal trainers: the effect of acoustic structure on an animal's response. In: Bateson PPG, Klopfer PH (eds) Human understanding and animal awareness. Plenum Press, New York

Menke C (1986) Auswirkungen von Streß auf das Verhalten einer schwarzbunten Milchviehherde unter Zuhilfenahme der elektronischen Datenverarbeitung. Diploma thesis, Gesamthochschule Kassel

Moberg GP (2000) Biological response to stress: implications for animal welfare. In: Moberg GP, Mench JA (eds) Biology of animal stress. CAB International, Oxon

Moreira TF, Nicolino RR, Meneses RM et al (2019) Risk factors associated with lameness and hoof lesions in pasture-based dairy cattle systems in Southeast Brazil. J Dairy Sci 102:10369–10378

Munksgaard L, de Passillé AM, Rushen J et al (1997) Discrimination of people by dairy cows based on handling. J Dairy Sci 80:1106–1112

Munksgaard L, de Passillé AM, Rushen J et al (2001) Dairy cows' fear of people: social learning, milk yield and behaviour at milking. Appl Anim Behav Sci 73:15–26

Murphey R, Duarte FAM (1983) Calf control by voice command in a Brazilian dairy. Appl Anim Ethol 11:7–18

Murphey RM, Duarte FAM, Penedo MCT (1980) Approachability of bovine cattle in pastures - breed comparisons and a breed X treatment analysis. Behav Genet 10:171–181

Murphey RM, Duarte FAM, Penedo MCT (1981) Responses of cattle to humans in open spaces - breed comparisons and approach-avoidance relationships. Behav Genet 11:37–48

Neumann F, Steinbeck K (1971) Hormonale Beeinflussung des Verhaltens. Klin Wochenschr 49:790–806

Nowak R, Boivin X (2015) Filial attachment in sheep: similarities and differences between ewe-lamb and human-lamb relationships. Appl Anim Behav Sci 164:12–28

Padilla de la Torre M, Briefer EF, Reader T et al (2015) Acoustic analysis of cattle (Bos taurus) mother-offspring contact calls from a source-filter theory perspective. Appl Anim Behav Sci 163:58–68

Pajor EA, Rushen J, de Passille AMB (2000) Aversion learning techniques to evaluate dairy cattle handling practices. Appl Anim Behav Sci 69:89–102

Pajor EA, Rushen J, de Passille AMB (2003) Dairy cattle's choice of handling treatments in a Y-maze. Appl Anim Behav Sci 80:93–107

Paul ES, Serpell JA (1993) Childhood pet keeping and humane attitudes in young adulthood. Anim Welf 2:321–337

Ploeg van der JD (1994) Styles of farming: an introductory note on concepts and methodology. In: van der Ploeg JD, Long A (eds) Born from within: practice and perspectives of endogenous rural development. Van Gorcum, Assen

Pongrácz P, Molnár C, Miklósi Á et al (2005) Human listeners are able to classify dog (*Canis familiaris*) barks recorded in different situations. J Comp Psychol 119:136–144

Pongrácz P, Molnár C, Miklósi Á (2006) Acoustic parameters of dog barks carry emotional information for humans. Appl Anim Behav Sci 100:228–240
Price EO, Wallach SJR (1990) Physical isolation of hand-reared Hereford bulls increases their aggressiveness toward humans. Appl Anim Behav Sci 27:263–267
Probst JK, Neff AS, Leiber F et al (2012) Gentle touching in early life reduces avoidance distance and slaughter stress in beef cattle. Appl Anim Behav Sci 139:42–49
Rault J-L (2016) Effects of positive and negative human contacts and intranasal oxytocin on cerebrospinal fluid oxytocin. Psychoneuroendocrinology 69:60–66
Rault JL, Waiblinger S, Boivin X et al (2020) The power of a positive human-animal relationship for animal welfare. Front Vet Sci 7:590867
Reinhardt V (1980) Untersuchungen zum Sozialverhalten des Rindes. Birkhäuser Verlag TH, Basel
Reinhardt V, Reinhardt A (1981) Cohesive relationships in a cattle herd (Bos indicus). Behaviour 77:121–151
Reinhardt C, Reinhardt A, Reinhardt V (1986) Social behaviour and reproductive performance in semi-wild Scottish Highland cattle. Appl Anim Behav Sci 15:125–136
Reiter D (2017) The influence of the milker's behaviour on social behaviour of dairy cows. Diploma thesis, University of Veterinary Medicine, Vienna
Renger H (1975) Aggressives Verhalten von Bullen dem Menschen gegenüber. Dissertation vet. med., Ludwig-Maximilians-Universität München
Rørvang MV, Jensen MB, Nielsen BL (2017) Development of test for determining olfactory investigation of complex odours in cattle. Appl Anim Behav Sci 196:84–90
Rouha-Mülleder C, Iben C, Wagner E et al (2009) Relative importance of factors influencing the prevalence of lameness in Austrian cubicle loose-housed dairy cows. Prev Vet Med 92:123–133
Rouha-Mülleder C, Palme R, Waiblinger S (2010) Assessment of animal welfare in 80 dairy cow herds in cubicle housing - animal health and other animal-related parameters. Wiener Tierarztliche Monatsschrift 97:231–241
Rousing T, Waiblinger S (2004) Evaluation of on-farm methods for testing the human-animal relationship in dairy herds with cubicle loose housing systems – test-retest and inter-observer reliability and consistency to familiarity of test person. Appl Anim Behav Sci 85:215–231
Ruis M, Coleman G, Waiblinger S et al (2010) A multimedia-based cognitive-behavioural intervention program improves attitudes and handling behaviours of stockpeople in livestock farming. Adv Anim Biosci 1(01):175
Rushen J, Munksgaard L, de Passillé AM et al (1998) Location of handling and dairy cows' responses to people. Appl Anim Behav Sci 55:259–267
Rushen J, de Passillé AM, Munksgaard L (1999) Fear of people by cows and effects on milk yield, behavior, and heart rate at milking. J Dairy Sci 82:720–727
Rushen J, Munksgaard L, Marnet PG et al (2001) Human contact and the effects of acute stress on cows at milking. Appl Anim Behav Sci 73:1–14
Rybarczyk P, Koba Y, Rushen J et al (2001) Can cows discriminate people by their faces? Appl Anim Behav Sci 74:175–189
Sachser N, Kaiser S, Hennessy MB (2013) Behavioural profiles are shaped by social experience: when, how and why. Philos Trans R Soc B Biol Sci 368:20120344
Sambraus HH, Sambraus D (1975) Prägung von Nutztieren auf Menschen. Z Tierpsychol 38:1–17
Sato S, Sako S, Maeda A (1991) Social licking patterns in cattle (*Bos taurus*) - influence of environmental and social factors. Appl Anim Behav Sci 32:3–12
Schlichting MC (1974) Untersuchungen über die Fähigkeiten des Betreuungspersonals in der Milchviehhaltung und dessen Einfluss auf die Milchleistung. KTBL-Schrift, Göttingen
Schmied C, Boivin X, Waiblinger S (2005) Ethogramm des sozialen Leckens beim Rind: Untersuchungen in einer Mutterkuhherde. In: Einschütz K (ed) Aktuelle Arbeiten zur artgemäßen Tierhaltung 2005 (KTBL-Schrift 441). Kuratorium für Technik und Bauwesen in der Landwirtschaft, pp 68–92
Schmied C, Boivin X, Waiblinger S (2008a) Stroking different body regions of dairy cows: effects on avoidance and approach behavior towards a human. J Dairy Sci 91:596–605

Schmied C, Waiblinger S, Scharl T et al (2008b) Stroking of different body regions by a human: effects on behaviour and heart rate of dairy cows. Appl Anim Behav Sci 109:25–38
Schmied C, Boivin X, Scala S et al (2010) Effect of previous stroking on reactions to a veterinary procedure: behaviour and heart rate of dairy cows. Interact Stud 11:467–481
Schütz KE, Hawke M, Waas JR et al (2012) Effects of human handling during early rearing on the behaviour of dairy calves. Anim Welf 21:19–26
Schwabe CW (1978) How man first learnt to heal. Cattle, priests, and progress in medicine. University of Minnesota Press, Minneapolis
Scott JP (1958) Aggression. The University of Chicago Press, Chicago
Scott JP (1962) Critical periods in behavioral development. Science 138:949–958
Seabrook MF (1972) A study to determine the influence of the herdsman's personality on milk yield. J Agric Labour Sci 1:45–59
Seabrook MF (1984) The psychological interaction between the stockman and his animals and its influence on performance of pigs and dairy cows. Vet Rec 115:84–87
Seabrook MF (1986) The relationship between man and animals in managed systems. In: Cole DJA, Brander GC (eds) Bioindustrial ecosystems. Elsevier, Amsterdam
Seabrook MF (2001) The effect of the operational environment and operating protocols on the attitudes and behaviour of employed stockpersons. In: Hovi M, Bouilhol M (eds) Proceedings of the 3rd NAHWOA workshop, human–animal relationship: stockmanship and housing in organic livestock systems, 21–24 October 2000, Network for Animal Health and Welfare in Organic Agriculture (NAHWOA), pp 23–32
Seabrook MF, Bartle NC (1992) Human factors. In: Phillips C, Piggins D (eds) Farm animals and the environment. CAB International, Wallingford
Špinka M (2019) Animal agency, animal awareness and animal welfare. Anim Welf 28:11–20
Spoolder HAM, Waiblinger S (2009) Pigs and humans. In: Marchant-Forde JN (ed) The welfare of pigs. Springer
Stafford K, Mellor D (2005) Dehorning and disbudding distress and its alleviation in calves. Vet J 169:337–349
Stephansen RS, Fogh A, Norberg E (2018) Genetic parameters for handling and milking temperament in Danish first-parity Holstein cows. J Dairy Sci 101:11033–11039
Tallet C, Spinka M, Maruscáková I et al (2010) Human perception of vocalizations of domestic piglets and modulation by experience with domestic pigs (*Sus scrofa*). J Comp Psychol 124:81–91
Taschke A (1995) Ethologische, physiologische und histologische Untersuchungen zur Schmerzbelastung der Rinder bei der Enthornung. Dissertation, Institut für Nutztierwissenschaften der ETH Zürich
Taylor AA, Davis H (1998) Individual humans as discriminative stimuli for cattle (Bos taurus). Appl Anim Behav Sci 58:13–21
Taylor-Clarke M, Jacobsen P, Haggard P (2004) Keeping the world a constant size: object constancy in human touch. Nat Neurosci 7:219–220
Toates FM (2001) Biological psychology: an integrative approach. Pearson Education Limited, Harlow, Essex
Tucker CB, Mintline EM, Banuelos J et al (2014) Pain sensitivity and healing of hot-iron cattle brands. J Anim Sci 92:5674–5682
Unshelm J (1988) Introductory remarks. In: Zayan R, Dantzer R (eds) Social stress in domestic animals. Kluwer Academic Publishers, Dordrecht
Uvnäs-Moberg K (1997) Oxytocin linked antistress effects - the relaxation and growth response. Acta Physiol Scand Suppl 161:38–42
Uvnäs-Moberg K (2004) Role of oxytocin in antistress and well-being. J Psychosom Res 56:595–595
van Reenen CG, van der Werf JTN, Bruckmaier RM et al (2002) Individual differences in behavioral and physiological responsiveness of primiparous dairy cows to machine milking. J Dairy Sci 85:2551–2561

Wagner K, Barth K, Palme R et al (2012) Integration into the dairy cow herd: long-term effects of mother contact during the first twelve weeks of life. Appl Anim Behav Sci 141:117–129

Waiblinger S (1996) Die Mensch-Tier-Beziehung bei der Laufstallhaltung von behornten Milchkühen. Reihe Tierhaltung Bd. 24, Gesamthochschule Kassel, Witzenhausen

Waiblinger S (2012) Die Bedeutung der Veterinärmedizin für den Tierschutz. In: Grimm H, Otterstedt C (eds) Das Tier an sich? – Disziplinen übergreifender Perspektiven für neue Wege im wissenschaftsbasierten Tierschutz. Vandenhoeck & Ruprecht, Göttingen

Waiblinger S (2017) Human-animal relations. In: Jensen P (ed) The ethology of domestic animals: an introductory text, 3rd edn. CABI, Wallingford

Waiblinger S (2019) Agricultural animals. In: Hosey G, Melfi V (eds) Anthrozoology. Human-animal interactions in domesticated and wild animals. Oxford University Press, New York

Waiblinger S, Menke C (1999) Influence of herd size on human-cow relationships. Anthrozoös 12:240–247

Waiblinger S, Rouha-Mülleder C (2015) Stress bei Milchkühen in verschiedenen Melkstandtypen - Ergebnisse einer Praxisuntersuchung. In: Savary P, Schick M (eds) 5. Tänikoner Melktechniktagung: Nachhaltige Milchgewinnung, 18–19 März 2015 Agroscope, Tänikon, Schweiz, pp 23–26

Waiblinger S, Spoolder HAM (2007) Quality of stockpersonship. In: Velarde A, Geers R (eds) On farm monitoring of pig welfare. Wageningen Academic Publishers, Wageningen

Waiblinger S, Baars T, Menke C (2001) Understanding the cow - the central role of human-animal relationship in keeping horned dairy cows in loose housing. In: Hovi M, Bouilhol M (eds) Proceedings of the 3rd NAHWOA workshop, human–animal relationship: stockmanship and housing in organic livestock systems, 21–24 October 2000. Network for Animal Health and Welfare in Organic Agriculture (NAHWOA)

Waiblinger S, Menke C, Coleman G (2002) The relationship between attitudes, personal characteristics and behaviour of stockpeople and subsequent behaviour and production of dairy cows. Appl Anim Behav Sci 79:195–219

Waiblinger S, Menke C, Coleman G (2003a) Mensch-Tier-Interaktionen beim Melken: Einflussfaktoren und Auswirkungen auf Verhalten und Milchleistung der Kühe. In: Aktuelle Arbeiten zur angewandten Ethologie 2002, KTBL-Schrift 418. KTBL, 125–133

Waiblinger S, Menke C, Fölsch DW (2003b) Influences on the avoidance and approach behaviour of dairy cows towards humans on 35 farms. Appl Anim Behav Sci 84:23–39

Waiblinger S, Menke C, Korff J et al (2004) Previous handling and gentle interactions affect behaviour and heart rate of dairy cows during a veterinary procedure. Appl Anim Behav Sci 85:31–42

Waiblinger S, Boivin X, Pedersen V et al (2006a) Assessing the human-animal relationship in farmed species: a critical review. Appl Anim Behav Sci 101:185–242

Waiblinger S, Mülleder C, Menke C (2006b) The human-animal relationship and animal health management. Proceedings of the Organic Congress, 30.-31.5.2006, 2006b Odense, Denmark. Odense, Denmark, pp 498–499

Waiblinger S, Mülleder C, Menke C et al (2006c) How do farmers' attitudes impact on animal welfare? The relationship of attitudes to housing design and management on dairy cow farms. In: Amat M, Mariotti V (eds) The importance of attitudes, values and economics to the welfare and conservation of animals. Proceedings of the 15th annual conference of the international society for anthrozoology, 5–6 October 2006, Barcelona, pp 55–56

Waiblinger S, Schmied C, Minero M et al (2007) Variation in Austrian and Italian dairy farmers' attitudes and handling practices and consequences for handling stress. Report for the EEC-project Welfare Quality®

Waiblinger S, Wagner K, Hillmann E et al (2020) Short- and long-term effects of rearing dairy calves with contact to their mother on their reactions towards humans. J Dairy Res 87:148–153

Watts JM, Stookey JM (2000) Vocal behaviour in cattle: the animal's commentary on its biological processes and welfare. Appl Anim Behav Sci 67:15–33

Weltagrarbericht 2009. http://www.weltagrarbericht.de/themen-des-weltagrarberichts/fleisch-und-futtermittel.html. Accessed 2022/04/01

Wenz C, Laube R-B (1989) Kraftinteraktionen zwischen Tierpfleger und Besamungsbullen beim Führvorgang einschließlich des Absamprozeßes. International Youth Camp on Applied Ethology of Farm Animals, 1989 Gödöllő, Ungarn, pp 410–433

Wildridge AM, Thomson PC, Garcia SC et al (2020) Transitioning from conventional to automatic milking: effects on the human-animal relationship. J Dairy Sci 103:1608–1619. https://doi.org/10.3168/jds.2019-16658

Willson DW, Baier FS, Grandin T (2021) An observational field study on the effects of changes in shadow contrasts and noise on cattle movement in a small abattoir. Meat Sci 179:108539

Windschnurer I, Schmied C, Boivin X et al (2008) Reliability and inter-test relationship of tests for on-farm assessment of dairy cows' relationship to humans. Appl Anim Behav Sci 114:37–53

Windschnurer I, Barth K, Waiblinger S (2009a) Can stroking during milking decrease avoidance distances of cows towards humans? Anim Welf 18:507–513

Windschnurer I, Boivin X, Waiblinger S (2009b) Reliability of an avoidance distance test for the assessment of animals' responsiveness to humans and a preliminary investigation of its association with farmers' attitudes on bull fattening farms. Appl Anim Behav Sci 117:117–127

Windschnurer I, Boivin X, Coleman G et al (2010) Modifying attitudes and behaviour towards dairy cattle by multimedia-based cognitive-behavioural intervention. Proceedings of the 44th congress of the international society for applied ethology (ISAE), 4–7 August 2010, Uppsala, p 52

Zimmermann A, Bai L, Ginty DD (2014) The gentle touch receptors of mammalian skin. Science 346:950–954

# Welfare at Calving and of the Growing Animals

# 10

Margit Bak Jensen and Katy Proudfoot

## Contents



**Abstract**

Dairy cows and their calves face several challenges around parturition and in the early life of the calf that impact their welfare. There is an increasing public awareness of some of these challenges, including those that begin before birth as the cow prepares for labour and continue until the calf is weaned from milk. Researchers have recognised that these challenges exist and have begun to define

M. B. Jensen (✉)
Aarhus University, Aarhus, Denmark
e-mail: margitbak.jensen@anivet.au.dk

K. Proudfoot
The Atlantic Veterinary College, Charlottetown, PE, Canada



M. Haskell (ed.), *Cattle Welfare in Dairy and Beef Systems*, Animal Welfare 23,
https://doi.org/10.1007/978-3-031-21020-4_10

the key animal welfare issues for the cow and calf during this time period. In this chapter, we review the experience of the cow around the time of calving, the effect of prolonged maternal contact on the dam and her calf, and social housing for young calves. Next, we discuss the welfare of youngstock post-weaning and of growing cattle, although this topic has received less research attention. We end the chapter with a discussion about advances and future challenges in animal welfare for the peri-parturient cow and her calf, as well as the growing animal.

**Keywords**

Maternal behaviour · Social behaviour · Housing · Milk-feeding management · Cow · Calf · Young stock

## 10.1 Introduction

The time around parturition can be very challenging for the welfare of both cow and calf. The challenges include the processes of labour, giving birth and recovery for the cow, and of adjusting to post-natal life for the calf. There is an increasing public awareness of some of these challenges. For instance, surveys reveal public concerns about the early separation of dam and calf (Ventura et al. 2013; Hötzel et al. 2017) and the individual housing of calves is one of the foci of the European Citizen Initiative 'end the cage age' which was registered by the EU Commission (Anonymous 2018). In addition to this public awareness, researchers have also identified animal welfare challenges that begin before birth as the cow prepares for labour and continue until the calf is weaned from milk. Researchers have recognised that these challenges exist and have begun to define the key animal welfare issues, including the impact of commonly used housing and management practices on the animals' biological health, affective experiences, and the ability to live a reasonably natural life (Fraser et al. 1997). In this chapter, we review the experience of the cow around the time of calving, the effect of prolonged maternal contact on the dam and her calf, and social housing for young calves. Next, we discuss the welfare of youngstock post-weaning and of growing cattle, although this topic has received less research attention. We end the chapter with a discussion about advances and future challenges in animal welfare for the peri-parturient cow and her calf, as well as the growing animal.

## 10.2 Around the Time of Calving

Researchers and other stakeholders are increasingly recognising that dairy cows are vulnerable around the time of calving. For example, parturition itself is considered painful (Mainau and Manteca 2011) and is often followed by health problems (Ingvartsen 2006) as well as a high cow mortality (Thomsen et al. 2004). A survey of veterinarians reported that a difficult labour ('dystocia') was believed to be one of the most painful experiences in adult dairy cattle and their calves (Huxley and Whay 2006). In addition to pain and health-related outcomes, more recent emphasis has also been put on allowing for the natural behaviour of dairy cows before calving, such as the motivation to seek seclusion during labour (Rørvang et al. 2018a). Due to this research, several countries now have recommendations on housing and management around the time of calving to allow for a more natural birth. For instance, in Canada it is recommended that cows calve in an individual maternity pen or a spacious group maternity pen (Anonymous 2009), while in Denmark (Anonymous 2017) legislation prescribes that cows must calve in an individual pen (unless they calve whilst at pasture). However, both individual and group maternity pens have their own challenges. Here, we discuss the natural behaviour of dairy cows as they approach labour, and how this information can be used to better manage and house cows during this period.

### 10.2.1 The Calving Environment

Studies using wild ungulates and cattle kept in a natural setting provide valuable insight into the most appropriate calving environment for dairy cows kept in commercial facilities. For example, Lidfors et al. (1994) recorded the characteristics of calving sites for both dairy and beef cattle housed outdoors and found that cattle were attracted to calving sites at high altitudes with dry and soft materials on the ground, and with some trees nearby and cover overhead. Cattle generally avoided areas of open pasture, and they separated from their group mates on the day of calving compared to previous days. Studies with beef cattle have shown that pregnant animals prefer a degree of isolation from herd mates before giving birth but form preferential associations with other maternal cows (those with calves of similar age) after calving (Finger et al. 2014; Swain et al. 2015). In wild ungulates, characteristics of a calving location have been less explored, but there is some evidence that in some species the dam will separate from herd mates as calving approaches (bison: Lott and Galland 1985; elk: Barbknecht et al. 2011). However, in the natural environment, the motivation to isolate from the herd is likely dependent upon available resources and predation pressure. The function of the isolation-seeking behaviour is likely a strategy to reduce mis-mothering (see Rørvang et al. 2018a for a detailed review).

The majority of cows in commercial dairy farms across North America, Europe, and other countries are kept indoors to calve, which may limit their ability to seek a secluded and desirable calving area (Jensen and Tolstrup 2021). Research using cows kept indoors in individual maternity pens to give birth has found that many cows will seek a secluded place to give birth if it is provided (Proudfoot et al. 2014a, b). For example, Proudfoot et al. (2014b) attached plywood barriers to approximately one half of an individual pen to allow cows a secluded 'corner' to give birth in if they chose (Fig. 10.1a). Approximately 80% of the cows calved in this corner that provided visual isolation from herd mates, suggesting that such an area in an

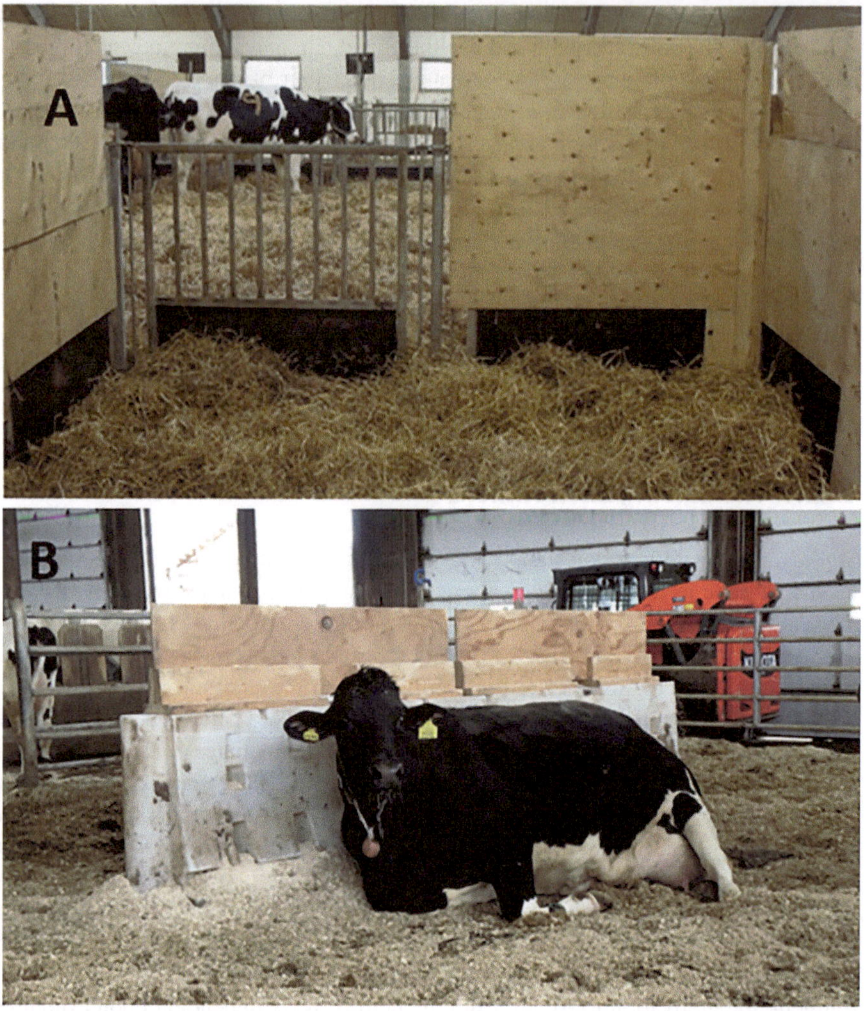


**Fig. 10.1** Methods for allowing cows to find seclusion at calving in individual (**a**; Proudfoot et al. 2014b) and group (**b**; Creutzinger et al. 2021a) pens. Photos courtesy of Katy Proudfoot

individual pen is perceived by cows as an opportunity to hide. This is an easy alteration to make to existing individual calving pens for many dairy producers.

Providing cows with an opportunity to seek visual isolation from herd mates when kept in a group pen during calving may be more complicated and impacted by social competition. For example, two studies created seclusion opportunities by fitting similar 'L'-shaped enclosures along the outside of a group maternity pen but used a different ratio of cows to enclosures (either one cow/enclosure or two cows/enclosure). The study providing one cow/enclosure used multiparous cows and found that approximately 50% of the cows were in the enclosure at calving (Rørvang et al. 2018b), whereas the study including two cows/enclosure used both primiparous and multiparous cows and found that only 10% of the cows calved in the enclosure (Jensen and Rørvang 2018). Additionally, Rørvang et al. (2018b) discovered that bolder cows with a higher social rank were most likely to use the enclosures to give birth. Thus, social competition for a resource allowing cows to hide likely impacts a cow's ability to seek seclusion during labour. An alternative approach would be to use a single-sided wall in the middle of a group maternity pen, allowing more than one cow to use the space at one time (Creutzinger et al. 2021a; Fig. 10.1b). This type of single-sided hide may help cows seek seclusion in a group setting and may also reduce the length of labour (Creutzinger et al. 2021b). Researchers are encouraged to continue to assess practical methods for allowing cows to separate themselves from other cows in group calving pens.

Some research using cows housed indoors has also evaluated the degree of seclusion cows may be motivated to seek in individual and group pens. For example, Rørvang et al. (2017) provided individually housed cows with three options for seclusion (tall and narrow [1.8 × 1.5 m], low and wide [1.0 × 2.5 m], and tall and wide [1.8 × 2.5 m]) but found no specific preference. However, cows that calved in the treatment with the highest amount of seclusion (tall and wide) were the most likely to have the longest labour. Although the cause-and-effect relationship between these factors is unknown, the authors speculate that cows with prolonged labour may be more motivated to seek seclusion due to greater restlessness and discomfort. In the same group pen with enclosures described above (Rørvang et al. 2018b), authors also tested whether cows would prefer extra protection from herd mates in the form of a push gate at the entry of the enclosure. In half of the pens, cows were trained to push through a gate to access the enclosures, whereas in the other half of the pens the enclosures had an open door where any cow could enter and exit freely. The gated pens allowed cows to enter and exit, but once a cow entered the enclosure another cow could not follow. However, with both types of pens, only about 50% of the cows calved in an enclosure; thus, cows in this study did not prefer extra protection, suggesting that cows may not have been able to combine a learnt response with their motivation to isolate.

The research to date on isolation behaviour of indoor-housed dairy cows has focused mainly on providing a physical space for cows to hide at parturition (e.g., Proudfoot et al. 2014b; Rørvang et al. 2017, 2018b). However, in a natural setting, cows and other ungulates also distance themselves from herd mates. For example, beef cattle on rangeland have been found to walk between 380 and 1250 m away from herd mates to give birth (Kiley-Worthington and de la Plain 1983; Flörcke and Grandin 2014). In one study where cows were kept indoors in pairs, cows increased the amount of time they spent in a different area of the pen as calving approached, though it is unclear whether the calving cow or her partner drove this response (Proudfoot et al. 2014a). Research is still needed to determine how cows utilise their space before they give birth and if additional lying space would be beneficial for animals kept in groups during parturition.

In addition to having a secluded place to give birth, cows preferred soft and dry places to give birth in a more natural setting (Lidfors et al. 1994). Although limited research has focused on the comfort of indoor calving pens, there is some evidence that cows prefer to give birth on a layer of sand and straw compared to a layer of rubber and straw (Campler et al. 2014). If the straw bedding is not on a substrate that allows it to drain, it may become wet and dirty quickly, which may impact the cow's comfort as well as her and her calf's risk of becoming ill after birth (Frank and Kaneene 1993; Elbers et al. 1998). However, there is evidence that straw bedding may be superior to other substrates in providing for cow health, as cows were found to be at higher risk of endometritis when kept on sand alone, sawdust, paper, or a combination of these materials, compared to straw (Cheong et al. 2011). Thus, indoor calving areas should have a thick layer of straw (30–46 cm; Cook and Nordlund 2004) ideally on top of a layer of sand (Campler et al. 2014). Individual calving areas should be completely cleaned (e.g., straw removed and replaced) between calvings, and group calving areas should have fresh bedding added daily and be completely replaced at least every 3–4 weeks (Cook and Nordlund 2004).

### 10.2.2 Behaviour and Dystocia

Dairy cows show distinctive changes in their behaviour as calving approaches and labour progresses. Cows undergo three stages of labour: in Stage 1, the cow's cervix begins to dilate, her pelvic ligaments relax, she begins to experience myometrial contractions, and her calf is moved into position for delivery; in Stage 2, the calf is pushed through the birth canal through forces in the uterus (myometrial contractions) and the abdomen (abdominal contractions); finally, Stage 3 occurs after birth when the cow's uterus undergoes involution (Noakes 2001; Jackson 2004). All stages of labour are likely associated with various levels of pain (Mainau and Manteca 2011) and are associated with changes in hormones (von Keyserlingk and Weary 2007); both pain and hormones likely contribute to changes in behaviour documented in cows during labour.

On the day of calving, cows eat about 30% less than on previous days (Schirmann et al. 2013), spend less time eating (Jensen 2012; Miedema et al. 2011), less time ruminating (Schirmann et al. 2013), and less time drinking water (Jensen 2012). Cows also show an increase in 'restlessness' on the day of calving compared to previous days, characterised by more transitions from standing to lying ('lying bouts'; Huzzey et al. 2005) due to more lying bouts occurring during the final 4–6 h before calving (Miedema et al. 2011; Jensen 2012). The frequency of lying bouts is especially high for cows giving birth for the first time compared to older animals (Neave et al. 2017). Cows also raise their tails, sniff and lick at the ground, turn their heads towards their abdomens, and increasingly spend time lying in lateral recumbency as calving gets nearer (e.g., in the 4–6 h before calving; Jensen 2012). These changes in behaviour provide insight into the cow's experience during labour but may also be useful for dairy farmers to monitor calving progress using technologies such as accelerometers and rumination collars (Borchers et al. 2017). More work is encouraged to develop sensitive algorithms for these technologies to detect early calving behaviours.

A cow's behaviour during labour is also impacted by her ability to give birth naturally and without assistance from farm staff or a veterinarian. 'Dystocia' is defined as a difficult birth resulting from prolonged or severely assisted calving (Mee 2008) and is considered to be one of the most painful experiences of dairy cattle (Huxley and Whay 2006). Some countries collect producer reports of dystocia incidence; for example, in the USA producers report that 4.7% of their cows experience dystocia (USDA 2016), and in Denmark, producers report that 1.8% of Holstein Frisian cows with purebred heifer calves and 1.9% of cows with bull calves experience dystocia (Byskov 2012). However, as these percentages are based on producer records, they may underestimate the true incidence or prevalence rate.

There are many risk factors for dystocia, including the breed of the dam and sire, calf sex and size, and position of the calf (e.g., malposition; reviewed by Mee 2008). Cows with dystocia have also been shown to change their behaviour compared to those with unassisted calvings; for example, cows with dystocia eat less, ruminate less, and have more lying bouts on the day of calving compared to those that calve without assistance (Proudfoot et al. 2009; Kovács et al. 2017). In addition, cows with dystocia rubbed against a wall more frequently, discharged more urine, and rubbed against the floor more often than those that go on to have a natural birth (Wehrend et al. 2006). After giving birth, cows with assisted births show some behavioural signs of pain compared to those calving naturally, such as a reduction in self-grooming behaviour (Barrier et al. 2012). As they become more commercially available, these behavioural changes on the day of calving may be detected by technologies (e.g., accelerometers and rumination collars; Kovács et al. 2017), which will allow dairy producers to identify cows with prolonged labour and a high risk for dystocia.

Not only does dystocia impact the behaviour and comfort of cows, but it also reduces both the cow and her offspring's health and survival after birth. Cows that require assistance during labour are more likely to experience retained placenta, vulvovaginal laceration, mastitis, and lameness after calving (Peeler et al. 1994; Kovács et al. 2016). Calves born to a dystocia are more likely to be stillborn and experience lower vitality, more physiological stress (measured as plasma cortisol), and lower passive immune transfer immediately after calving (Peeler et al. 1994; Lombard et al. 2007; Barrier et al. 2013). Likely due to lower immunity, calves born to assisted calvings are also more likely to become ill and die later in life compared to those born without assistance (Lombard et al. 2007; Barrier et al. 2013). Interestingly, the timing of an intervention during calving can impact stillbirths and the cow's risk of illness after calving. Kovács et al. (2016) found that the highest risk of stillbirth occurred when cows were inappropriately assisted during labour (e.g., premature pulling of the calf). In the same study, calves born after inappropriate assistance also had lower vitality scores. Thus, farm staff should be trained in recognising appropriate signs when a cow needs assistance during labour, and to only assist when it is required (e.g., see a detailed description of obstetric intervention by Schuenemann et al. 2011).

### 10.2.3 Housing and Management of the Dam and Her Newborn

Just as studies using more natural settings provide insights into cow housing and management before calving, this type of work can also inform on appropriate indoor facilities post-calving. Researchers studying a feral cow herd in Italy found that cows and calves stayed near each other in the first few days of life; the calves stayed in a hidden area while the cow grazed nearby (Vitale et al. 1986). As the calves became older, the dam and her offspring joined the herd, allowing both to socialise with other cows and calves. This type of setting differs dramatically from the housing and management of dairy cows and their newborn calves on most commercial dairy farms in North America and Europe. The dam and calf are usually separated immediately after birth (e.g., within an hour to up to 24 h) or, less commonly, are kept together for a short period (e.g., a week or two). Newer, alternative management practices are arising that allow prolonged contact between the dam and her calf, mimicking more natural settings. The housing and management, as well as early behaviour of the dam and her calf when kept together for hours to a few weeks, will be discussed in this section, and more prolonged contact (e.g., the entire milk-feeding period) will be covered in detail in the next section (Sect. 10.3).

When indoor-housed dairy cows are given the opportunity to stay with their calves for a short period after calving, they spend the majority of the first few hours after birth focused on their offspring. Jensen (2012) kept the dam with her calf for 24 h after birth in an individual pen and found that the dam spent over 50% of her time licking and sniffing the calf in the first hour after birth. However, the interaction between the dam and her calf is impacted by the cow's housing and social environment. Cows that give birth in individual pens (e.g., as in the Jensen 2012 study) can direct their attention to their newborn without distraction, while cows that give birth in group pens may endure interference from other cows. For example, cows kept in groups spent less time licking their calves in the 6 h after calving compared to those kept in individual pens, likely because other cows in the group were also licking the newborn (Edwards 1983). It has also been reported that about 30% of calves born in group pens suckled a cow other than their mother (Edwards 1983; Illman and Spinka 1993). This 'mis-mothering' behaviour can be reduced by providing the cow and her newborn a secluded space after calving (Jensen et al. 2019). Interference from other cows may be stressful to the new mother and may also adversely affect the quality of the colostrum of cows being nursed before giving birth which may increase the risk of calf mortality (Reimus et al. 2020). Thus, it is recommended that cows either give birth in individual pens—to which they are moved well in time before calving—or, if calving in groups, the cow and her offspring should have the opportunity to seclude themselves or be quickly moved to a protected space to limit mis-mothering. However, moving the cow after she has entered labour should be avoided as this has been found to prolong the second stage of labour (Proudfoot et al. 2013).

If the cow and calf are separated, care should be taken to ensure that both are moved into low-stress environments. Calves benefit from social housing in early life (covered in more detail in Sect. 10.4), and cows may benefit from being moved into smaller groups with less competition for resources. In commercial settings, cows are sometimes moved directly into the milking herd, which may include high competition for resources due to overstocking. Alternatively, cows may be moved into designated post-calving pens (e.g., 'fresh' pens) where they are given special care for a few weeks before re-entering the milking herd. Researchers have found reductions in competition when cows are housed in smaller groups after calving compared to larger groups (6 vs. 24 cows per pen; Jensen and Proudfoot 2017). Campler et al. (2019) found that cows kept in a straw-bedded, individual maternity pen for two extra days after calving had higher lying times and lower feeding rates (a sign of reduced competition) compared to those moved directly into a group free stall pen after giving birth. Primiparous animals calving for the first time may especially benefit from a 'fresh' pen. Østergaard et al. (2010) found that primiparous cows moved into a fresh pen with other post-parturient animals for 1 month had improved health and productivity compared with those moved into a lactating group. Thus, if the cow and calf are to be separated soon after birth, the dam benefits from being in a more comfortable pen with less competition where she can recover from labour.

### 10.2.4 Dam and Calf Response to Early Separation

Traditionally, it has been recommended that dairy calves are separated from their dams immediately after calving (e.g., within a few hours to 24 h). A low response to separation is often used as an argument for early separation, and both the dam and her calf show more behavioural signs of stress when the pair is kept together longer (from days up to weeks) before separation (Stěhulová et al. 2008; Flower and Weary 2001). These findings are not surprising, as in nature the survival of the calf depends on care provided by its dam, and the calf is likely developing a stronger bond with the dam during the first days and weeks.

In addition to acute stress, separation of the dam and calf has been found to impact cognitive components of the calves' emotional states. For example, researchers have found that separation from the dam causes signals of distress, such as vocalisations, restlessness, and reduced feed intake (reviewed by Weary et al. 2008). More recently, Daros et al. (2014) used a cognitive bias test to assess the impact of maternal separation on the emotional experience of 42-day-old calves. Calves were trained to discriminate between red and white colours displayed on a computer monitor; each colour indicated either a reward (an allotment of milk) or a punishment (a 'timeout'). Before and after separation from the dam, the calves were then given a set of ambiguous colours on the screen (intermediate between red and white). Researchers found that after separation from the dam, calves were more likely to respond to the ambiguous cues as if they were negative compared to before separation. These results suggest that separation from the dam causes a negative emotional experience in young calves.

## 10.3 Unweaned Calves Kept with the Dam

There is growing social pressure to keep dairy cows with their calves for longer than what has been traditionally considered common practice (e.g., within 24 h of birth). For example, when researchers in Canada asked participants: 'should dairy calves be separated from the cow within the first few hours after birth?'; authors found that 76% of the participants that were not associated with the dairy industry said 'No' to this question, whereas individuals associated with the industry had a range of responses (e.g., 33% of dairy farmers, 40% of dairy professionals, and 0% of veterinarians said 'No'; Ventura et al. 2013). The recommendation to remove the calf from the dam soon after birth was initially based on concerns over the health and welfare of both animals (see Beaver et al. 2019 for a review of these arguments). However, there is limited research showing any evidence of the health benefits of early separation and insufficient research assessing the welfare impacts of early separation (see detailed companion reviews by Beaver et al. 2019, and Meagher et al. 2019). Together, these findings have led to increased interest in developing systems that allow prolonged contact between the pair.

### 10.3.1 Maternal Contact

Over the last 30 years, studies have assessed the impact of various types of contact on the dam and calf. For the purposes of this chapter, we will focus on three types of contact described in detail by Johnsen et al. (2016) and Sirovnik et al. (2020): (1) 'free contact' systems where the dam and calf have unrestricted access to each other including the ability to suckle/nurse, (2) 'restricted suckling' systems which allow the dam and calf to have short daily contact only to nurse, (3) 'half day contact' systems which allow the dam and calf to be housed together throughout the day or night. Regardless of contact type, studies have also assessed differences in dam and calf outcomes when the pair is separated early compared with varying durations of contact (reviewed by Beaver et al. 2019 and Meagher et al. 2019).

#### 10.3.1.1 Effects of Maternal Contact on Health, Behaviour, and Productivity

In a systematic review of the literature, Beaver et al. (2019) found that there was no consistent evidence that cows and calves were more likely to become ill after calving if they were kept together. For example, there is no strong evidence that keeping cows and calves together increases the risk of Johne's disease in calves (e.g., Nielsen and Toft 2011; Donat et al. 2016), which is often used as a justification for early separation. Conversely, a majority of studies have shown that cows allowed to nurse their calves are less likely to be diagnosed with mastitis after calving (e.g., Fröberg et al. 2008; González-Sedano et al. 2010), and calves allowed to suckle are less likely to get scours in early life (e.g., Weary and Chua 2000; Wagenaar and Langhout 2007). Thus, there are no consistent health risks to keeping cows and calves together after calving and some potential health benefits (e.g., reduced mastitis and scours) of allowing calves to suckle from their dams.

In a second systematic review of the literature, Meagher et al. (2019) describe evidence that early cow-calf contact can cause stress during separation but can also have positive impacts on the social behaviour and growth of calves. For example, Meagher et al. (2019) reported that the majority of studies included in their review showed an increase in calf growth during the suckling period when calves had contact with their dam (e.g., Flower and Weary 2001; Kisac et al. 2011). Stěhulová et al. (2008) found that calves kept with their dams for 7 days habituated more quickly to a novel situation compared to those separated at 1 and 4 days after birth and moved into individual calf pens. Duve et al. (2012) compared the social behaviour of calves kept individually, in pairs and with their dam for 4 weeks after calving. Calves kept with their dams showed the best adaptation to a stressor (restraint), followed by those kept in pairs and those kept individually. In addition, calves kept with their dams or in pairs spent more time engaged in play behaviour (suggested to indicate a positive affective state; Boissy et al. 2007) compared to those kept individually (Duve et al. 2012). These results suggest that cows may be providing their calves with social support that is superior to the social support that can be provided by their peers (Rault 2012). Although separation is more stressful for both the dam and calf with increased periods of contact, there are social and growth benefits of allowing some early contact between dams and calves.

#### 10.3.1.2 Challenges of Dam Rearing

Although there is evidence that keeping young calves with their dams have some benefits, more research is still needed to address common challenges with this housing approach. The two main challenges are housing and management during the contact phase and defining weaning strategies that limit the inevitable stress of separation.

Most dairy facilities are built around the concept of early separation by housing the dairy cows in one area of the barn (e.g., free stall or bedded pack) and the calves in their own area (e.g., nursery). Thus, transitioning to a housing system where the cows and calves are together for the full or partial day may be a challenge. Some studies have used innovative housing strategies using existing farm facilities. For example, one study allowing full contact kept the cow and calf together in a free stall pen with access to an automatic milking system (Fröberg and Lidfors 2009). Others have added a separate 'creep' area for calves directly adjacent to a free stall pen (Roth et al. 2009; Wagner et al. 2013). Wagner et al. (2013) constructed a 'selection gate' that provided calves with access to their dams at certain times of day, allowing the producer more flexibility with when calves and cows had contact. Finally, some studies have used deep-bedded straw packs with a concrete loafing area and a separate calf creep area (Johnsen et al. 2015). More research is encouraged to determine other options for easily transitioning a current dairy facility to allow contact and to develop new, innovative housing strategies designed specifically to allow for dam–calf contact.

A second challenge is developing weaning strategies that limit the stress of separation. Regardless of when the dam and calf are separated, both show behavioural indicators of acute stress after separation. It may be argued that the benefits of dam contact should outweigh the increased stress at separation the longer dam and calf have been together, but these costs and benefits are not easily comparable. Instead, measures may be taken to minimise stress at separation by gradually weakening the maternal bond and the calf's dependence on the dam's milk as recently reviewed in Jensen (2018). This is because, in addition to separation from the cow, withdrawal from access to milk as part of the weaning process can be stressful for the calf due to hunger (Thomas et al. 2001; Budzynska and Weary 2008). One method used to reduce the stress of separation is to first prevent nursing before physically reducing contact. For example, Johnsen et al. (2015) allowed dams and calves to have visual and physical contact, but prevented nursing by partial separation (a fence) before separation 8 weeks after calving, and found that calves showed a reduced vocal response at separation compared with calves that could hear but not see or touch the dam. A further study by Johnsen et al. (2018) found that calves without nutritional dependence on the dam (e.g., fed supplemental milk and were not allowed to suckle) showed fewer behavioural responses to separation compared to calves that could suckle from the dam. Another method for separating the stressors at weaning is the use of a nose flap on calves; these devices allow for continued cow-calf contact without the opportunity for nursing (Loberg et al. 2008). Thus, care should be taken to ensure that stress is reduced during separation, regardless of how long the pair has had contact.

There are several other challenges associated with systems that allow dam–calf contact that require further research. For example, a sustainable system will allow the dairy farmer to make sufficient profit from milk to remain in business (reviewed by Johnsen et al. 2016). Thus, research is still needed to ensure that cows are producing sufficient saleable milk without compromising their health. For example, when the calf and machine are both extracting milk from the cow, she is at risk of losing excessive weight (Margerison et al. 2002). Moreover, there is also evidence that the milk ejection response can be impaired in dams that are nursing their calves, resulting in lower milk yield during machine milking (e.g., De Passillé et al. 2008; Barth 2020). More research is encouraged to help mitigate these challenges while ensuring the health of dams and their calves kept in these systems.

## 10.4 Unweaned Calves Housed with Peers

### 10.4.1 Peer Contact (Pair or Group Housing)

Artificially reared, milk-fed calves are traditionally housed individually, and surveys show that most milk-fed calves in Canada (Vasseur et al. 2010 (70%); Winder et al. 2018 (63%)) and Brazil (Hötzel et al. 2014 (70%)) are housed in individual pens or hunches. Especially in European countries like Germany, the Netherlands, Sweden, and Denmark, housing milk-fed calves in groups is becoming more common (Marcé et al. 2010). This change is in part driven by reduced labour cost, especially when using automated milk-feeding systems (Kung et al. 1997; Medrano-Galarza et al. 2017). The main arguments against group housing have been increased risk of diarrhoea and respiratory disease, but research suggests several advantages in terms of welfare of housing calves in small groups.

Calves are typically fed a low milk allowance to stimulate concentrate intake and to facilitate early weaning off milk, mainly to reduce costs of feed. This management practice also fits well with individual housing pre-weaning. However, this milk-feeding practice renders the calves hungry and does not take advantage of their capacity for growth (Rosenberger et al. 2017). When calves are housed with peers, competition for milk, or access to a milk feeder, is a challenge, especially if the milk allowance is low. Another challenge is abnormal cross-sucking among calves, and both competition and cross-sucking require special preventive measures in milk-feeding management when group housing is practised.

#### 10.4.1.1 Effects on Behaviour, Production, Health

**Behaviour**

Under semi-natural conditions, the calf's first social contact is with the dam, but after a few weeks of life calves associate increasingly with other calves (Wood-Gush et al. 1984; Vitale et al. 1986). Under commercial conditions, milk-fed dairy calves' opportunity to associate with other calves is limited if they are housed individually, but some types of individual pens allow physical contact between calves housed in neighbouring pens. Individually housed calves use this opportunity to sniff and lick the neighbouring calf's head through partitions from a young age, although the level of social behaviour performed was less than that of pair-housed calves that could also push and butt each other as well as perform social play behaviour (Duve and Jensen 2012). The opportunity to interact fully reduces calves' fearfulness in novel social situations. Calves housed in pairs or small groups more readily approached and interacted with an unfamiliar calf in a standard social test than individually housed calves (De Paula Vieira et al. 2012; Jensen et al. 1997). The level of social contact possible between individual pens may be auditory, visual, and physical. To test the effect of each mode of contact on social behaviour, calves were tested in a social interaction test. Calves with only auditory contact were the most fearful, pair-housed calves were the least fearful, while individually housed calves with physical contact were intermediate (visual contact in addition to auditory contact had a minimal effect on calves' behaviour; Jensen and Larsen 2014). Therefore, pair housing was superior to all types of individual pens, and those individual pens with opportunity for physical contact between neighbouring calves were better than individual pens without any opportunity for contact.

Although individually housed calves were more reluctant to approach and interact with unfamiliar calves, once they had made contact, these calves engaged in more agonistic social interactions (Duve and Jensen 2011; De Paula Vieira et al. 2012). This behaviour may be due to an inability to respond appropriately during social interactions, suggesting that social skills are only developed through full social interaction, i.e. during social housing. Cognitive skills are also improved by social housing. No difference in speed of learning an operant task between pair-housed calves and individually housed calves was detected in a simple learning task (Gaillard et al. 2014; Bučková et al. 2019), but reversal learning was superior in pair-housed calves (Gaillard et al. 2014) and in calves housed in dam-calf groups (Meagher et al. 2015) compared to individually housed calves. Calves housed in dam-calf groups were also less neophobic (Costa et al. 2014), and thus more likely to rapidly adjust to novel foods and other changes in their environment.

Emotional state can change how animals perceive the world, and pair-housed calves judged their chance of success higher than individually housed calves when presented with ambiguous cues in a judgement bias test (Bučková et al. 2019), indicating more positive affective states. Social housing with peers also reduced calves' responses to stressors and thus provided social buffering (i.e., reducing the negative effects of stressful experiences; Rault 2012) when calves were separated from the group (Færevik et al. 2006), subjected to blood sampling (Duve et al. 2012) and during the period of weaning from milk (De Paula Vieira et al. 2010; Bolt et al. 2017). Thus, social housing appears not only to provide opportunity for positive experiences but also mitigates effects of negative experience.

**Production**

Housing calves in pairs or small groups stimulated solid food intake in the pre-weaning period in a range of studies, including calves fed a low milk allowance (e.g., Hepola et al. 2006; Tapki 2007), calves fed a higher milk allowance (e.g., Costa et al. 2015; Whalin et al. 2018), and ad libitum-fed calves (e.g., De Paula Vieira et al. 2010; Overvest et al. 2018). However, Jensen et al. (2015) only found a positive effect of pair housing on concentrate intake when calves were fed a high milk allowance, compared to when they were fed a low milk allowance, and some studies failed to find any effect of social housing when offering milk ad libitum (Chua et al. 2002) and a restricted allowance of 6 l of milk per day (Jensen and Larsen 2014; Bolt et al. 2017). Calves consume very little solid food before 4 weeks of age and although early social housing may stimulate calves to eat concentrate and hay at an early age, any effect on intake may not be detectable until calves begin to eat a substantial amount. For instance, Miller-Cushon and DeVries (2016) and Mahendran et al. (2021) did not find any detectable differences in concentrate intake between individually and pair-housed calves until after 4 weeks of age. Thus, the period of measurement may help explain some of the differences between studies. Especially for calves fed high volumes of milk, low intakes of concentrate is a challenge to weaning off milk, but pair and group housing, and the resulting social facilitation of feeding, may ease the transition from milk to solid food. Pairs of calves with simultaneous access to concentrates ate more (Miller-Cushon et al. 2014), emphasising that calves have to be able to eat at the same time for social facilitation to stimulate increased solid food intake.

**Health**

The main argument against group housing calves is a risk of poor health, and some studies find calf health to be better in individual than group housing (Gulliksen et al.

2009; Karle et al. 2019). However, group size is an important determinant of group-housed calves' health. Calves in groups of more than eight calves had a higher risk of respiratory disease (Svensson et al. 2003; Svensson and Liberg 2006; Karle et al. 2019), and thus group size should be kept small for health reasons. Calves managed in stable groups (all-in-all-out) had a higher daily gain and a lower incidence of disease than calves housed in dynamic groups (continuous introduction; Pedersen et al. 2009), and the effect of group size on health may depend on age variation within the group and on management.

#### 10.4.1.2 Challenges of Social Housing (Cross-Sucking, Competition, Regrouping)

**Cross-Sucking**

Calves housed in groups may direct their sucking behaviour towards other calves' heads or bodies. This abnormal behaviour is termed cross-sucking and is most intense in the first 10–20 min after milk ingestion (Lidfors 1993). The risk of cross-sucking can be reduced by offering the milk via a teat, in teat buckets or automated teat-feeders. When the milk is offered in teat buckets, the calves spend more time ingesting the milk, they suck the teat after the milk is ingested, and they perform less cross-sucking after the ingestion of milk compared to when milk is offered in buckets or troughs (reviewed by De Passillé 2001; Jensen 2003; Fig. 10.2). It is important to leave the teat buckets with the calves for approx. 20 min after the milk is drunk to efficiently reduce cross-sucking (Jung and Lidfors 2001). The use of dry teats in combination with feeding milk in open buckets is also recommended to prevent cross-sucking (De Passillé and Caza 1997), but dry teats were more attractive when dipped in milk (Jung and Lidfors 2001) and are likely used more if placed directly above the milk bucket. For instance, dry teats placed away from the teat bucket in individual hutches did not affect the time calves spent sucking on pen fixtures (Pempek et al. 2017). Therefore, feeding the milk via a teat by using a teat bucket is preferable and more likely to prevent cross-sucking than offering dry teats when a manual milk-feeding system is used.

When computer-controlled milk feeders are used, the milk is also delivered via a teat, typically to one calf in the group at a time. To ensure that the calves' behavioural need to suck is met, the teat must be available for the calf for the duration of the sucking motivation (20 min; Lidfors 1993) and not be retracted as soon as the milk is drunk. Competition for access to the milk feeder can compromise calves' access to suck the teat after milk ingestion. For instance, with 24 calves per feeder, calves were disturbed by other calves for 50% of the time they spent in the feeder compared to 10% with 12 calves per feeder, which resulted in less time spent in both sucking milk and non-nutritive sucking after milk ingestion (Jensen 2004) and thus a higher risk of cross-sucking on other calves.

**Competition**

Even in small groups, calves compete for their milk (Jensen and Budde 2006). Among calves fed milk restrictively via teat buckets, placing a barrier between the teats separating calves' heads and shoulders prevented displacements and prevented

**Fig. 10.2** When calves are fed milk via a teat they spend more time ingesting the milk, they suck on the teat after milk ingestion and perform less abnormal cross-sucking. Top: calves fed milk in open buckets cross-sucking each other after milk ingestion. Bottom: calves fed via a teat (a teat-bar with six teats for six calves). Jensen and Budde (2006). Photos courtesy of Margit Bak Jensen and Marlene Budde

calves from drinking from each other's teat buckets (Jensen et al. 2008). When calves were offered milk via teats connected to a common trough (one teat per calf), barriers between teats reduced displacements. However, there was a significant variation in milk intake within the group, as drinking speed determined milk intake, and thus barriers did not ensure an even distribution of milk among the calves as intended (Nielsen et al. 2008a). Among ad libitum-fed calves, providing more than one teat per calf reduced displacements from teats and can thus reduce behavioural competition (von Keyserlingk et al. 2004). When computer-controlled milk feeding is used, calves cannot steal other calves' milk allowance, but there may be competition for feeder access. When housed in groups of 24 with one feeder, calves were subject to more displacements from the feeder than when housed in groups of 12 with one feeder (Jensen 2004). The effect of group size on competition likely impacts the youngest calves more than older calves. Among calves fed via a computer-controlled feeder, calves introduced at 6 days of age were less active and required more assistance to learn to use the milk feeder than calves introduced at 14 days of age (Rasmussen et al. 2006; Jensen 2007). In accordance with this, Fujiwara et al. (2014) found that the younger the calf was at introduction to the group pen (between 5 and 14 days old), the longer it took before it had its first milk meal. Comparing calves introduced at 1 and 5 days of age, Medrano-Galarza et al. (2018) also found that younger calves took longer to learn to use the milk feeder than older calves. The larger the group size, the more likely it is that the older calves cope better with the transition to group housing and automatic feeding. It is not recommended that young calves are introduced into larger groups of older calves.

**Group Composition and Regrouping**

Calves are often reared in groups of same-age calves. Researchers have found that calves reared with an older companion had enhanced food intake (De Paula Vieira et al. 2012), likely due to learning from older animals. However, when housed under commercial stocking densities of space and access to feed, social competition affects younger individuals more than older ones, particularly for access to feeders and feed. For instance, in age-heterogeneous groups of weaned calves, the younger calves gained less weight than similar-aged calves in age-homogeneous groups (Færevik et al. 2010), and therefore it may be advantageous to maintain animals in a homogeneous group, especially if there is competition for resources.

Calves form preferential social relationships with calves of the same age (Raussi et al. 2010), and being housed in a stable group enables calves to stay with familiar and preferred partners, which provide social support (reviewed by Rault 2012). A way to maintain familiarity in groups of calves is to pair calves when few days old, create small groups from these pairs during the milk-feeding period and to combine these small groups into larger groups after weaning. Avoiding competition for resources, especially when group composition changes, may be a way to reduce the negative effects of regrouping and thereby improve animal welfare.

#### 10.4.1.3 Milk-Feeding Management

Dairy calves depend on milk to cover their nutritional needs until their rumen is sufficiently developed to digest enough solids. Feeding calves too little milk results in hunger and poor growth, poor feed efficiency, and increased incidence of disease (Khan et al. 2011; Costa et al. 2019). The daily ad libitum intake of Holstein calves corresponds to 10–12 l of whole milk and approx. 20% of calf body weight (Jasper and Weary 2002; von Keyserlingk et al. 2004; Sweeney et al. 2010). Calves restricted to half of this amount (approx. 5 l/day of replacer or whole milk) fed via a computer-controlled milk feeder had more than twice as many unrewarded visits to the feeder than calves offered 8 l/day or more (Jensen and Holm 2003; Jensen 2004; Nielsen et al. 2008b; De Paula Vieira et al. 2008). A high number of unrewarded visits (e.g., visits after the calf has consumed its allowance) is due to the calves attempting to get additional milk and illustrates that they are hungry (Jensen and Holm 2003; Jensen 2004; De Paula Vieira et al. 2008). Another sign of hunger is more frequent and louder vocalisation of low-fed calves compared to calves offered a milk allowance corresponding to their desired intake (Thomas et al. 2001). Prevention of hunger is a widely recognised welfare criteria (Welfare Quality 2009), and continued focus on fulfilling calves' nutritional needs throughout to weaning is warranted.

Restricted milk feeding (at approx. 10% of body weight) stimulates early concentrate intake as calves attempt to compensate for the lack of energy from milk. Unfortunately, calves consume little concentrate during the first 4 weeks of life (Khan et al. 2011), and restricted milk feeding results in reduced daily gain compared to calves offered higher milk allowances. On the other hand, a higher daily gain was found in calves fed milk replacer at 8 l/day vs. 5 l/day (Jensen 2006), in calves fed 12 l. vs. 6 l/day whole milk (Rosenberger et al. 2017), in calves fed an allowance corresponding to 20% vs. 10% of body weight (Khan et al. 2007), and in calves fed ad libitum vs an allowance corresponding to 10% of body weight (Jasper and Weary 2002). Indicating a more positive emotional state (Boissy et al. 2007) when fed higher milk allowances, calves fed high milk allowances performed more locomotor play behaviour than low-fed calves (Krachun et al. 2010). Long-term effects of high milk allowances on production and health are indicated by a younger age at first breeding (Raeth-Knight et al. 2009) and higher milk yield as adult cows (Soberon and Van Amburgh 2013).

**Weaning Off High Milk Allowances**

One problem with feeding high milk allowances is to get the calves weaned off milk because their concentrate intake is low. However, concentrate intake can be stimulated in high-fed calves through social facilitation when they are housed in pairs or small groups (see Sect. 10.4.1.1) and by gradually weaning them over a longer period (Jensen 2006; Khan et al. 2007; Sweeney et al. 2010). Current recommendations are that milk allowance should be high early in life, and weaning should not be initiated before the calves are able to eat enough concentrate to substitute the lost milk (Sweeney et al. 2010). One way of managing this is to offer high milk (20% body weight) during the first 3–4 weeks, followed by low milk (10% body weight) until gradual weaning at 7–8 weeks of age (Khan et al. 2007). Another approach is to adjust initiation and completion of weaning to the individual calf's' intake of solid feed (De Passillé and Rushen 2012).

**Final Remarks**

Social housing has positive effects on calf welfare and production, but requires suitable group management, including housing in pairs or small groups, as well as milk-feeding practices to reduce cross-sucking and competition. Feeding higher levels of milk results in improved performance but requires that the transition to solid feed is facilitated by gradual weaning methods. Several positive effects of social housing on calves' learning ability and social skills have been found, but future research should investigate the effects of early social housing on long-term effects on the behaviour and welfare of cattle

## 10.5 Housing and Management of Growing Animals

The welfare of growing cattle has reviewed less research interest, but the main concerns for animal welfare have centred around low space allowances, unsuitable lying surfaces, and aggression.

### 10.5.1 Effects of Housing

In Europe and North America, most dairy cows are housed in free stall barns (Eurostat; Dairy 2014). However, there is little data on how growing animals, including replacement heifers (post-weaning to pre-calving) and growing animals reared for meat (bulls and heifers post-weaning to pre-slaughter), are housed. When cows are housed in free stalls, it is recommended to house replacement heifers in free stall barns, at least for some time before calving to facilitate the transition to the dairy herd. Other typical housing types for young stock include group pens which are deep bedded with straw and pens with fully or partially slatted concrete floors. In North America, animals reared for meat may also be kept in feedlots (Endres and Schwartzkopf-Genswein 2018).

Studies that compared various types of housing (often with different space allowances) found that growing animals housed in group pens with deep-bedded straw packs had fewer problems changing position from standing to lying compared to animals housed in group pens with harder lying surfaces. For instance, in on-farm studies including 450–600 kg young bulls (various dairy-, beef-, and cross-breeds) the number of lying bouts increased with increasing softness of the floor in group pens, while interruptions of the lying down and getting up sequence were highest among young bulls on concrete slatted floors and were not reduced by overlaying the concrete slats with rubber (Gygax et al. 2007; Absmanner et al. 2009). In these studies, the daily lying time was not affected by the softness of the flooring, but bulls lay down for longer periods with fewer overall lying bouts on harder floors than they did on softer floors. The longer lying bouts may reflect a reluctance to perform the standing to lying transition, which likely involves concussion of the knees on the hard concrete flooring. Longer lying periods on a hard surface may explain a higher prevalence of joint injuries and swellings among bulls on the concrete slatted floors compared to straw bedding, while the prevalence among bulls kept in pens with rubber-covered slats was intermediate (Schulze Westerath et al. 2007). Among replacement heifers, housing in slatted floor pens with more than seven animals per pen from 3 to 7 months of age was associated with a 1.7-fold increase in risk of culling in first lactation compared to housing in bedded group pens (Hultgren and Svensson 2009). However, animals generally had more space in straw-bedded pens than in fully slatted concrete floor pens, and it is unclear whether the above differences are due to the hard floor in addition to the limited space.

In many European countries, the housing of growing animals in pens with fully slatted floors became widespread in the 1980s. The space allowance was typically low, and research showed that a reduction in space allowance from approx. 4.0 to

1.5 m$^2$/animal reduced lying time and increased aggression and physiological stress reactions in young stock of both sexes in the weight interval 250–500 kg (reviewed by Ingvartsen and Andersen 1992). This type of housing also adversely affected the growth of the animals, and a meta-analysis showed that reduced space allowance from 4.7 to 1.5 m$^2$/animal changed feed intake, daily gain, and feed conversion ratio to 92, 81, and 115%, respectively, in bulls and steers weighing 250–500 kg (Ingvartsen and Andersen 1992). These pens were also used for replacement heifers in some countries, and similar effects of space allowance were found on lying and social behaviour (Hindhede et al. 1996). In addition, an epidemiological study showed that housing replacement heifers in pens with fully slatted floors from day 90 to conception was associated with a lower milk production in first lactation compared to being housed in deep-bedded pens (Svensson and Hultgren 2008).

Several attempts have been made to make the concrete slatted floors more suitable as a lying surface by covering them with rubber. Researchers showed that bulls in pens with rubber-covered slats had more social interactions (Brscic et al. 2015) and fewer interruptions of lying down sequences compared to bulls is corresponding pens with concrete slats (Brscic et al. 2015; Graunke et al. 2011). In addition, hoof health was generally better among bulls in pens with rubber-covered slats (Graunke et al. 2011). However, a preference experiment showed that young bulls preferred a straw-bedded floor over the rubber-covered floor. They ranked flooring in a designated resting area from most to least preferred as follows: a solid floor with straw bedding, a solid floor with sawdust bedding, a solid floor covered with a rubber mat, a slatted floor with rubber coverage, and a concrete slatted floor (Lowe et al. 2001). The preference of young bulls for a slatted floor with rubber coverage over a concrete slatted floor was confirmed by Platz et al. (2007), who also found that bulls on slats covered with rubber had more lying bouts than bulls in corresponding concrete slatted floor pens (3.0 m$^2$/animal). Thus, overlaying concrete slats with rubber in fully slatted floor pens reduces some of the problems growing animals have lying down and getting up due to better traction and some softness; however, it does not reduce the problems to the same extent as straw bedding.

Modifying slatted floor pens by increasing the pen size and space allowance and establishing a deep straw bed in half of the pen to provide a suitable lying area is a way to make these pens more suitable. For instance, in pens with 3.0 m$^2$/heifer (300–400 kg), providing a 1.5 m$^2$/animal area with deep straw resulted in more lying periods compared to a fully slatted floor heifers clearly preferred the straw-bedded area for lying, but 1.5 m$^2$/animal was not sufficient for all animals to lie down at the same time (Hindhede et al. 1996). However, increasing the straw-bedded area from 1.8 to 2.7 and 3.6 m$^2$/animal resulted in a higher level of synchronised lying (Mogensen et al. 1997) and a lower occurrence of aggression (Nielsen et al. 1997). The risk of infectious hoof diseases (e.g., heel horn erosion) is higher in wet bedding, and increasing the bedded lying area from 1.8 m$^2$/animal to 2.7 and 3.6 m$^2$/animal reduced the prevalence of heel horn erosion (Mogensen et al. 1997), likely because the straw bedding was dryer at the higher space allowance. Finally, providing a straw-bedded lying area may also improve production, as, among heifers at a space allowance of 6 m$^2$/animal, heifers housed in straw-bedded pens had a greater daily weight gain and feed conversion ratio than heifers housed in fully slatted floor pens on concrete slats (Keane et al. 2017).

In dairy cows, it is well documented that animals lie down for longer in free stalls that are deep bedded with straw, sawdust, or sand than in free stalls with mattresses with minimal bedding (e.g., Tucker et al. 2003; Calamari et al. 2009); moreover, there are fewer hock injuries among cows in free stalls with a deep bedding of straw, sawdust, or sand than in enclosures with mattresses with minimal bedding (Weary and Taszkun 2000; Wechsler et al. 2000; Vokey et al. 2001; Livesey et al. 2002). However, there is limited research on this aspect in growing animals. Older studies found that placing rubber mats in free stalls with concrete floor increased heifers' use of stalls (O'Connel et al. 1993), increased their lying time, and reduced the prevalence of sole haemorrhage after calving (Leonard et al. 1994). In addition, a four-fold increase (amounts not given) of the amount of straw bedding added to concrete-surfaced free stalls improved hoof health among heifers (Colam-Ainsworth et al. 1989), while adding minimal bedding to rubber mats in free stalls did not reduce injuries among young bulls (Schulze Westerath et al. 2007). Increased lying time and reduced prevalence of injuries is likely achieved by adding a substantial amount of bedding to rubber mats or mattresses for growing animals, but to the best of our knowledge, there are no studies on the effect of various amounts of bedding in free stalls for growing animals. Regarding the alleys in free stall barns, the effect of a rubber cover of a concrete floor has not been investigated in growing animals. However, in dairy cows, rubber on the floor results in a more natural gait, e.g., longer steps (Telezhenko et al. 2007), indicating that the animal walks with more ease on the rubber floor. Heifers (approx. 200 kg) reduced their lying time considerably during the first day after they were moved from deep bedding to a pen with free stalls, and approx. 20% of the time lying down was spent lying in the alley (von Keyserlingk et al. 2011). This rejection of free stalls decreased during subsequent days, and housing pregnant replacement heifers in a barn with free stalls reduces rejection of free stalls after calving (O'Connel et al. 1993). Rejection of free stalls is a risk of culling, and research on how to ease transitions between housing is encouraged.

## 10.6 Advances and Challenges in Animal Welfare

### 10.6.1 Advances

Traditionally, research on parturient dairy cows has focused on identifying and treating common diseases that commonly occur after calving. Over the last 10 years, researchers have developed an understanding of maternal dairy cow behaviour before giving birth. More recently, researchers have shifted their focus to creating post-calving environments where the dam and her calf can be housed together. These advancements have significantly improved our understanding of dairy cow welfare, specifically around the natural behaviour of cows and their calves. Dairy producers are recommended to provide sufficient space and resources in maternity pens to allow cows the opportunity to seclude themselves if they choose.

Social housing of calves that are separated from the dam at birth has also received considerable research interest during the past 20 years. This research has shown positive effects of social housing on calf welfare and production and has provided solutions to the challenges of competition over milk when calves are socially housed. The move towards group housing of dairy calves in the milk-feeding period is largely supported by these advances. Furthermore, advancements in research into the affective experiences and cognitive abilities of milk-fed calves have significantly improved our understanding of the importance of the social environment for the dairy calf.

### 10.6.2 Challenges

Despite the advancements in knowledge about the natural maternal behaviour of dairy cows, there remains a key gap in our understanding of dam and calf affective states. Measuring affect in animals has traditionally been a challenge; however, new research tools have been developed to assess animal emotions. For example, cognitive bias testing is one method that has been used to measure emotional valence (i.e., positive and negative affect) in animals. Cognitive bias testing has already been used to assess the effect of maternal separation from calves; researchers found evidence that separation may lead to negative judgement bias in calves. Research is also still needed to understand the negative impact of labour and dystocia on pain post-calving as well as practical methods for mitigating pain. More research on the impact of various housing practices before and after giving birth on affective states in dams and calves is warranted.

Research is also still needed to create practical recommendations for providing cow-calf contact after birth. Social science research can help guide this line of research, as these new systems should be aligned with public values about animal welfare (Ventura et al. 2013; Beaver et al. 2020) and be in compliance with the UN Sustainable Development Goal of 'responsible consumption' (Keeling et al. 2019). Additionally, providing farm animals with opportunities for positive experiences is an area of increasing interest (Mellor 2016), and research is needed into the identification and validation of behavioural indicators of how animals experience various situations (e.g., de Oliveira and Keeling 2018) as well as into how the balance of positive and negative experiences affects how the animals feel overall (Webb et al. 2019). Research in the natural sciences can then help create practical systems that allow for cow-calf contact, including housing, management, and weaning practices that reduce animal distress and provide opportunity for positive experiences.

The effect of housing and management on welfare of growing animals has received little research focus. Growing animals are increasingly housed in group pens with free stalls, yet a majority of the current research has focused on pens with concrete slatted floors with or without a designated lying area. There is a lack of research on how the lying surface and the amount of bedding affect behaviour and welfare in replacement heifers and fattening animals of different weight classes in free stall housing. Furthermore, there is a lack of research on how competition for feeding and lying spaces is associated with resource availability in growing animals. Research on dairy cows suggests that they prefer to compete and share with certain individuals over others. Sharing of resources may be facilitated by early social experiences. Therefore, the effect of group stability and the potential beneficial effect of letting growing animals choose their own social company should be explored in future research. Cattle are often regrouped many times from birth to calving, or slaughter, which leads to aggression. Allowing the animals to maintain social bonds when new groups are formed may buffer the negative effects of the regrouping, and other management practices, but research is needed to understand the formation of preferential social relationships and their impact on cattle welfare.

## References

Absmanner E, Rouha-Mulleder C, Scharl T, Leisch F, Troxler J (2009) Effects of different housing systems on the behaviour of beef bulls—an on-farm assessment on Austrian farms. Appl Anim Behav Sci 118:12–19

Anonymous (2009) Code of Practice for the care and handling of farm animals - Dairy Cattle. Dairy Farmers of Canada and the National Farm Animal Care Council. ISBN 979-0-9812240-0-8 (book)

Anonymous (2017) Law nr 520, 26/05/2010; Ministerial order number 58 of 11/1/2017. https://www.retsinformation.dk

Anonymous (2018) Commission Decision (EU) 2018/1222 of 5 September 2018 on the proposed citizens' initiative entitled 'End the Cage Age' (notified under document C(2018) 5829). https://eur-lex.europa.eu/legal-content/EN/TXT/HTML/?uri=CELEX:32018D1222&from=EN

Barbknecht AE, Fairbanks WS, Rogerson JD, Maichak EJ, Scurlock BM, Meadows LL (2011) Elk parturition site selection at local and landscape scales. J Wildl Manage 75:646–654

Barrier AC, Ruelle E, Haskell MJ, Dwyer CM (2012) Effect of a difficult calving on the vigour of the calf, the onset of maternal behaviour, and some behavioural indicators of pain in the dam. Prev Vet Med 103:248–256. https://doi.org/10.1016/j.prevetmed.2011.09.001

Barrier AC, Haskell MJ, Birch S, Bagnall A, Bell DJ, Dickinson J, Macrae AI, Dwyer CM (2013) The impact of dystocia on dairy calf health, welfare, performance and survival. Vet J 195:86–90

Barth K (2020) Effects of suckling on milk yield and milk composition of dairy cows in cow–calf contact systems. J Dairy Res 87(S1):133–137

Beaver A, Meagher RK, von Keyserlingk MA, Weary DM (2019) Invited review: A systematic review of the effects of early separation on dairy cow and calf health. J Dairy Sci 102:5784–5810

Beaver A, Proudfoot K, von Keyserlingk MAG (2020) Symposium review: considerations for the future of dairy cattle housing: an animal welfare perspective. J Dairy Sci 103:5746–5758

Boissy A, Manteuffel G, Jensen MB, Moe RO, Spruijt B, Keeling L, Winckler C, Forkman B, Dimitrov I, Langbein J, Bakken M, Viessier I, Aubert A (2007) Assessment of positive emotions in animals to improve their welfare. Physiol Behav 92:375–397

Bolt SL, Boyland NK, Mlynski DT, James R, Croft DP (2017) Pair housing of dairy calves and age at pairing: effects on weaning stress, health, production and social networks. PLoS One 12(1):e0166926. https://doi.org/10.1371/journal.pone.0166926

Borchers MR, Chang YM, Proudfoot KL, Wadsworth BA, Stone AE, Bewley JM (2017) Machine-learning-based calving prediction from activity, lying, and ruminating behaviors in dairy cattle. J Dairy Sci 100:5664–5674

Brscic M, Ricci R, Prevedello P, Lonardi C, De Nardi R, Contiero B, Gottardo F, Cozzi G (2015) Synthetic rubber surface as an alternative to concrete to improve welfare and performance of finishing beef cattle reared on fully slatted flooring. Animal 9:1386–1392

Bučková K, Špinka M, Hintze S (2019) Pair housing makes calves more optimistic. Scientific Reports 9:20246. https://doi.org/10.1038/s41598-019-56798-w

Budzynska M, Weary DM (2008) Weaning distress in dairy calves: Effects of alternative weaning procedures. Appl Anim Behav Sci 112:33–39

Byskov K (2012) Consequences for the dairy producer—meat production without subsidies (In Danish; 'Konsekvens for mælkeproducenten—Kødproduktion uden støtte'). 21 pp. Downloaded 30-09-2020 from http://projektfinansiering.vfl.dk/Kvaegafgiftsfonden/2012/Koedproduktionudenstoette/Sider/Koedproduktion-uden-stoette-PSP-2160_67b8656c-06c0-4c3d-a2fd-7fe4adcf47e1.pdf

Calamari L, Calegari F, Stefanini L (2009) Effect of different free stall surfaces on behavioural, productive and metabolic parameters in dairy cows. Appl Anim Behav Sci 120:9–17

Campler M, Munksgaard L, Jensen MB, Weary DM, von Keyserlingk MAG (2014) Short communication: Flooring preferences of dairy cows at calving. J Dairy Sci 97:892–896. https://doi.org/10.3168/jds.2013-7253

Campler MR, Munksgaard L, Jensen MB (2019) The effect of transition cow housing on lying and feeding behavior in Holstein dairy cows. J Dairy Sci 102:7398–7407

Cheong SH, Nydam DV, Galvão KN, Crosier BM, Gilbert RO (2011) Cow-level and herd-level risk factors for subclinical endometritis in lactating Holstein cows. J Dairy Sci 94:762–770. https://doi.org/10.3168/jds.2010-3439

Chua B, Coenen E, van Delen J, Weary DM (2002) Effects of pair versus individual housing on the behavior and performance of dairy calves. J Dairy Sci 85:360–364

Colam-Ainsworth P, Lunn GA, Thomas RC, Eddy RG (1989) Behaviour of cows in cubicles and its possible relationship with laminitis in replacement dairy heifers. Vet Rec 125:573–575

Cook NB, Nordlund KV (2004) Behavioral needs of the transition cow and considerations for special needs facility design. Vet Clin North Am Food Anim Pract 20:495–520. https://doi.org/10.1016/j.cvfa.2004.06.011

Costa JHC, Daros RR, von Keyserlingk MAG, Weary DM (2014) Complex social housing reduces food neophobia in dairy calves. J Dairy Sci 97:7804–7810

Costa JHC, Meagher RK, von Keyserlingk MAG, Weary DM (2015) Early pair housing increases solid feed intake and weight gains in dairy calves. J Dairy Sci 98:6381–6386

Costa JHC, Cantor MC, Adderley NA, Neave HW (2019) Key animal welfare issues in commercially raised dairy calves: social environment, nutrition, and painful procedures. Can J Anim Sci 99:649–660. https://doi-org.ez.statsbiblioteket.dk/10.1139/cjas-2019-003

Creutzinger KC, Dann HM, Moraes LE, Krawczel PD, Proudfoot KL (2021a) Effects of prepartum stocking density and a blind on physiological biomarkers, health, and hygiene of transition Holstein dairy cows. J Dairy Sci 104:886–898

Creutzinger KC, Dann HM, Krawczel PD, Moraes LE, Pairis-Garcia MD, Proudfoot KL (2021b) The effect of stocking density and a blind on the behavior of Holstein dairy cows in group maternity pens. Part II: Labor length, lying behavior, and social behavior. J Dairy Sci 104(6):7122–7134

Dairy (2014) Dairy Cattle Management Practices in the United States, 2014. Accessed Jul. 12, 2019. https://www.aphis.usda.gov/animalhealth/nahms/dairy/downloads/dairy14/Dairy14drPartI1.pdf

Daros RR, Costa JH, von Keyserlingk MA, Hötzel MJ, Weary DM (2014) Separation from the dam causes negative judgement bias in dairy calves. PLoS One 9(5):e98429. https://doi.org/10.1371/journal.pone.0098429

de Oliveira D, Keeling LJ (2018) Routine activities and emotion in the life of dairy cows: Integrating body language into an affective state framework. PLoS ONE 13(5):e0195674. https://doi.org/10.1371/journal.pone.0195674

De Passillé AM (2001) Sucking motivation and related problems in calves. Appl Anim Behav Sci 72:175–187

De Passillé AM, Caza N (1997) Cross-sucking by calves occurs after meals and is reduced when calves suck a dry teat. J Dairy Sci 80(Suppl 1):229
De Passillé AM, Rushen J (2012) Adjusting the weaning age of calves fed by automated feeders according to individual intakes of solid feed. J Dairy Sci 95:5292–5298
De Passillé AM, Marnet P-G, Lapierre H, Rushen J (2008) Effects of twice-daily nursing on milk ejection and milk yield during nursing and milking in dairy cows. J Dairy Sci 91:1416–1422
De Paula Vieira A, Guesdon V, De Passillé AM, von Keyserlingk MAG, Weary DM (2008) Behavioural indicators of hunger in dairy calves. Appl Anim Behav Sci 109:180–189
De Paula Vieira A, von Keyserlingk MAG, Weary DM (2010) Effects of pair versus single housing on performance and behaviour of dairy calves before and after weaning from milk. J Dairy Sci 93:3079–3085
De Paula Vieira A, von Keyserlingk MAG, Weary DM (2012) Presence of an older weaned companion influences feeding behaviour and improves performance of dairy calves before and after weaning from milk. J Dairy Sci 95:3218–3224
Donat K, Schmidt M, Köhler H, Sauter-Louis C (2016) Management of the calving pen is a crucial factor for paratuberculosis control in large dairy herds. J Dairy Sci 99:3744–3752
Duve LR, Jensen MB (2011) The level of social contact affects social behaviour in pre-weaned dairy calves. Appl Anim Behav Sci 135:34–43
Duve LR, Jensen MB (2012) Social behavior of young dairy calves housed with limited or full social contact with a peer. J Dairy Sci 95:1–10
Duve LR, Weary DM, Halekoh U, Jensen MB (2012) The effects of social contact and milk allowance on responses to handling, play, and social behavior in young dairy calves. J Dairy Sci 95:6571–6581
Edwards SA (1983) The behaviour of dairy cows and their newborn calves in individual or group housing. Appl Anim Ethol 10:191–198
Elbers ARW, Miltenburg JD, De Lange D, Crauwels APP, Barkema HW, Schukken YH (1998) Risk factors for clinical mastitis in a random sample of dairy herds from the southern part of the Netherlands. J Dairy Sci 81:420–426. https://doi.org/10.3168/jds.S0022-0302(98)75592-4
Endres MI, Schwartzkopf-Genswein K (2018) Overview of cattle production systems. In: Tucker C (ed) Advances in cattle welfare. Woodhead, Duxford, pp 1–26. 250 pp
Færevik G, Jensen MB, Bøe KE (2006) Dairy calves social preferences and the significance of a companion during separation from the group. Appl Anim Behav Sci 99:205–221
Færevik G, Jensen MB, Bøe KE (2010) The effect of group composition and age on social behavior and competition in groups of weaned dairy calves. J Dairy Sci 93:4274–4279
Finger A, Patison KP, Heath BM, Swain DL (2014) Changes in the group associations of free-ranging beef cows at calving. Anim Prod Sci 54:270–276
Flörcke C, Grandin T (2014) Separation behaviour for parturition of red Angus beef cows. Open J Anim Sci 4:43–50
Flower FC, Weary DM (2001) Effects of early separation on the dairy cow and calf: 2. Separation at 1 day and 2 weeks after birth. Appl Anim Behav Sci 70:275–284
Frank NA, Kaneene JB (1993) Management risk factors associated with calf diarrhea in Michigan dairy herds. J Dairy Sci 76:1313–1323. https://doi.org/10.3168/jds.S0022-0302(93)77462-7
Fraser D, Weary DM, Pajor EA, Milligan BN (1997) A scientific conception of animal welfare that reflects ethical concerns. Anim Welfare 6:187–205
Fröberg S, Lidfors L (2009) Behaviour of dairy calves suckling the dam in a barn with automatic milking or being fed milk substitute from an automatic feeder in a group pen. Appl Anim Behav Sci 117:150–158
Fröberg S, Gratte E, Svennersten-Sjaunja K, Olsson I, Berg C, Orihuela A, Galina CS, García B, Lidfors L (2008) Effect of suckling ('restricted suckling') on dairy cows' udder health and milk let-down and their calves' weight gain, feed intake and behaviour. Appl Anim Behav Sci 113:1–14
Fujiwara M, Rushen J, De Passillé AM (2014) Dairy calves' adaptation to group housing with automated feeders. Appl Anim Behav Sci 158:1–7

Gaillard C, Meagher RK, von Keyserlingk MAG, Weary DM (2014) Social housing improves dairy calves' performance in two cognitive tests. PLoS One 9(2):e90205. https://doi.org/10.1371/journal.pone.0090205

González-Sedano M, Marın-Mejıa B, Maranto MI, de Magalhaes-Labarthe AL, Alonso-Dıaz MA (2010) Effect of residual calf suckling on clinical and sub-clinical infections of mastitis in dual-purpose cows: Epidemiological measurements. Res Vet Sci 89:362–366

Graunke KL, Telezhenko E, Hessle A, Bergsten C, Loberg JM (2011) Does rubber flooring improve welfare and production in growing bulls in fully slatted floor pens? Anim Welfare 20:173–183

Gulliksen SM, Lie KI, Løkken T, Østerås O (2009) Calf mortality in Norwegian dairy herds. J Dairy Sci 92:2782–2795

Gygax L, Mayer C, Schulze Westerath H, Friedli K, Wechler B (2007) On-farm assessment of the lying behaviour of finishing bulls kept in housing systems with different floor qualities. Anim Welfare 16:205–208

Hepola H, Hänninen L, Pursiainen P, Tuure VM, Syrjälä-Qvist L, Pyykkönen M, Saloniemi H (2006) Feed intake and oral behavior of dairy calves housed individually or in groups in warm or could buildings. Livest Sci 105:94–104

Hindhede J, Sørensen JT, Jensen MB, Krohn CC (1996) Effect of space allowance, access to bedding and flock size in slatted floor systems on production and health of heifers. Acta Agric Scand A Anim Sci 46:46–53

Hötzel MJ, Longo C, Balcao LF, Cardoso CS, Costa JHC (2014) A survey of management practices that influence performance and welfare of dairy calves reared in Southern Brazil. PLoS One 9(12):e114995. https://doi.org/10.1371/journal.pone.0114995

Hötzel MJ, Cardoso CS, Roslindo A, von Keyserlingk MAG (2017) Citizens' views on the practices of zero-grazing and cow-calf separation in the dairy industry: does providing information increase acceptability? J Dairy Sci 100:4150–4160

Hultgren J, Svensson C (2009) Heifer rearing conditions affect length of productive life in Swedish dairy cows. Prev Vet Med 89:255–264

Huxley JN, Whay HR (2006) Current attitudes of cattle practitioners to pain and the use of analgesics in cattle. Vet Rec 159:662–668

Huzzey JM, von Keyserlingk MAG, Weary DM (2005) Changes in feeding, drinking, and standing behavior of dairy cows during the transition period. J Dairy Sci 88:2454–2461. https://doi.org/10.3168/jds.S0022-0302(05)72923-4

Illman G, Spinka M (1993) Maternal behaviour of dairy heifers and sucking of their newborn calves in group housing. Appl Anim Behav Sci 36:91–98

Ingvartsen KL, Andersen HR (1992) Space allowance and type of housing for growing cattle: a review of performance and possible relations to neuroendocrine function. Acta Agric Scand Sect A Anim Sci 43:65–80

Ingvartsen KL (2006) Feeding- and management-related diseases in the transition cow: physiological adaptations around calving and strategies to reduce feeding-related diseases. Anim Feed Sci Technol 126:175–213

Jackson P (2004) Normal birth. In: Jackson PGG (ed) Handbook of veterinary obstetrics, 2nd edn. Saunders, Philadelphia, PA, pp 1–12

Jasper J, Weary D (2002) Effects of ad libitum milk intake on dairy calves. J Dairy Sci 85:3054–3058

Jensen MB (2003) The effects of feeding method, milk allowance and social factors on milk feeding behaviour and cross-sucking in group housed dairy calves. Appl Anim Behav Sci 80:191–206

Jensen MB (2004) Computer controlled milk feeding of dairy calves: The effects of number of calves per feeder and number of milk portions on use of feeder and social behavior. J Dairy Sci 87:3428–3438

Jensen MB (2006) Computer-controlled milk feeding of group-housed calves, the effect of milk allowance and weaning type. J Dairy Sci 89:201–206

Jensen MB (2007) Age at introduction to the group affects dairy calves' use of a computer-controlled milk feeder. Appl Anim Behav Sci 107:22–31

Jensen MB (2012) Behaviour around the time of calving in dairy cows. Appl Anim Behav Sci 139:195–202. https://doi.org/10.1016/j.applanim.2012.04.002

Jensen MB (2018) The role of social behavior in cattle welfare. In: Tucker C (ed) Advances in cattle welfare. Woodhead, Duxford, pp 123–155

Jensen MB, Budde M (2006) The effect of milk feeding method and group size on feeding behavior and cross-sucking in group-housed dairy calves. J Dairy Sci 89:4778–4783

Jensen MB, Holm L (2003) The effect of milk flow rate and milk allowance on feeding related behaviour in dairy calves fed by computer controlled milk feeders. Appl Anim Behav Sci 82:87–100

Jensen MB, Larsen LE (2014) Effects of level of social contact on dairy calf behavior and health. J Dairy Sci 97:5035–5044

Jensen MB, Proudfoot KL (2017) Effect of group size and health status on behavior and feed intake of multiparous dairy cows in early lactation. J Dairy Sci 100:9759–9768

Jensen MB, Rørvang MV (2018) The degree of visual cover and location of birth fluids affect dairy cows' choice of calving site. J Dairy Sci 101:9483–9492. https://doi.org/10.3168/jds.2018-14724

Jensen MB, Tolstrup RB (2021) A survey on management and housing of peri-parturient dairy cows and their calves. Animal 15:100388

Jensen MB, Vestergaard KS, Krohn CC, Munksgaard L (1997) Effect of single versus group housing and space allowance on responses of calves during open-field test. Appl Anim Behav Sci 54:109–121

Jensen MB, De Passillé AM, von Keyserlingk MAG, Rushen J (2008) A barrier can reduce competition over teats in pair-housed milk-fed calves. J Dairy Sci 91:1607–1613

Jensen MB, Duve LR, Weary DM (2015) Pair housing and enhanced milk allowance increase play behavior and improve performance in dairy calves. J Dairy Sci 98:2568–2575

Jensen MB, Herskin MS, Rørvang MV (2019) Secluded maternity areas for parturient dairy cows offer protection from herd members. J Dairy Sci 102:5492–5500

Johnsen JF, Ellingsen K, Grøndahl AM, Bøe KE, Lidfors L, Mejdell CM (2015) The effect of physical contact between dairy cows and calves during separation on their post-separation behavioural response. Appl Anim Behav Sci 166:11–19

Johnsen JF, Zipp KA, Kälber T, De Passillé AM, Knierim U, Barth K, Mejdell CM (2016) Is rearing calves with the dam a feasible option for dairy farms?—Current and future research. Appl Anim Behav Sci 181:1–11

Johnsen JF, Mejdell CM, Beaver A, De Passillé AM, Rushen J, Weary DM (2018) Behavioural responses to cow-calf separation: the effect of nutritional dependence. Appl Anim Behav Sci 201:1–6

Jung J, Lidfors L (2001) Effects of amount of milk, milk flow and access to a rubber teal on cross-sucking and non-nutritive sucking in dairy calves. Appl Anim Behav Sci 72:201–213

Karle BM, Maier GU, Love WJ, Dubrovsky SA, Williams DR, Anderson RJ, Van Eenennaam AL, Lehenbauer TW, Aly SS (2019) Regional management practices and prevalence of bovine respiratory disease in California's preweaned dairy calves. J Dairy Sci 102:7583–7596

Keane MP, McGee M, O'Riordan EG, Kelly AK, Earley B (2017) Effect of space allowance and floor type on performance, welfare and physiological measurements of finishing beef heifers. Animal 11:2285–2294

Keeling L, Tunón H, Olmos Antillón G, Berg C, Jones M, Stuardo L, Swanson J, Wallenbeck A, Winckler C, Blokhuis H (2019) Animal welfare and the united nations sustainable development goals. Front Vet Sci 6:336. https://doi.org/10.3389/fvets.2019.00336

Khan MA, Lee HJ, Lee WS, Kim HS, Ki KS, Hur TY, Suh GH, Kang SJ, Choi YJ (2007) Structural growth, rumen development and metabolic and immune responses of Holstein male calves fed milk through step-down and conventional methods. J Dairy Sci 90:3376–3387

Khan MA, Weary DM, von Keyserlingk MAG (2011) Invited review: Effects of milk ration on solid feed intake, weaning and performance in dairy heifers. J Dairy Sci 94:1071–1081

Kiley-Worthington M, de la Plain S (1983) The behaviour of beef suckler cattle. Tierhaltung, 14. Birkhäuser, Basel

Kisac P, Brouček J, Uhrinčať M, Hanus A (2011) Effect of weaning calves from mother at different ages on their growth and milk yield of mothers. Czech J Anim Sci 56:261–268

Kovács L, Kézér FL, Szenci O (2016) Effect of calving process on the outcomes of delivery and postpartum health of dairy cows with unassisted and assisted calvings. J Dairy Sci 99:7568–7573

Kovács L, Kézér FL, Ruff F, Szenci O (2017) Rumination time and reticuloruminal temperature as possible predictors of dystocia in dairy cows. J Dairy Sci 100:1568–1579

Krachun C, Rushen J, De Passillé AM (2010) Play behaviour in dairy calves is reduced by weaning and by a low energy intake. Appl Anim Behav Sci 122:71–76

Kung L, Demarco S, Siebenson LN, Joyner E, Haenlein GFW, Morris RM (1997) An evaluation of two management systems for rearing calves fed milk replacer. J Dairy Sci 80:2529–2533

Leonard N, O'Connel J, O'Farrel K (1994) The effects different housing design on behaviour and foot le-sions in Friesian heifers. Vet Rec 134:490–494

Lidfors LM (1993) Cross-sucking in group-housed dairy calves before and after weaning off milk. Appl Anim Behav Sci 38:15–24

Lidfors LM, Moran D, Jung J, Jensen P, Castren H (1994) Behaviour at calving and choice of calving place in cattle kept in different environments. Appl Anim Behav Sci 42:11–28

Livesey CT, Marsh C, Metcalf JA, Laven RA (2002) Hock injuries in cattle kept in straw yards or cubicles with rubber mats or mattresses. Vet Rec 150:677–679

Loberg JM, Hernandez CE, Thierfelder T, Jensen MB, Berg C, Lidfors L (2008) Weaning and separation in two steps—a way to decrease stress in dairy calves suckled by foster cows. Appl Anim Behav Sci 111:222–234

Lombard JE, Garry FB, Tomlinson SM, Garber LP (2007) Impacts of dystocia on health and survival of dairy calves. J Dairy Sci 90:1751–1760

Lott DF, Galland JC (1985) Parturition in American bison: Precocity and systematic variation in cow isolation. Z Tierpsychol 69:66–71

Lowe DE, Steen RWJ, Beattie VE (2001) Preferences of housed finishing beef cattle for different floor types. Anim Welfare 10:395–404

Mahendran SA, Wathes DC, Booth RE, Blackie N (2021) The health and behavioural effects of individual versus pair housing of calves at different ages on a UK commercial dairy farm. Animals 11:612. https://doi.org/10.3390/ani11030612

Mainau E, Manteca X (2011) Pain and discomfort caused by parturition in cows and sows. Appl Anim Behav Sci 135:241–251. https://doi.org/10.1016/j.applanim.2011.10.020

Marcé C, Guatteo R, Bareille N, Fourichon C (2010) Dairy calf housing systems across Europe and risk for calf infectious diseases. Animal 4:1588–1596

Margerison JK, Phillips CJC, Preston TR (2002) Restricted suckling of tropical dairy cows by their own calf or other cows' calves. J Anim Sci 80:1663–1670

Meagher RK, Daros RR, Costa JHC, von Keyserlingk MAG, Hötzel MJ, Weary DM (2015) Effects of degree and timing of social housing on reversal learning and response to novel objects in dairy calves. PLoS One 10(8):e0132828. https://doi.org/10.1371/journal.pone.0132828

Meagher RK, Beaver A, Weary DM, von Keyserlingk MA (2019) Invited review: A systematic review of the effects of prolonged cow–calf contact on behavior, welfare, and productivity. J Dairy Sci 102:5765–5783

Medrano-Galarza C, LeBlanc SJ, Jones-Bitton A, DeVries TJ, Rushen J, De Passillé AM, Haley DB (2017) Can J Anim Sci 98:250–259

Medrano-Galarza C, LeBlanc SJ, DeVries TJ, Jones-Bitton A, Rushen J, De Passillé AM, Endres MI, Haley DB (2018) Effect of age of introduction to an automated milk feeder on calf learning and performance and labor requirements. J Dairy Sci 101:9371–9384

Mee JF (2008) Prevalence and risk factors for dystocia in dairy cattle: A review. Vet J 176:93–101. https://doi.org/10.1016/j.tvjl.2007.12.032

Mellor DJ (2016) Updating animal welfare thinking: moving beyond the "five freedoms" towards "a life worth living". Animals 6:21

Miedema HM, Cockram MS, Dwyer CM, Macrae AI (2011) Behavioural predictors of the start of normal and dystocic calving in dairy cows and heifers. Appl Anim Behav Sci 132:14–19. https://doi.org/10.1016/j.applanim.2011.03.003

Miller-Cushon EK, DeVries TJ (2016) Effect of social housing on the development of feeding behavior and social feeding preferences of dairy calves. J Dairy Sci 99:1406–1417

Miller-Cushon EK, Bergeron R, Leslie KE, Mason GJ, DeVries TJ (2014) Competition during the milk-feeding stage influences the development of feeding behavior of pair-housed dairy calves. J Dairy Sci 97:6450–6462

Mogensen L, Nielsen LH, Hindhede J, Sørensen JT, Krohn CC (1997) Effect of space allowance in deep bedding systems on resting behaviour, production, and health of dairy heifers. Acta Agric Scand A Anim Sci 47:178–186

Neave HW, Lomb J, von Keyserlingk MAG, Behnam-Shabahang A, Weary DM (2017) Parity differences in the behavior of transition dairy cows. J Dairy Sci 100:548–561. https://doi.org/10.3168/jds.2016-10987

Nielsen SS, Toft N (2011) Effect of management practices on paratuberculosis prevalence in Danish dairy herds. J Dairy Sci 94:1849–1857

Nielsen LH, Mogensen L, Krohn C, Hindhede J, Sørensen JT (1997) Resting and social behaviour of dairy heifers housed in slatted floor pens with different sized bedded lying areas. Appl Anim Behav Sci 54:307–316

Nielsen PP, Jensen MB, Lidfors L (2008a) The effects of teat bar design and weaning method on behavior, intake, and gain of dairy calves. J Dairy Sci 91:2423–2432

Nielsen PP, Jensen MB, Lidfors L (2008b) Milk allowance and weaning method affect the use of a computer controlled milk feeder and the development of cross-sucking in dairy calves. Appl Anim Behav Sci 109:223–237

Noakes DE (2001) Parturition and the care of parturient animals. In: Noakes DE, Parkinson TJ, England GCW (eds) Arthur's veterinary reproduction and obstetrics. Saunders, Philadelphia, PA, pp 155–187

O'Connel JM, Giller PS, Meaney WJ (1993) Wealing training and cubicle usage as heifers. Appl Anim Behav Sci 37:185–195

Østergaard S, Thomsen P, Burow E (2010) Separate housing for one month after calving improves production and health in primiparous cows but not in multiparous. J Dairy Sci 93:3403–3890

Overvest MA, Crossley RE, Miller-Cushon EK, DeVries TJ (2018) Social housing influences the behavior and feed intake of dairy calves during weaning. J Dairy Sci 101:8123–8134

Pedersen RE, Sørensen JT, Skjøth F, Hindhede J, Nielsen TR (2009) How milk-fed dairy calves perform in stable versus dynamic groups. Lives Sci 121:215–218

Peeler EJ, Otte MJ, Esslemont RJ (1994) Inter-relationships of periparturient diseases in dairy cows. Vet Rec 134:129–132

Pempek JA, Eastridge ML, Proudfoot KL (2017) The effect of a furnished individual hutch pre-weaning on calf behavior, response to novelty, and growth. J Dairy Sci 100:4807–4817

Platz S, Ahrens F, Bahrs E, Nuske S, Erhard MH (2007) Association between floor type and behaviour, skin lesions and claw dimensions in group-housed fattening bulls. Prev Vet Med 80:209–221

Proudfoot KL, Huzzey JM, Von Keyserlingk MAG (2009) The effect of dystocia on the dry matter intake and behavior of Holstein cows. J Dairy Sci 92:4937–4944

Proudfoot KL, Jensen MB, Heegaard PMH, von Keyserlingk MAG (2013) Effect of moving dairy cows at different stages of labor on behavior during parturition. J Dairy Sci 96:1638–1646

Proudfoot KL, Weary DM, Von Keyserlingk MAG (2014a) Maternal isolation behavior of Holstein dairy cows kept indoors. J Dairy Sci 92:277–281

Proudfoot KL, Jensen MB, Weary DM, Von Keyserlingk MAG (2014b) Dairy cows seek isolation at calving and when ill. J Dairy Sci 97:2731–2739

Raeth-Knight M, Chester-Jones H, Hayes S, Linn J, Larson R (2009) Impact of conventional or intensive milk replacer programs on Holstein heifer performance through six months of age and during first lactation. J Dairy Sci 92:799–809

Rasmussen L, Jensen MB, Jeppesen LL (2006) The effect of age at introduction and number of milk-portions on calves' responses to integration into a dynamic group of dairy calves fed by computer controlled milk feeder. Appl Anim Behav Sci 100:153–163

Rault JL (2012) Friends with benefits: Social support and its relevance for farm animal welfare. Appl Anim Behav Sci 136:1–14
Raussi S, Niskanen S, Siivonen J, Hänninen L, Hepola H, Jauhiainen L, Veissier I (2010) The formation of preferential relationships at early age in cattle. Behav Process 84:726–731
Reimus K, Alvåsen K, Emanuelson U, Viltrop A, Mõtus K (2020) Herd-level risk factors for cow and calf on-farm mortality in Estonian dairy herds. Acta Vet Scand 62:1–15
Rørvang MV, Herskin MS, Jensen MB (2017) Dairy cows with prolonged calving seek additional isolation. J Dairy Sci 100:2967–2975
Rørvang MV, Nielsen BL, Herskin MS, Jensen MB (2018a) Prepartum maternal behavior of domesticated cattle: a comparison with managed, feral, and wild ungulates. Front Vet Sci 5:1–11. https://doi.org/10.3389/fvets.2018.00045
Rørvang MV, Herskin MS, Jensen MB (2018b) The motivation-based calving facility: Social and cognitive factors influence isolation seeking behaviour of Holstein dairy cows at calving. PLoS One 13:e0191128. https://doi.org/10.1371/journal.pone.0191128
Rosenberger K, Costa JHC, Neave HW, von Keyserlingk MAG, Weary DM (2017) The effect of milk allowance on behavior and weight gains in dairy calves. J Dairy Sci 100:504–512
Roth BA, Barth K, Gygax L, Hillmann E (2009) Influence of artificial vs. mother-bonded rearing on sucking behaviour, health and weight gain in calves. Appl Anim Behav Sci 119:143–150
Schirmann K, Chapinal N, Weary DM, Vickers L, von Keyserlingk MAG (2013) Short communication: Rumination and feeding behavior before and after calving in dairy cows. 2013. J Dairy Sci 96:7088–7092. https://doi.org/10.3168/jds.2013-7023
Schuenemann GM, Nieto I, Bas S, Galvão KN, Workman J (2011) Assessment of calving progress and reference times for obstetric intervention during dystocia in Holstein dairy cows. J Dairy Sci 94:5494–5501. https://doi.org/10.3168/jds.2011-4436
Schulze Westerath H, Gygax L, Mayer C, Wechler B (2007) Leg lesions and cleanliness of finishing bulls kept in housing systems with different lying area surfaces. Vet J 174:77–85
Sirovnik J, Barth K, de Oliveira D, Ferneborg S, Haskell MJ, Hillmann E, Jensen MB, Mejdell CM, Napolitano F, Vaarst M, Verwer CM, Waiblinger S, Zipp KA, Johnsen JF (2020) Methodological terminology and definitions for research and discussion of cow-calf contact systems. J Dairy Res 87(S1):108–114
Soberon F, Van Amburgh ME (2013) Lactation biology symposium: The effect of nutrient intake from milk or milk replacer of preweaned dairy calves on lactation milk yield as adults: A meta-analysis of current data. J Dairy Sci 91:706–712
Stěhulová I, Lidfors L, Špinka M (2008) Response of dairy cows and calves to early separation: Effect of calf age and visual and auditory contact after separation. Appl Anim Behav Sci 110:144–165
Svensson C, Hultgren J (2008) Associations between housing, management, and morbidity during rearing and subsequent first-lactation milk production of dairy cows in Southwest Sweden. J Dairy Sci 91:1510–1518
Svensson C, Liberg P (2006) The effect of group size on health and growth rate of Swedish dairy calves in pens with automatic milk feeders. Prev Vet Med 73:43–53
Svensson C, Lundborg K, Emanuelson U, Olsson SO (2003) Morbidity in Swedish dairy calves from birth to 90 days of age and individual calf-level risk factors for infectious diseases. Prev Vet Med 58:179–197
Swain DL, Patison KP, Heath BM, Bishop-Hurley GJ, Finger A (2015) Pregnant cattle associations and links to maternal reciprocity. Appl Anim Behav Sci 168:10–17
Sweeney BC, Rushen J, Weary DM, De Passillé AM (2010) Duration of weaning, starter intake and weight gain of dairy calves fed large amounts of milk. J Dairy Sci 93:148–152
Tapki I (2007) Effects of individual or combined housing systems on behavioural and growth responses of dairy calves. Acta Agric Scand A Anim Sci 57:55–60
Telezhenko E, Lidfors L, Bergsten C (2007) Dairy cow preferences for soft or hard flooring when standing or walking. J Dairy Sci 90:3716–3724
Thomas TJ, Weary DM, Appleby MC (2001) Newborn and 5-week old calves vocalise in response to milk deprivation. Appl Anim Behav Sci 74:165–173

Thomsen PT, Kjeldsen AM, Sørensen JT, Houe H (2004) Mortality (including euthanasia) among Danish dairy cows (1990–2001). Prev Vet Med 62:19–33
Tucker CB, Weary DM, Fraser D (2003) Effects of three types of free-stall surfaces on preferences and stall usage by dairy cows. J Dairy Sci 86:521–529
USDA (2016) Dairy cattle management practices in the United States (2014) USDA–Anim. Plant Health Insp. Serv.–Vet. Serv., Center Epidemiol. Anim. Health, Natl. Anim. Health Monit. Syst., Fort Collins, CO. #692.0216 (2016)
Vasseur E, Borderas F, Cue RI, Lefebvre D, Pellerin D, Rushen J, Wade KM, De Passillé AM (2010) A survey of dairy calf management practices in Canada that affect animal welfare. J Dairy Sci 93:1307–1316
Ventura BA, Von Keyserlingk MAG, Schuppli CA, Weary DM (2013) Views on contentious practices in dairy farming: The case of early cow-calf separation. J Dairy Sci 96:6105–6116
Vitale AF, Tenucci M, Papiri M, Lovari S (1986) Social behaviour of calves of semi-wild Maremma Cattle (Bos primigenius taurus). Appl Anim Behav Sci 26:217–231
Vokey FJ, Guard CL, Erb HN, Galton DM (2001) Effects of alley and stall surfaces on indices of claw and leg health in dairy cattle housed in a free-stall barn. J Dairy Sci 84:2686–2699
von Keyserlingk MAG, Weary DM (2007) Maternal behavior in cattle. Horm Behav 52:106–113. https://doi.org/10.1016/j.yhbeh.2007.03.015
von Keyserlingk MAG, Brusius L, Weary DM (2004) Competition for teats and feeding behavior by group housed dairy calves. J Dairy Sci 87:4190–4194
von Keyserlingk MAG, Cunha GE, Fregonesi JA, Weary DM (2011) Introducing heifers to freestall housing. J Dairy Sci 94:1900–1190
Wagenaar JPTM, Langhout J (2007) Practical implications of increasing 'natural living' through suckling systems in organic dairy calf rearing. NJAS Wageningen J Life Sci 54(4):375–386
Wagner K, Barth K, Hillmann E, Palme R, Futschik A, Waiblinger S (2013) Mother rearing of dairy calves: Reactions to isolation and to confrontation with an unfamiliar conspecific in a new environment. Appl Anim Behav Sci 147:43–54
Weary DM, Chua B (2000) Effects of early separation on the dairy cow and calf: 1. Separation at 6 h, 1 day and 4 days after birth. Appl Anim Behav Sci 69:177–188
Weary DM, Taszkun I (2000) Hock lesions and free-stall design. J Dairy Sci 83:697–702
Weary D, Jasper J, Hötzel MJ (2008) Understanding weaning distress. Appl Anim Behav Sci 110:24–41
Webb LE, Veenhoven R, Harfeld JL, Jensen MB (2019) What is animal happiness? Ann N Y Acad Sci 1438:62–76
Wechsler B, Schaub J, Friedli K, Hauser R (2000) Behaviour and leg injuries in dairy cows kept in cubicle systems with straw bedding or soft lying mats. Appl Anim Behav Sci 69:189–197
Wehrend A, Hofmann E, Failing K, Bostedt H (2006) Behaviour during the first stage of labour in cattle: Influence of parity and dystocia. Appl Anim Behav Sci 100:164–170
Welfare Quality (2009) Welfare Quality assessment protocol for cattle. Accessed Fb. 14, 2021. http://www.welfarequality.net/media/1121/wqr11.pdf
Whalin LJ, Weary DM, von Keyserlingk MAG (2018) Short communication: Pair housing dairy calves in modified calf hutches. J Dairy Sci 101:5428–5433
Winder CB, Bauman CA, Duffield TF, Barkema HW, Keefe GP, Dubuc J, Uehlinger F, Kelton DF (2018) Canadian National Dairy Study: Heifer calf management. J Dairy Sci 101:10565–10579
Wood-Gush DGM, Dennison SGC, Carson K, Hunt KA (1984) The early behaviour of suckler calves in the field. Biol Behav 9:295–306

# Precision Livestock Farming Technologies for Dairy and Beef Production

# 11

Annabelle Beaver and S. Mark Rutter

## Contents



**Abstract**

Precision Livestock Farming (PLF) describes the use of technology within livestock systems to monitor animals, their products, and the environment. A main aim of PLF technologies is to provide continuous individual-animal data to farmers, which can then be used to inform management decisions and improve production and resource-use efficiency. For dairy and beef cattle, PLF technologies have historically been used to provide data on oestrus (specifically for dairy systems), production parameters, and animal health. As the sophistication of PLF technologies increases, so too does their capacity to measure more complex outcomes, such as indicators of animal welfare. This chapter provides an overview of the use of PLF technologies within dairy and beef cattle systems and con-

A. Beaver (✉) · S. M. Rutter
Harper Adams University, Newport, Shropshire, UK
e-mail: ABeaver@harper-adams.ac.uk



M. Haskell (ed.), *Cattle Welfare in Dairy and Beef Systems*, Animal Welfare 23,
https://doi.org/10.1007/978-3-031-21020-4_11

cludes with a discussion on the implications of PLF for cattle welfare. Current and future PLF technologies are described, using the framework of the Five Freedoms to guide the discussion.

**Keywords**

Precision livestock farming · Cattle sensors · Cattle Welfare · Continuous monitoring · Targeted treatment

## 11.1 Introduction

The digital revolution has impacted most aspects of modern life, and agriculture is no exception. The biggest impact so far in farming has been in arable production, where the term Precision Farming was first coined to describe the use of sensor-derived and other data to help target resources (e.g. fertiliser, pesticides) to where they are needed (Whelan and McBratney 2000). The same approach when applied in animal production systems is called Precision Livestock Farming, which is typically abbreviated to PLF (Berckmans 2017). PLF systems aim to manage individual animals by continuously monitoring their production, reproductive status, health, physiology, environmental impact, and welfare using a range of sensors and real-time data analysis, alerting farmers to any animals that require their attention (Berckmans 2017). It is important to emphasise that this is an aim, and the degree to which *individual* animals are *continuously* monitored varies between different animal production systems. However, the PLF systems used with intensively managed dairy cows arguably come the closest to achieving this overall aim of continuous individual-animal monitoring (Rutter 2012), as discussed in the next section.

It is worth noting that much of the PLF research discussed in this chapter involves developing and testing technologies for eventual application on commercial farms. However, we also discuss technologies that are used for research purposes (e.g. to assess pain in cattle), with the ultimate goal of informing practices to improve welfare, rather than for use on commercial farms. The line between these objectives is not clear-cut, as some technology that is developed for research purposes may also have future or indirect on-farm applications.

## 11.2 The Use of PLF Technologies Within Dairy Systems

### 11.2.1 A Brief History of Dairy PLF Technologies, with a Focus on Oestrus Detection

The precursors of contemporary dairy PLF technologies date back to the 1970s. The first automated systems developed for use on dairy farms were individual electronic milk meters that were installed in the milking parlour and were designed to keep pace with ever-increasing herd sizes and rising milk yields (see Mottram 2016). In the 1980s, the use of artificial insemination meant that automatic, behaviour-based oestrus detection systems followed suit. Research from that era (e.g. Claus et al. 1983) determined that the inability of farm staff to accurately detect oestrus by visual observation alone was a significant barrier to achieving higher rates of conception, highlighting the need for development of automated systems to increase detection sensitivity. Additionally, artificial insemination, which was already widely in use in the 1970s (Foote 2002), depended upon considerably more labour and managerial expertise than was required for natural service. When oestrous behaviours do occur, they can be quite distinctive (e.g. increased activity level, mounting and standing to be mounted; Roelofs et al. 2010) and readily detected by single-sensor systems; thus, oestrous behaviours were prime candidates for these early PLF prototypes.

Heat detection, however, has remained a primary factor in limiting dairy cattle reproductive performance, as 35% of cows demonstrate no overt behavioural signs of oestrus (Palmer et al. 2010). Moreover, behavioural signs of oestrus have become more discreet over time, and the duration of oestrus has become progressively shorter (Lucy 2007; Dobson et al. 2008). Alongside these challenges, the repertoire of PLF technologies to measure oestrus has expanded. Currently, a variety of sensor types are available, with the majority used to detect changes in activity by means of animal-mounted tri-axial accelerometers (Reith and Hoy 2018): activity meters fastened around the neck can measure three-dimensional neck movements, while leg-mounted devices can record the number of steps and quantify lying time. These capabilities provide scope to detect restlessness, increased activity level, and decreased eating time, which are characteristic behaviours associated with oestrus (Roelofs et al. 2010). Changes in activity levels recorded by these accelerometers provide a reasonably accurate prediction of ovulation time (Stevenson et al. 2014).

An accelerometer can also be mounted on the ear of a cow (designed to fit around the radio-frequency identification (RFID) tag) or in the form of a rumen bolus (capsule that is administered orally into the rumen). The boluses can detect behavioural changes associated with oestrus (Knight 2020) and alert the farmer to the appropriate insemination window. Further, milk progesterone (P4) concentration, which is often considered a gold-standard in evaluation of reproduction status, can be measured using biosensors (e.g. the Herd Navigator Lattec I/S, Hillerød, Denmark) (Saint-Dizier and Chastant-Maillard 2012). Video camera software also shows promise for use in oestrus detection. Camera-based PLF technology can be used to detect the removal of paint affixed to the sacrococcygeal region to identify standing heat, the behavioural signpost for primary oestrus (Alawneh et al. 2006). Secondary oestrous behaviours (e.g. flehmen and chin resting) can be identified using automated image analysis; the software has been designed to automatically remove sequences of inactivity, greatly reducing the observation time required (see Saint-Dizier and Chastant-Maillard 2012). Figure 11.1 illustrates the range of behavioural and physiological data that can be collected by various sensors designed for use on commercial dairy farms.

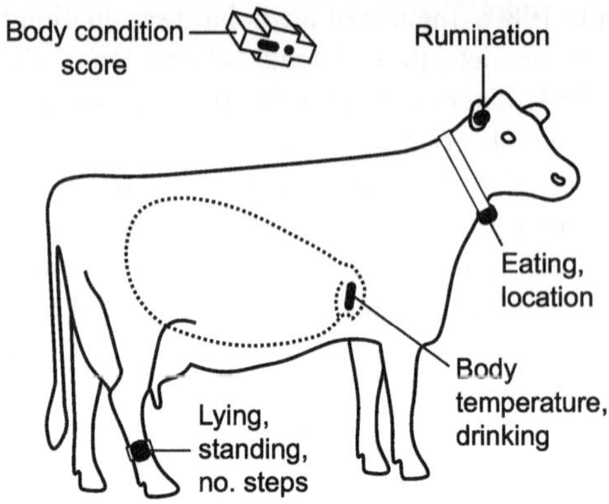


**Fig. 11.1** An illustration of the various behavioural and physiological parameters that can currently be measured on commercial dairy farms using a variety of sensors

### 11.2.2 Use of Dairy PLF Technology for Investigating Health and Other Outcomes

As the suite of available technologies grows to include more sophisticated methods of data collection, the outcome measures available for study have concurrently expanded. For instance, options now exist to measure the onset of calving using tail-mounted sensors (tri-axial accelerometers or inclinometers), vaginal temperature detection boluses, and reticulorumen boluses (as reviewed by van Erp-Van der Kooij and Rutter 2020). More recently, PLF technologies in dairy systems have broadened to include the domain of dairy cattle health monitoring, including detection of lameness and transition cow disease.

There are a variety of automated methods used to detect lameness in dairy cattle. Early diagnosis of lameness is particularly challenging based upon visual observation alone (Alsaaod et al. 2019). Pressure sensors in the floor (e.g. GaitWise, ILVO, Merelbeke, Belgium) provide data on hoof placement and leg pressure: detection of stride-to-stride inconsistencies can allow for the diagnosis of lameness in its early stages (Van Nuffel et al. 2015). Further, leg-mounted accelerometers are able to detect important behavioural differences related to lameness, such as step count and lying time (Thorup et al. 2017).

Accelerometers are not only useful in lameness detection but can predict and identify a variety of other diseases and conditions in dairy cattle. Prior to the onset of many significant transition diseases, physical activity is known to be diminished, as is rumination behaviour (see review by Dittrich et al. 2019). Rumination and activity changes (monitored via neck-mounted electronic tags) are effective in identifying metabolic and digestive disorders associated with calving, such as ketosis, displaced abomasum (Stangaferro et al. 2016a), and severe mastitis and metritis (Stangaferro et al. 2016b, c). Activity data from leg-mounted accelerometers have also recently been used to differentiate cattle with and without antibodies to *Mycobacterium avium* subsp. *paratuberculosis* (MAP; the causative agent of ruminant Johne's disease). In addition to demonstrating differences in lying patterns (i.e.

number and duration of lying bouts), antibody-positive animals were found to spend less time lying down during peak lactation (Charlton et al. 2019).

A consistent pattern across a variety of prominent production diseases is a decrease in feeding time. Changes in feeding behaviours have been observed for cows diagnosed with hypocalcaemia, ketosis, metritis, mastitis, and lameness, sometimes even prior to diagnosis of clinical disease onset (see review by Dittrich et al. 2019). Neck-mounted accelerometers can be used to estimate grazing time (Oudshoorn et al. 2013). Eating and drinking data can also be extracted from automatic feed systems (such as Insentec, Hokofarm Group, Marknesse, the Netherlands) that are able to identify individual cows via RFID; there is a growing body of research harnessing this technology to predict disease based upon feeding patterns (e.g. metritis (Huzzey et al. 2007); ketosis (Goldhawk et al. 2009); claw horn lesions (Proudfoot et al. 2010); and endometritis (Thompson et al. 2019)). To date, the use of data from these automatic bins has mainly been applied in a research context; automatic drinkers may be more financially feasible than feed bins due to a lower bin-to-cow ratio. However, the models based upon drinking behaviour have not always been as successful (e.g. Sahar et al. 2020).

Technology to monitor social interactions has also been used to assess cow welfare. In the 1980s, Rutter et al. (1987) used a microcomputer to create a matrix of interactions to map out replacements (i.e. where one cow replaces another) at the feed face. The automatically generated dominance matrix was found to correspond exactly to the matrix obtained from visual observation, with a high correlation between observed and automatically collected angular dominance values. Changes in cow social interactions have been shown to be associated with cow health issues. For example, pre-calving social interactions have been shown to be important predictors of transition disease (Proudfoot et al. 2018), and information about social interactions can prove useful in identifying animals at greater risk of developing disease after calving. A renewed interest in this topic has led to the harnessing of data from automatic feed bins to understand social dynamics. Huzzey et al. (2014) and McDonald et al. (2019) have developed algorithms from automated feed and water bins, respectively, to identify agonistic interactions between cattle. This information, in turn, has been successfully used to predict transition disease.

PLF technologies have also been used for monitoring the health of dairy calves, although not to the same extent as in adult cattle. Acoustic analysis has previously been used to monitor coughing in calves (Ferrari et al. 2010), which is of particular interest given that respiratory infection is a leading cause of death in dairy calves worldwide (Svensson et al. 2006; Windeyer et al. 2014). For calves, accelerometers have also been used to measure lying duration, activity level (i.e. number of steps), and rumination, which can be important indicators of disease. For example, prior to diagnosis with bovine respiratory disease (BRD), calves showed fewer lying bouts and had a lower step count for 3 days after diagnosis (Swartz et al. 2017). Other studies have found accelerometer data useful for the observation of changes in lying behaviour for calves with a variety of conditions such as navel inflammation (Studds et al. 2018), Salmonellosis (Lowe et al. 2019), or experimental infection with *Mannheimia haemolytica* (Hixson et al. 2018).

Automated calf feeders (Fig. 11.2) are particularly useful for understanding the feeding behaviour of dairy calves, in addition to making health-related inferences. Ear-tag transponders linked to the feeders are used to adjust individual-animal intake or track intake data over time. Changes in feeding patterns can be used to understand health outcomes; altered drinking speeds (Knauer et al. 2017) and decreases in unrewarded feeder visits (Svensson and Jensen 2007) have also been associated with BRD diagnosis. However, there may be an interaction between feeding behaviour and management practices (e.g. milk allowance) on the efficacy of disease prediction. For instance, if calves are fed restricted milk, there may be reduced sensitivity for timely detection of behavioural changes associated with disease, such as reduced milk intake. However, other behavioural changes associated with illness, such as decreases in unrewarded visits to the milk feeder, appear to be consistent for calves fed varying milk allowances (as reviewed by Costa et al. 2021).

Finally, at the herd level, there are a variety of PLF technologies available to assist with farm maintenance by automating management practices that have traditionally been done by hand. Tasks such as milking, cleaning, and feed provision have been automated through the introduction of automatic milking systems (AMS), robot scrapers to remove manure, robots that push the feed and robots that mix and deliver mixed forage-based rations. There are also systems available to monitor environmental variables and thus control climatic conditions (see review by van Erp-Van der Kooij and Rutter 2020). In the current environment, adoption of PLF solutions on dairy farms can help address growing shortages in skilled labour and promote evidence-based, data-driven management decisions (Eastwood et al. 2012; Pham and Stack 2018).

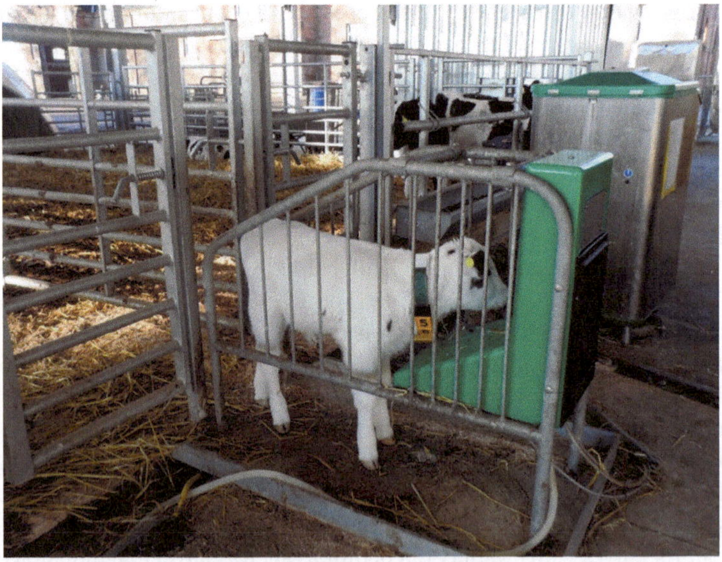

**Fig. 11.2** An automated calf feeder can control and monitor the volume and timing of milk for individual calves. Photograph courtesy of SRUC

## 11.3 The Use of PLF Technologies Within Beef Systems

The use of PLF technologies is markedly less common within beef compared to dairy systems (Richeson et al. 2018), and these technologies have generally been used to tackle different issues. Natural mating still predominates in most beef production systems (Baruselli et al. 2018), which removes the need for technology to facilitate oestrus detection. However, beef systems are increasingly utilising electronic identification (EID), with many countries making its use mandatory (Hossain and Quaddus 2013).

RFID technology has been used in conjunction with an electronic weigh scale for remote weighing of beef cattle to obtain liveweight or monitor weight gain (González et al. 2014). A limitation to EID is that animals must walk close (i.e. within approximately 0.5 m) of the RFID reader panel (Morris et al. 2012). Thus, these remote weighing systems are often fitted within a short raceway linking important resources (e.g. between a grazing area and a water point). Recently, a walk-over weighing system was used to develop an algorithm to detect calving date in beef cattle (Menzies et al. 2017), building upon earlier work by Aldridge et al. (2016); however, further research is needed in larger herds across differing environments to assess the accuracy of the model. Eventually, this technology could potentially increase reproductive efficiency and thus profitability.

Furthermore, Miller et al. (2019) discuss promising results from 3D imaging technology, paired with machine-learning algorithms, to predict carcass characteristics of live beef cattle. They highlight the inefficiency of visual observation or weighing alone in the selection of finishing cattle for slaughter, which can result in large numbers of carcasses failing to meet target conformation grades at the abattoir. They conclude that 3D imaging technique may be used effectively to predict liveweight, saleable meat yield, and carcass conformation grades in live animals, thus potentially reducing inefficiency in beef production systems (Miller et al. 2019).

Pedigree matchmaker (PMM) has been refined for use in sheep and involves matching ewes with their lambs using EID (Morris et al. 2012). Pedigree information is obtained by estimating the probability that ewe-lamb pairs are related based upon the frequency that a given lamb is observed to follow a particular dam. This method obviates more costly or labour-intensive options such as DNA testing or tagging at birth (Richards and Atkins 2007). A similar system has proved useful in suckler beef systems, with 90% of maternal pairs correctly matched based upon the number of times the animals were recorded together (i.e. the 'half-weight method'; Menzies et al. 2018).

Wearable PLF technology has been used to some extent to measure health-related outcomes in beef cows and calves, with a focus on respiratory disease. In calves, White et al. (2016) used a remote early disease identification system (Precision Animal Solutions, LLC, Canton, MO) which links to a real-time location system (Smartbow, MKWE, Vienna, Austria) with the goal of diagnosing BRD. The researchers achieved higher positive and negative predictive values compared to visual observation alone, suggesting that remote monitoring may offer a more accurate BRD diagnosis in beef calves. Electronic monitoring systems have also been used to recording eating and drinking behaviour (Buhman et al. 2005) or proximity to the feed face (Quimby et al. 2011) with the aim of differentiating between feedlot calves with and without bovine respiratory disease (BRD). Similarly, Pillen et al. (2016) used pedometers to distinguish feedlot cattle with BRD by monitoring changes in standing time and step count; differences were evident prior to diagnosis. Steers with severe clinical illness from inoculation with *Mycoplasma bovis* were also shown to travel shorter distances and spend less time near the feed bunk (White et al. 2012). Finally, Toaff-Rosenstein et al. (2016) determined that rectal temperature loggers (TidbiT v2, Onset Computer Corporation, Bourne, MD, USA) and leg-mounted accelerometers (HOBO Pendant-G, Onset Computer Corporation, Bourne, MD, USA) were able to differentiate beef steers challenged with bovine respiratory syncytial virus from an unchallenged cohort.

However, while accelerometers have been used extensively in the dairy industry, they have not been commonly used in the beef sector, and their use has mainly addressed the behaviour of beef cattle after castration. Several studies have concluded that surgical castration (compared to other techniques such as banding) led to increases in standing time and a reduced step count (White et al. 2008; Petherick et al. 2014; Roberts et al. 2018), although multiple potential interpretations may account for this difference. These data could ultimately be used to inform farmers of best practice when it comes to castration method. Accelerometers (such as the IceRobotics IceTag, Queensferry, UK) have also been researched in the context of extensive beef systems, but some research has shown that the sensor faces difficulty in differentiating grazing from standing (Ungar et al. 2018).

A number of advanced technologies have recently been introduced to bolster sustainability and improve grazing management, with applications in the grass-fed beef sector. Virtual fencing, which will be discussed in more detail in subsequent sections, makes use of animal-mounted training collars and is gaining traction in parts of the world where extensive livestock grazing is common (Umstatter 2011; Jachowski et al. 2014). Another collar-mounted device, EGrazor makes use of solar-powered sensors to estimate pasture intake from behavioural data (CSIRO Robotics and Autonomous Systems Group, Queensland, Australia; Appelqvist et al. 2022). Once the system is fitted and deployed, minimal labour is needed and cattle may return to their normal grazing routine. The group has also developed a next-generation smart ear tag (Ceres Tag) with the goal of assisting farmers with tracking of their herds by means of geolocation (see CSIRO n.d.; Appelqvist et al. 2022). Although sentinel monitoring, whereby only certain animals within the herd are tagged, can improve cost-effectiveness of large-scale livestock monitoring, individual health issues and injury in un-monitored animals would not be detected.

Technologies to determine animal location have evolved greatly over the last half century, with global positioning-based tracking systems now commercially available (Maroto-Molina et al. 2019). For free-range livestock, GPS tracking solutions need not rely upon a ground-communication station and can provide data on an animal's movements and the extent to which it interacts with resources (e.g. water points, mineral blocks, etc.) as well as with other tracked animals. However, several limitations such as wireless signal impairments and financial cost (Bhakta et al.

2019) impair the uptake of these systems on commercial farms. Maroto-Molina et al. (2019) attempted to overcome some of these limitations by proposing a low-cost IoT (Internet of Things)-based solution, using several GPS collars and Bluetooth tags connected to an energy-efficient low power wide area (LPWA) network; the system proved promising for both beef cattle and sheep.

## 11.4 Welfare Implications of Precision Livestock Farming

### 11.4.1 Current and Future Implications of Precision Livestock Farming on Cattle Welfare

Beginning with milk yield and oestrus, and expanding to health conditions such lameness and transition diseases, the capacity of PLF systems to measure more complex outcomes has developed alongside the increasing sophistication of these technologies. Most recently, PLF systems have expanded their data collection capacities into the animal welfare domain (van Erp-Van der Kooij and Rutter 2020). If health is considered to be multifactorial, welfare is arguably more complex still, as the entire concept of health and biological functioning is just one of several facets of animal welfare (Fraser et al. 1997).

For the purposes of this chapter, we will use the framework of the Five Freedoms and associated five needs (FAWC 2009) to discuss how available technology could be harnessed to assess animal welfare, and how existing and emerging technology could potentially be adapted to improve the welfare of cattle. As the Five Freedoms have been historically critiqued for their emphasis on mitigating negative welfare, we will extend the discussion to include considerations for how PLF technologies may be used to promote positive affective states in cattle (Mellor 2016; Lawrence et al. 2019). We will also consider precision technology used for other livestock species and its applicability to the dairy and beef sectors.

#### 11.4.1.1 Freedom from Hunger and Thirst: By Ready Access to Fresh Water and a Diet to Maintain Full Health and Vigour

In a literal sense, freedom from hunger and thirst can be evaluated directly using PLF systems to measure eating and drinking time of individual animals, in addition to quantity consumed. Automatic feed and water bins not only provide these data but also offer insight into parameters such as frequency of bin visits, number of bins visited per day, and number of bins visited per meal. As feeding time does not perfectly correlate with quantity consumed, an understanding of the interactions between some of these other dimensions provides a more complete picture of feeding dynamics (Sahar et al. 2020). These data can in turn be integrated to determine whether the animal's nutritional needs are being met. Further, using the algorithm developed by Huzzey et al. (2014), agonistic interactions can be identified from PLF data, offering insights into the social structure of the herd and any social factors that may have led to reductions in feeding or drinking time.

Computer vision systems are also being investigated as potential lower-cost methods of measuring individual-animal feed intake. Bezen et al. (2020) installed an RGB-D (Red, Green, Blue, Depth) camera and used the images to train deep Convolutional Neural Network (CNN) models; although these methods require further large-scale validation, the preliminary results have been promising.

A 'rumen' temperature bolus (e.g. smaXtec Classic Bolus, Graz, Austria) can detect when an animal drinks, as the water (which will be below body temperature in most situations) passes directly into the reticulum where the sensor is located, resulting in a temporary reduction in the temperature reading (Vázquez-Diosdado et al. 2019). Importantly, the demand for certain resources, such as fresh water, is influenced by temperature and humidity, which is quantified using indices that incorporate both parameters (temperature-humidity indices (THI)). Higher temperature and humidity lead to increased thirst, which can in turn increase competition for the drinker and frequency of agonistic interactions (McDonald et al. 2019). These agonistic interactions involve displacements from the water trough, in which a cow's drinking behaviour is interrupted by another animal. Thus, in order to provide freedom from thirst, particularly for subordinate animals, the temperature and humidity within the barn should be regulated. There are automated systems available to monitor these environmental variables (e.g. via Tinytag Plus 2 loggers, Gemini Data Loggers, Chichester, UK, or HOBO Pro dataloggers, Onset Computer Corporation, Bourne, MD, USA), to appraise the severity of heat stress, and even to record sweating rate in cattle (e.g. the Evapo-meter (Delfin Technologies Ltd, Kuopio, Finland)) (Ji et al. 2017). PLF systems to directly control climate conditions are in use elsewhere in the livestock industry and are in development for use on dairy farms (see van Erp-Van der Kooij and Rutter 2020).

Extensive beef systems may be particularly good candidates for technology aimed at addressing the freedom from hunger and thirst, given the potential for limited availability of feed and water and increased difficulty for stockpeople to keep a close eye on the animals. However, it is much more challenging to measure feed intake and feeding behaviour in pasture-based systems. Technologies exist for this purpose (e.g. pressure sensors and accelerometers), but, to date, they are mainly applied at the research level (Maroto Molina et al. 2020). RFID systems can be employed to better understand the optimal number and distance between water points, in addition to flagging individuals who have not visited a waterer within a relevant time frame (Williams et al. 2019). Advanced systems have integrated multiple data streams from RFID, flow meters, and water sensors to concurrently track water levels, temperature, quality, and individual consumption. If there is insufficient water available, for instance, the system will issue an alert (Tang et al. 2021)

Virtual fencing, which is already commercially available (Boviguard, Agrifence, Gloucester, UK), could be used to reduce thirst and hunger by automatically shepherding cattle towards water points, or to areas of high vegetation (as discussed by Rutter 2014). This system could be used in combination with existing PLF technologies that track vegetation growth and availability (Schellberg et al. 2008). Physical fencing of livestock can result in some negative environmental impacts on wildlife, such as population fragmentation, changes in community structure, and the prevention of movement including migrations (Gadd 2012). Although more research is needed, virtual fencing may therefore confer some environmental benefits and enhance conservation efforts (Riesch et al. 2022) including the exclusion of agricultural animals from environmentally sensitive areas (Campbell et al. 2020). Virtual fencing makes use of a dynamic virtual boundary determined through selected geographical coordinates. When an animal strays close to the boundary, it is typically warned using an auditory cue; if the animal then attempts to traverse the boundary, an electric shock is given (Umstatter et al. 2015). As this technique may be considered to raise other welfare issues (Lee et al. 2018; Lomax et al. 2019), the virtual fencing should ideally be implemented through the use of positive reinforcement where possible (Rutter 2014).

#### 11.4.1.2 Freedom from Pain, Injury, and Disease: By Prevention or Rapid Diagnosis and Treatment

As previously discussed, PLF systems have been adopted for use in detecting and predicting important diseases and conditions in dairy cattle. Lameness is arguably the most significant welfare challenge facing the modern dairy cow and is known to be a painful condition (Shearer et al. 2013). Alsaaod et al. (2019) note that even trained observers underestimate the prevalence of moderately lame animals, based upon visual observation alone, thus missing the opportunity for early intervention or treatment. In addition to the PLF technologies previously mentioned to detect lameness, there are several types of vision-based systems in development to facilitate lameness detection (Kang et al. 2021). For example, Song et al. (2008) trialled a 2D vision-based trackway system, obtaining a strong correlation between automatically captured and manually labelled hoof-location data. The main limitation of image processing techniques is the difficulty of differentiating the walking cow from a complex background. Thus, other researchers (e.g. Jiang et al. 2019) have attempted to use deep learning algorithms to extract cow lameness features from naturalistic backgrounds. 3D computer vision detection is also in development, which can improve sensitivity (Viazzi et al. 2014) and more accurately identify the back postures of the animals (Pezzuolo et al. 2018). Finally, infrared thermography has been used as a diagnostic tool for detecting hoof temperature differences and has successfully differentiated between cattle with and without hoof lesions (Stokes et al. 2012)

Neonatal calf mortality is a significant issue in extensive beef systems (Bunter et al. 2014), and technology is in development to increase remote monitoring capabilities for cows and their calves. The Pedigree Matchmaker (PMM) technique, if adopted more widely in beef systems, could facilitate more rapid intervention by means of its ability to track to movements of cow-calf pairs. In sheep, PMM data have not only been used to obtain pedigree information, but also to research lamb and ewe behaviour traits to map the level of association between the ewe and offspring (Brown et al. 2011). In cattle, algorithms could be developed to alert the farmer to abnormal reductions in 'close reads' (number of reads within 5 s of the dam), average time between reads for calf and dam, or drops in the number of times the reader is activated (particularly if located at an important resource such as a water point). A telemetric monitoring system is also in development to identify both the time and location of calving in extensive beef herds, consisting of a transmitter, a terrestrial receiver, and a central location server (Stephen et al. 2019). Optimisation of this system could lead to swift intervention and more feasible monitoring of cow-calf pairs. Similarly, global positioning systems or drones may be useful for cattle monitoring on pasture and the tracking of cow-calf pairs (see review by Beaver et al. 2020)

Camera-based approaches are also useful for health monitoring and disease diagnosis. For instance, cattle body condition score (BCS) can be assessed regularly using a BCS camera (such as DeLaval, Tumba, Sweden; Hallén Sandgren and Emanuelson 2016), which is based upon 3D imaging. This technology has the potential to provide more precision than visual assessment alone, although some studies have found its accuracy is diminished when BCS is outside midrange values (Mullins et al. 2019). Near-infrared spectroscopic sensing systems have been used to measure Somatic Cell Count (SCC) in milk, and when incorporated into an automatic milking system, have been able to assess milk quality of individual animals (Kawasaki et al. 2008). Advanced monitoring methods have been applied towards understanding pig welfare, and these techniques could also be applied to the dairy context. Fernández-Carrión et al. (2017) used an optical flow algorithm to translate motion capture of pigs with the African Swine Fever virus into digital models; changes in activity were detected prior to the onset of clinical disease symptoms. As discussed in Sect. 11.2.2, changes in activity level are a hallmark of many significant diseases in dairy cattle and applying similar advanced monitoring methods to dairy contexts has the potential to enhance identification, prediction, and prevention of cattle disease.

#### 11.4.1.3 Freedom from Discomfort: By Providing an Appropriate Environment Including Shelter and a Comfortable Resting Area

Although lacking a consistent definition in the literature, 'cow comfort' and methods to improve it on dairy farms have received much attention by both regional and national animal welfare organisations (see Beaver et al. 2021a). As the antonym of discomfort, comfort ties directly back to the third of the Five Freedoms. Lying behaviour is sometimes used in the evaluation of comfort (e.g. Haley et al. 2000), and parameters such as time spent lying down or number of lying bouts can be easily measured using tri-axial accelerometers (e.g. IceRobotics CowAlert, Queensferry, UK; Fig. 11.3). Computer vision-based systems could be employed to more specifically assess cow positioning within cubicles (Porto et al. 2013), which may be useful in evaluating collisions with stall hardware and body position while lying down. Several studies have identified differences in time spent kneeling, number of unfulfilled intentions to lie down, or lying outside the lying area as indicators of reduced comfort (Krohn and Munksgaard 1993; Popescu et al. 2013), and these parameters could also be evaluated using vison-based systems.

Dairy cattle prefer lying in open spaces and will take advantage of the space allowance to assume more expansive postures, such as lying with the limbs

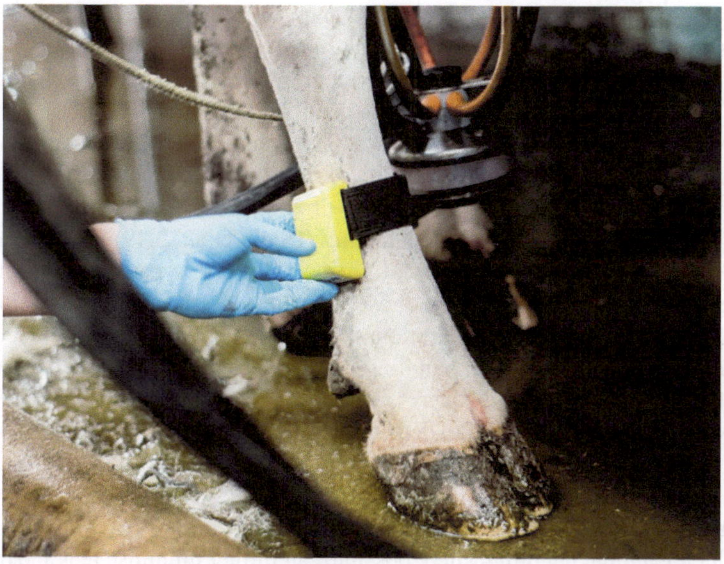

**Fig. 11.3** A dairy cow wearing an IceRobotics Ice-Qube on a rear leg. Incorporating a triple-axis accelerometer, this sensor can determine when a cow is lying down, standing and, during walking, the number of steps taken. As part of the on-farm CowAlert system, it can alert when the cow is in oestrus as well as alert any changes in locomotion associated with lameness. Photograph courtesy of IceRobotics, Queensferry, Scotland

outstretched (Beaver et al. 2021b). Recent research has also shown that cattle will compromise on their preferred bedding substrate in order to access open lying spaces (Shewbridge Carter et al. 2021); however, the provision of open lying spaces is sometimes not considered on dairy farms due to a reduction in cleanliness. Certain PLF technologies such as automatic scrapers can facilitate the adoption of open lying areas on dairy farms. For instance, the High Welfare Floor (Newman et al. 2018) allows for the separation of urine (which filters through the floor into a Permavoid drainage layer) from faeces, which remain atop the textile and can be collected by a specialised automated scraper system.

As the Five Freedoms are not mutually exclusive entities, temperature control systems are again highly relevant when it comes to providing freedom from discomfort. There is a significant untapped potential for existing PLF technologies to link with climate control systems (Jukan et al. 2017), such as through the provision of a retractable roof that responds to changes in THI, wind speed, and weather conditions. Virtual fencing could also be used to direct animals towards shelter when inclement weather is approaching (Rutter 2014).

#### 11.4.1.4 Freedom from Fear and Distress: By Ensuring Conditions and Treatment Which Avoid Mental Suffering

For beef cattle, remote sensing can be used to measure physiological responses that may be indicative of acute pre-slaughter stress. Thermal infrared (TIR) imagery is a promising means to detect changes in eye and skin temperature as an indicator of acute stress prior to slaughter, which has mostly been investigated in pigs (e.g. Weschenfelder et al. 2013). Infrared thermography has also been implemented to detect fear-related responses during cattle handling (Stewart et al. 2008), which can in turn be used to understand which procedures or actions are particularly aversive. More recent research in cattle has made use of infrared thermograms of the cows' eyes and forelimbs in different situations, to provide insight into behavioural indicators of anxiety, including right-side laterality (Uddin et al. 2019), flight speed, and restlessness in the crush (Uddin et al. 2021). TIR offers particular promise for remote-image data collection, and thus contactless assessment, of physiological parameters such as heart and respiration rate (Jorquera-Chavez et al. 2019), thereby reducing stress associated with handling. Other types of vison-based technologies have also been applied to study distress behaviour, mostly in pigs. Specifically, optical flow analysis has been used to detect abnormal movement (i.e. tripping or trampling) of pigs in the slaughterhouse (Gronskyte et al. 2016). This type of analysis would certainly be of value in the beef sector as well, to better understand behaviours associated with pre-slaughter stress.

Reducing animal handling where possible can further minimise fear and distress. Cattle in extensive systems can be directed towards handling facilities using virtual fencing technology rather than corralled by farm workers (Rutter 2014). Moreover, the use of AMS in dairy systems can reduce the need for handling and has been associated with improved human-animal relationships as evidenced by decreased stress responses to handling and reduced avoidance distances (Wildridge et al. 2020). Technology such as AMS and automated feeding systems for calves also have the ability to provide animals with agency in their environments, a key component to promoting positive welfare (Špinka 2019). Further, cameras mounted in AMS waiting areas have previously been used in combination with machine-learning algorithms to automatically detect agonistic behaviours (Guzhva et al. 2016). Knowledge of herd dynamics can facilitate the creation of management groups based on knowledge of social interactions, which can minimise distress associated with negative social interactions.

With respect to the prediction of calving time, the clear welfare application of these sensors is to reduce pain and injury associated with dystocia, through prompt intervention, if necessary. The subtler welfare application is the ability to monitor cows from a distance. Self-isolation behaviour prior to calving has been observed in domesticated cattle in various housing situations (Lidfors et al. 1994; Proudfoot et al. 2014), and fear of humans may be perpetuated by disturbance during calving (des Roches et al. 2016). The ability to monitor cows from a distance, which could be achieved through a combination of sensors and vison-based approaches, could minimise potential fear during an already stressful event.

Finally, Meen et al. (2015) suggest that sound analysis of vocalisations in dairy cattle could represent a promising welfare indicator, by automatically detecting differences in frequencies of calls to distinguish positive and negative affective state and alerting farmers to a welfare compromise. Vocal patterns have been correlated with arousal and valence in other farmed species such as pigs and goats but have not yet been explored in depth in cattle (see review by Green et al. 2018). In one study, Green et al. (2021) investigated vocalisation patterns of cows after separation from their calves. When the call type was ambiguous, narrow-band spectrograms were assessed by analysing fundamental frequency and stability of the vocalisations across the duration of each call. Kinematic diagrams also highlighted variation in vocal patterns according to cow emotional state, suggesting that sound analysis may prove promising for on-farm cattle welfare assessment (Green et al. 2021).

#### 11.4.1.5 Freedom to Express Normal Behaviour: By Providing Sufficient Space, Proper Facilities, and Company of the Animals' Own Kind

Precision technologies within the agricultural context routinely harness behavioural data, usually with the goal of reaching conclusions about aspects of heath (e.g. transition disease) and biological functioning (e.g. oestrus detection); however, these technologies have rarely been used to assess behaviour as the primary outcome measure (Beaver et al. 2020). The Freedom to express normal behaviour arose out of concerns over behavioural restriction (Brambell 1965), and thus mirrors the other freedoms with its focus on negative affect, at least in its original sense. Yet, the wording of this freedom, as the only freedom 'to', presents a unique invitation to explore positive affect, which is increasingly incorporated into the understanding of animal welfare (Mellor 2016; Lawrence et al. 2019).

Proximity loggers have been implemented to detect positive social behaviours such as social grooming, which can then be used to understand social network structures in dairy herds (Boyland et al. 2016). The use of a veterinary telemedicine system (i.e. electronic communication to remotely deliver health information) can facilitate pasture access for cattle (Warren et al. 2003). Improved access to pasture would allow cattle the space to perform normal grazing and social behaviours.

On commercial farms, calves are often fed a restricted amount of milk twice per day, by bottle or bucket (USDA 2016). This feeding pattern stands in contrast to the behaviour of semi-wild cattle, observed to nurse the dam up to 12 times per day during the first week of life (Vitale et al. 1986). Technologies such as automated calf feeders can promote normal behaviour by facilitating management of group-housed dairy calves and allowing calves to mimic natural suckling patterns by adjusting meal frequency and quantity consumed.

### 11.4.2 Potential Risks to Cattle Welfare from Precision Livestock Farming

It should be noted that PLF technologies are designed to facilitate rather than replace good stockmanship (Rutter 2017), and the successful integration of PLF systems into normal farm practice will depend upon vigilance in maintaining its functionality. If PLF instead supplants visual observation, there is a risk of negative welfare implications, particularly as technology does not always function properly and requires the knowledge and experience of stockpeople to refine implementation and interpret results.

As noted by Wurtz et al. (2019), much of the new and existing research into PLF systems, such as automated tracking of animals, does not build upon previous studies. New research groups entering the PLF research arena may instead start from scratch and end up recreating similar technologies to previous groups. This situation may occur because publications from the preceding research do not include enough detail to allow replication, or the algorithms created are subject to intellectual property rights restrictions. Wurtz et al. (2019) recommend, in particular, that reporting standards for machine vision literature be improved to facilitate efforts of future researchers to build upon the provided evidence. This observation suggests that some PLF research may not be fully in alignment with the 3 Rs (i.e. Reduction, Refinement, and Replacement; (NC3Rs 2020)). This repetition, if avoidable, could potentially affect cattle welfare, in the event that invasive techniques are used to measure physiological parameters, or if animals are subjected to stressful conditions or experimental infection (e.g. where disease is induced or monitored without intervention to allow detection technologies to be developed). As Jukan et al. (2017) suggest, an ultimate aim should be to integrate relevant technologies into a centralised database to share information and best practices. Jukan et al. (2017) also recommend that researchers consider the possibility of building on research conducted with other species or in other contexts.

As discussed in the present chapter, PLF technologies for dairy systems were originally developed to detect behaviours associated with oestrus, and technologies are still widely in use for this purpose. In terms of animal welfare, these technologies can arguably improve welfare by identifying fertility compromises that could be linked to disease or other underlying issues such as chronic stress (Walker et al. 2008). However, the extent to which fertility is a reliable welfare metric has been widely debated, with some suggesting that improvements to fertility could theoretically reduce welfare by exposing the animal to the myriad welfare risks associated with calving, such as dystocia and transition disease (Ritter et al. 2019). Moreover, the ability to detect oestrus without the need to rely on visual observation may reduce the incentive for producers to provide cattle with freedom of movement or suitable environments to express their full range of oestrous activities, potentially impinging on their ability to express important normal behaviour.

There is surprisingly little research addressing how wearable technology itself influences cow behaviour. Certain contemporary studies make particular note of how lightweight the technology has become, and authors postulate that the technology itself will not influence or restrict behaviour when worn, due to its diminutive size and weight (e.g. Saitoh and Kato 2021). Indeed, sensors have become progressively smaller, lighter, and less obtrusive over the last several decades (Maroto-Molina et al. 2019); for instance, Roberts et al. (1995) described an early differential GPS (DGPS) device for animal location monitoring that weighed approximately 2.5 kg. However, we call for further research directly addressing affects of wearable technology on cattle behaviour (of both the wearer and herdmates) and potential consequences to animal welfare.

Finally, the general public has expressed concern over intensification in farming (Spooner et al. 2014) because it deviates from the 'naturalness' element evoked by small family-run farms (Boogaard et al. 2010; Gieseke et al. 2018). Although herd size is an unreliable predictor of animal welfare, citizens disapprove of this facet of intensification (Gieseke et al. 2018). Larger dairy farms are more likely to implement PLF technologies (Gargiulo et al. 2018), which may further reduce public acceptance of these technologies. It remains to be seen whether technology can be incorporated into the lay public's definition of naturalness, and whether younger generations may accept a broader definition of naturalness that includes technology (see Beaver et al. 2020). Preliminary research has been conducted on public attitudes towards digital farming technologies in Germany (Pfeiffer et al. 2020) but further research is needed to explore this topic in full. Although public acceptance does not directly influence the welfare of cattle on farm, retailers have a prominent voice in which animal welfare standards are prioritised and enforced. These retailers, in turn, are significantly influenced by consumer views (Grandin 2014).

## 11.5 Summary

A variety of precision livestock farming technologies are already being used on commercial farms to help improve the efficiency of production, monitor animal health, and alert the farmer to any animals requiring their attention. Although the majority of PLF technologies currently target dairy production, systems for beef production are increasing in availability. Whilst these systems generally help to improve animal welfare, farmers need to be aware of the limitations of these systems and need to be able to manage their animals using more traditional management when technologies fail. PLF systems are not intended to replace the skills of expert stock people but instead provide a tool to help them manage their animals more effectively.

## References

Aldridge MN, Lee SJ, Taylor JD, Popplewell GI, Job FR, Pitchford WS, Aldridge MN, Lee SJ, Taylor JD, Popplewell GI, Job FR, Pitchford WS (2016) The use of walk over weigh to predict calving date in extensively managed beef herds. Anim Prod Sci 57:583–591. https://doi.org/10.1071/AN15172

Alawneh JI, Williamson NB, Bailey D (2006) Comparison of a camera-software system and typical farm management for detecting oestrus in dairy cattle at pasture. N Z Vet J 54(2):73–77

Alsaaod M, Fadul M, Steiner A (2019) Automatic lameness detection in cattle. Vet J 246:35–44. https://doi.org/10.1016/J.TVJL.2019.01.005

Appelqvist I, De Wet H, Marusic S, Janakievski F (2022) Industry 4.0 and the impact on the agrifood industry. In: Food engineering innovations across the food supply chain. Academic, Cambridge, MA, pp 343–356

Baruselli PS, Ferreira RM, Filho MFS, Bó GA (2018) Review: Using artificial insemination v. natural service in beef herds. Animal 12:s45–s52. https://doi.org/10.1017/S175173111800054X

Beaver A, Proudfoot KL, von Keyserlingk MAG (2020) Symposium review: Considerations for the future of dairy cattle housing: An animal welfare perspective. J Dairy Sci 103:5746–5758. https://doi.org/10.3168/JDS.2019-17804

Beaver A, Strazhnik E, von Keyserlingk MAG, Weary DM (2021a) The freestall reimagined: Effects on stall hygiene and space usage in dairy cattle. Animals 11:1711. https://doi.org/10.3390/ANI11061711

Beaver A, Weary DM, von Keyserlingk MAG (2021b) Invited review: The welfare of dairy cattle housed in tiestalls compared to less-restrictive housing types: A systematic review. J Dairy Sci. https://doi.org/10.3168/JDS.2020-19609

Berckmans D (2017) General introduction to precision livestock farming. Anim Front 7:6–11. https://doi.org/10.2527/AF.2017.0102

Bezen R, Edan Y, Halachmi I (2020) Computer vision system for measuring individual cow feed intake using RGB-D camera and deep learning algorithms. Comput Electron Agric 172:105345. https://doi.org/10.1016/J.COMPAG.2020.105345

Bhakta I, Phadikar S, Majumder K (2019) State-of-the-art technologies in precision agriculture: a systematic review. J Sci Food Agric 99:4878–4888. https://doi.org/10.1002/JSFA.9693

Boogaard BK, Bock BB, Oosting SJ, Wiskerke JSC, van der Zijpp AJ (2010) Social acceptance of dairy farming: the ambivalence between the two faces of modernity. J Agric Environ Ethics 24:259–282. https://doi.org/10.1007/S10806-010-9256-4

Boyland NK, Mlynski DT, James R, Brent LJN, Croft DP (2016) The social network structure of a dynamic group of dairy cows: From individual to group level patterns. Appl Anim Behav Sci 174:1–10. https://doi.org/10.1016/J.APPLANIM.2015.11.016

Brambell FWR (1965) Report of the technical committee to enquire into the welfare of animals kept under intensive livestock husbandry systems (Command Rep. 2836). Her Majesty's Stationery Office, London

Brown DJ, Swan AA, Mortimer ML (2011) Pedigree matchmaker: can it tell us more than just pedigree? Proc Assoc Advmt Anim Breed Genet 19:231–234

Buhman MJ, Perino LJ, Galyean ML, Wittum TE, Montgomery TH, Swingle RS (2005) Association between changes in eating and drinking behaviors and respiratory tract disease in newly arrived calves at a feedlot. Am J Vet Res 61:1163–1168. https://doi.org/10.2460/ajvr.2000.61.1163

Bunter KL, Johnston DJ, Wolcott ML, Fordyce G (2014) Factors associated with calf mortality in tropically adapted beef breeds managed in extensive Australian production systems. Anim Prod Sci 54:25–36. https://doi.org/10.1071/AN12421

Campbell DL, Ouzman J, Mowat D, Lea JM, Lee C, Llewellyn RS (2020) Virtual fencing technology excludes beef cattle from an environmentally sensitive area. Animals 10:1069. https://doi.org/10.3390/ani10061069

Charlton GL, Bleach ECL, Rutter SM (2019) Cows with paratuberculosis (Johne's disease) alter their lying behavior around peak lactation. J Dairy Sci 102:11328–11336. https://doi.org/10.3168/JDS.2019-16854

Claus R, Karg H, Zwiauer D, von Butler I, Pirchner F, Rattenberger E (1983) Analysis of factors influencing reproductive performance of the dairy cow by progesterone assay in milk-fat. Br Vet J 139:29–37. https://doi.org/10.1016/S0007-1935(17)30586-9

Costa JHC, Cantor MC, Neave HW (2021) Symposium review: Precision technologies for dairy calves and management applications. J Dairy Sci 104:1203–1219. https://doi.org/10.3168/JDS.2019-17885

CSIRO (n.d.) Ceres Tag: smart ear tags for livestock. https://www.csiro.au/en/research/animals/livestock/ceres-tag

des Roches ADB, Veissier I, Boivin X, Gilot-Fromont E, Mounier L (2016) A prospective exploration of farm, farmer, and animal characteristics in human-animal relationships: An epidemiological survey. J Dairy Sci 99:5573–5585. https://doi.org/10.3168/JDS.2015-10633

Dittrich I, Gertz M, Krieter J (2019) Alterations in sick dairy cows' daily behavioural patterns. Heliyon 5:e02902. https://doi.org/10.1016/J.HELIYON.2019.E02902

Dobson H, Walker SL, Morris MJ, Routly JE, Smith RF (2008) Why is it getting more difficult to successfully artificially inseminate dairy cows? Animal 2:1104–1111. https://doi.org/10.1017/S175173110800236X

Eastwood CR, Chapman DF, Paine MS (2012) Networks of practice for co-construction of agricultural decision support systems: Case studies of precision dairy farms in Australia. Agric Syst 108:10–18. https://doi.org/10.1016/J.AGSY.2011.12.005

FAWC (2009) Farm animal welfare in Great Britain: past, present and future. Trends Food Sci Technol 70

Fernández-Carrión E, Martínez-Avilés M, Ivorra B, Martínez-López B, Ramos ÁM, Sánchez-Vizcaíno JM (2017) Motion-based video monitoring for early detection of livestock diseases: The case of African swine fever. PLoS One 12:e0183793. https://doi.org/10.1371/JOURNAL.PONE.0183793

Ferrari S, Piccinini R, Silva M, Exadaktylos V, Berckmans D, Guarino M (2010) Cough sound description in relation to respiratory diseases in dairy calves. Prev Vet Med 96:276–280. https://doi.org/10.1016/J.PREVETMED.2010.06.013

Foote RH (2002) The history of artificial insemination: Selected notes and notables1. J Anim Sci 80:1–10. https://doi.org/10.2527/animalsci2002.80e-suppl_21a

Fraser D, Weary D, Pajor E, Milligan B (1997) A scientific conception of animal welfare that reflects ethical concerns. Anim Welfare 6(3):187–205

Gadd ME (2012) Barriers, the beef industry and unnatural selection: a review of the impact of veterinary fencing on mammals in southern Africa. In: Somers M, Hayward M (eds) Fencing for conservation. Springer, New York, pp 153–186. https://doi.org/10.1007/978-1-4614-0902-1_9

Gargiulo JI, Eastwood CR, Garcia SC, Lyons NA (2018) Dairy farmers with larger herd sizes adopt more precision dairy technologies. J Dairy Sci 101:5466–5473. https://doi.org/10.3168/JDS.2017-13324

Gieseke D, Lambertz C, Gauly M (2018) Relationship between herd size and measures of animal welfare on dairy cattle farms with freestall housing in Germany. J Dairy Sci 101:7397–7411. https://doi.org/10.3168/JDS.2017-14232

Goldhawk C, Chapinal N, Veira DM, Weary DM, von Keyserlingk MAG (2009) Prepartum feeding behavior is an early indicator of subclinical ketosis. J Dairy Sci 92:4971–4977. https://doi.org/10.3168/JDS.2009-2242

González LA, Bishop-Hurley G, Henry D, Charmley E, González LA, Bishop-Hurley G, Henry D, Charmley E (2014) Wireless sensor networks to study, monitor and manage cattle in grazing systems. Anim Prod Sci 54:1687–1693. https://doi.org/10.1071/AN14368

Grandin T (2014) Animal welfare and society concerns finding the missing link. Meat Sci 98:461–469. https://doi.org/10.1016/J.MEATSCI.2014.05.011

Green AC, Johnston IN, Clark CEF (2018) Invited review: The evolution of cattle bioacoustics and application for advanced dairy systems. Animal 12:1250–1259

Green AC, Lidfors LM, Lomax S, Favaro L, Clark CEF (2021) Vocal production in postpartum dairy cows: Temporal organization and association with maternal and stress behaviors. J Dairy Sci 104:826–838

Gronskyte R, Clemmensen LH, Hviid MS, Kulahci M (2016) Monitoring pig movement at the slaughterhouse using optical flow and modified angular histograms. Biosyst Eng 141:19–30. https://doi.org/10.1016/J.BIOSYSTEMSENG.2015.10.002

Guzhva O, Ardö H, Herlin A, Nilsson M, Åström K, Bergsten C (2016) Feasibility study for the implementation of an automatic system for the detection of social interactions in the waiting area of automatic milking stations by using a video surveillance system. Comput Electron Agric 127:506–509. https://doi.org/10.1016/J.COMPAG.2016.07.010

Haley DB, Rushen J, De Passillé AM (2000) Behavioural indicators of cow comfort: Activity and resting behaviour of dairy cows in two types of housing. Can J Anim Sci 80:257–263. https://doi.org/10.4141/A99-084

Hallén Sandgren C, Emanuelson U (2016) Consistency of measurements from an automatic body condition scoring camera. In: Kamphuis C, Steeneveld W (eds) Precision dairy farming. Wageningen Academic, Leeuwarden, pp 285–290

Hixson CL, Krawczel PD, Caldwell JM, Miller-Cushon EK (2018) Behavioral changes in group-housed dairy calves infected with Mannheimia haemolytica. J Dairy Sci 101:10351–10360. https://doi.org/10.3168/JDS.2018-14832

Hossain MA, Quaddus M (2013) Does mandatory pressure increase RFID adoption? A case study of Western Australian livestock industry. IFIP Adv Inf Commun Technol 402:184–202. https://doi.org/10.1007/978-3-642-38862-0_12

Huzzey JM, Veira DM, Weary DM, Von Keyserlingk MAG (2007) Prepartum behavior and dry matter intake identify dairy cows at risk for metritis. J Dairy Sci 90:3220–3233. https://doi.org/10.3168/JDS.2006-807

Huzzey JM, Weary DM, Tiau BYF, Von Keyserlingk MAG (2014) Short communication: Automatic detection of social competition using an electronic feeding system. J Dairy Sci 97:2953–2958. https://doi.org/10.3168/JDS.2013-7434

Jachowski DS, Slotow R, Millspaugh JJ (2014) Good virtual fences make good neighbors: opportunities for conservation. Anim Conserv 17:187–196. https://doi.org/10.1111/acv.12082

Ji B, Banhazi T, Wang C, Li B (2017) PLF technologies: Model development for solving heat stress problems on dairy farms. In: Precision livestock farming 2017—Papers presented at the 8th European conference on precision livestock farming, ECPLF 2017, pp 821–836

Jiang B, Wu Q, Yin X, Wu D, Song H, He D (2019) FLYOLOv3 deep learning for key parts of dairy cow body detection. Comput Electron Agric 166:104982. https://doi.org/10.1016/J.COMPAG.2019.104982

Jorquera-Chavez M, Fuentes S, Dunshea FR, Warner RD, Poblete T, Jongman EC (2019) Modelling and validation of computer vision techniques to assess heart rate, eye temperature, ear-base temperature and respiration rate in cattle. Animals 9:1089. https://doi.org/10.3390/ANI9121089

Jukan A, Masip-Bruin X, Amla N (2017) Smart computing and sensing technologies for animal welfare: A systematic review. ACM Comput Surv 50. https://doi.org/10.1145/3041960

Kang X, Zhang XD, Liu G (2021) A review: development of computer vision-based lameness detection for dairy cows and discussion of the practical applications. Sensors 21:753. https://doi.org/10.3390/S21030753

Kawasaki M, Kawamura S, Tsukahara M, Morita S, Komiya M, Natsuga M (2008) Near-infrared spectroscopic sensing system for on-line milk quality assessment in a milking robot. Comput Electron Agric 63(1):22–27

Knauer WA, Godden SM, Dietrich A, James RE (2017) The association between daily average feeding behaviors and morbidity in automatically fed group-housed preweaned dairy calves. J Dairy Sci 100:5642–5652. https://doi.org/10.3168/JDS.2016-12372

Knight CH (2020) Review: Sensor techniques in ruminants: more than fitness trackers. Animal 14:s187–s195. https://doi.org/10.1017/S1751731119003276

Krohn CC, Munksgaard L (1993) Behaviour of dairy cows kept in extensive (loose housing/pasture) or intensive (tie stall) environments II. Lying and lying-down behaviour. Appl Anim Behav Sci 37:1–16. https://doi.org/10.1016/0168-1591(93)90066-X

Lawrence AB, Vigors B, Sandøe P (2019) What is so positive about positive animal welfare?—a critical review of the literature. Animals 9:783. https://doi.org/10.3390/ANI9100783

Lee C, Colditz IG, Campbell DL (2018) A framework to assess the impact of new animal management technologies on welfare: A case study of virtual fencing. Front Vet Sci 5:187. https://doi.org/10.3389/fvets.2018.00187

Lidfors LM, Moran D, Jung J, Jensen P, Castren H (1994) Behaviour at calving and choice of calving place in cattle kept in different environments. Appl Anim Behav Sci 42:11–28. https://doi.org/10.1016/0168-1591(94)90003-5

Lomax S, Colusso P, Clark CE (2019) Does virtual fencing work for grazing dairy cattle? Animals 9:429. https://doi.org/10.3390/ani9070429

Lowe GL, Sutherland MA, Waas JR, Schaefer AL, Cox NR, Stewart M (2019) Physiological and behavioral responses as indicators for early disease detection in dairy calves. J Dairy Sci 102:5389–5402. https://doi.org/10.3168/JDS.2018-15701

Lucy MC (2007) Fertility in high-producing dairy cows: reasons for decline and corrective strategies for sustainable improvement. Soc Reprod Fertil Suppl 64:237–254. https://doi.org/10.5661/rdr-vi-237

Maroto Molina FM, Marín CCP, Moreno LM, Buendía EIA, Marín DCP (2020) Welfare Quality® for dairy cows: towards a sensor-based assessment. J Dairy Res 87:28–33. https://doi.org/10.1017/S002202992000045X

Maroto-Molina F, Navarro-García J, Príncipe-Aguirre K, Gómez-Maqueda I, Guerrero-Ginel JE, Garrido-Varo A, Pérez-Marín DC (2019) A low-cost IOT-based system to monitor the location of a whole herd. Sensors (Switzerland) 19. https://doi.org/10.3390/s19102298

McDonald PV, von Keyserlingk MAG, Weary DM (2019) Technical note: Using an electronic drinker to monitor competition in dairy cows. J Dairy Sci 102:3495–3500. https://doi.org/10.3168/JDS.2018-15585

Meen GH, Schellekens MA, Slegers MHM, Leenders NLG, van Erp-van der Kooij E, Noldus LPJJ (2015) Sound analysis in dairy cattle vocalisation as a potential welfare monitor. Comput Electron Agric 118:111–115. https://doi.org/10.1016/J.COMPAG.2015.08.028

Mellor DJ (2016) Updating animal welfare thinking: moving beyond the "five freedoms" towards "a life worth living". Animals 6:21. https://doi.org/10.3390/ANI6030021

Menzies D, Patison KP, Corbet NJ, Swain DL, Menzies D, Patison KP, Corbet NJ, Swain DL (2017) Using Walk-over-Weighing technology for parturition date determination in beef cattle. Anim Prod Sci 58:1743–1750. https://doi.org/10.1071/AN16694

Menzies D, Patison KP, Corbet NJ, Swain DL (2018) Using temporal associations to determine maternal parentage in extensive beef herds. Anim Prod Sci 58:943–949. https://doi.org/10.1071/AN16450

Miller GA, Hyslop JJ, Barclay D, Edwards A, Thomson W, Duthie CA (2019) Using 3D Imaging and Machine Learning to Predict Liveweight and Carcass Characteristics of Live Finishing Beef Cattle. Front Sustain Food Syst 3:30. https://doi.org/10.3389/fsufs.2019.00030

Morris JE, Cronin GM, Bush RD (2012) Improving sheep production and welfare in extensive systems through precision sheep management. Anim Prod Sci 52:665–670. https://doi.org/10.1071/AN11097

Mottram T (2016) Animal board invited review: Precision livestock farming for dairy cows with a focus on oestrus detection. Animal 10:1575–1584. https://doi.org/10.1017/S1751731115002517

Mullins IL, Truman CM, Campler MR, Bewley JM, Costa JHC (2019) Validation of a commercial automated body condition scoring system on a commercial dairy farm. Animal 9:287. https://doi.org/10.3390/ANI9060287

NC3Rs (2020) The 3Rs | NC3Rs. Accessed 23 July 2021. https://nc3rs.org.uk/the-3rs

Newman AP, Mbanaso F, Nnadi EO, Fontaneda LÁS, Shuttleworth AB (2018) Drainage of animal housing units for maximum animal welfare and environmental control studies on microbiological safety and drainage behaviour. In: World environmental and water resources congress 2018: Groundwater, sustainability, and hydro-climate/climate change—selected papers from the world environmental and water resources congress 2018. American Society of Civil Engineers (ASCE), pp 14–26

Oudshoorn FW, Cornou C, Hellwing ALF, Hansen HH, Munksgaard L, Lund P, Kristensen T (2013) Estimation of grass intake on pasture for dairy cows using tightly and loosely mounted di- and tri-axial accelerometers combined with bite count. Comput Electron Agric 99:227–235. https://doi.org/10.1016/J.COMPAG.2013.09.013

Palmer MA, Olmos G, Boyle LA, Mee JF (2010) Estrus detection and estrus characteristics in housed and pastured Holstein–Friesian cows. Theriogenology 74:255–264. https://doi.org/10.1016/J.THERIOGENOLOGY.2010.02.009

Petherick JC, Small AH, Mayer DG, Colditz IG, Ferguson DM, Stafford KJ (2014) A comparison of welfare outcomes for weaner and mature Bos indicus bulls surgically or tension band castrated with or without analgesia: 1. Behavioural responses. Appl Anim Behav Sci 157:23–34. https://doi.org/10.1016/J.APPLANIM.2014.05.003

Pezzuolo A, Guarino M, Sartori L, Marinello F (2018) A feasibility study on the use of a structured light depth-camera for three-dimensional body measurements of dairy cows in free-stall barns. Sensors 18:673. https://doi.org/10.3390/S18020673

Pfeiffer J, Gabriel A, Gandorfer M (2020) Understanding the public attitudinal acceptance of digital farming technologies: a nationwide survey in Germany. Agric Hum Values 38:107–128. https://doi.org/10.1007/S10460-020-10145-2

Pham X, Stack M (2018) How data analytics is transforming agriculture. Bus Horiz 61:125–133. https://doi.org/10.1016/J.BUSHOR.2017.09.011

Pillen JL, Pinedo PJ, Ives SE, Covey TL, Naikare HK, Richeson JT (2016) Alteration of activity variables relative to clinical diagnosis of bovine respiratory disease in newly received feedlot cattle. Bov Pract 50:1–8

Popescu S, Borda C, Diugan EA, Spinu M, Groza IS, Sandru CD (2013) Dairy cows welfare quality in tie-stall housing system with or without access to exercise. Acta Vet Scand 55:1–11. https://doi.org/10.1186/1751-0147-55-43

Porto SMC, Arcidiacono C, Anguzza U, Cascone G (2013) A computer vision-based system for the automatic detection of lying behaviour of dairy cows in free-stall barns. Biosyst Eng 115:184–194. https://doi.org/10.1016/J.BIOSYSTEMSENG.2013.03.002

Proudfoot KL, Weary DM, von Keyserlingk MAG (2010) Behavior during transition differs for cows diagnosed with claw horn lesions in mid lactation. J Dairy Sci 93:3970–3978. https://doi.org/10.3168/JDS.2009-2767

Proudfoot KL, Jensen MB, Weary DM, Von Keyserlingk MAG (2014) Dairy cows seek isolation at calving and when ill. J Dairy Sci 97:2731–2739. https://doi.org/10.3168/JDS.2013-7274

Proudfoot KL, Weary DM, LeBlanc SJ, Mamedova LK, von Keyserlingk MAG (2018) Exposure to an unpredictable and competitive social environment affects behavior and health of transition dairy cows. J Dairy Sci 101:9309–9320. https://doi.org/10.3168/JDS.2017-14115

Quimby WF, Sowell BF, Bowman JGP, Branine ME, Hubbert ME, Sherwood HW (2011) Application of feeding behaviour to predict morbidity of newly received calves in a commercial feedlot. Can J Anim Sci 81:315–320. https://doi.org/10.4141/A00-098

Reith S, Hoy S (2018) Review: Behavioral signs of estrus and the potential of fully automated systems for detection of estrus in dairy cattle. Animal 12:398–407. https://doi.org/10.1017/S1751731117001975

Richards JS, Atkins K (2007) Determining pedigree by association in Merino flocks. Proc Assoc Adv Anim Breed Genet 17:403–406

Richeson JT, Lawrence TE, White BJ (2018) Using advanced technologies to quantify beef cattle behavior. Transl Anim Sci 2:223–229. https://doi.org/10.1093/tas/txy004

Riesch F, Komainda M, Horn J, Isselstein J (2022) Real-world applications for virtual fences–what are potential benefits for conservation? In: International grassland congress proceedings, pp 1–4

Ritter C, Beaver A, von Keyserlingk MAG (2019) The complex relationship between welfare and reproduction in cattle. Reprod Domest Anim 54:29–37. https://doi.org/10.1111/RDA.13464

Roberts G, Williams A, Last JD, Penning PD, Rutter SM (1995) A low-power postprocessed DGPS system for logging the locations of sheep on hill pastures. Navigation 42:327–336. https://doi.org/10.1002/j.2161-4296.1995.tb01894.x

Roberts SL, Powell JG, Hughes HD, Richeson JT (2018) Effect of castration method and analgesia on inflammation, behavior, growth performance, and carcass traits in feedlot cattle. J Anim Sci 96:66–75. https://doi.org/10.1093/JAS/SKX022

Roelofs J, López-Gatius F, Hunter RHF, van Eerdenburg FJCM, Hanzen C (2010) When is a cow in estrus? Clinical and practical aspects. Theriogenology 74:327–344. https://doi.org/10.1016/J.THERIOGENOLOGY.2010.02.016

Rutter SM (2012) A "smart" future for ruminant livestock production? Cattle Pract 20:186–193

Rutter SM (2014) Smart technologies for detecting animal welfare status and delivering health remedies for rangeland systems. Rev Sci Tech Off Int Epiz 33:181–187

Rutter SM (2017) Advanced livestock management solutions. Adv Sheep Welf 245–261. https://doi.org/10.1016/B978-0-08-100718-1.00013-3

Rutter SM, Jackson DA, Johnson CL, Forbes JM (1987) Automatically recorded competitive feeding behaviour as a measure of social dominance in dairy cows. Appl Anim Behav Sci 17:41–50. https://doi.org/10.1016/0168-1591(87)90006-2

Sahar MW, Beaver A, von Keyserlingk MAG, Weary DM (2020) Predicting disease in transition dairy cattle based on behaviors measured before calving. Animals 10:928. https://doi.org/10.3390/ANI10060928

Saint-Dizier M, Chastant-Maillard S (2012) Towards an automated detection of oestrus in dairy cattle. Reprod Domest Anim 47:1056–1061. https://doi.org/10.1111/j.1439-0531.2011.01971.x

Saitoh T, Kato Y (2021) Evaluation of wearable cameras for monitoring and analyzing calf behavior: A preliminary study. Animals 11:2622. https://doi.org/10.3390/ani11092622

Schellberg J, Hill MJ, Gerhards R, Rothmund M, Braun M (2008) Precision agriculture on grassland: Applications, perspectives and constraints. Eur J Agron 29:59–71. https://doi.org/10.1016/J.EJA.2008.05.005

Shearer JK, Stock ML, Van Amstel SR, Coetzee JF (2013) Assessment and management of pain associated with lameness in cattle. Vet Clin North Am Food Anim Pract 29:135–156. https://doi.org/10.1016/j.cvfa.2012.11.012

Shewbridge Carter L, Rutter SM, Ball D, Gibbons J, Haskell MJ (2021) Dairy cow trade-off preference for 2 different lying qualities: Lying surface and lying space. J Dairy Sci 104:862–873. https://doi.org/10.3168/jds.2020-18781

Song X, Leroy T, Vranken E, Maertens W, Sonck B, Berckmans D (2008) Automatic detection of lameness in dairy cattle—Vision-based trackway analysis in cow's locomotion. Comput Electron Agric 64:39–44. https://doi.org/10.1016/J.COMPAG.2008.05.016

Špinka M (2019) Animal agency, animal awareness and animal welfare. Anim Welf 28:11–20. https://doi.org/10.7120/09627286.28.1.011

Spooner JM, Schuppli CA, Fraser D (2014) Attitudes of Canadian citizens toward farm animal welfare: A qualitative study. Livest Sci 163:150–158. https://doi.org/10.1016/J.LIVSCI.2014.02.011

Stangaferro ML, Wijma R, Caixeta LS, Al-Abri MA, Giordano JO (2016a) Use of rumination and activity monitoring for the identification of dairy cows with health disorders: Part I. Metabolic and digestive disorders. J Dairy Sci 99:7395–7410. https://doi.org/10.3168/JDS.2016-10907

Stangaferro ML, Wijma R, Caixeta LS, Al-Abri MA, Giordano JO (2016b) Use of rumination and activity monitoring for the identification of dairy cows with health disorders: Part II. Mastitis. J Dairy Sci 99:7411–7421. https://doi.org/10.3168/JDS.2016-10908

Stangaferro ML, Wijma R, Caixeta LS, Al-Abri MA, Giordano JO (2016c) Use of rumination and activity monitoring for the identification of dairy cows with health disorders: Part III. Metritis. J Dairy Sci 99:7422–7433. https://doi.org/10.3168/JDS.2016-11352

Stephen C, Menzies D, Swain D, Patison K, Corbet N, Norman S (2019) Telemetric monitoring of calving using a novel calf alert device. Clin Theriogenol 11

Stevenson JS, Hill SL, Nebel RL, DeJarnette JM (2014) Ovulation timing and conception risk after automated activity monitoring in lactating dairy cows. J Dairy Sci 97:4296–4308. https://doi.org/10.3168/JDS.2013-7873

Stewart M, Schaefer AL, Haley DB, Colyn J, Cook NJ, Stafford KJ, Webster JR (2008) Infrared thermography as a non-invasive method for detecting fear-related responses of cattle to handling procedures. Anim Welf 17:387–393

Stokes JE, Leach KA, Main DCJ, Whay HR (2012) An investigation into the use of infrared thermography (IRT) as a rapid diagnostic tool for foot lesions in dairy cattle. Vet J 193:674–678. https://doi.org/10.1016/J.TVJL.2012.06.052

Studds MJ, Deikun LL, Sorter DE, Pempek JA, Proudfoot KL (2018) Short communication: The effect of diarrhea and navel inflammation on the lying behavior of veal calves. J Dairy Sci 101:11251–11255. https://doi.org/10.3168/JDS.2018-15003

Svensson C, Jensen MB (2007) Short communication: identification of diseased calves by use of data from automatic milk feeders. J Dairy Sci 90:994–997. https://doi.org/10.3168/JDS.S0022-0302(07)71584-9

Svensson C, Hultgren J, Oltenacu PA (2006) Morbidity in 3–7-month-old dairy calves in south-western Sweden, and risk factors for diarrhoea and respiratory disease. Prev Vet Med 74:162–179. https://doi.org/10.1016/J.PREVETMED.2005.11.008

Swartz TH, Findlay AN, Petersson-Wolfe CS (2017) Short communication: Automated detection of behavioral changes from respiratory disease in pre-weaned calves. J Dairy Sci 100:9273–9278. https://doi.org/10.3168/JDS.2016-12280

Tang W, Biglari A, Ebarb R, Pickett T, Smallidge S, Ward M (2021) A smart sensing system of water quality and intake monitoring for livestock and wild animals. Sensors 21:2885. https://doi.org/10.3390/s21082885

Thompson A, Proudfoot KL, Franks B, von Keyserlingk MAG (2019) Social environment and individual differences in feeding behavior are associated with risk of endometritis in dairy cows. Animals 9:828. https://doi.org/10.3390/ANI9100828

Thorup VM, Charlton GL, Gauld C, Bleach E, Rutter SM (2017) The largest reduction in activity due to lameness occurs during early lactation in multiparous dairy cows. In: Lameness in ruminants, pp 342–344

Toaff-Rosenstein RL, Gershwin LJ, Zanella AJ, Tucker CB (2016) The sickness response in steers with induced bovine respiratory disease before and after treatment with a non-steroidal anti-inflammatory drug. Appl Anim Behav Sci 181:49–62

Uddin J, Phillips CJ, Goma AA, McNeill DM (2019) Relationships between infrared temperature and laterality. Appl Anim Behav Sci 220:104855. https://doi.org/10.1016/j.applanim.2019.104855

Uddin J, Phillips CJ, Auboeuf M, McNeill DM (2021) Relationships between body temperatures and behaviours in lactating dairy cows. Appl Anim Behav Sci 241:105359. https://doi.org/10.1016/j.applanim.2021.105359

Umstatter C (2011) The evolution of virtual fences: A review. Comput Electron Agric 75(1):10–22

Umstatter C, Morgan-Davies J, Waterhouse T (2015) Cattle Responses to a Type of Virtual Fence. Rangel Ecol Manag 68:100–107. https://doi.org/10.1016/J.RAMA.2014.12.004

Ungar ED, Nevo Y, Baram H, Arieli A (2018) Evaluation of the IceTag leg sensor and its derivative models to predict behaviour, using beef cattle on rangeland. J Neurosci Methods 300:127–137. https://doi.org/10.1016/J.JNEUMETH.2017.06.001

USDA (2016) Dairy cattle management practices in the United States, 2014. Accessed April 7, 2021. https://www.aphis.usda.gov/animal_health/nahms/dairy/downloads/dairy14/Dairy14_dr_PartIII.pdf

van Erp-Van der Kooij E, Rutter SM (2020) Using precision farming to improve animal welfare. CAB Rev Perspect Agric Vet Sci Nutr Nat Resour 15:1–10. https://doi.org/10.1079/PAVSNNR202015051

Van Nuffel A, Zwertvaegher I, Van Weyenberg S, Pastell M, Thorup VM, Bahr C, Sonck B, Saeys W (2015) Lameness detection in dairy cows: Part 2. Use of sensors to automatically register changes in locomotion or behavior. Animals 5:861–885. https://doi.org/10.3390/ANI5030388

Vázquez-Diosdado JA, Miguel-Pacheco GG, Plant B, Dottorini T, Green M, Kaler J (2019) Developing and evaluating threshold-based algorithms to detect drinking behavior in dairy cows using reticulorumen temperature. J Dairy Sci 102:10471–10482. https://doi.org/10.3168/JDS.2019-16442

Viazzi S, Bahr C, Van Hertem T, Schlageter-Tello A, Romanini CEB, Halachmi I, Lokhorst C, Berckmans D (2014) Comparison of a three-dimensional and two-dimensional camera system for automated measurement of back posture in dairy cows. Comput Electron Agric 100:139–147. https://doi.org/10.1016/J.COMPAG.2013.11.005

Vitale AF, Tenucci M, Papini M, Lovari S (1986) Social behaviour of the calves of semi-wild Maremma cattle, Bos primigenius taurus. Appl Anim Behav Sci 16:217–231. https://doi.org/10.1016/0168-1591(86)90115-2

Walker SL, Smith RF, Jones DN, Routly JE, Dobson H (2008) Chronic stress, hormone profiles and estrus intensity in dairy cattle. Horm Behav 53:493–501. https://doi.org/10.1016/J.YHBEH.2007.12.003

Warren S, Nagl L, Schmitz R, Yao J, Hildreth T, Erickson H, Poole D, Andresen D (2003) A distributed infrastructure for veterinary telemedicine. Annu Int Conf IEEE Eng Med Biol Proc 2:1394–1397. https://doi.org/10.1109/IEMBS.2003.1279573

Weschenfelder AV, Saucier L, Maldague X, Rocha LM, Schaefer AL, Faucitano L (2013) Use of infrared ocular thermography to assess physiological conditions of pigs prior to slaughter and predict pork quality variation. Meat Sci 95:616–620. https://doi.org/10.1016/J.MEATSCI.2013.06.003

Whelan BM, McBratney AB (2000) The "null hypothesis" of precision agriculture management. Precis Agric 2:265–279. https://doi.org/10.1023/A:1011838806489

White BJ, Coetzee JF, Renter DG, Babcock AH, Thomson DU, Andresen D (2008) Evaluation of two-dimensional accelerometers to monitor behavior of beef calves after castration. Am J Vet Res 69:1005–1012. https://doi.org/10.2460/AJVR.69.8.1005

White BJ, Anderson DE, Renter DG, Larson RL, Mosier DA, Kelly LL, Theurer ME, Robért BD, Walz ML (2012) Clinical, behavioral, and pulmonary changes in calves following inoculation with Mycoplasma bovis. Am J Vet Res 73:490–497. https://doi.org/10.2460/AJVR.73.4.490

White BJ, Goehl DR, Amrine DE, Booker C, Wildman B, Perrett T (2016) Bayesian evaluation of clinical diagnostic test characteristics of visual observations and remote monitoring to diagnose bovine respiratory disease in beef calves. Prev Vet Med 126:74–80. https://doi.org/10.1016/J.PREVETMED.2016.01.027

Wildridge AM, Thomson PC, Garcia SC, Jongman EC, Kerrisk KL (2020) Transitioning from conventional to automatic milking: Effects on the human-animal relationship. J Dairy Sci 103:1608–1619. https://doi.org/10.3168/JDS.2019-16658

Williams LR, Fox DR, Bishop-Hurley GJ, Swain DL (2019) Use of radio frequency identification (RFID) technology to record grazing beef cattle water point use. Comput Electron Agric 156:193–202. https://doi.org/10.1016/J.COMPAG.2018.11.025

Windeyer MC, Leslie KE, Godden SM, Hodgins DC, Lissemore KD, LeBlanc SJ (2014) Factors associated with morbidity, mortality, and growth of dairy heifer calves up to 3 months of age. Prev Vet Med 113:231–240. https://doi.org/10.1016/J.PREVETMED.2013.10.019

Wurtz K, Camerlink I, D'Eath RB, Fernández AP, Norton T, Steibel J, Siegford J (2019) Recording behaviour of indoor-housed farm animals automatically using machine vision technology: A systematic review. PLoS One 14:e0226669. https://doi.org/10.1371/JOURNAL.PONE.0226669

# Strategies and Tools for Genetic Selection in Dairy Cattle and Their Application to Improving Animal Welfare

# 12

Jennie E. Pryce, Christa Egger-Danner, and Geoff Simm

## Contents



**Abstract**

Genetic improvement of farm animals, especially selection within breeds focussed on high production and efficiency, is often cited as a potential threat to animal welfare. However, many animal welfare issues can be addressed, at least partially, by animal breeding and genetics. In this chapter, we explore the relationship between genetic selection and animal welfare, the strategies and tools for genetic improvement and how they can contribute to improved animal welfare. A growing public awareness of animal welfare and environmental issues has led to breeding goals being broadened beyond farmer profitability. As animal

J. E. Pryce (✉)
Agriculture Victoria Research, Bundoora, VIC, Australia

School of Applied Systems Biology, La Trobe University, Bundoora, VIC, Australia
e-mail: jennie.pryce@agriculture.vic.gov.au

C. Egger-Danner
ZuchtData EDV-Dienstleistungen GmbH, Vienna, Austria

G. Simm
Global Academy of Agriculture and Food Systems, Royal (Dick) School of Veterinary Studies, University of Edinburgh, Easter Bush Campus, Midlothian, UK



M. Haskell (ed.), *Cattle Welfare in Dairy and Beef Systems*, Animal Welfare 23,
https://doi.org/10.1007/978-3-031-21020-4_12

welfare and behaviour are complex and multi-factorial, so the emergence of selection indices that include a large number of traits to optimise animal welfare in a way that is consistent with enterprise sustainability for the farmer is necessary. This trend is likely to continue and will be aided by the advent of new technologies for measuring animal welfare in conjunction with DNA-based predictions of genetic merit (genomic selection). The dairy cattle industry has been exemplary for the application of genomic selection, in addition to enabling selection decisions to be made earlier in life, it can be used to select for traits where it was not possible to select for previously. These include important welfare-related traits, such as improved disease resistance and heat tolerance. Dairy cattle breeding is a very international activity with just a few breeding companies dominating the market in semen for the most numerous breeds, especially the Holstein. Consequently, genetic diversity within breeds is diminishing and although genetic gain has been significant, the rate of inbreeding now presents itself as a threat to the future success of breeding programmes. A greater emphasis on diversity in breeding programmes and the traits under selection is needed as major themes in research and application. Innovation in methods to measure these new traits, (e.g. molecular phenotyping, sensor development, digitalisation data science, etc.) could dramatically transform selection for animal welfare, as these technologies can enable large-scale objective measurements of animal behaviours. In addition to animal-based outcome measures, factors like housing, feeding, specific management practices pose other risks to welfare. Risk factors and their interactions have an impact on the development of diseases or other challenges to welfare. Collaborative efforts between animal behaviour scientists, geneticists, engineers, data scientists, and others will potentially provide solutions to these challenges.

**Keywords**

Selection index · Dairy genetics · Genomic selection · Breeding goal · Resilience · Inbreeding

## 12.1 Introduction

The study of welfare is focused on improving the lives of animals (Von Keyserlingk and Weary 2017) and encompasses the health and functional fitness of animals in addition to promoting positive psychological states. Consideration of animal welfare is an important part of designing breeding programmes for ethical and commercial reasons; it is clearly important to animals themselves, to farmers, and to many consumers, and needs to be considered in designing programmes that are resilient and forward-looking. The three main challenges in designing welfare-friendly breeding programmes are: (1) defining what to improve (referred to as breeding goals by animal breeders) and the welfare indicators to use (referred to as selection criteria by animal breeders); (2) accessing measurements on large numbers of animals in a cost-effective way that can be used for genetic selection purposes to reduce the risks; and (3) developing and validating approaches to assess emotional states (Weary et al. 2017) with breeding programmes in mind.

Directives such as the OIE (World Organisation for Animal Health) Terrestrial Animal Code (OIE 2022) that aims to implement improvement of standards of worldwide animal health and welfare from a veterinary point of view, have highlighted the importance of genetic selection for animal welfare; for example 'individual animals within a breed should be selected to propagate offspring that exhibit traits beneficial to animal health and welfare by promoting robustness and longevity. These include resistance to infectious and production related diseases, ease of calving, fertility, body condition score and temperament'. Breeding activities facilitate long-term permanent and cumulative improvement of welfare, whereas improved management is faster in the short term, but might not be sustainable or permanent.

While selection between breeds and crossbreeding is likely to have impacts on animal welfare, the focus of this chapter is mainly to consider within-breed options including: (1) essential principles of genetic selection; (2) the expansion of dairy cattle breeding objectives to include traits associated with animal welfare; (3) animal welfare in breeding decisions including how technological advances and collaboration are key components of success in this area. The aim of this chapter is to provide examples of where breeding solutions have been applied in the past and thoughts on where this approach might be especially useful in the future, rather than cataloguing an exhaustive list of examples of potential animal welfare solutions applied to the field of animal breeding.

**Glossary Box (After Simm et al. 2021)**

Additive genetic effects—the influence on an animal's genotype or phenotype due to genes that act together in a relatively linear and cumulative manner. Estimated breeding values (EBVs) are used to estimate the aggregate effect of such genes on traits of interest.

AI or Artificial insemination—deposition of semen into the reproductive tract of a female animal—usually after earlier semen collection, dilution, freezing and storage, and subsequent thawing. It allows elite males to produce many more offspring than by natural mating.

Breeding goal—the set of traits which a breeding programme is intended to improve.

Breeding programme—the set of activities associated with breeding future generations of animals, including choice of breeding objective and selection criteria, recording, genotyping and genetic evaluation of animals, selection and mating of animals, monitoring genetic gain and inbreeding, etc.

BLUP: Best linear unbiased prediction—statistical procedure for estimating breeding values. It is applied under several sets of assumptions or models which account for different relationships between animals. BLUP estimates environmental effects and breeding values simultaneously, often for multiple traits, and so disentangles genetics from management, feeding, etc., more effectively, and leads to more accurate estimates of breeding value than other methods.

EBV: estimated breeding value—an estimate of the additive genetic merit of an animal, derived from performance records from the animal itself and/or its relatives, and their pedigree relationships; increasingly also uses genomic information.

Genome wide association studies—establish relationships between genetic markers (genotypes, usually SNPs) and animal performance in traits of interest; this information can be used to improve understanding of the genetic control of traits and can also be included in genomic selection to identify other animals with the most favourable combination of genetic markers.

Genomic selection—selection of breeding animals based on the use of genome-wide genetic markers (usually SNPs) to estimate breeding values. The relationships among genetic marker genotypes and animal phenotypes are first measured in a 'reference population', in order to estimate breeding values of selection candidates from genotypes using only genotypes, or a combination of genotypes and performance records.

Genotype—the set of genes/alleles that an animal inherits—may refer to a pair of alleles at a specific locus/site in the genome, or to the collective effect of many loci affecting a trait of interest.

Heritability—that fraction of the total phenotypic variation that is due to additive genetic variation; the proportion of superiority of parents that gets passed on to offspring.

Heterosis/hybrid vigour—the advantage in performance of crossbred animals over the mid-parent mean for the trait of interest.

Introgression—introduction of a new gene of interest (e.g. for polledness), usually via crossing with another breed carrying that gene, followed by backcrossing to the original breed while ensuring breeding animals carry the gene of interest. Some such changes are now possible via gene editing, though this is largely still at an experimental stage.

Multi-trait—refers to simultaneous estimation of genetic parameters or breeding values for multiple traits, or simultaneous selection for multiple traits.

Non-additive genetic effects—the influence on an animal's genotype or phenotype due to genes that act non-additively, e.g. show dominance, where the presence of a dominant allele partially or completely masks the effect of a recessive allele at the same locus; or epistasis, where the effects of a gene at one locus are influenced by the genotype at another locus.

Phenotype—is an observable or measurable trait such as stature, milk volume, temperament. The phenotype is a result of the animal's genotype and its 'environment' (essentially all non-genetic influences). The relative importance of these is measured by the heritability.

Qualitative traits—traits that are often under the control of single genes (e.g. coat colour, polledness, many genetic disorders) that fall into discrete classes.

Quantitative traits—traits affected by genes at many different loci (polygenic), as well as by non-genetic factors like feeding and management (often termed 'environmental' effects). The performance of animals in quantitative traits tends to show continuous variation.

Selection criteria (auxiliary traits)—the set of measurements on which selection is based; these may be the same as breeding goal traits, where these can be measured directly in candidates for selection, or proxies for these traits.

Selection index or total merit index—An overall score of genetic merit allowing optimal selection for multiple traits—with the emphasis on each breeding goal trait usually depending on its relative economic value and the scope for genetic improvement (which depends on the additive genetic variance in that trait and covariance with other traits under selection). Index scores can be derived directly from (multi-trait) BLUP EBVs for breeding goal traits and their economic values. Examples in dairy cattle include £PLI in the UK and Balanced Performance Index (BPI) in Australia.

SNP—single nucleotide polymorphisms are commonly used genetic markers arising from variation that occurs at a single nucleotide (A, C, G or T) within the DNA sequence. Single Nucleotide Polymorphism (SNP) 'chips' are available for most domestic species, that allow detection of variants at 10s–100s of thousands of SNP loci dispersed evenly across the genome.

Trait—an animal characteristic of interest in breeding programmes that can be classified or measured and subjected to selection.

Threshold model (TM)—a type of statistical model used in the estimation of genetic parameters and breeding values for traits that are influenced by many genes but that have a limited number of categories, and an underlying normal distribution of liability, e.g. presence or absence of disease, or a small number of scores indicating severity of disease or degree of calving difficulty.

## 12.2 The Principles of Genetic Selection

In this section, we outline some of the key concepts and strategies in livestock breeding relevant to dairy cattle welfare. For a fuller description, see Simm et al. (2021). Traditionally, there have been three main strategies for the genetic improvement of farmed livestock: (i) selection between breeds or strains, (ii) selection within breeds or strains, and (iii) crossbreeding. Newer molecular genetic tools are beginning to augment these strategies enhancing existing selection approaches via 'genomic selection' (now in widespread commercial use), and allowing the base sequence of genes to be altered in a targeted manner via gene editing (subject to tight regulation in most countries). The principles behind each of these strategies and how they can be implemented to improved welfare will be discussed below.

***Selection Between Breeds or Strains*** For genetic improvement strategies to be effective, it is important to decide what the important traits are (the 'breeding goal'). Historically, scientists and breeders have focussed on traits with the highest economic importance (e.g. milk yield and milk composition), though dairy farmers have long been concerned with the functional fitness of cows, often assessed via the proxy of conformation or 'type' scoring (see, e.g., Miglior et al. 2017). There is a growing need to consider other traits related to animal welfare and environmental impact that may not be properly recognised by their economic values alone. It is logical to choose the most appropriate breed or cross, based on its performance in this set of traits. Selection between breeds or strains can achieve dramatic and rapid 'one off' genetic change when there are large genetic differences between populations. Further improvement depends on selection within the chosen breed or strain.

Crossbreeding involves mating animals of different breeds, lines, or species, for a range of reasons including: (i) improving system efficiency by crossing 'complementary' breeds that excel in different traits—for instance, crossing of *Bos taurus* breeds selected for high production with *Bos indicus* breeds showing high heat and disease tolerance in the tropics; (ii) 'grading up' to a new breed or strain—as has happened often over the last few decades in the dairy sectors of many countries; (iv) as an intermediate step in the creation of a new synthetic or composite breed; (vi) to introduce a single gene for a favourable characteristic, such as polledness—the absence of horns—to an existing breed ('introgression'), or (v) to exploit heterosis or hybrid vigour—the advantage in performance above the mid-parent mean often seen in crosses, and widely applied in some pastoral dairy industries such as that in New Zealand (Lopez-Villalobos et al. 2000).

Selection within breeds involves comparing animals of that breed and identifying preferred animals to become parents of the next generation. When repeated in each generation, this produces cumulative changes in successive generations, as seen in the dairy sector of many countries.

***Genetic Variation*** There are many traits of interest in farmed animals under the control of single genes (e.g. coat colour, polledness, many genetic disorders). These are often termed qualitative traits, if they fall into discrete classes. Many other traits of interest in animals are affected by genes at many different loci (polygenic), as well as by non-genetic factors like feeding and management (often termed 'environmental' effects). Although classical Mendelian segregation is at work at each of these loci, it is difficult to distinguish different phenotypes. Instead, the performance of animals tends to show continuous variation. Often the performance of animals follows a normal distribution, and is measured on some scale, hence these are termed quantitative traits.

For quantitative traits it is useful to think of an animal's phenotype being comprised of its genotype (which can be further subdivided into an additive genetic component, or 'breeding value', and a non-additive genetic component) and an environmental component. Modern methods of livestock improvement attempt to disentangle these components as far as possible through the application of statistical methods such as linear models, best linear unbiased prediction (BLUP), etc. (see glossary). Selection between and within breeds acts largely on additive genetic merit, while crossbreeding may be used to benefit from additive or non-additive genetic differences between animals, or both of these.

Many of the 'tools' used in within-breed selection rest on properties of this normal distribution of performance. For example, the variance in performance in a group of animals can be split into additive genetic, non-additive genetic and environmental components. This allows comparisons of the relative importance of these different sources of variation, and is useful when deciding on a strategy for genetic improvement, and for predicting responses to selection. An important related measure is the heritability of the trait—defined as the ratio of additive genetic variation to total phenotypic variation in the trait of interest. Put simply, it tells us the relative influence of genetics (nature) and environment (nurture) on traits of interest. The profitability and public acceptability of livestock enterprises depends on an increasing number of animal characteristics, and it is important to know how these are related. Phenotypic or genetic correlations (derived from variances and covariances) are used to quantify the association between observed performance or breeding values, respectively, in pairs of traits. It is worth mentioning here the special case, relevant to selection for some welfare-related traits, especially disease traits, where we record the presence or absence of disease, or a limited number of categories of severity, rather than the continuous scale we see in many other traits, but there is an underlying normal distribution of 'liability'. These traits require a particular type of statistical model known as a 'threshold model' to derive genetic parameters and estimate breeding values, but respond to selection in just the same way.

***Breeding Programmes*** Effective selection within breeds increases the average level of additive genetic merit or breeding value of the population in the traits concerned. The key steps in a breeding programme are shown in Fig. 12.1 and include: (i) defining the breeding goal (the set of traits we wish to improve); (ii) deriving

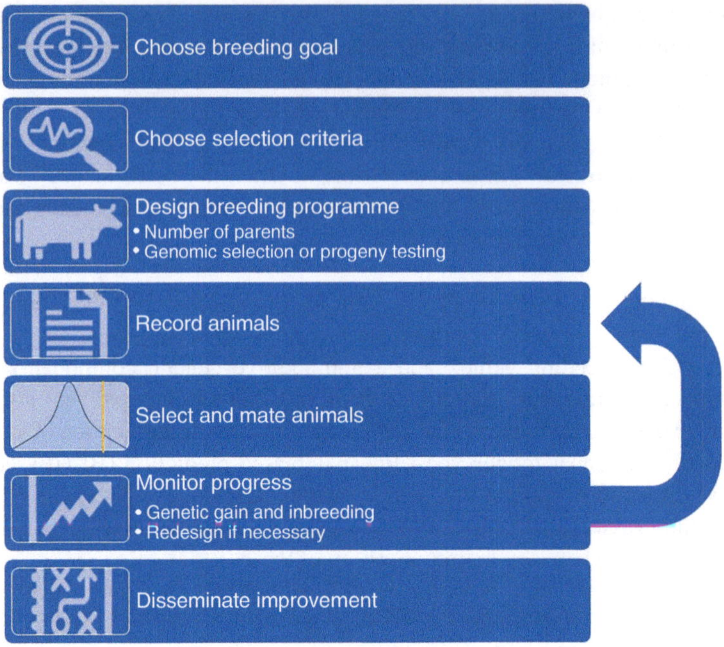


**Fig. 12.1** Steps involved in within-breed improvement programmes based on objective performance (after Harris et al. (1984), Simm et al. (2021))

relative economic values for breeding goal traits—this helps optimise the weighting on different traits in a multi-trait selection index; (iii) deciding on the selection criterion (the traits we measure as candidates for selection—these may be breeding goal traits themselves, or proxies for these, e.g. if breeding goal traits cannot be measured directly because they are expensive, expressed in one sex only, or expressed late in life); (iv) estimating 'genetic parameters' for the breeding goal traits and selection criteria—especially the phenotypic and genetic variances for key traits and the covariances among them, and the heritabilities and correlations derived from these (co)variances; (v) designing the breeding programme, e.g. deciding on the numbers of males and females to be selected annually, to achieve a balance between maximising genetic gain and minimising levels of inbreeding; (vi) implementing the programme, i.e. doing the routine recording, genetic evaluation (estimating breeding values of candidates for selection), and mating of animals; and (vii) monitoring progress and redesigning the programme where necessary, e.g. if there are unforeseen consequences of selection, or markets change.

For many traits that are associated with animal welfare (e.g. disease resistance, calving ease, thermal comfort), while the heritability is low (i.e. genetic variation is proportionately small when compared to the non-genetic variation) the genetic variation that exists may still be relatively high in absolute terms (meaning that there are large genetic differences between some individuals and families). On top of this, selection of farm animals can only be effective when the traits of interest—or alternative selection criteria, or correlated traits—are measurable and accurately recorded. A further complication is that, even within a herd, animals do not necessarily face an equal disease challenge, so interpretation of disease records is complex (Bishop and Woolliams 2010). In addition to error, other non-genetic sources of variation include feeding, climate, chance events, and other unknown effects and these may dominate the measurable variation of many traits. Together, these often contribute to low heritability estimates for many animal health and welfare traits.

***Genetic and Genomic Evaluations*** Estimated breeding values (EBVs) are used to identify the best candidates for breeding. Best linear unbiased prediction (BLUP) is a very widely used statistical technique that disentangles genetic from environmental effects in the best possible way, and so produces the most accurate EBVs. Conventionally, BLUP uses performance records from related animals to increase the accuracy of EBVs. The more records, and the closer the relationships of recorded animals to the target animal, the more accurate the EBVs. Until recently, most dairy cattle breeding programmes have been based on structured progeny testing of young AI bulls, with daughters' performance for a wide range of traits being recorded in commercial herds. Very accurate EBVs can be produced for bulls with many hundreds of daughters recorded. In the last decade or so, the practice of genomic selection has virtually supplanted planned progeny testing in many industrialised countries—we discuss this later. Typically, breeding values for dairy cattle are estimated nationally by genetic evaluation units that are often part of government ministries, breed societies, universities, or research institutes. For many years, INTERBULL—a subcommittee of the International Committee for Animal Recording (ICAR)—has provided guidance on, and helped harmonise approaches to, genetic evaluation, as well as providing international evaluations that combine information optimally from multiple countries.

A range of molecular genetic tools is enhancing our ability to select for desired performance or inherited disease status. Increasingly, automated methods are available for detecting genetic polymorphisms (variations in the bases present at particular sites on the chromosome—these variations exist within coding regions of genes, but also in many other parts of the genome). Single Nucleotide Polymorphism (SNP) 'chips' are available for most domestic species, which allow detection of variants at 10s–100s of thousands of SNP loci dispersed across the genome. This in turn allows whole genome association studies, where particular sequences of SNPs identify segments of the genome associated with a trait, e.g. high milk yield, or disease resistance. Once such associations have been established, SNP information can be used in the so-called genomic selection to identify other animals with this favourable combination of SNPs (see Fig. 12.2). This allows earlier estimation of breeding values (as genotypes can be obtained directly on candidates for selection much sooner than performance records), higher accuracy of EBVs (especially when genomic and performance records are combined), or both. Within the last decade, breeding programmes have changed from using progeny testing to genomic

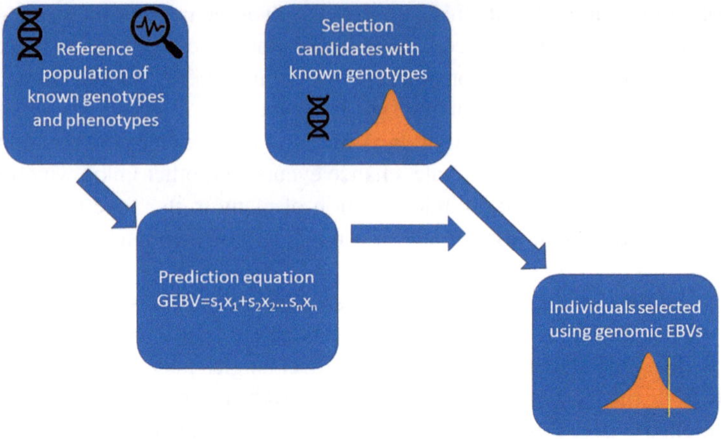


**Fig. 12.2** Genomic prediction using a reference population of known phenotypes and genotypes is used to generate a genomic prediction equation which is applied to genotyped animals. The best animals are selected for breeding using the genomic breeding values derived from this equation. (After Goddard and Hayes (2009), Eggen (2012))

selection in many countries, where the best bulls mated to the best females are young bulls selected based on their genomic EBV. Genomic selection has transformed livestock breeding internationally because, in addition to enabling selection decisions to be made earlier in life, genomic selection can be used to select for traits that were not accessible before, including important welfare traits, such as improved disease resistance, resilience to climate variability and thermal stress, etc.

***Rates of Genetic Gain*** Annual rates of response to selection in polygenic traits depend on four main factors: (i) the selection intensity achieved (i.e. the superiority of selected parents above the mean), (ii) the accuracy with which genetic merit in the trait of interest is predicted (accuracy of estimating breeding values), (iii) the amount of additive genetic variation in the trait of interest, and (iv) the generation interval (the average age of parents when their offspring are born). Generally speaking, the higher the selection intensity, accuracy, and genetic variation, and the lower the generation interval, the higher the annual rate of genetic improvement. Breeders have most control over the selection intensity and generation interval (but both within biological limits) and—at least at a national level—choice of method to estimate breeding values.

Rates of genetic gain in production traits, fertility, longevity, and udder health have increased substantially since the introduction of genomic selection; largely driven by reduced generation intervals (García-Ruiz et al. 2016). Dedicated female reference populations that have entire herds of genotyped cows with these measurements recorded are a valuable source of information for these new traits. In addition to national genetic evaluation units, commercial companies are also developing their own genomic predictions for health traits through use of health data collected on cows that are genotyped.

Livestock breeding industries in industrialised nations often have a pyramid structure, with elite or nucleus breeders at the top, one or more middle tiers of pure-bred or crossbred multipliers, and a final tier of commercial herds or flocks, or end users. Pig, poultry, and dairy cattle breeding operations in many countries are dominated by a relatively small number of international breeding companies who supply breeding stock to commercial producers. Because of the widespread use of artificial insemination (AI) in dairy cattle breeding, breeding companies supply semen from elite dairy bulls, with most elite cows owned by individual farmers. AI also allows commercial dairy farmers to directly access elite genetic material, bypassing the multiplier tiers present in other sectors.

Genetic improvement, including selection between breeds, crossing, and within-breed selection has led to dramatic changes in the performance of dairy cattle over the last 70 years or so (Simm et al. 2021). The development and widespread adoption of technologies for semen collection, freezing, and artificial insemination (AI) in dairy cattle has both enabled effective genetic improvement in many countries through progeny testing, and—together with related embryo transfer technologies—led to international exchange of genetic material, and dairy cattle breeding becoming a truly international endeavour.

Genetic selection in domesticated species has been practised with a great deal of success and has focussed primarily on improving traits that have market value or are associated with reducing costs of production. For example, Cole and VanRaden (2018) showed around a 300 kg increase in fat yield for US Holstein cows born between 1957 and 2015. From the 200 kg/year base in 1957, genetics and management/feeding each representing 28% of the gain. A major challenge now is to extend this approach to characteristics, like those associated with animal welfare and environmental impact of livestock, that have high societal value but low or hidden current market value.

## 12.3 Including Animal Welfare in Dairy Cattle Breeding Objectives

Animal welfare is an area of science that generally includes the measurement of multiple indicators to assess the physical, behavioural, and emotional state of the individual (Broom 1991). Some of these states are difficult to quantify or measure objectively. Animal breeding, on the other hand, relies on objective measurements, although breeding values are often developed as a by-product of data primarily collected for farm management decisions. For example, milk production breeding values are generally estimated using pedigree data, genomic data, and data collected from routine milk-recording. Fertility breeding values use mating and pregnancy test data either recorded by farmers or professional service providers; health breeding values largely use clinically recorded data (as diagnosed by veterinarians or farmers), while longevity EBVs use data on herd-entry and exit dates. Other examples include claw health recorded by professional hoof trimmers, conformation scoring, auction sales, slaughterhouse data, etc. There are many other examples of breeding values that are by-products of recording for another purpose. Egger-Danner et al. (2015) describe the potential sources of data and their uses. Typically, a genetic evaluation unit will produce a set of >30 breeding values for different traits.

***Single-Trait Selection*** For many years, selection focused on milk production traits and conformation in many countries. Conformation, or the appearance of cows has for a long time been regarded by producers as helping to ensure that their cattle are productive and long-lasting, in addition to taking honours in the show-ring or pedigree sales (Miglior et al. 2017). In the late 1990s, it became clear that an undesirable consequence of narrow, production-orientated selection criteria was a reduction in health and fertility. The decline in fertility, in particular, has been well documented (Lucy 2001; Berry et al. 2014). But, there was also evidence that there were unfavourable genetic correlations between production traits and other animal welfare traits, (e.g. mastitis resistance, lameness, reproductive and metabolic disorders), which were starting to deteriorate (Rauw et al. 1998). This led to a large number of studies focused on determining the extent of genetic control of health traits, generally through the use of clinical observations of disease. The heritability estimates from these studies show that generally the genetic control is small (Table 12.1), yet there is sufficient genetic variation to make genetic progress.

***Selecting for Health Traits*** Before the consequences of narrow dairy breeding objectives described above were widely understood, the Nordic countries already had a long history of recording and providing genetic evaluations of health traits. For example, in Norway, veterinary treatments had to be registered on an individual basis from 1975 (Heringstad and Østerås 2013), with similar schemes being established in Denmark, Finland, and Sweden through the 1980s. In addition to the Nordic countries, routine genetic evaluations of mastitis have been in place in Austria and Germany since 2010, and in France and Canada from 2012 (Egger-Danner et al. 2015), with many others following. Valuable lessons that have been learnt by dairy geneticists and others about the dangers of narrow breeding goals. In addition to selection on fertility, disease resistance traits have become key areas where breeding values are being developed for future breeding goals.

Pain or injury associated with injury or disease of the feet or legs often causes the animal to alter the way it walks to avoid putting weight on the affected limb or limbs. This behavioural expression of pain is what we know as 'lameness' or 'altered locomotion'. As well as affecting the way the animal walks, poor foot and leg health adversely affects feeding, ruminating, and lying behaviour among others (Whay and Shearer 2017). There are many methods of scoring the degree of lameness but in dairy cattle, lameness scoring is considered to be an important welfare indicator (Table 12.2) with feet and leg problems being common. For example Van Der Waaij et al. (2005) estimating that 70% of cows in the Netherlands have at least one hoof

**Table 12.1** Ranges of heritability estimates of udder health, lameness and claw disorders, and metabolic diseases summarised from 3 recent review papers

| | Model | Range in heritability estimate | Review paper |
|---|---|---|---|
| **Udder health** | | | Egger-Danner et al. (2015) |
| Clinical mastitis | | 0.02–0.09 | |
| Improved SCC | | 0.01–0.17 | |
| Electrical conductivity | | 0.12–0.36 | |
| Pathogen information | | 0.04–0.09 | |
| **Lameness and claw disorders** | | | Heringstad et al. (2018) |
| Digital dermatitis/interdigital dermatitis | LM | 0.01–0.11 | |
| | TM | 0.09–0.20 | |
| Heel horn erosion | LM | 0.03–0.07 | |
| | TM | 0.09 | |
| Interdigital hyperplasia | LM | 0.01–0.14 | |
| | TM | 0.19–0.39 | |
| Sole haemorrhage | LM | 0.02–0.08 | |
| | TM | 0.07–0.09 | |
| Sole ulcer | LM | 0.01–0.12 | |
| | TM | 0.07–0.18 | |
| White line disease | LM | 0.01–0.09 | |
| | TM | 0.06–0.10 | |
| Lameness | LM | 0.02–0.10 | |
| | TM | 0.02–0.15 | |
| Locomotion | LM | 0.03–0.11 | |
| **Metabolic diseases** | | | Pryce et al. (2016) |
| Ketosis | LM | 0.01–0.08 | |
| | TM | 0.02–0.16 | |
| Milk fever | LM | 0.01–0.08 | |
| | TM | 0.09–0.18 | |
| Displaced abomasum | LM | 0–0.08 | |
| | TM | 0.12–0.32 | |
| Tetany | LM | 0.004 | |
| | TM | 0.02–0.05 | |

LM means linear model and TM means threshold model

issue. Although management and housing play a key part in controlling lameness, genetic improvement is a strategy worthy of consideration. Genetic improvement of hoof health, as a general trait, can be achieved through data collected from hoof trimming, veterinary treatments, or on-farm databases. Lameness scoring can be used as in auxiliary trait for prediction of claw health (Heringstad et al. 2018).

Another promising approach is to develop breeding values for different types of lameness, as there is evidence to suggest that heritabilities vary between claw

**Table 12.2** Top 10 welfare indicators from ICAR survey (adapted from Haskell et al. (2019): https://www.icar.org/Documents/Prague-2019/Presentations/02%20-%20Marie%20Haskell.pdf

| Welfare indicator | No. scoring |
|---|---|
| Body condition score | 28 |
| Lameness in loose-housed cows | 24 |
| Diarrhoea | 18 |
| Temperament | 16 |
| Skin alterations, swellings, or injuries | 16 |
| Lameness in tie-stalls | 16 |
| Existing records | 16 |
| Cleanliness | 15 |
| Claw trimmer data | 13 |
| Hampered respiration | 11 |

diseases recorded by hoof trimmers (Ødegård et al. 2013; Buch et al. 2011). This requires the development of preferably national databases of accurate and consistent data records. In fact, there has been a lot of effort recently to harmonize recording of claw disorder, e.g. the ICAR Claw Health Atlas (Egger-Danner et al. 2014). The increase in electronic capture of data has enabled the assembly of much more in the way of clinical observations of disease, with many farmers keeping electronic records as evidence for quality assurance programmes.

***Predictor Traits*** Many traits that are currently evaluated are correlated, so selection for one breeding value can have favourable (or unfavourable) effects on other traits. Perhaps the best example of this is selection for mastitis resistance using somatic cell count. Many countries first introduced mastitis EBVs through a proxy trait of somatic cell count (SCC), which can be considered to be the cow's immune response to infection. When SCC is high, the cow is responding to a likely infection in the udder. Since then including farm or veterinary records on mastitis observations has become more prominent in the development of breeding values for mastitis.

Heringstad et al. (2006) showed that selection against mastitis leads to favourable correlated responses to selection in other diseases, such as ketosis and retained placenta, indicating the existence of a general robustness or reduced liability to disease. Selecting for general disease resistance, or immunity is also becoming popular. For example, in a study by De la Paz (2008) comparing cows with high and low antibody and cell-mediated immune response, high responders had a decreased risk of disease occurrence for several diseases, including mastitis, ketosis, metritis, and retained placenta. The heritability of response to an immunity challenge is high enough to justify selection (Thompson-Crispi et al. 2012). In fact, selection tools for immunity are available commercially. Cows identified as high responders based on estimated breeding values for cell and antibody-mediated immune responses were found to have half the disease occurrence compared with low responders (Thompson-Crispi et al. 2012).

Body condition score (BCS) is often considered to be an indicator of hunger, reduced fertility (Banos et al. 2004), or metabolic disease (Pryce et al. 2016), and therefore a welfare indicator. A survey by the International Committee on Animal Recording on the use of welfare indicators showed that body condition score and lameness were the most popular (Table 12.2). Body condition score is often part of the suite of conformation traits scored by breed societies, or it is part of quality assurance systems and is only recorded once a year and only from a subset of animals in the herd. Thus, evaluating changes that may indicate a change in welfare is not easy. However, BCS is reasonably heritable (Pryce and Harris 2006) and already considered as part of the breeding objective in countries such as New Zealand (Zhang and Amer 2021) with the justification that the costs associated are incurred through having to replenish body reserves mobilised in lactation, especially if cows being thinner leads to earlier drying off dates and less days in milk. For more information, see DairyNZ (2022).

***Resilience*** An area of growing interest is resilience, which could have positive implications for animal welfare. Resilience is defined as 'the capacity of an animal to be minimally affected by disturbances, or to rapidly return to the state pertained before exposure to a disturbance' (Colditz and Hine 2016). A disturbance can be physical (disease, temperature) or emotional (e.g. negative interaction with humans, novel environments, social stressors) (Berghof et al. 2019). In terms of animal welfare, it is likely that an animal with better resilience will have a less negative experience during the disturbance than one with lower resilience.

This area of research has led to novel ways to calculate traits of interest, for example the variance of a trait under a particular challenge may describe the impact of a disturbance on individuals in a population, so the animals with least variation may be more resilient to their environment than the others (Berghof et al. 2019). Using daily milk yields from automatic milking systems (AMS), Elgersma et al. (2018) showed that cows with low within-cow variation in milk yield had genetically less disease and greater longevity. Furthermore, for cows that have genetically the same level of milk yield, those with less variable milk yield have better health, longevity, and fertility, and a higher BCS than those with more variable milk yield (Poppe et al. 2021). Following on from this, Poppe et al. (2021) proposed that AMS can be used to identify cows that have low within-cow variability in milk yield (deviations from the expected lactation curve) and fast recovery after a challenge event, and that these are likely to be the most resilient.

Another example of a measure of resilience is the reduction in yield (i.e. slope) after passing a temperature and humidity index (THI) threshold (Fig. 12.3). Cows that are more tolerant to heat have less steep slopes (i.e. a smaller reduction in production as temperature and humidity rise). Nguyen et al. (2016) used the decline in milk, fat, and protein yields as THI increases as indicators of heat stress. The study found that using high density SNP genotypes, heat tolerance genomic breeding values can be predicted at the accuracy of 0.42–0.61. Genomically predicted heat susceptible and predicted heat tolerant animals show reduced milk yield losses, rectal

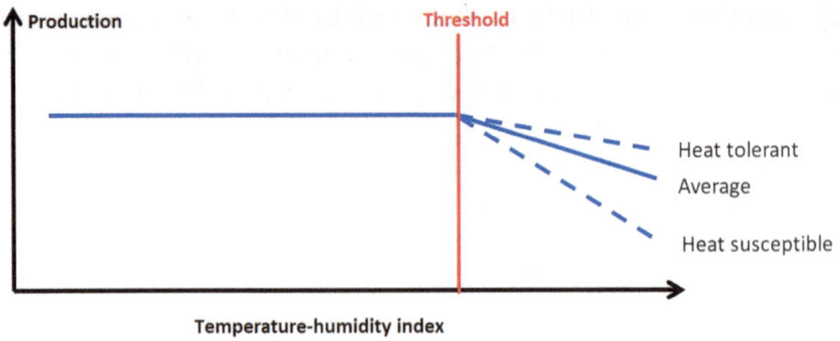


**Fig. 12.3** Heat tolerance defined as the slope of decline in milk, fat, or protein yield when temperature and humidity exceed a thermoneutral threshold

and intra-vaginal temperatures when experiencing a mild simulated heat wave (Garner et al. 2016). Clearly, traits associated with resilience are a growing area that could produce new solutions to breeding for improved animal welfare.

***Selection Indices*** As we have shown, there are now many traits that can and should be included in breeding programmes. Being able to select for many traits simultaneously has led to a considerable amount of work developing multi-trait selection indices, building on the approach developed by Hazel (1943). The idea is that farmers can focus on a single index score when selecting breeding animals, instead of selecting for multiple EBVs simultaneously.

A selection index starts with the identification of the breeding goal, which is often net farm profit broadly representing at least the following categories: milk production, type, longevity, udder health, fertility, other traits (Egger-Danner et al. 2015). The breeding goal is calculated as the sum of each EBV multiplied by a weight, which is usually based on the economic or perceived value of the trait. A selection index for n traits can be written as:

$$\text{Index} = b_1\text{EBV}_1 + b_2\text{EBV}_2 + b_3\text{EBV}_3 + \ldots + b_m\text{EBV}_n,$$

where the *b*-values are the weights to each of the EBVs. Selecting on this index gives the highest selection response in genetic merit such as ranking for highest profit.

***Selection Indices for Improved Animal Welfare*** Currently, animal welfare is rarely considered in deriving the weights to apply to traits in a selection index, with these typically based solely on economic parameters. For example, the cost of disease is generally used to calculate economic values for disease resistance traits. It is challenging to appropriately define the weights from an animal welfare perspective, as it is adding a subjective layer to an objective process (i.e. we need to put a value on the animal's experience of a disease or other welfare issue), as animal welfare is an ethical concept requiring societal input. However, if these challenges can be overcome, then a framework exists to devise non-market values to apply to traits under selection (Nielsen et al. 2005).

The thinking to date takes into consideration consumer willingness to pay for aspects of traits that have perceived societal or animal welfare value. It is also possible to devise indices that have either a desired outcome, or selection response, or restrict the change in a trait. Other aspects affecting breeding objectives will some become more important, for example, as the growing human population places more pressure on limited resources and global changes leading to hotter and drier—or otherwise more extreme—conditions in which to manage livestock, there is also a need to recognise increased consumer awareness of animal welfare and farming conditions. So, future breeding goals need to adapt to these considerations by including economic, societal, and environmental considerations simultaneously (Boichard and Brochard 2012; Martin-Collado et al. 2015).

An approach that has gained some traction in existing breeding programmes is placing additional emphasis on traits perceived to be associated with improved animal welfare. Martin-Collado et al. (2015) used the '1000 minds' methodology to add objectivity to perceived non-market values through a survey, where questions on perceived values are assessed through a series of comparisons that are of similar actual value. The idea being that if opinions are canvassed from many farmers (hence the '1000 minds' name), then the comparative value of a trait to groups of farmers with similar philosophies can be quantified. This approach was the foundation to determine farmer preferences for national selection indices being developed in Australia. Although farmer preferences were the focus of the research by Martin-Collado et al. (2015), it was clear that animal welfare and improving the functional ability of dairy cows were at the forefront of farmers' desires for future generations of cattle. To provide selection tools that give farmers of different philosophies an index that best suited their needs, 3 indices were released for Australian dairy breeders in 2015 to use in selection decisions; one index focused on profitability, another had more emphasis on health and fertility traits, while the final index was more focused on conformation traits (Byrne et al. 2016). At around the same time the total merit index (TMI) used for selecting Fleckvieh and Brown Swiss cattle in Austria and Germany was updated to include farmer preferences (Fuerst-Waltl et al. 2016). In this process more emphasis was placed on fitness traits.

Indices or sub-indices that focus entirely on animal welfare traits may start to emerge, especially as a vehicle to capture the complexity of animal welfare. For example, if we want to achieve favourable selection responses in psychological state in addition to health and productivity, then the first task is to identify selection criteria. While the health and productivity part of our breeding objective may have an economic dimension, and therefore be at least partially captured in our current selection indices, the positive psychological response is much more challenging and is only starting to emerge as a potential selection criterion.

## 12.4 Other Considerations

***The Interaction with Environment*** An animal's performance is a result of the genes it inherits and the environment in which it is kept, including climatic factors and a host of management factors. Livestock keepers have long been aware of the fact that some breeds or strains perform better in some environments than others—termed a genotype x environment (GxE) interaction. There is good evidence, for instance, that *Bos taurus* dairy breeds highly selected for production in temperate climates often perform worse (e.g. lower production, higher disease incidence, shorter herdlife) than tropically-adapted *Bos indicus* breeds, or crosses with these, in tropical environments with extreme thermal, nutritional, and disease challenges (see Simm et al. 2021). GxE interactions also lead to differences in the ranking of sires within a breed, in different production systems (Fig. 12.4). Hence, it is crucially important for animal welfare that appropriate breeds or crosses are chosen for particular environments and systems when GxE interactions are present. Evidence to inform such decisions may come from experimental studies or analysis of industry data.

Although risk factors such as stocking density, stall size, etc., pose direct issues for animal welfare, it is how animals deal with these risk factors that often has a genetic component. Moreover, there is reason to believe that substantial GxE interactions exist, i.e. that the ranking of sires or families for welfare indicators differs between environments at the herd level. To date, most GxE studies have explored interactions between countries. However, Zwald et al. (2003) found that the heritabilities of production traits in colder climates within the USA were lower than in hot

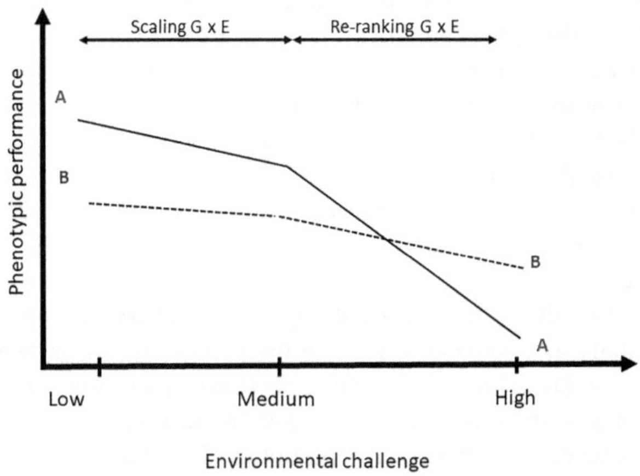


**Fig. 12.4** The relative performance of two animals (or breeds/crosses) changes across three envionmental challenge levels (low, medium, and high). In low and medium challenge environments, Animal A outperforms animal B. However, as the challenge level increases, A's performance deteriorates rapidly, and there is a cross over in rankings. In the highest challenge environment, Animal B performs the best

climates (0.26 and 0.39, respectively), and the genetic correlation between these two groups was 0.66, implying that heat stress may play a role in genotype by environment interactions. Understanding of the interaction between genetics and the environment needs to grow, in order for researchers to attempt to predict performance or risk in different systems. The use of machine learning methods, i.e. learning from patterns in data, lends itself disentangling this information and enabling predictions of risk (Lasser et al. 2021) which should partly help to improve management across systems/environments.

***Inbreeding*** Although conserving genetic diversity is not itself a welfare concern, the consequences of it are. Inbreeding arises when the parents of an individual are genetically related. More specifically it is close to zero when there are no common ancestors and increases if the parents are closely related. The consensus is that inbreeding depression is the result of the 'load' of deleterious recessive genes arising from common ancestors. Fitness traits such as fertility (McParland et al. 2007) and health (Baes et al. 2019) are especially sensitive to inbreeding. Inbreeding also leads to the manifestation of genetic diseases (arising from single deleterious mutations). Inbreeding is especially important in dairy cattle because AI and genomic selection allow very intense selection of males, and AI facilitates rapid international dissemination of genes from favoured sires. Moreover, the concentration of breeding decisions in a few global companies in a highly competitive market arguably leads to a focus on shorter term marketability rather than longer term sustainable use of genetic variation.

The downside to accelerating rate of genetic gain (largely through genomic selection) has been the impact it has had on effective population size, especially in popular dairy breeds, such as the Holstein (Makanjuola et al. 2020). Effective population size is a measure that accounts for the number of breeding males and females in a population, not just the overall population size and current estimates are 43–66 in the Holstein breed (Makanjuola et al. 2020). This provides a better indicator of genetic variability in a population, and changes in this over time. There has been an increase in the rate of inbreeding reported in most dairy populations (Vanraden et al. 2011; Makanjuola et al. 2020) with evidence that the rate of inbreeding has accelerated since the introduction of genomic selection (Doublet et al. 2019). Conserving diversity and genetic variation is important to maintain dairy genetic resources and reduce the consequences of inbreeding, such as inbreeding depression in fitness traits.

As inbreeding increases, the risk of homozygous deleterious recessive mutations existing also increases. A mutation is a change in the nucleotide sequence of the genome and most are harmless and rare, and some are positive—in fact, mutations give rise to genetic variation. With the widespread international use of selected sires via AI, heterozygote 'carriers' of such recessive genes can spread quickly in the population before affected homozygous descendants appear and are detected. There are examples of genetic diseases that arise as single mutations, such as Complex Vertebral Malformation (CVM), Bovine Leucocyte Adhesion Deficiency (BLAD), and Deficiency of Uridine Monophosphate Synthase (DUMPS) in Holsteins. Most of these diseases are the result of reasonably recent (rare) mutations. For example, complex vertebral malformation, or CVM, can be traced to two former elite Holstein sires. Because of their widespread use, the sires appeared on both sides of the pedigree of affected calves (Agerholm et al. 2001). More recently, a mutation affecting calf survival has been identified and is associated with cholesterol deficiency leading to emaciated calves that fail to thrive and presents a serious animal welfare issue (Kipp et al. 2016). At its peak, 13% of registered Canadian Holsteins calves were affected. The occurrence of these diseases highlights the importance of managing rates of inbreeding, which arises as a result of the co-occurrence of common ancestor(s).

Genomic data can be used to control or monitor inbreeding in a population by quantifying genomic relationships between animals in addition to estimating inbreeding depression (Baes et al. 2019; Bjelland et al. 2013). One of the advantages in using genomic, rather than pedigree relationships, is that it is a more accurate estimate of identity by descent, because it does not suffer from lack of depth of pedigree data and pedigree errors. The use of genomic metrics should allow breeders to improve management of the risks associated with inbreeding, allowing better evaluation of the trade-offs between the genetic value of the progeny and the undesirable side effects associated with inbreeding (Baes et al. 2019).

## 12.5 New Technologies, Traits, and Methods

The way an animal interacts with its environment affects its welfare and detrimental responses could loosely be described as negative experiences or "stress". Brito et al. (2020) described how aspects of the hypothalamic–pituitary–adrenal (HPA) and sympathetic–adrenal–medullary (SAM) systems have genetic variation and could be useful targets. For example, glucocorticoid concentrations (cortisol and corticosterone) may be indicative of stress (König and May 2019). An alternative approach is to consider behavioural traits instead. For example, in their review, Haskell et al. (2014) cited 4 studies where cortisol was higher in excitable animals than calm animals, with the measurable behaviour being temperament. A 'good' temperament in the dairying context is often described as a calm response to being milked and docility at handling (Haskell et al. 2014). Temperament is often included in genetic evaluations and is evaluated using farmer recorded scores on a scale such as 1–5 and could become a trait for consideration in more welfare-focused selection indices.

***Sensors*** Emerging technologies are likely to facilitate the development of breeding values for behaviours. These could include cameras, microphones (for vocalisations), body temperature sensors, or accelerometers (Brito et al. 2020; see also Chap. 11 in this volume). For example, wearable sensors using accelerometers can provide massive quantities of longitudinal data that can help in defining traits associated with cow comfort, such as the amount of time spent ruminating, lying, feeding, walking, etc. (Bikker et al. 2014). One of the advantages of embracing sensor defined phenotypes is that they can be used to detect the risk of disease, or compromises in animal welfare, before clinical disease has occurred. While this is clearly advantageous for management purposes, it could be argued that the primary objective of a health breeding value is to select against the occurrence of clinical cases. So, although sensor data are likely to become a valuable addition to estimating breeding values into the future, especially as a way to collect very large amounts of objective data, sensor data could also be used to record clinical manifestations of disease.

From a technical perspective, incorporating data from multiple sources is a well-established process in animal breeding through the implementation of multi-trait models. One advantage of sensor systems or cameras is that they can provide continuous measurements. Development of EBVs using this potentially massive source of data is largely in its infancy (Cole et al. 2020) and there are still many hurdles to overcome, including differences between systems, lack of uniformity across devices, and lack of integration with national databases leading to disconnected data silos (Lasser et al. 2021). So, to date, the contribution of sensor data to EBVs is under-explored.

***Mid-Infrared Spectroscopy*** One of the most promising ways of evaluating sub-clinical disease is the mid-infrared (MIR) analysis of milk samples. MIR is used routinely to quantify the fat, protein, casein, lactose, and urea concentration of milk in milk-recording programmes. Many farmers participate in these programmes as they use the data for management decisions. So again, the MIR data is a by-product of an alternative use of these data. MIR analysis of milk has been used to predict other milk characteristics such as fatty acid composition, milk protein composition, milk coagulation properties, milk acidity, mineral composition, and ketone bodies (De Marchi et al. 2014). Recently, Luke et al. (2019b) showed that biomarkers of early lactation disease (measured in serum) are predictable using MIR analysis. Furthermore, many of these biomarkers are also heritable and show promise for genomic selection (Luke et al. 2019a; Van Den Berg et al. 2021). It seems likely that as we gain further understanding of the potential value of these measurements, breeding of dairy cattle will be further transformed.

***Gene Editing*** Gene editing allows genetic material to be added, removed, or altered at specific locations on the genome. Gene editing is especially useful for traits where a small number of edits are required. Two examples of how this technology can be used to improve animal and human welfare are in selecting for hornless (polled) cattle or breeding cows that are tolerant to heat stress.

Successful inclusion of the polled allele into dairy breeds would eliminate horns in dairy cattle (Mueller et al. 2019). Most dairy heifers are disbudded or dehorned at an early age and although it is a standard management practice, without the use of appropriate anaesthetics and analgesics it can be very painful (Stock et al. 2013) and it is increasingly scrutinised by the public as a potential welfare issue. In *Bos taurus* breeds polledness is controlled at a single locus, with the polled allele dominant to the horned allele. Therefore, mating a homozygous polled bull (PP) to a herd of horned cows (hh) will result in all the offspring being polled (Ph). If a bull is heterozygous (Ph) and the cows are horned (hh), then half the offspring will be polled (Ph). Two mutations that prevent development of horns in certain breeds of cattle have been mapped on the bovine genome (Medugorac et al. 2012) and these have become targets for gene editing (Fahrenkrug and Carlson 2014).

The so-called slick mutation is also a target for gene editing. The slick mutation is an adaptation to heat found naturally in Senepol cattle, where it appears to be associated with the type of coat, being slick or smooth, with some possible alteration to sweating ability (Davis et al. 2017). Breeding programmes in some countries including Puerto Rico and the USA have already started incorporating the slick mutation in Holstein cattle without the use of gene editing (Carabaño et al. 2019; Hansen 2020), which will enable the use of the desirable mutation with little to no background genetics from the donor breed.

Gene editing has the potential to be used in a more extensive way to provide a combination of desirable characteristics such as thermotolerance, disease resistance, and polledness, although the successful editing of large numbers of alleles at different loci has not been reported yet (Van Eenennaam 2019), but could become a reality. However, gene editing could largely complement traditional breeding techniques and programmes (Van Eenennaam 2019).

## 12.6 Putting It All Together

To make the most of these opportunities, expertise from many different disciplines is required. We need think-tanks of animal scientists and technologists to think collaboratively on measurements that could be useful for genetic evaluation purposes. For example, to capture animal welfare adequately in breeding programmes we need to include measurements that capture an animal's emotional state in addition to physical state. It is likely that some of the technologies we have discussed will lend themselves to capturing some of this information, however, working with engineers and data scientists to generate records of the welfare indicators we wish to measure is a priority area for the future. One approach is to first test and validate these measurements in research herds and then take the most promising to more numerous genotyped populations to develop genomic prediction equations that can be used to select individuals. Finally, we need to understand the genetic correlations between these new selection criteria and the traits we currently select for. Providing the new measurements pass these stages, the final step is taking the measurements and genotypes to genetic evaluations. This can be either public (i.e. for all farmers to use), or for proprietary marketing by large independent breeding companies and then marketed exclusively on their bulls.

Financial incentives or penalties on the basis of genetic merit for welfare associated characteristics, such as inbreeding metrics or horned/polled status, could also be considered, with checks made using genotyping, as extent of inbreeding and presence of specific genotypes can be accurately quantified using genomic data. If these approaches are useful, they could even complement, or to some extent replace farm audits. There are also opportunities for supermarkets to insist on their suppliers having welfare-focused breeding programmes.

## 12.7 Concluding Statements

Dairy cattle breeding has a mixed past with regard to animal welfare. Narrow breeding goals that focused almost entirely on milk production traits have been detrimental to many animal traits associated with welfare. Increasing awareness of welfare issues and the broadening of breeding goals to include animal welfare traits are welcome changes. Advances in understanding the genetics of welfare-related traits, and technologies and methodologies for recording and evaluation of these can help prevent welfare problems and allow active pursuit of better welfare. Collaborative efforts between animal behaviour scientists, geneticists, engineers, data scientists, and others will potentially provide solutions to these challenges.

**Acknowledgements** Jennie Pryce thanks DairyBio (a programme funded by Dairy Australia, the State Government of Victoria and The Gardiner Foundation, Melbourne) for financial support and colleagues from Agriculture Victoria for inspirational conversations.

Christa Egger-Danner was supported by the COMET project D4Dairy—Digitalisation, Data Integration, Detec[1]tion, Decision Support in Dairying in Austria (Vienna). These projects are supported by BMK (Federal Minis[1]try Republic of Austria Climate Action, Environment, Energy, Mobility, Innovation and Technology), BMDW (Federal Ministry Republic of Austria Digital and Eco[1]nomic Affairs, Vienna), the province of Lower Austria, and the city of Vienna in the framework of COMET—Competence Centers for Excellent Technologies. The COMET programme is handled by the FFG (Austrian Research Promotion Agency, Vienna).

## References

Agerholm JS, Bendixen C, Andersen O, Arnbjerg J (2001) Complex vertebral malformation in Holstein calves. J Vet Diagn Investig 13:283–289

Baes CF, Makanjuola BO, Miglior F, Marras G, Howard JT, Fleming A, Maltecca C (2019) Symposium review: the genomic architecture of inbreeding: how homozygosity affects health and performance. J Dairy Sci 102:2807–2817

Banos G, Brotherstone S, Coffey MP (2004) Evaluation of body condition score measured throughout lactation as an indicator of fertility in dairy cattle. J Dairy Sci 87:2669–2676

Berghof TVL, Poppe M, Mulder HA (2019) Opportunities to improve resilience in animal breeding programs. Front Genet 9

Berry D, Wall E, Pryce J (2014) Genetics and genomics of reproductive performance in dairy and beef cattle. Animal 8:105–121

Bikker JP, Van Laar H, Rump P, Doorenbos J, Van Meurs K, Griffioen GM, Dijkstra J (2014) Technical note: evaluation of an ear-attached movement sensor to record cow feeding behavior and activity. J Dairy Sci 97:2974–2979

Bishop SC, Woolliams JA (2010) On the genetic interpretation of disease data. PLoS One 5:e8940

Bjelland DW, Weigel KA, Vukasinovic N, Nkrumah JD (2013) Evaluation of inbreeding depression in Holstein cattle using whole-genome SNP markers and alternative measures of genomic inbreeding. J Dairy Sci 96:4697–4706
Boichard D, Brochard M (2012) New phenotypes for new breeding goals in dairy cattle. Animal 6:544–550
Brito LF, Oliveira HR, Mcconn BR, Schinckel AP, Arrazola A, Marchant-Forde JN, Johnson JS (2020) Large-scale phenotyping of livestock welfare in commercial production systems: A new frontier in animal breeding. Front Genet 11:793–793
Broom DM (1991) Animal welfare: concepts and measurement2. J Anim Sci 69:4167–4175
Buch LH, Sørensen AC, Lassen J, Berg P, Eriksson J-Å, Jakobsen J, Sørensen MK (2011) Hygiene-related and feed-related hoof diseases show different patterns of genetic correlations to clinical mastitis and female fertility. J Dairy Sci 94:1540–1551
Byrne TJ, Santos BFS, Amer PR, Martin-Collado D, Pryce JE, Axford M (2016) New breeding objectives and selection indices for the Australian dairy industry. J Dairy Sci 99:8146–8167
Carabaño MJ, Ramón M, Menéndez-Buxadera A, Molina A, Díaz C (2019) Selecting for heat tolerance. Anim Front 9:62–68
Colditz IG, Hine BC (2016) Resilience in farm animals: biology, management, breeding and implications for animal welfare. Anim Prod Sci 56:1961–1983
Cole JB, Vanraden PM (2018) Symposium review: possibilities in an age of genomics: the future of selection indices. J Dairy Sci 101:3686–3701
Cole JB, Eaglen SAE, Maltecca C, Mulder HA, Pryce JE (2020) The future of phenomics in dairy cattle breeding. Anim Front 10:37–44
DairyNZ (2022) Economic values (EV'S) are an estimate of the dollar value to a NZ dairy farmer of a unit change in each trait. https://www.dairynz.co.nz/animal/animal-evaluation/interpreting-the-info/economic-values/. Accessed 8 December 2022
Davis SR, Spelman RJ, Littlejohn MD (2017) Breeding and genetics symposium: breeding heat tolerant dairy cattle: the case for introgression of the "slick" prolactin receptor variant into Bos taurus dairy breeds. J Anim Sci 95:1788–1800
De La Paz JM (2008) Using antibody and cell-mediated immune response to test antigens in periparturient dairy cows as a measure of disease resistance. University of Florida, Gainesville, FL
De Marchi M, Toffanin V, Cassandro M, Penasa M (2014) Invited review: mid-infrared spectroscopy as phenotyping tool for milk traits. J Dairy Sci 97:1171–1186
Doublet A-C, Croiseau P, Fritz S, Michenet A, Hozé C, Danchin-Burge C, Laloë D, Restoux G (2019) The impact of genomic selection on genetic diversity and genetic gain in three French dairy cattle breeds. Genet Sel Evol 51:52
Eggen A (2012) The development and application of genomic selection as a new breeding paradigm. Anim Front 2:10–15
Egger-Danner C, Nielsen P, Fiedler A, Müller K, Fjeldaas T, Döpfer D, Daniel V, Bergsten C, Cramer G, Christen A (2014) ICAR claw health atlas. ICAR Technical Series
Egger-Danner C, Cole J, Pryce J, Gengler N, Heringstad B, Bradley A, Stock K (2015) Invited review: overview of new traits and phenotyping strategies in dairy cattle with a focus on functional traits. Animal 9:191–207
Elgersma GG, De Jong G, Van Der Linde R, Mulder HA (2018) Fluctuations in milk yield are heritable and can be used as a resilience indicator to breed healthy cows. J Dairy Sci 101:1240–1250
Fahrenkrug SC, Carlson DF (2014) Hornless livestock. Google Patents
Fuerst-Waltl B, Fuerst C, Obritzhauser W, Egger-Danner C (2016) Sustainable breeding objectives and possible selection response: finding the balance between economics and breeders' preferences. J Dairy Sci 99:9796–9809
García-Ruiz A, Cole JB, Vanraden PM, Wiggans GR, Ruiz-López FJ, Van Tassell CP (2016) Changes in genetic selection differentials and generation intervals in US Holstein dairy cattle as a result of genomic selection. Proc Natl Acad Sci 113:E3995–E4004
Garner JB, Douglas ML, Williams SRO, Wales WJ, Marett LC, Nguyen TTT, Reich CM, Hayes BJ (2016) Genomic selection improves heat tolerance in dairy cattle. Sci Rep 6:34114

Goddard ME, Hayes BJ (2009) Mapping genes for complex traits in domestic animals and their use in breeding programmes. Nat Rev Genet 10:381–391

Hansen PJ (2020) Prospects for gene introgression or gene editing as a strategy for reduction of the impact of heat stress on production and reproduction in cattle. Theriogenology 154:190–202

Harris DL, Arboleda CR, Stewart TS (1984) Animal breeding programs: a systematic approach to their design. AGRIS

Haskell MJ, Simm G, Turner SP (2014) Genetic selection for temperament traits in dairy and beef cattle. Front Genet 5

Haskell MJ, Vasseur E, Mosconi C, Egger-Danner C (2019) Mapping the global use of welfare indicators for dairy cows. In: Proceedings of the 43rd ICAR Conference, Prague 2019

Hazel LN (1943) The genetic basis for constructing selection indexes. Genetics 28:476–490

Heringstad B, Østerås O (2013) More than 30 years of health recording in Norway. ICAR Technical Series no. 17, 39

Heringstad B, Gianola D, Chang Y, Ødegård J, Klemetsdal G (2006) Genetic associations between clinical mastitis and somatic cell score in early first-lactation cows. J Dairy Sci 89:2236–2244

Heringstad B, Egger-Danner C, Charfeddine N, Pryce JE, Stock KF, Kofler J, Sogstad AM, Holzhauer M, Fiedler A, Müller K, Nielsen P, Thomas G, Gengler N, De Jong G, Ødegård C, Malchiodi F, Miglior F, Alsaaod M, Cole JB (2018) Invited review: genetics and claw health: opportunities to enhance claw health by genetic selection. J Dairy Sci 101:4801–4821

Kipp S, Segelke D, Schierenbeck S, Reinhardt F, Reents R, Wurmser C, Pausch H, Fries R, Thaller G, Tetens J, Pott J, Haas D, Raddatz BB, Hewicker-Trautwein M, Proios I, Schmicke M, Grünberg W (2016) Identification of a haplotype associated with cholesterol deficiency and increased juvenile mortality in Holstein cattle. J Dairy Sci 99:8915–8931

König S, May K (2019) Invited review: phenotyping strategies and quantitative-genetic background of resistance, tolerance and resilience associated traits in dairy cattle. Animal 13:897–908

Lasser J, Matzhold C, Egger-Danner C, Fuerst-Waltl B, Steininger F, Wittek T, Klimek P (2021) Integrating diverse data sources to predict disease risk in dairy cattle. bioRxiv. https://doi.org/10.1101/2021.03.25.436798

Lopez-Villalobos N, Garrick DJ, Holmes CW, Blair HT, Spelman RJ (2000) Profitabilities of some mating systems for dairy herds in New Zealand. J Dairy Sci 83:144–153

Lucy M (2001) Reproductive loss in high-producing dairy cattle: where will it end? J Dairy Sci 84:1277–1293

Luke TDW, Nguyen TTT, Rochfort S, Wales WJ, Richardson CM, Abdelsayed M, Pryce JE (2019a) Genomic prediction of serum biomarkers of health in early lactation. J Dairy Sci 102:11142–11152

Luke TDW, Rochfort S, Wales WJ, Bonfatti V, Marett L, Pryce JE (2019b) Metabolic profiling of early-lactation dairy cows using milk mid-infrared spectra. J Dairy Sci 102:1747–1760

Makanjuola BO, Miglior F, Abdalla EA, Maltecca C, Schenkel FS, Baes CF (2020) Effect of genomic selection on rate of inbreeding and coancestry and effective population size of Holstein and Jersey cattle populations. J Dairy Sci 103:5183–5199

Martin-Collado D, Byrne T, Amer P, Santos B, Axford M, Pryce J (2015) Analyzing the heterogeneity of farmers' preferences for improvements in dairy cow traits using farmer typologies. J Dairy Sci 98:4148–4161

McParland S, Kearney JF, Rath M, Berry DP (2007) Inbreeding effects on milk production, calving performance, fertility, and conformation in Irish Holstein-Friesians. J Dairy Sci 90:4411–4419

Medugorac I, Seichter D, Graf A, Russ I, Blum H, Göpel KH, Rothammer S, Förster M, Krebs S (2012) Bovine polledness–an autosomal dominant trait with allelic heterogeneity. PLoS One 7:e39477

Miglior F, Fleming A, Malchiodi F, Brito LF, Martin P, Baes CF (2017) A 100-year review: identification and genetic selection of economically important traits in dairy cattle. J Dairy Sci 100:10251–10271

Mueller ML, Cole JB, Sonstegard TS, Van Eenennaam AL (2019) Comparison of gene editing versus conventional breeding to introgress the POLLED allele into the US dairy cattle population. J Dairy Sci 102:4215–4226

Nguyen TTT, Bowman PJ, Haile-Mariam M, Pryce JE, Hayes BJ (2016) Genomic selection for tolerance to heat stress in Australian dairy cattle. J Dairy Sci 99:2849–2862
Nielsen H-M, Christensen LG, Groen A (2005) Derivation of sustainable breeding goals for dairy cattle using selection index theory. J Dairy Sci 88:1882–1890
Ødegård C, Svendsen M, Heringstad B (2013) Genetic analyses of claw health in Norwegian red cows. J Dairy Sci 96:7274–7283
OIE (2022) Terrestrial code online access. https://www.oie.int/en/what-we-do/standards/codes-and-manuals/terrestrial-code-online-access/. Accessed 8 December 2022
Poppe M, Bonekamp G, Van Pelt ML, Mulder HA (2021) Genetic analysis of resilience indicators based on milk yield records in different lactations and at different lactation stages. J Dairy Sci 104:1967–1981
Pryce JE, Harris BL (2006) Genetics of body condition score in New Zealand dairy cows. J Dairy Sci 89:4424–4432
Pryce JE, Parker Gaddis KL, Koeck A, Bastin C, Abdelsayed M, Gengler N, Miglior F, Heringstad B, Egger-Danner C, Stock KF, Bradley AJ, Cole JB (2016) Invited review: opportunities for genetic improvement of metabolic diseases. J Dairy Sci 99:6855–6873
Rauw W, Kanis E, Noordhuizen-Stassen E, Grommers F (1998) Undesirable side effects of selection for high production efficiency in farm animals: a review. Livest Prod Sci 56:15–33
Simm G, Pollott G, Mrode R, Houston R, Marshall K (2021) Genetic improvement of farmed animals. CABI, Wallingford
Stock ML, Baldridge SL, Griffin D, Coetzee JF (2013) Bovine dehorning: assessing pain and providing analgesic management. Vet Clin N Am Food Anim Pract 29:103–133. https://doi.org/10.1016/j.cvfa.2012.11.001
Thompson-Crispi K, Hine B, Quinton M, Miglior F, Mallard B (2012) Short communication: association of disease incidence and adaptive immune response in Holstein dairy cows. J Dairy Sci 95:3888–3893
Van Den Berg I, Ho PN, Luke TDW, Haile-Mariam M, Bolormaa S, Pryce JE (2021) The use of milk mid-infrared spectroscopy to improve genomic prediction accuracy of serum biomarkers. J Dairy Sci 104:2008–2017
Van Der Waaij E, Holzhauer M, Ellen E, Kamphuis C, De Jong G (2005) Genetic parameters for claw disorders in Dutch dairy cattle and correlations with conformation traits. J Dairy Sci 88:3672–3678
Van Eenennaam AL (2019) Application of genome editing in farm animals: cattle. Transgenic Res 28:93–100
Vanraden PM, Olson KM, Wiggans GR, Cole JB, Tooker ME (2011) Genomic inbreeding and relationships among Holsteins, jerseys, and Brown Swiss. J Dairy Sci 94:5673–5682
Von Keyserlingk MAG, Weary DM (2017) A 100-year review: animal welfare in the journal of dairy science—the first 100 years. J Dairy Sci 100:10432–10444
Weary DM, Droege P, Braithwaite VA (2017) Chapter two - Behavioral evidence of felt emotions: approaches, inferences, and refinements. In: Naguib M, Podos J, Simmons LW, Barrett L, Healy SD, Zuk M (eds) Advances in the study of behavior. Academic Press, New York
Whay HR, Shearer JK (2017) The impact of lameness on welfare of the dairy cow. Vet Clin North Am Food Anim Pract 33:153–164
Zhang X, Amer P (2021) A new selection index percent emphasis method using subindex weights and genetic evaluation accuracy. J Dairy Sci 104:5827–5842
Zwald NR, Weigel KA, Fikse WF, Rekaya R (2003) Identification of factors that cause genotype by environment interaction between herds of Holstein cattle in seventeen countries. J Dairy Sci 86:1009–1018

# Part IV

# Cattle welfare: culture and sustainability interactions

# The Sustainability of Cattle Production Systems

# 13

Donald M. Broom

## Contents



D. M. Broom (✉)
Department of Veterinary Medicine and St Catharine's College, University of Cambridge, Cambridge, UK
e-mail: dmb16@cam.ac.uk



M. Haskell (ed.), *Cattle Welfare in Dairy and Beef Systems*, Animal Welfare 23,
https://doi.org/10.1007/978-3-031-21020-4_13

**Abstract**

The welfare of animals used for food production is a key part of the sustainability of any system. However, animal welfare should be considered along with adverse effects on a wide range of aspects of human welfare and the environment. All of these components should be included in sustainability evaluation and a scoring method based on scientific information should be used. Cattle production has a major advantage because cattle eat leaves that humans cannot eat. However, the microflora in the cattle gut produces greenhouse gases. Analysis of beef-production systems shows that the sustainability of the best systems is very much better than that of the worst systems. The least sustainable are extensive grazing that causes land degradation and the use of feedlots or indoor housing with grain feeding. The most sustainable beef-production systems are semi-intensive silvopastoral systems and well-managed pasture-fed beef from areas where crop production is uneconomic.

A number of key welfare issues affect the dairy industry including lameness, mastitis, reproductive disorders, and aspects of calf management. Urgent changes that are needed include reversing genetic selection for high milk yield per cow and avoiding high feed intakes that cause metabolic pressure. Other major topics concerning the sustainability of dairy production are: minimising grain use, feeding high-protein forage plants, reducing greenhouse gas emissions by changing diet, and improving labelling and traceability.

**Keywords**

Sustainability assessment · Land and water usage · Biodiversity · Animal welfare · Greenhouse gas · Human-edible feed · Feedlot · Forage-based systems · Silvopastoral · Beef · Dairy cows · Calves

## 13.1 Introduction

The sustainability of human management of land and livestock is important to an increasing proportion of consumers. The welfare of animals used for, or affected by, this management is a key part of that sustainability. In the past, for the production of food there was a 'push' economy in which producers had the major influence on how products were produced. More recently, there has been an increasing change to a 'pull' economy where consumers require transparency about production methods and refuse to buy if they do not consider products to be sustainable and hence of high quality (Broom 2014, 2017a, 2022b). The demand 'pull' has become more detailed in its requirements than previously assumed by some economists (Kim and Lee 2009; Antonelli and Gehringer 2015). There is currently much negative publicity about beef and dairy products because of sustainability issues, some correct and some unjustified, so good quality information about sustainability is important to the industry and to the public. In relation to usage of world resources, it is better if those plant materials that can be eaten by humans are eaten by humans rather than being fed to animals that humans then eat. Some of the public have an aesthetic dislike of eating animals but others argue that they do not want animals to be killed in order to produce human food. However, a very large number of animals are killed during plant production, as mentioned by Broom (2018b, 2022b). These include soil-dwelling animals, animals killed by pesticides, animals killed during harvest, and animals killed while food is stored. Indeed, some plant production methods kill many more animals than some cattle production methods.

The meaning of the term 'sustainable' is now much wider than it was in the past (Herrero et al. 2012; Broom 2017a, b). Examples of reasons why systems have been called unsustainable include when there was no market for a product, when a resource was depleted and became unavailable to the system, for example water or an essential nutrient, or when a system product, such as dioxin, prevented the functioning of other systems. One or more of these reasons, or several other negative impacts, as discussed below, can result in a system being unsustainable. A system or procedure is sustainable if it is acceptable now and if its expected future effects are acceptable, in particular in relation to resource availability, consequences of functioning, and morality of action (Broom 2014). Governments, other agencies, and the general public use ethical evaluations of available information, ideally information of good scientific quality, when deciding on what is acceptable (Bañon Gomis et al. 2011).

Sustainability has many components. Factors that might make a food-production system unsustainable include: adverse effects on human welfare, including on health; poor welfare of production animals; inefficient usage of world resources; harmful environmental effects, such as greenhouse gas production, water pollution including by nitrogen and phosphorus, low biodiversity or destruction of natural ecosystems, reduced carbon sequestration, unacceptable genetic modification, not being 'fair trade' (in that producers in poor countries are not properly rewarded), insufficient job satisfaction for those working in the industry; and damage to rural communities (Broom 2017b, 2021a, b). Decisions about the sustainability of any product, system, or action will depend on trade-offs among the components and comparisons will usually result in a hierarchy of perceived sustainability values where there are alternatives (Marshall and Toffel 2005; Pope et al. 2017).

Some of the studies of sustainability consider only one, or a small number, of its components (Smith et al. 2013) and statements about a product may be based on evidence concerning a limited number of aspects of sustainability. For example, when considering only greenhouse gas production, pig and poultry meat may appear more sustainable than meat derived from pasture feeding or all red meat may be considered equally unsustainable (Steinfeld et al. 2006; Clonan et al. 2015; Siegrist and Hartmann 2019), but neither is true when all components are considered. Sustainability evaluations of beef production have often been limited to feedlot systems and inefficient extensive systems, but other systems have a much higher sustainability value (Broom 2021b). Dairy systems that differ in feeding and housing methods have been compared, but also without a consideration of all aspects that contribute to sustainability (Broom 2021a). Failure to consider all components of sustainability can affect whether decisions made in regard to production systems are effective and fair.

The major part of this paper concerns how the various cattle production methods can have negative or positive effects on components of sustainability, including animal welfare. A separate area of study, which is not reviewed here, is how different methods might need to be developed and used in future to cope with the consequences of climate change, such as the hotter and drier conditions which livestock may have to face. This area is considered by other authors (e.g. Pryce and de Haas 2018). In Sect. 13.2, components of sustainability of cattle production are discussed. The results for beef-production methods of a simple but comprehensive method for sustainability assessment and decision-making are described in Sect. 13.3. The sustainability of some aspects of dairy production methods are then discussed in Sect. 13.4.

## 13.2 Cattle Production: Sustainability Components

### 13.2.1 Considering All Components

The assessment of the sustainability of agricultural production systems should include all components of sustainability. Attempts to do this have included the use of life-cycle analysis (e.g. Ciambrone 2018) and measurement of system externalities methodologies (Balmford et al. 2012, 2018). Some life-cycle analyses for animal production do not consider all environmental aspects of sustainability, for example life-cycle analysis methodology has often not considered the extent to which land used for beef production could be used for other human food production (de Vries et al. 2015). Other components sometimes neglected, for example by those focussing solely on carbon footprint, are animal welfare, aspects of human welfare, and water use.

### 13.2.2 Human Welfare: Health

Health is an important part of the welfare of human and non-human animals. The impact of beef and other meat on human health is positive if intake is moderate and if little processed meat is consumed (Wagemakers et al. 2009). The *n*-3 series polyunsaturated fatty acids (PUFAs), recommended for a healthy human diet, are at higher concentrations in beef from forage-fed cattle than from grain-fed cattle (Nuernberg et al. 2005; Warren et al. 2008; Daley et al. 2010). Antimicrobial resistance (AMR) is one of the greatest current threats to human health (Broom 2020a). More antibiotics are used in systems with high densities of cattle (Sneeringer et al. 2015) and the disposal system for urine and faeces from these systems may also promote the development of antimicrobial-resistant pathogens.

### 13.2.3 Cattle Welfare and the Welfare of Other Species Affected by Cattle Production

Ninety-five percent of a large sample of EU citizens said that farm animal welfare is important to them and affects their purchasing (EU D.G. Health and Food Safety 2016; Broom 2020b). Much scientific information is available about cattle welfare (Phillips 2018; Webster 2018; Broom 2022a) and in other chapters of this book. Disease problems have a large effect on calf welfare and are worse at high densities of animals (Webster 1994; EFSA 2006). Disease can also be worse at high density in adult animals, for example in most feedlot and indoor-rearing systems (Schneider et al. 2009; Magrin et al. 2018). Lack of food in areas with degraded pastures is a major cause of poor welfare. There are negative effects on cattle welfare if there is high metabolic pressure on the animals, sometimes because of too high a proportion of concentrate feed in the diet, or if there is lack of access to grazing areas.

The term welfare refers to all animals, some that are sentient and some that are not. If cattle are foraging for their food, usually on pasture, their foraging area may have been produced by removal of original habitat with impact on the welfare of the animals that lived there. Cattle moving around can cause harm to small animals but usually take some care to minimise the more obvious harms. Otherwise, foraging cattle have a small negative impact and some positive impacts on the welfare of other animals. If the cattle are fed on imported food, as with almost all plant production by humans, there will be some mortality of other animals and some negative effects on welfare. The animals negatively affected include: those displaced for arable production, those harmed by soil cultivation and planting methods, those harmed by chemicals used to control or accidentally affect plants or animals in or near arable fields, those harmed during harvesting, those harmed by transport methods, and those harmed during storage of plant material. Hence if cattle are fed grain, and to a much smaller extent if they are fed imported forage, there are some negative effects on the welfare of other animals. Cattle products and by-products may have positive or negative effects on the welfare of non-human animal species.

### 13.2.4 Cattle Production and Efficiency of Use of World Resources: Land Usage

When plant-derived food that humans could eat is fed to livestock, a proportion of the energy in the plant food is effectively wasted compared with direct consumption by humans. Some land used for producing farm animal feed could have been used for human food production (de Boer and van Ittersum 2018). However, a large proportion of land used for producing forage eaten by farmed animals does not have the highly fertile soils and favourable rainfall to allow arable crops to be grown. Herbivores such as cattle that eat forage plants are important in relation to use of world resources, as compared with pigs or poultry which compete with humans for food (Broom et al. 2013; de Vries et al. 2015; Broom 2017b, 2018b). This assessment of the efficiency of land usage in the world should take account of the proportions of livestock feed components that are human-edible, including crop residues and food industry by-products (Wilkinson 2011). Feed conversion by cattle is sometimes said to be less efficient than that by pigs and poultry (Gerber et al. 2015). However, ruminants use 5.9 kg of human-edible feed/kg of protein output, whereas monogastrics need 15.8 kg/kg (Mottet et al. 2017). Meat production from intensively kept pigs and broiler chickens uses less human-edible feed per kg boneless meat than grain-fed cattle from feedlot systems but much more than ruminants fed on forage plants only (Mottet et al. 2017). Hence, for cattle that are kept intensively for part of the year, it is generally better to feed conserved herbage (which is typically grown on land unsuitable for arable cropping) rather than grain during the season with little forage growth. Good levels of productivity are possible in forage-only systems. Cattle on semi-intensive silvopastoral systems containing pasture plus leaves of high-protein shrubs or trees (see Sect. 13.3 for a more detailed description of the forages used in silvopastoral and the image in Fig. 13.1) in upland, temperate conditions in Colombia were 1.5–4 times more productive per unit of land used than cattle fed on fertilised pasture only, hence no supplementary feed was required (Murgueitio et al. 2008; Broom et al. 2013). Similar productivity can be obtained from lowland permanent pasture in temperate conditions (Orr et al. 2019). The incorporation of protein-rich plants into the pasture facilitates this. Much land used for extensive pasture that is not degraded could not be used for other human food production (Halder 2013) so the efficiency of usage is high. Efficiency of land use where pasture is degraded is reduced according to the degree of degradation.

### 13.2.5 Cattle Production and Efficiency of Use of World Resources: Land Area

A comparison of the land area required for meat or dairy production systems should include land occupied by animals, land to produce their feed, and land to process meat. Broom (2019b) found that the land to produce 1 tonne of beef is: extensive unmodified pasture: 27 ha, feedlots with pre-feedlot extensive conditions: 21 ha,

**Fig. 13.1** Cattle in a semi-intensive silvopastoral system in a warm temperate mountain area in Colombia. They are choosing to feed on *Leucaena leucocephala*. (Photograph with permission of C. Cuartas)

feedlots with pre-feedlot irrigated pasture conditions: 9 ha, fertilised irrigated pasture: 10 ha, and semi-intensive silvopastoral systems: 2 ha. If the diet includes a high proportion of concentrates, the total land use per unit of beef production is higher. The land area component is negative where there is degraded land. Any farming method that results in land degradation is considered unacceptable by some consumers (Özgüner et al. 2012; Jendoubi et al. 2020). Greater meat production per hectare of land means more land available for other purposes such as conservation (Balmford et al. 2018).

### 13.2.6 Cattle Production and Efficiency of Use of World Resources: Amount of Water Used

The water footprint of human food production (Mekonnen and Hoekstra 2010) is increasingly important, especially in regions or seasons where there may be water shortages. In a comparison of beef-production systems (Broom 2019b), only conserved water in farm reservoirs, human water supplies, or from rivers and streams was considered. This is the water that requires human time, energy, and financial cost to obtain. Rainfall is only included if it forms part of the conserved water. Conserved water used per kilogram of beef was: feedlots with irrigated pasture pre-feedlot: 673 l, feedlots with extensive pasture pre-feedlot: 553 l, fertilised irrigated pasture: 411 l, extensive unmodified pasture: 155 l, and semi-intensive silvopastoral systems: 87 l.

### 13.2.7 Cattle Production and Harmful Environmental Effects: Greenhouse Gas Production

Methane from cattle has an important global warming effect but the FAO (Steinfeld et al. 2006) exaggerated the global warming effects of methane from cattle because they did not fully consider the relatively short duration of methane in the atmosphere (Allen et al. 2018). Also, the data for $CO_{2e}$ (carbon dioxide equivalent) in cattle production came from extensive grazing and feedlot systems, but other systems have much lower outputs. The calculations included: $CO_2$ from fossil fuel use, $CH_4$ and $N_2O$ from manure, $CH_4$ from rumen fermentation, and $CO_{2e}$ from the cultivation and transport of feed (de Vries and de Boer 2010). That study obtained values for $CO_{2e}$ per kg of meat of 3.9–10 kg for pork, 3.7–6.9 kg for chicken, and 14 to 32 kg for beef. The methanogens in the gut of cattle, principally Archaea such as *Methanobrevibacter*, can be reduced by using dietary supplements and by including certain plant species in the available forage. These measures can reduce methane production per unit of product (Patra et al. 2017; Broom 2021a, b). Reducing methane production in cattle may be more difficult in the most extensive systems.

Feedlots and semi-intensive silvopastoral systems, which have high protein intake and high yield per unit area (Murgueitio et al. 2008), produce less greenhouse gas per unit of product than lower yielding systems. For example, they can have up to four times lower $CO_{2e}$ per tonne of beef than relatively inefficient extensive pasture systems (Balmford et al. 2018). If pasture is degraded, $CO_{2e}$ produced per tonne of product is even higher. Semi-intensive silvopastoral systems produce 1.6–1.8 times less methane than fertilised pasture or efficient unfertilised pasture systems (Murgueitio et al. 2008; Broom et al. 2013). Where food supplementation is used, the greenhouse gas cost is higher, partly because tillage and other soil structure damage result in greenhouse gas production (Vellinga et al. 2004; Nawaz et al. 2017).

### 13.2.8 Cattle Production and Harmful Environmental Effects: Water Pollution, N/P Cycle Disruption

Artificial fertiliser, which can cause nitrate pollution and interfere with world nitrogen (N) and phosphorus (P) cycles (Rockström et al. 2009), is used in some feed-production systems and some pasture systems but not on unfertilised extensive pasture or semi-intensive silvopastoral systems. Water pollution by ammonia and other pollutants from manure can be minimised by capturing and processing them (Nader et al. 1998) but while high concentrations of animals increase the risk of such pollution, localisation of animals in a small area can mean that there are better opportunities to capture N and P and minimise the loss of these elements during storage and spreading of manure. However, Eghball and Power (1994) found that in feedlots in the USA, 50% of the nitrogen was lost, mainly by runoff, ammonia volatilisation and denitrification, before the manure was removed from the feedlot, and only 25% was incorporated into the soil. Much of the lost nitrogen and some of the lost phosphorus can enter waterways (Gerber et al. 2015). Systems with no fertiliser usage produce less water pollution. Water pollution is greatest when there is a high density of animals and when concentrates are fed (Roche et al. 2013).

### 13.2.9 Cattle Production and Harmful Environmental Effects: Biodiversity

Biodiversity is defined as the variety of animal and plant life in an environment and can be measured as the number of species, the number of rare species, or as a function of number of species and number of individuals of each in an area. It is a matter of much public concern (Skogen et al. 2018) and can be evaluated in farmed and unmodified environments. Biodiversity is reduced when natural vegetation is cleared to produce pasture (Koellner et al. 2013). Degraded land has lower biodiversity than undamaged agricultural land but can regenerate to produce biodiverse environments (Plieninger and Gaertner 2011). Farmland with trees and areas where natural vegetation is left at the margins of cultivated areas are more biodiverse than when every part is cultivated. Pasture resulting from monoculture has lower biodiversity than semi-natural pasture or herbal leys (Luoto et al. 2003; Goh and Bruce 2005). The use of herbicides and pesticides on cultivated land can have a large impact on local and global biodiversity, hence the biodiversity of farmland is now lower than it has ever been (Benton et al. 2003; Batáry et al. 2020). The food plants of insects, birds, and mammals disappear (Butler et al. 2007) and seed dressings and crop sprays kill very large numbers of non-target insects and other animals (Geiger et al. 2010). Since farmland occupies such a high proportion of the earth's land surface, many people would say that it is a crime of enormous significance that mean biodiversity has declined more in the last 20 years than at any time in human history.

Biodiversity is much higher when shrubs and trees as well as pasture plants are present (Fischer et al. 2010). Silvopastoral systems have much higher numbers of insects and birds than pasture areas in the same locality (Burgess 1999; Fajardo et al. 2008; Múnera et al. 2008; Rivera et al. 2008; McAdam and McEvoy 2009). Crops such as maize and soya have less biodiversity than pasture (Cremene et al. 2005; Zabel et al. 2019), especially when herbicides are used to encourage only one plant species, and extensive pasture systems have higher biodiversity than irrigated pasture. Where cattle are housed in a feedlot, or indoors on slatted floors, and the cattle feed is mainly maize, soya, or other crop plants, the overall consequence is a large negative impact on biodiversity. Systems that rear the cattle on extensive pasture for the first months of life slightly reduce this negative effect.

### 13.2.10 Cattle Production and Harmful Environmental Effects: Carbon Sequestration

The conversion of land from its original state to farmland or for other human use affects the amount of carbon sequestered in the soil and in the biomass above ground. Cattle production systems with a greater standing crop of plants, for example silvopastoral systems which are used with both beef and dairy cattle, also have a much greater depth of soil than grass-only systems and therefore have a lower risk of soil erosion (Molina et al. 2008). The greater soil depth also means that they act as a carbon sink and sequester more carbon (Schmidinger and Stehfest 2012). Permanent pasture is better in this respect than frequently renewed pasture and both are better than land used for crop production (Garnett 2009). Pasture degradation reduces carbon sequestration (Maia et al. 2009). When compared with the total carbon sequestration capacity of temperate and tropical native forest, there is a reduction of 72% for arable land, 56–69% for pasture, and 44–46% for agroforestry (Tzilivakis et al. 2019; Toru and Kibret 2019).

### 13.2.11 Cattle Production and Genetic Modification

Consumers who will not buy products associated with genetic modification (GM) because of perceived risks associated with GM or because they do not agree with genetic modification of organisms, show more antipathy to modifying animals or using mixtures of animal and non-animal cells, than to modifying plants or microorganisms (Knight 2008; Hudson et al. 2015). Gene editing is a form of genetic modification, while cloning involves human interference in biological processes but is not GM (Broom 2014, 2018a). Adverse effects of GM on human health, animal welfare, or the environment are unacceptable to most people so are unsustainable (Frewer et al. 2014). Food from a genetically modified plant or animal could contain allergenic proteins or could have unforeseen negative environmental impacts (EFSA GMO Panel 2010), or the welfare of animals used by humans may be poor because of the procedures involved or due to genetic changes in the modified individuals. Animal products are not accepted by some consumers if the feed of those animals was genetically modified, hence 'no GM feed' labels. Legislation is in place in most countries requiring testing before general release of any product of genetic modification, including gene editing. Such testing should involve a full range of aspects of sustainability (Broom 2018a). Cattle producers, aware of consumer attitudes, usually express reluctance to use genetically modified or cloned animals due to the risk to their sales. Although no GM animals are in current use in the cattle industries, production companies that use the most intensive systems may be more likely to use GM in the future. This would further increase consumer hostility to products from cattle products.

### 13.2.12 Cattle Production and Fair Trade

Many consumers consider that failure to properly reward the producers of food in poorer countries is morally wrong. Hence products like coffee, cocoa, and fruit have a well-established 'Fair Trade' labelling system (Nicholls and Opal 2005; Broom and Johnson 2019). Fair-trade labels could be more widely used for beef and dairy products.

### 13.2.13 Cattle Production and Worker Satisfaction

Working with animals in rural environments can be a rewarding experience, so work with cattle is generally regarded as desirable (Viljoen and Wiskerke 2012). On semi-intensive silvopastoral farms in Colombia and Mexico, whose standards of animal welfare and positive environmental impact are high, workers like the work and continue employment on those farms for longer than people who work on conventional farms (Calle et al. 2009). Worker satisfaction may be lower in other cattle production environments but carefully controlled investigation comparing systems is needed (Calvo-Lorenzo 2018). Worker satisfaction is likely to be better when sustainable systems are used.

### 13.2.14 Cattle Production and Rural Community Preservation

Agricultural and social system changes can lead to rural communities declining and disappearing and there is public pressure for government action to safeguard such communities. Subsidies to preserve rural communities in the European Union have reduced migration to cities and keeping populations in rural areas is seen as a major success of the EU Common Agricultural Policy (Gray 2000; Broom 2010). Consumers often prefer beef and dairy products produced by small rural communities, which might confer some advantage for more extensive production systems.

## 13.3 Sustainability of Global Beef-Production Systems

This section presents a comparison of the sustainability of beef-production systems. The components of sustainability and the relevant published information about beef that are described in Sect. 13.2 were used to produce Table 13.1, which is modified after Broom (2021a, b). Comparisons of the sustainability of animal production systems should be relevant across the whole world but regional analyses of what is practicable can also be affected by local physical conditions and the different kinds of pasture usage in temperate and tropical situations. The product might come from temperate, tropical, or sub-tropical sources. However, the breeds used in the various systems are selected by producers according to factors such as growth rate in local conditions, ability to adapt to expected environmental temperatures, and vulnerability to disease, for example tick-borne diseases. The transport of the product is a factor affecting sustainability and could lead to differences between protein sources. While local production would involve less fossil fuel use than imports from distant places, there would not necessarily be substantial variation in sustainability associated with the production system.

In comparing systems for sustainability of cattle production it is best to assume that appropriate breeds were used for all data investigated. There is an overlap between beef and dairy production (Lowe and Gereffi 2009) so the origin of calves

**Table 13.1** Sustainability components and score for beef-production systems. Numbers indicate negative scores; zero-tolerance (Z) for some consumers is indicated where there is evidence for it. Other components that do not alter the scoring are not included in this Table. (Modified after Broom 2021a, b)

| | Sustainability components | | | | | | | | | |
|---|---|---|---|---|---|---|---|---|---|---|
| | | | World resources | | | | | | | |
| Beef-production systems | Human health | Animal welfare | Land usage | Land area | Water used | Greenhouse gas production | Water pollution; N/P cycle disruption | Biodiversity | Carbon sequestration | Total scores |
| Extensive pasture degraded | −1 | −3 | −3 | −5, Z | −1 | −5, Z | 0 | −4 | −4 | **−26, ZZ** |
| Extensive pasture not degraded | −1 | 0 | 0 | −3 | −1 | −3 | 0 | −2 | −2 | **−12** |
| Fertilised irrigated pasture + concentrates | −1 | 0 | −4 | −3 | −4 | −2 | −2 | −4 | −3 | **−23** |
| Fertilised irrigated pasture no concentrates | −1 | 0 | −2 | −2 | −3 | −2 | −1 | −3 | −2 | **−16** |
| Fertilised irrigated pasture to feedlot | −2 | −2, Z | −4 | −2 | −4 | −1 | −2 | −4 | −4 | **−25, Z** |
| Extensive pasture to feedlot | −2 | −2, Z | −2 | −4 | −4 | −2 | −2 | −3 | −4 | **−25, Z** |
| Fertilised irrigated pasture to indoor | −2 | −3, Z | −4 | −2 | −4 | −1 | −2 | −4 | −4 | **−26, Z** |
| Extensive pasture to indoor | −2 | −3, Z | −2 | −4 | −4 | −3 | −2 | −3 | −3 | **−26, Z** |
| Indoor rearing to indoor | −2 | −4, Z | −4 | −3 | −4 | −1 | −3 | −4 | −4 | **−29, Z** |
| Semi-intensive silvopastoral | −1 | 0 | 0 | −1 | 0 | −1 | 0 | −1 | −1 | **−5** |

is an important variable affecting aspects of sustainability (de Vries et al. 2015). Another important variable is whether or not the system complies with organic standards but taking account of this is difficult due to the great variation in both organic and non-organic systems. Similar problems apply to comparisons of high-roughage and low-roughage systems. For example, roughage effects on greenhouse gas production differ depending on whether the feed is conserved forage or fresh plants grazed or browsed from a variety of species, and whether the fresh or conserved feed is from high-nutrient or low-nutrient sources (Mogensen et al. 2015). Figures for the carbon sequestration capacity of land in comparison with forests (see Sect. 13.2.10 for details) are used rather than the cost of compensation for reduction in carbon sequestration (Lubowski et al. 2006). Neither organic standard nor level of roughage is specified in the comparison of beef-production systems below and the dairy or beef origin of calves is considered in Sect. 13.4.

In Table 13.1, each component of sustainability is scored from 0, meaning no negative effect, to −5, indicating a very negative effect. If some consumers have been reported to avoid the product altogether because of a sustainability component, this is indicated by a Z in Table 13.1. The most widely used system for beef production world-wide is the extensive unmodified pasture system used in temperate, tropical, upland, and lowland areas. The cattle are reared throughout their lives on pasture, initially with their mothers and then in age-related groups. This kind of system was called extensive grazing management by Allen et al. (2018). The term 'unmodified' is used to distinguish this system from fertilised pasture and silvopastoral systems that are also extensive and does not mean that livestock do not alter the pasture. The manure from the animals is usually left on the land. No artificial fertiliser is used and the land is not irrigated. The level of degradation of the pasture and soil depends on how much of the plant material is removed during grazing. The land after extensive grazing may still have leaves on the plants or the plants may be reduced to stalks and roots or there may be exposed soil (Davidson et al. 2008). There are great differences in sustainability between systems where the pasture and soil become degraded and those where the pasture recovers fully and rapidly after a period with cattle on it. For this analysis, degraded means that more than half of what would normally be pasture with photosynthetic cover is exposed soil or non-photosynthetic.

Another beef-production system is fertilised pasture, irrigated when necessary, and either with or without use of additional concentrate feed. Supplementary pasture-derived or concentrate food is used to increase cattle growth rates or when pasture growth is insufficient to feed the animals during a dry period or during the cold seasons in temperate areas. Two further kinds of systems involve a final growth period in feedlots or inside buildings, often on slatted floors. When young, the cattle may be kept on fertilised, irrigated pasture, kept on extensive pasture, or housed and then transported to feedlots or indoor housing and fed high levels of concentrates for the last few months of life. A fifth system, increasingly used throughout the life of the cattle in tropical and sub-tropical countries, is the semi-intensive silvopastoral system that utilises, in addition to pasture plants, shrubs with edible leaves such as the high-protein leguminous shrub *Leucaena leucocephala*, together with trees which may also have edible leaves (Murgueitio et al. 2008, 2015; Ku Vera et al. 2011; Broom et al. 2013). The trees can provide shade and those with edible leaves provide nutrients when pasture yield is low. Rotational management minimises damage to forage plants and the term 'semi-intensive' reflects the higher density of animals than in pasture-only systems. As with other systems, the plant species used vary according to climate (Peri et al. 2016; Broom 2017b, Pachas et al. 2018; Radrizzani et al. 2019).

The factors taken into account when scoring each of the components of sustainability shown in Table 13.1 are summarised here but described in detail in Broom (2021a, b).

Using a method that scores all components of sustainability for which there are differences among the systems compared should help in decision-making by individual consumers, production and consumer organisations, and those making government policy. Table 13.1 shows that systems may be negative in one aspect of sustainability but less negative in others. Beef farming using feedlots was assessed as better than extensive grazing without pasture degradation in greenhouse gas production and in total land area used but feedlots were worse with regard to animal welfare, land usage, and amount of water used. Perhaps the most important result was that the most sustainable beef-production systems are much better than the worst. The systems which the total scores in Table 13.1 show to be the least sustainable are those involving indoor housing on slatted floors with concentrate feed (scores of −26 to −29) (Roath and Krueger 1982; Broom and Kirkden 2004; Duff and Galyean 2007; Hall 2008), extensive grazing where the pasture becomes degraded (score −26) (Hall 2008), and the two feedlot systems (score −25) (Köhler 1993). The zero-tolerance indicators for systems that are completely unacceptable to some consumers identify degraded pastures with two unacceptable components and indoor-housed cattle and cattle kept in feedlots as unacceptable in relation to animal welfare. In the near future, all systems using feeding of grain and soya to cattle may be regarded as unsustainable on biodiversity grounds. In some regions, systems with high water use may also become unacceptable. The systems whose sustainability scores were the best were: semi-intensive silvopastoral system (−5), followed by extensive grazing on non-degraded pastures (−12) and fertilised irrigated pasture grazing with no concentrate feeding (−16). There was insufficient information available for cell-cultured meat, but the best of these methodologies is likely to have few negative scores.

## 13.4 Sustainability of Dairy Production Systems

### 13.4.1 Dairy Systems

The concept of sustainability, described above, is the same for dairy and beef products but the greater diversity of dairy products means that they may be unsustainable because of consumer rejection of all or only some dairy products. The various systems of dairy farm management have parallels with some of the beef systems mentioned above and are reviewed by Webster (2018) and Endres (2021). Therefore, the information presented above is not repeated in this section and details that apply only to dairy systems are discussed here. Dairy systems differ from beef systems because of the constraint that the lactating cows have to be milked. The breeds and growth patterns are also different but the biology and needs of the animals are almost identical.

### 13.4.2 Dairy Production and Human Health

The effects of dairy products on consumer health have major consequences for the future survival and success of dairy farming. While some people may decide to avoid dairy products, most can now access good information about how to manage the levels of saturated fats in their diet and blood cholesterol levels, to avoid obesity and how to take advantage of the nutritional and health benefits of dairy products (Tunick and Van Hekken 2015). However, since dairy systems use more antibiotics than beef systems (Hommerich et al. 2019), antimicrobial resistance is more likely to be promoted by dairy production than by beef production. Thus, it can be argued that overall, dairy has a worse effect on human health.

### 13.4.3 Dairy Production and Cattle Welfare

The discussion in Sect. 13.2.3 above, about cattle production and effects on the welfare of species other than cattle and humans, is also relevant to dairy cattle. Some dairy cow herds have such severe welfare problems that consumers are already avoiding dairy products for this reason. The problems include lameness, mastitis, impaired reproduction, inability to show normal behavioural and physiological responses, infectious diseases, and injury (see review by EFSA 2009). Lameness, mastitis, and impaired reproduction are worse in high-producing dairy cows (Schukken et al. 2005; Oltenacu and Broom 2010). Many dairy cows are at or beyond a production level that is metabolically damaging to them (Knaus 2009) and the problems are worse in cows treated with bovine somatotrophin which increases milk yield and metabolic pressure (EU SCAHAW 1999). Since dairy cattle have been selected for high milk production to too great an extent, there is an urgent need for selection factors to be changed to reduce production per cow. This has been implemented by some countries (see the chapter by Pryce et al.), and further uptake of this approach will help to protect the future of the industry (Broom 2021a). In addition to the increase in genetic potential for yield, feeding energy-dense, highly fermentable diets to optimise milk production exacerbates the metabolic disorders leading to lameness, mastitis, and reproductive problems and hence causes poor welfare. Grain feeding can lead to digestive disorders such as rumen acidosis which has been estimated to have a prevalence rate of up to 60%, depending on the stage of lactation (Penner 2018). Diets should be changed throughout the world in order to alleviate the problems for the cows. There has been consolidation of production into a few, large, resource-intensive farming operations (Scott and Gooch 2017). The problems are not necessarily worse in large dairy herds than in small ones; all can change genetic strains and diet and give adequate individual care to the cows (Broom 2013; Robbins et al. 2016).

The management system for the cows (pasture-based or indoor housing) will affect welfare. In some extensive grazing systems, pasture degradation and subsequent starvation is a major welfare issue for dairy cattle. Inadequate housing that does not meet the needs of the animals is a major welfare problem for some dairy cattle. The welfare of some dairy cows and many dairy bulls is poor because of long-term tethering or other close confinement and lack of access to a grazing area. These welfare challenges are especially important for consumers, and hence for the future of the industry. Dairy calves may be reared in individual crates or pens and may show indicators of poor welfare including stereotypies; difficulties in standing, lying, and grooming; excessive grooming of the front of the body with the ingestion of much hair and the formation of hairballs in the gut; and substantial adverse reactions to walking and to transport (Broom and Leaver 1978; Veissier et al. 1994; Broom 1996, 2022a; Boe and Faerevik 2003; Gaillard et al. 2014; Phillips 2018). In contrast, group housing, if well managed, improves feeding, health, development of behaviour and cognition (Miller-Cushon et al. 2018). Dairy calves are subject to earlier maternal deprivation than beef calves and, for many, indoor-rearing conditions in the earliest part of life.

Dairy and beef calves may also be subjected to practices that cause pain including castration and horn-disbudding. The extent of the pain can be measured using pain-related behaviours and pain can be prevented using anaesthesia and analgesia (Stafford and Mellor 2005; Stilwell et al. 2008a, b, 2009, 2010, 2012, 2019). There is public concern about pain in livestock. Because of the very poor welfare of most high-producing cows and of many dairy calves used for veal production or as herd-replacers, dairy cattle welfare is worse than that of beef cattle.

### 13.4.4 Dairy Production and Land Usage

Dairy cattle can consume material that humans cannot consume (e.g. grass) and convert it to milk and hence milk products. Since land in many parts of the world is not suitable for arable crop production but can support pasture, properly managed extensive, pasture-based dairy production can be an efficient and sustainable use of available world resources (Marshall and Collins 2018). However, while some dairy farming systems use very little grain or soya, others use large quantities of concentrates so are less sustainable. Grasses continue to dominate feed resources for livestock production globally but feed from sources such as grain may represent around 30–40% of the diet of dairy cattle in some regions such as North America, which is characterised by more intensive forms of production (Blümmel et al. 2018). Grain and soya used as cattle feed is less sustainable because such feeding competes for land and other resources with crop production for human food (Herrero et al. 2010; Broom 2018b; Balmford et al. 2018). Such practices are likely to disappear in the near future because of consumer pressure, competing markets for grain, and consequent government action.

### 13.4.5 Dairy Production and Land Area Required

Dairy production uses a little more land than beef production because dairy calves are usually fed materials produced from crops during their early development whilst a high proportion of lactating beef cows are pasture fed. Housed cattle fed on highly digestible crop-derived feeds have high growth rates and feed use efficiencies. However, as for beef cattle, the area of land required for these dairy intensive systems is high when land used for feed production is taken into account (Broom 2019a; Woolf 2020).

Systems that have a high level of dairy production per unit of land make more land available for other uses such as nature reserves (Balmford et al. 2018). This important fact emphasises the necessity for comprehensive sustainability assessment. Systems where there is high production per unit of land (litres of milk/ha), and where this calculation takes into account the land needed to produce all the food for the cattle, should be preferred to inefficient use of land. However, such a preference is correct only when all aspects of sustainability are considered in the comparison of the different systems. The smallest area of land per unit of production, as for beef production, has been shown in some circumstances to be semi-intensive silvopastoral systems (Murgueitio et al. 2008, 2015; Broom et al. 2013).

### 13.4.6 Dairy Production and Water Usage

Because dairy cows produce large quantities of milk, they generally use more conserved water then beef cows and more of this water typically comes from sources that has been purified for human use. The greatest water usage, as it is in beef systems, is in systems where pasture or crops are irrigated and the least is in semi-intensive silvopastoral systems because of their efficient water-holding capacity in soil and in plant biomass (Broom et al. 2013; Broom 2019a, b).

### 13.4.7 Dairy Production, Greenhouse Gas Production, and Pollution by Nitrogen and Phosphorus

Greenhouse gas production from dairy cattle is high where arable crops are used to feed the cows. Stoll-Kleemann and O'Riordan (2015) estimated that 45% of greenhouse gas production from cattle production comes from feed production and processing, with 39% from cow digestion and 10% from their faeces. The use of artificial fertilisers on crops or pasture increases greenhouse gas production and nitrogenous and phosphorus pollution. As with land area per unit of production, greenhouse gas per unit of production is lower if the animals are high producing than if they are low producing. Some studies have suggested that dairy production systems using intensive dairy systems have lower emissions than extensive systems (e.g. FAO 2010). However, all greenhouse gas emissions associated with the production of the feed should be properly considered in such a calculation and the sustainability metric should be calculated considering all components. One likely consequence of such calculations is that extensive grazing with some degree of land degradation is likely to be unacceptable to the public because of its consequences for climate change. The management of urine and faeces from dairy cattle can allow processing to reduce nitrogenous compounds acting as greenhouse gases and water pollutants. The work of Dirksen et al. (2020) showing that cattle can be trained to localise urine and faeces deposition is important because this could facilitate treatment of nitrogenous waste and hence reduce greenhouse gas output per unit of product.

### 13.4.8 Dairy Production and Biodiversity Loss

Biodiversity loss and carbon sequestration loss are greater in most dairy production systems than in extensive beef-production systems. This is because growing the plants that produce supplementary feed, ploughing land for reseeding pasture, and producing and using artificial fertiliser all reduce biodiversity and carbon sequestration. However, destruction of forest for farming has more frequently occurred in order to farm beef cattle than to farm dairy cattle.

### 13.4.9 Dairy Production and Other Sustainability Components

Genetic modification of cattle for dairy production has not been commercially important so far but any attempt to use GM techniques to produce additional chemicals in milk should be subject to careful assessment of impact on cattle welfare and all other sustainability components. Both milk and beef production can be fair trade. Worker satisfaction is usually greater when there is more direct contact between the workers and individual cattle. Tasks such as milking and daily movement for milking tend to promote human interactions with cattle and consequently, higher job satisfaction. Rural communities often rely on dairy or beef production or both. Indeed, many rural communities depend on mixed use cattle. The general trend of specialisation of breeds for dairy or beef can usefully be reversed in many rural communities.

### 13.4.10 How to Implement Sustainable Dairy Production

Much dairy production research is still aimed at increasing productivity without considering the necessity of maintaining or creating a sustainable industry. However, some research has focussed on how to optimise extensive systems to increase grazing efficiency and optimise product quality while reducing greenhouse gases such as methane, protecting biodiversity and maintaining good animal welfare, including animal health (Woodfield and Judson 2018). There are some serious welfare problems at present, including lameness, mastitis, and aspects of calf management, but in several systems, dairy cows and calves can be produced with very good welfare. Retailers and all who buy dairy products should insist on seeing evidence for high welfare standards. In relation to evidence of welfare and environmental impacts, transparency is essential. Consumers need to obtain accurate information. The best dairy production systems use resources that cannot easily be used in any other way for human food production, have no negative impacts on animal welfare, and have no greater negative impact on the environment than wholly plant-based alternatives. However, other widely used practices in dairy production are not sustainable, so much change in the industry is needed in order for it to survive the current and expected increase in consumer concern over these practices.

Research also shows how to change ruminant diets so as to reduce the rumen population of methane-producing microorganisms. Van Wijk (2020) emphasises the potential advantages of integrated crop–livestock systems in combatting climate change and improving resilience in agricultural production. This includes agroforestry practices such as silvopastoralism (Moreno and Rolo 2019). Many different kinds of land can produce more edible plant material and more milk is produced on dairy farms when pasture-only systems are replaced by trees and shrubs with edible, high-protein leaves as well as pasture plants (Murgueitio et al. 2008; Broom et al. 2013; Broom 2017b). For example, in some parts of the world, dairy cattle could be managed on pastures with shrubs such as *Leucaena leucocephala* (Radrizzani et al. 2019). Where there is an unfavourable growing season, cut leafy branches of trees such as ramón (*Brosimum alicastrum)* can be an important source of food (Ku-Vera et al. 2013). Other benefits are that biodiversity is much higher in semi-intensive silvopastoral systems than in pasture-only systems. One consequence of this is that there are more tick and fly predators so there is less cattle disease. Another consequence is the potential to better conserve forest resources (Burgess 1999; Múnera et al. 2008; Rivera et al. 2008; McAdam and McEvoy 2009; Sutherland et al. 2018).

Although no carefully evaluated sustainability scoring of different dairy production systems has yet been conducted, it is clear that sustainable dairy production in the future will involve avoiding the worst negative impacts on animal welfare, world resource usage, and biodiversity reduction. In order to achieve sustainability, it is necessary to change the genetic selection programmes and feeding regimes for dairy cows so as to reduce the high levels of milk production per individual and reduce lameness, mastitis, and reproductive disorders. The changes in selection and feeding to improve welfare will also involve ceasing to feed large quantities of grain and soya which will improve usage of world resources, biodiversity, and water use. Increased and improved use of semi-intensive silvopastoral systems and other high-protein forage will also have these effects. Dietary change to reduce greenhouse gas production will be important to consumers and to meet already stated government targets. Avoidance of close confinement of calves will be necessary in all countries if the veal and dairy industries are to survive increased consumer scrutiny. With well-controlled labelling and traceability, consumers can choose to buy dairy products that are high-welfare, have low greenhouse gas emission, are indicated as fair trade and are otherwise sustainable.

## 13.5 Conclusions

1. Sustainability is a wide-ranging concept with many components, of which cattle welfare is one, but all of which should be taken into account in a sustainability scoring system based on the available scientific literature.
2. The analysis presented here makes it clear that some widespread practices in beef and dairy production are not sustainable and hence major changes in these industries are urgently needed if they are to survive. Sustainable systems do currently exist and all of the changes mentioned are feasible, but changes need to occur in the near future to ensure a sustainable future for these industries.
3. The great advantage that cattle production has over other animal production systems is that the major components of the diet of cattle can be plant material that humans cannot eat. The future of animal production will involve increasing the proportion of animals that eat leaves or other primary products of photosynthesis.
4. Sustainability components that can be assessed for cattle production systems include human health, animal welfare, land usage, land area, amount of water used, greenhouse gas production, water pollution, N/P cycle disruption, biodiversity loss, and carbon sequestration. Analysis of the effects of cattle production should take account of all effects including those above and, for example, effects on the welfare and biodiversity of species affected by producing food for cattle.
5. The results of sustainability scoring for beef production show that there is a wide range in sustainability across the systems, with the best systems very much better than the worst systems.
6. The least sustainable beef-production systems were extensive grazing that causes land degradation and the use of feedlots or indoor housing with grain feeding.
7. Semi-intensive silvopastoral systems are the most sustainable beef-production systems. Well-managed pasture-fed beef from land where crop production is uneconomic is also sustainable.
8. The dairy industry urgently requires system change. The poor welfare of high-producing dairy cows, because of lameness, mastitis, and reproductive disorders, and of closely confined calves is likely to lead to the avoidance by many consumers of all dairy products in the near future. Other major topics concerning the sustainability of dairy production are minimising grain use, feeding high-protein leaves of shrubs and trees, changing diet to reduce methane output, and improving labelling and traceability.

## References


Allen MR, Shine KP, Fuglestvedt JS et al (2018) A solution to the misrepresentations of $CO_2$-equivalent emissions of short-lived climate pollutants under ambitious mitigation. npj Clim Atmos Sci 1:1–8

Antonelli C, Gehringer A (2015) The competent demand-pull hypothesis. In: Crespi F, Quatraro F (eds) The economics of knowledge, innovation and systemic technology policy. Routledge, London, pp 48–69

Balmford A, Green R, Phalan B (2012) What conservationists need to know about farming. Proc R Soc B Biol Sci 279(1739):2714–2724. https://doi.org/10.1098/rspb.2012.0515

Balmford A, Amano T, Bartlett H et al (2018) The environmental costs and benefits of high-yield farming. Nat Sustain 1:477–485. https://doi.org/10.1038/s41893-018-0138-5

Bañon Gomis AJ, Guillén Parra M, Hoffman WM et al (2011) Rethinking the concept of sustainability. Bus Soc Rev 116:171–191

Batáry P, Báldi A, Ekroos J, Gallé R, Grass I, Tscharntke T (2020) Biologia Futura: landscape perspectives on farmland biodiversity conservation. Biologia Futura 71(1-2):9–18

Benton TG, Vickery JA, Wilson JD (2003) Farmland biodiversity: is habitat heterogeneity the key? Trends Ecol Evol 18:182–188

Blümmel M, Muller A, Schader C et al (2018) The use and abuse of cereals, legumes and crop residues in rations for dairy cattle. In: Webster J (ed) Achieving sustainable production of milk Volume 3: Dairy herd management and welfare. Burleigh Dodds Science Publishing, Cambridge, UK

Boe KE, Faerevik G (2003) Grouping and social preferences in calves, heifers and cows. Appl Anim Behav Sci 80:175–190

Broom DM (1996) Scientific research on veal calf welfare. In: Veal perspectives to the year 2000. Proceedings of international symposium, Le Mans, France. Fédération de la Vitellerie Francaise, Paris, pp 147–153

Broom DM (2010) Animal welfare: an aspect of care, sustainability, and food quality required by the public. J Vet Med Educ 37:83–88. https://doi.org/10.3138/jvme.37.1.83

Broom DM (2013) Cow welfare and herd size: towards a sustainable dairy industry. Cattle Pract 21:169–173

Broom DM (2014) Sentience and animal welfare. CABI, Wallingford, p 200

Broom DM (2017a) Animal welfare in the European Union. European Parliament Policy Department, Citizen's Rights and Constitutional Affairs, Brussels, p 75. https://doi.org/10.2861/891355. http://www.europarl.europa.eu/RegData/etudes/STUD/2017/583114/IPOL_STU(2017)583114_EN.pdf

Broom DM (2017b) Components of sustainable animal production and the use of silvopastoral systems. Revista Brasileira Zootecnia 46:683–688. https://doi.org/10.1590/S1806-92902017000800009

Broom DM (2018a) Animal welfare and the brave new world of modifying animals. In: Grandin T, Whiting M (eds) Are we pushing animals to their biological limits? CABI, Wallingford, pp 172–180

Broom DM (2018b) The scientific basis for action on animal welfare and other aspects of sustainability. In: D'Silva J, McKenna C (eds) Farming, food and nature: respecting animals, people and the environment. Routledge, New York, pp 93–100

Broom DM (2019a) Land and water usage in beef production systems. Animals 9(6):286. https://doi.org/10.3390/ani9060286

Broom DM (2019b) Bem-estar bovino e atitudes públicas em relação ao tema. In: Quaresma M, Silva D, Quintas H, Romão R (eds) Livro de Resumos das XX Jornadas das Associação Portuguesa de Buiatria. Associação Portuguesa de Buiatria, Lisbon, pp 48–49

Broom DM (2020a) The necessity of human attitude change and methods of avoiding pandemics. Anim Sentience 2020:369

Broom DM (2020b) EU regulations and the current position of animal welfare. In: Vosough Ahmadi D (ed) The economics of farm animal welfare: theory, evidence and policy. CABI, Wallingford, pp 147–155
Broom DM (2021a) Dairy cattle welfare and other aspects of sustainability. In: Endres M (ed) Understanding the behaviour and improving the welfare of dairy cattle. Burleigh Dodds Science Publishing, Cambridge, UK, pp 1–13. https://doi.org/10.19103/AS.2020.0084.01
Broom DM (2021b) A method for assessing sustainability, with beef production as an example. Biol Rev 96:1836–1853. https://doi.org/10.1111/brv.12726
Broom DM (2022a) Broom and Fraser's domestic animal behaviour and welfare, 6th edn. CABI, Wallingford, p 545
Broom DM (2022b) Sustainable livestock feeding and management: what changes are needed? In: Proceedings of the Animal Nutrition Conference of Canada, Saskatoon. Ottawa, Canada, Animal Nutrition Association of Canada, pp 87–98
Broom DM, Johnson KG (2019) Stress and animal welfare: key issues in the biology of humans and other animals, 2nd edn. Springer Nature, Cham, Switzerland, p 230
Broom DM, Kirkden RD (2004) Welfare, stress, behavior, and pathophysiology. In: Dunlop RH, Malbert C-H (eds) Veterinary pathophysiology. Blackwell, Ames, Iowa, pp 337–369
Broom DM, Leaver JD (1978) The effects of group-housing or partial isolation on later social behaviour of calves. Anim Behav 26:1255–1263
Broom DM, Galindo FA, Murgueitio E (2013) Sustainable, efficient livestock production with high biodiversity and good welfare for animals. Proc R Soc B 280:20132025. https://doi.org/10.1098/rspb.2013.2025
Burgess PJ (1999) Effects of agroforestry on farm biodiversity in the UK. Scott For 53:24–27
Butler SJ, Vicker JA, Norris K (2007) Farmland biodiversity and the footprint of agriculture. Science 315:381–384
Calle A, Montagnini F, Zuluaga AF (2009) Farmers' perceptions of silvopastoral system production in Quindio, Colombia. Bois et Forets des Tropiques 300:79–94
Calvo-Lorenzo MS (2018) Practical approach and outlook regarding animal welfare concerns related to beef feedlot production. In: American Association of Bovine Practitioners Proceedings of the Annual Conference, pp 140–148
Ciambrone DF (2018) Environmental life cycle analysis. CRC Press, Boca Raton, FL, p 160
Clonan A, Wilson P, Swift JA et al (2015) Red and processed meat consumption and purchasing behaviours and attitudes: impacts for human health, animal welfare and environmental sustainability. Public Health Nutr 18:2446–2456
Cremene C, Groza G, Rakosy L et al (2005) Alterations of steppe-like grasslands in Eastern Europe: a threat to regional biodiversity hotspots. Conserv Biol 19:1606–1618
Daley CA, Abbott A, Doyle PS et al (2010) A review of fatty acid profiles and antioxidant content in grass-fed and grain-fed beef. Nutr J 9:1–12
Davidson EA, Asner GP, Stone TA et al (2008) Objective indicators of pasture degradation from spectral mixture analysis of Landsat imagery. J Geophys Res 113:G00B03
de Boer IJ, van Ittersum MK (2018) Circularity in agricultural production. Animal production systems and Plant production systems. Wageningen University and Research. https://www.nutrientplatform.org/wp-content/uploads/2019/06/3939_Circularity-in-agricultural-production-012019final.pdf
de Vries M, de Boer IJM (2010) Comparing environmental impacts for livestock products: a review of life cycle assessments. Livest Sci 128:1–11
de Vries MD, van Middelaar CE, de Boer IJM (2015) Comparing environmental impacts of beef production systems: a review of life cycle assessments. Livest Sci 178:279–288. https://doi.org/10.1016/j.livsci.2015.06.020
Dirksen N, Langbein J, Matthews L et al (2020) Conditionability of 'voluntary' and 'reflexive-like' behaviors, with special reference to elimination behavior in cattle. Neurosci Biobehav Rev 115:5–12
Duff GC, Galyean ML (2007) Board-invited review: recent advances in management of highly stressed, newly received feedlot cattle. J Anim Sci 85:823–840

EFSA (2006) The risks of poor welfare in intensive calf farming systems: an update of the scientific veterinary committee report on the welfare of calves. EFSA J 366:1–36
EFSA (2009) Scientific report on the overall effects of farming systems on dairy cow welfare and disease. Annex EFSA J 1143:1–38
EFSA Panel on Genetically Modified Organisms (GMO Panel) (2010) Scientific opinion on the assessment of allergenicity of GM plants and microorganisms and derived food and feed. EFSA J 8(7):1700
Eghball B, Power JF (1994) Beef cattle feedlot manure management. J Soil Water Conserv 49:113–122
Endres M (ed) (2021) Understanding the behaviour and improving the welfare of dairy cattle. Burleigh Dodds Science Publishing, Cambridge, UK
EU D.G. Health and Food Safety (2016) Special Eurobarometer 442 Attitudes of Europeans towards Animal Welfare. European Commission, Brussels
EU SCAHAW (1999) EU scientific committee on animal health and animal welfare report on animal welfare aspects of the use of bovine somatotrophin. European Commission, Brussels
Fajardo ND, Johnston R, Neira LA (2008) Sistemas ganaderos amigos de los aves. In: Murgueitio E, Cuartas CA, Naranjo JF (eds) Ganadería del Futuro. Fundación CIPAV, Cali, Colombia, pp 171–203
FAO (2010) Greenhouse gas emissions from the dairy sector: a life cycle assessment. Food and Agriculture Organization of the United Nations, Rome, Italy
Fischer J, Zerger A, Gibbons P et al (2010) Tree decline and the future of Australian farmland biodiversity. Proc Natl Acad Sci 107:19597–19602
Frewer LJ, Coles D, Houdebine LM et al (2014) Attitudes towards genetically modified animals in food production. Br Food J 116:1296–1313
Gaillard C, Meagher RK, von Keyserlingk MAG et al (2014) Social housing improves dairy calves' performance in two cognitive tests. PLoS One 9:e90205
Garnett T (2009) Livestock-related greenhouse gas emissions: impacts and options for policy makers. Environ Sci Pol 12:491–503
Geiger F, Bengtsson J, Berendse F et al (2010) Persistent negative effects of pesticides on biodiversity and biological control potential on European farmland. Basic Appl Ecol 11:97–105
Gerber PJ, Mottet A, Opio CI et al (2015) Environmental impacts of beef production: review of challenges and perspectives for durability. Meat Sci 109:2–12
Goh KM, Bruce GE (2005) Comparison of biomass production and biological nitrogen fixation of multi-species pastures (mixed herb leys) with perennial ryegrass-white clover pasture with and without irrigation in Canterbury, New Zealand. Agric Ecosyst Environ 110:230–240
Gray J (2000) The common agricultural policy and the re-invention of the rural in the European Community. Sociol Rural 40:30–52
Halder JC (2013) Land suitability assessment for crop cultivation by using remote sensing and GIS. J Geogr Geol 5:65–74
Hall SJ (2008) A comparative analysis of the habitat of the extinct aurochs and other prehistoric mammals in Britain. Ecography 31:187–190
Herrero M, Thornton PK, Notenbaert AM, Wood S, Msangi S, Freeman HA, Bossio D, Dixon J, Peters M, van de Steeg J et al (2010) Smart investments in sustainable food production: Revisiting mixed croplivestock systems. Science 327:822–825. https://doi.org/10.1126/science.1183725
Herrero MT, Thornton PK, Notenbaert AMO, Msangi S, Wood S, Kruska RL, Dixon JA, Bossio DA, Steeg, JVD, Freeman HA, Li X (2012) Drivers of change in crop–livestock systems and their potential impacts on agro-ecosystems services and human wellbeing to 2030. A study commissioned by the CGIAR Systemwide Livestock Programme
Hommerich K, Ruddat I, Hartmann M et al (2019) Monitoring antibiotic usage in German dairy and beef cattle farms—a longitudinal analysis. Front Vet Sci 6:244
Hudson J, Caplanova A, Novak M (2015) Public attitudes to GM foods. The balancing of risks and gains. Appetite 92:303–313

Jendoubi D, Hossain MS, Giger M et al (2020) Local livelihoods and land users' perceptions of land degradation in Northwest Tunisia. Environ Dev 33:100507

Kim W, Lee JD (2009) Measuring the role of technology-push and demand-pull in the dynamic development of the semiconductor industry: the case of the global DRAM market. J Appl Econ 12:83–108

Knaus W (2009) Dairy cows trapped between performance demands and adaptability. J Sci Food Agric 89:1107–1114

Knight AJ (2008) Perceptions, knowledge and ethical concerns with GM foods and the GM process. Public Underst Sci 18:177–188

Köhler M (1993) Skeleton and habitat of recent and fossil ruminants. F. Pfeil, Munich

Koellner T, De Baan L, Beck T, Brandão M, Civit B, Goedkoop M, Margni M, Canals LM, Müller-Wenk R, Weidema B, Wittstock B (2013) Principles for life cycle inventories of land use on a global scale. Int J Life Cycle Assess 18:1203–1215

Ku Vera JC, Ruiz GA, Albores MS et al (2011) Alimentación de rumiantes en sistemas silvopastoriles intensivos: avances de investigación básica. In: Memorias, 3rd Congreso sobre Sistemas Silvopastoriles Intensivos para la Ganadería Sostenible del Siglo, vol XXI. Fundación Produce, Universidad Autónoma de Yucatán, Merida, Mexico, pp 8–16

Ku-Vera JC, Ayala-Burgos AJ, Solorio-Sánchez FJ et al (2013) Tropical tree foliages and shrubs as feed additives in ruminant rations. In: Nutritional strategies of animal feed additives. Nova Science Publishers, New York, pp 59–76

Lowe M, Gereffi G (2009) A value chain analysis of the US beef and dairy industries. Center on Globalization, Governance & Competitiveness, Duke University, pp 1–55

Lubowski RN, Plantinga AJ, Stavins RN (2006) Land-use change and carbon sinks: econometric estimation of the carbon sequestration supply function. J Environ Econ Manag 51:135–152

Luoto M, Rekolainen S, Aakkula J et al (2003) Loss of plant species richness and habitat connectivity in grasslands associated with agricultural change in Finland. AMBIO 32:447–452

Magrin L, Brscic M, Armato L et al (2018) An overview of claw disorders at slaughter in finishing beef cattle reared in intensive indoor systems through a cross-sectional study. Prev Vet Med 161:83–89

Maia SM, Ogle SM, Cerri CE et al (2009) Effect of grassland management on soil carbon sequestration in Rondônia and Mato Grosso states, Brazil. Geoderma 149:84–91

Marshall A, Collins R (eds) (2018) Improving grassland and pasture management in temperate agriculture. Burleigh Dodds Science Publishing, Cambridge, UK

Marshall JD, Toffel MW (2005) Framing the elusive concept of sustainability: a sustainability hierarchy. Environ Sci Technol 39:673–682

McAdam JH, McEvoy PM (2009) The potential for silvopastoralism to enhance biodiversity on grassland farms in Ireland. In: Rigueiro-Rodriguez A, McAdam J, Mosquera-Losada MR (eds) Agroforestry in Europe: current status and future prospects. Springer, Berlin, Germany, pp 343–356

Mekonnen MM, Hoekstra AY (2010) The green, blue and grey water footprint of farm animals and animal products, vol 1. UNESCO-IHE Institute for Water Education, Delft

Miller-Cushon E, Leslie K, DeVries T (2018) Ensuring the health and welfare of dairy calves and heifers. In: Webster J (ed) Achieving sustainable production of milk. Vol 3: Dairy herd management and welfare. Burleigh Dodds Science Publishing, Cambridge, UK, p 149

Mogensen L, Kristensen T, Nielsen NI et al (2015) Greenhouse gas emissions from beef production systems in Denmark and Sweden. Livest Sci 174:126–143

Molina CCH, Molina DCH, Molina EJ, Molina JP (2008) Carne, leche y medio ambiente en el Sistema silvopastoril intensivo con *Leucaena leucocephala* (Lam.) de Wit Mimosaceae. In: Murgueitio E, Cuartas CA, Naranjo JF (eds) Ganadería del Futuro. Fundación CIPAV, Cali, Colombia, pp 41–65

Moreno G, Rolo V (2019) Agroforestry practices: silvopastoralism. In: Mosquera-Losada M, Prabhu R (eds) Agroforestry for sustainable agriculture. Burleigh Dodds Science Publishing, Cambridge, UK

Mottet A, De Haan C, Falcucci A, Tempio G, Opio C, Gerber P (2017) Livestock: on our plates or eating at our table? A new analysis of the feed/food debate. Glob Food Sec 14:1–8

Múnera E, Bock BC, Bolivar DM, Botero JA (2008) Composición y estructura de la avifauna en diferentes hábitats en el departamento de Córdoba, Colombia. In: Murgueitio E, Cuartas CA, Naranjo JF (eds) Ganadería del Futuro. Fundación CIPAV, Cali, Colombia, pp 205–225

Murgueitio E, Cuartas CA, Naranjo JF (2008) Ganadería del Futuro. Fundación CIPAV, Cali Colombia

Murgueitio E, Flores MX, Calle Z, Chará JD, Barahona R, Molina CH, Uribe F (2015) Productividad en sistemas silvopastoriles intensivos en América Latina. In: Montagnini F, Somarriba E, Murgueitio E, Fassola H, Eibl B (eds) Sistemas Agroforestales. Funciones Productivas, Socioeconómicas y Ambientales. Serie Técnica Informe Técnico. CATIA, Turrialba, Costa Rica, pp 59–101

Nader G, Tate KW, Atwill R, Bushnell J (1998) Water quality effect of rangeland beef cattle excrement. Rangelands Arch 20:19–25

Nawaz A, Lal R, Shrestha RK, Farooq M (2017) Mulching affects soil properties and greenhouse gas emissions under long-term no-till and plough-till systems in Alfisol of central Ohio. Land Degrad Dev 28:673–681

Nicholls A, Opal C (2005) Fair trade. Sage, Thousand Oaks, CA

Nuernberg K, Dannenberger D, Nuernberg G, Ender K, Voigt J, Scollan ND, Wood JD, Nute GR, Richardson RI (2005) Effect of a grass-based and a concentrate feeding system on meat quality characteristics and fatty acid composition of longissimus muscle in different cattle breeds. Livest Prod Sci 94:137–147

Oltenacu PA, Broom DM (2010) The impact of genetic selection for increased milk yield on the welfare of dairy cows. Anim Welf 19(S):39–49

Orr RJ, Griffith BA, Rivero MJ, Lee MR (2019) Livestock performance for sheep and cattle grazing lowland permanent pasture: benchmarking potential of forage-based systems. Agronomy 9(2):101

Özgüner H, Eraslan Ş, Yilmaz SERAP (2012) Public perception of landscape restoration along a degraded urban streamside. Land Degrad Dev 23:24–33

Pachas ANA, Shelton HM, Lambrides CJ, Dalzell SA, Murtagh GJ (2018) Effect of tree density on competition between *Leucaena leucocephala* and *Chloris gayana* using a Nelder Wheel trial. I. Aboveground interactions. Crop Pasture Sci 69:419–429

Patra A, Park T, Kim M, Yu Z (2017) Rumen methanogens and mitigation of methane emission by anti-methanogenic compounds and substances. J Anim Sci Biotechnol 8(1):13. https://doi.org/10.1186/s40104-017-0145-9

Penner G (2018) Disorders of digestion and metabolism in dairy cattle: the case of subacute rumen acidosis. In: Webster J (ed) Achieving sustainable production of milk volume 3: dairy herd management and welfare. Burleigh Dodds Science Publishing, Cambridge, UK

Peri PL, Dube F, Varella AC (2016) Silvopastoral systems in the subtropical and temperate zones of South America: an overview. In: Silvopastoral systems in southern South America. Springer, Cham, Switzerland, pp 1–8

Phillips C (2018) Understanding the behaviour of dairy cattle. In: Webster J (ed) Achieving sustainable production of milk volume 3: dairy herd management and welfare. Burleigh Dodds Science Publishing, Cambridge, UK

Plieninger T, Gaertner M (2011) Harnessing degraded lands for biodiversity conservation. J Nat Conserv 19:18–23

Pope J, Bond A, Huge J, Morrison-Saunders A (2017) Reconceptualising sustainability assessment. Environ Impact Assess Rev 62:205–215

Pryce J, de Haas Y (2018) Genetic selection for dairy cow welfare and resilience to climate change. In: Webster J (ed) Achieving sustainable production of milk volume 3: dairy herd management and welfare. Burleigh Dodds Science Publishing, Cambridge, UK

Radrizzani A, Shelton HM, Kravchuk O, Dalzell SA (2019) Survey of long-term productivity and nutritional status of Leucaena leucocephala-grass pastures in subtropical Queensland. Anim Prod Sci 56(12):2064–2073. https://doi.org/10.1071/AN15084

Rivera LB, Botero M, Escobar S, Armbrecht I (2008) Diversidad de hormigas en sistemas ganaderos. In: Murgueitio E, Cuartas CA, Naranjo JF (eds) Ganadería del Futuro. Fundación CIPAV, Cali, Colombia, pp 228–243

Roath LR, Krueger WC (1982) Cattle grazing influence on a mountain riparian zone. Rangel Ecol Manag 35:100–103

Robbins JA, von Keyserlingk MAG, Fraser D, Weary DM (2016) Invited review: farm size and animal welfare. J Anim Sci 94:5439–5455

Roche LM, Kromschroeder L, Atwill ER, Dahlgren RA, Tate KW (2013) Water quality conditions associated with cattle grazing and recreation on national forest lands. PLoS One 8:e68127

Rockström J, Steffen W, Noone K, Persson Å, Chapin FS, Lambin EF, Lenton TM, Scheffer M, Folke C, Schellnhuber HJ, Nykvist B, De Wit CA, Hughes T, Van Der Leeuw S, Rodhe H et al (2009) A safe operating space for humanity. Nature 461(7263):472–475

Schmidinger K, Stehfest E (2012) Including $CO_2$ implications of land occupation in LCAs—method and example for livestock products. Int J Life Cycle Assess 17:962–972

Schneider MJ, Tait RG Jr, Busby WD, Reecy JM (2009) An evaluation of bovine respiratory disease complex in feedlot cattle: impact on performance and carcass traits using treatment records and lung lesion scores. J Anim Sci 87:1821–1827

Schukken YH, Tikofsky LL, Zadoks RN (2005) Environmental control for mastitis prevention, milk quality and food safety. In: Hogeveen H (ed) Mastitis in dairy production: current knowledge and future solutions. Wageningen Academic Publishers, Wageningen, The Netherlands, pp 109–114

Scott N, Gooch C (2017) 'Towards' sustainability of dairy farming: an over Invan Belzen. In: N. (ed) Achieving sustainable production of milk Volume 2: Safety, quality and sustainability. Burleigh Dodds Science Publishing, Cambridge

Siegrist M, Hartmann C (2019) Impact of sustainability perception on consumption of organic meat and meat substitutes. Appetite 132:196–202

Skogen K, Helland H, Kaltenborn B (2018) Concern about climate change, biodiversity loss, habitat degradation and landscape change: embedded in different packages of environmental concern? J Nat Conserv 44:1220

Smith P, Haberl H, Popp A, Erb KH, Lauk C, Harper R, Tubiello FN, de Siqueira Pinto A, Jafari M, Sohi S, Masera O, Böttcher H, Berndes G, Bustamante M, Ahammad H, Clark H, Dong H, Elsiddig EA, Mbow C, Ravindranath NH, Rice CW, Robledo Abad C, Romanovskaya A, Sperling F, Herrero M, House JI, Rose S (2013) How much land-based greenhouse gas mitigation can be achieved without compromising food security and environmental goals? Glob Chang Biol 19:2285–2302. https://doi.org/10.1111/gcb.12160

Sneeringer S, Macdonald JM, Key N, Mcbride WD, Mathews K (2015) Economics of antibiotic use in US livestock production. USDA, Economic Research Report, p 200

Stafford KJ, Mellor DJ (2005) Dehorning and disbudding distress and its alleviation in calves - a review. Vet J 169:337–349

Steinfeld H, Gerber P, Wassenaar T, Castel V, Rosales M, De Haan C (2006) Livestock's long shadow: environmental issues and options. FAO, Rome, Italy

Stilwell G, Lima MS, Broom DM (2008a) Effects of nonsteroidal anti-inflammatory drugs on long-term pain in calves castrated by use of an external clamping technique following epidural anaesthesia. Am J Vet Res 69:744–750

Stilwell G, Lima MS, Broom DM (2008b) Comparing plasma cortisol and behaviour of calves dehorned with caustic paste after non-steroidal-anti-inflammatory analgesia. Livest Sci 119:63–69

Stilwell G, Carvalho RC, Lima MS et al (2009) Effect of caustic paste disbudding, using local anaesthesia with and without analgesia, on behaviour and cortisol of calves. Appl Anim Behav Sci 116:35–44

Stilwell G, Carvalho RC, Carolino N, Lima MS, Broom DM (2010) Effect of hot-iron disbudding on behaviour and plasma cortisol of calves sedated with xylazine. Res Vet Sci 88:188–193

Stilwell G, Lima MS, Carvalho RC et al (2012) Effects of hot-iron disbudding, using regional anaesthesia with and without carprofen, on cortisol and behaviour of calves. Res Vet Sci 92:338–341

Stilwell GT, Ferrador AM, Santos MS et al (2019) Use of topical local anesthetics to control pain during treatment of hoof lesions in dairy cows. J Dairy Sci 102:6383–6390

Stoll-Kleemann S, O'Riordan T (2015) The sustainability challenges of our meat and dairy diets. Environ Sci Policy Sustain Dev 57:34–48

Sutherland WJ, Dicks LV, Everard M, Geneletti D (2018) Qualitative methods for ecologists and conservation scientists. Methods Ecol Evol 9:7–9

Toru T, Kibret K (2019) Carbon stock under major land use/land cover types of hades sub-watershed, eastern Ethiopia. Carbon Balance Manag 14:1–14

Tunick MH, Van Hekken DL (2015) Dairy products and health: recent insights. J Agric Food Chem 63:9381–9388

Tzilivakis J, Warner DJ, Holland JM (2019) Developing practical techniques for quantitative assessment of ecosystem services on farmland. Ecol Indic 106:105514

Van Wijk M (2020) The contribution of integrated crop-livestock systems in combatting climate change and improving resilience in agricultural production. In: Deryng D (ed) Climate change and agriculture. Burleigh Dodds Science Publishing, Cambridge, UK

Veissier I, Gesmier V, Le Neindre P, Gautier JY, Bertrand G (1994) The effects of rearing in individual crates on subsequent social behaviour of veal calves. Appl Anim Behav Sci 41:199–210

Vellinga TV, Van den Pol-van Dasselaar A, Kuikman PJ (2004) The impact of grassland ploughing on CO2 and N2O emissions in the Netherlands. Nutr Cycl Agroecosyst 70:33–45

Viljoen AJ, Wiskerke SC (eds) (2012) Sustainable food planning: evolving theory and practice. Wageningen Academic Publishers, Wageningen, The Netherlands

Wagemakers JJ, Prynne CJ, Stephen AM, Wadsworth ME (2009) Consumption of red or processed meat does not predict risk factors for coronary heart disease; results from a cohort of British adults in 1989 and 1999. Eur J Clin Nutr 63:303–311. https://doi.org/10.1038/sj.ejcn.1602954

Warren HE, Scollan ND, Enser M, Hughes SI, Richardson RI, Wood JD (2008) Effects of breed and a concentrate or grass silage diet on beef quality in cattle of 3 ages. I: animal performance, carcass quality and muscle fatty acid composition. Meat Sci 78:256–269

Webster J (1994) Animal welfare: a cool eye towards Eden. Blackwell, Oxford, UK

Webster J (ed) (2018) Achieving sustainable production of milk volume 3: dairy herd management and welfare. Burleigh Dodds Science Publishing, Cambridge, UK

Wilkinson JM (2011) Re-defining efficiency of feed use by livestock. Animal 5:1014–1022

Woodfield D, Judson H (2018) Balancing pasture productivity with environmental and animal health requirements. In: Marshall A, Collins R (eds) Improving grassland and pasture management in temperate agriculture. Burleigh Dodds Science Publishing, Cambridge, UK

Woolf J (2020) Quantifying the role of livestock in climate change. In: Deryng D (ed) Climate change and agriculture. Burleigh Dodds Science Publishing, Cambridge, UK

Zabel F, Delzeit R, Schneider JM, Seppelt R, Mauser W, Václavík T (2019) Global impacts of future cropland expansion and intensification on agricultural markets and biodiversity. Nat Commun 10:1–10

[illegible]

# The Sheltering of Unwanted Cows in India

# 14

Arvind Sharma, Uttara Kennedy, and Clive J. C. Phillips

## Contents




A. Sharma (✉)
Department of Animal Husbandry, Government of Himachal Pradesh, Veterinary Polyclinic, Shahpur, Himachal Pradesh, India

Post Doctoral Research Fellow (Adjunct), School of Veterinary Science, The University of Queensland, Gatton, Australia

School of Veterinary Science, The University of Queensland, Gatton, QLD, Australia
e-mail: arvind.sharma@uqconnect.edu.au

U. Kennedy
School of Veterinary Science, The University of Queensland, Gatton, QLD, Australia

RSPCA Queensland, Animal Care Campus, Wacol, QLD, Australia

C. J. C. Phillips
Curtin University Sustainability Policy (CUSP) Institute, Bentley, Perth, Australia

Estonian University of Life Sciences, Institute of Veterinary Medicine and Animal Sciences, Tartu, Estonia




M. Haskell (ed.), *Cattle Welfare in Dairy and Beef Systems*, Animal Welfare 23,
https://doi.org/10.1007/978-3-031-21020-4_14

**Abstract**

In India, the sheltering of old, infertile, and unproductive cows in traditional shelters, or gaushalas, for as long as they live, is a unique example of reverence for animal life. The cow is worshipped as a goddess in the predominantly Hindu society. There are more than 5000 cow shelters spread over the country, and these accommodate from 20 to 10,000 cows per shelter. Most of the shelters are funded by religious organisations, Hindu temple trusts and philanthropy, with some additional funding from the government. These shelters primarily hold abandoned street cows whose numbers have reached dangerous levels, causing a danger to humans with whom they share the roads. Health, housing and feeding management are major issues facing the cows in the shelters, and we recommend attention to isolation of diseased animals, treatment for endo- and ectoparasites, adequate ventilation, avoidance of overcrowding in sheds, proper flooring and space for resting and free movement. Challenges include shortages of fodder, human labour and veterinarians. To ensure the sustainability of cow shelters, welfare assessment protocols should be used to audit the shelters. With the provision of adequate measures to safeguard welfare, this system of preventing dairy cow wastage could be emulated in other countries, most of which have greater resources for this than India has.

**Keywords**

Cow shelters · Gaushala · Welfare assessment · Cow welfare · India · Animal ethics

## 14.1 Introduction

The intensification of the world's dairy industries has accelerated in the twenty-first century. In response to growing demand, particularly in Asia, cows are increasingly likely to be kept in industrial scale farms with tens of thousands of other cows. The most rapid transition is happening in China, where small owner-occupied farms are being replaced by government-sponsored farms with up to 30,000 cows in each. Usually the intensification of dairy farming (globally as well as in Asia) in terms of cows per farm is accompanied by permanent housing of the cows. This has generally been associated with a reduction in longevity of the cow. The building designs used for permanent housing in intensive dairy farming typically involve hard concrete flooring, exposure of the feet to pooled slurry and restricted space allowances, which exacerbate welfare issues such as foot disorders and lameness, leading to increased culling rates and reduced longevity of dairy cows (Bruijnis et al. 2013; Broom 2007; Haskell et al. 2006). Culling rates are about 40–50% in modern intensive units, i.e. the lactating or productive period for the cow is only 2–2.5 years on average before being slaughtered, directly or indirectly as a result of the high milk yields expected of her (Compton et al. 2017; De Vries 2017; Hadley et al. 2006). The high milk yield creates a significant nutrient deficit in early lactation, which reduces conception rates and predisposes cows to disease (Chiumia et al. 2012). Coupled with calving the cow for the first time at two years of age, whereas it was frequently three years in the past, the average modern dairy cow lives for only 4–5 years, which is a fraction of the natural lifespan of the species, estimated to be 20–25 years (e.g. Moran and Doyle 2015).

Poor fertility, a high risk of mastitis and lameness are together responsible for most cows lasting such a short time in the herd. In the last century, dairy cow replacement rates in the United Kingdom were typically 20–25%, i.e. the cows lasted 4–5 years in the herd (Esslemont et al. 1985; Fishwick and Russell 1955; Leaver 1982; Nix 1989; Russell 1974). Milk yield expectations were approximately one half of what is expected in intensive systems today, and disease and nutritional challenges to the dairy cow were correspondingly lower. In New Zealand, where dairy cow systems relied almost totally on grazed pasture rather than feeding conserved feeds indoors, the replacement rate was even lower in the 1950s, under 7%, i.e. cows lasted on average 14 years (McMeekan 1960). Thus, longevity expectations have declined significantly in modern milk production systems. Therefore, it is not surprising that public concern about intensive dairy farming systems has increased; not only regarding the short life of the cow, but also the pressure of constantly producing high yields of milk and destruction of male (bobby) calves at just a few days of age.

However, there are alternatives to this bleak picture of cattle premature wastage. In India, where there is considerable reverence for the life of animals, cattle that have reached the end of their productive lives on dairy farms are not immediately slaughtered, but given retirement in shelters traditionally referred to as 'gaushalas', for as long as they live. Cows in gaushalas are housed loose in sheds, or are tethered in some gaushalas, with either free access to adjacent open paddocks or restricted or no access to open areas. Water is provided (from natural sources or from automatically or manually filled troughs) and feed consists mainly of agricultural by-products. The number of cattle depends on the carrying capacity of the shelter, but ranges from 20 up to 10,000 animals. Overall management is provided by a trust composed of religious leaders and the businesspeople/traders who traditionally fund these shelters. A manager runs the daily routines and activities of the shelters, with salaried workers for each paddock. Vendors can be seen sitting in front of the gates of shelters selling bales of green fodder to the visitors who, out of a sense of religious duty, offer them to the cows daily on their way to work.

The average age of the animals in these gaushalas/shelters is 11 years, according to a recent survey (Sharma et al. 2019a), suggesting that many are living well into their teens. Not only that, the Indian public and government services support this sheltering activity, in an attempt to give cows the quality of life that the public believes they deserve. All of this is achieved in the 39th poorest country in the world (Alkire et al. 2019), with 1.4 billion people and 192 million cattle, the biggest cattle herd of any country worldwide (Anon 2019). It is now time for people in other countries in the world with large dairy industries to ask whether the Indian example could provide a model for their dairy industries to follow, so that cows do not have to endure a short, miserable life. This would probably cost the consumer considerably, but milk is sold at relatively low prices in most Western countries, compared to many other sources of nutrients for human consumption (Drewnowski 2010; Headey and Alderman 2019).

While the Indian gaushala cow 'retirement' system potentially represents a high welfare model of care for old, non-productive dairy cows that the world should take note of, there are strains appearing in the system as it occurs in its home country. Due to the burgeoning cattle population, the strict ban on cow slaughter in most states and political activism, there has been an increase in the number of rescued cows in shelters, leading to overcrowding and hence compromised welfare. The number of cattle in the streets has also increased due to vigilantism against cattle smuggling across state borders and the inability of cow shelters to house all of them. There are many welfare issues affecting these cows in shelters in the contemporary scenario that is far from the ideal traditional sheltering concept, and these need to be tackled through scientific management and intervention.

This chapter will firstly discuss the cultural and religious background in which the concept of cow sheltering exists in India. This will be followed by a discussion of the modern concept of cow shelters in India, housing, health, feeding and nutritional management and stockpersonship of sheltered cows. The current trends in the economics of running cow shelters will also be detailed in comparison to the traditional concept of financial management.

### 14.1.1 A Historical Perspective on Indian Symbolism of the Cow

To explore the historical origins of concepts of cow welfare, preservation of quality of life and, more broadly, the inclusion of animals in our moral sphere in the West, we should first return to the intellectual debate of a few centuries ago by great thinkers such as Voltaire, David Hume and Jeremy Bentham. All of these thinkers commented on the moral and ethical place of animals in our society (Jordan 2005). In more recent times, turning points such as the first rudimentary animal protection laws, the formation of the first societies for the protection of animals and subsequently the formation of the Brambell Committee and the Farm Animal Welfare Council to create the "Five Freedoms" have been well documented (RSPCA (Australia) n.d.; Duncan 2006; Keeling et al. 2011).

In contrast to the Western story of the animal welfare movement, the Indian subcontinent's relationship with its animals stems from a completely different historical background. Without an understanding of these deep philosophical and political roots, we cannot hope to understand the unique belief systems in this region. This belief system has provided the societal and ethical context that has led to the structure of the dairy industry in India being markedly different from that of the West. This belief system has resulted in an abhorrence of the killing of cows and the almost complete absence of a beef industry in a country with the largest cattle population in the world (Anon 2019). It is also the only part of the world that has large-scale, institutionalised preservation of dairy cattle past their productive life.

Reverence for the cow is a centrepiece of Hindu culture, the roots of which can be traced back to the Indus Valley civilization around 3000 BCE. Through the millennia, the cow has evolved to become one of the most important symbols of Hindu identity, often synonymous with religious or nationalistic sentiment and pride. The issue of protecting and revering the cow has been the cause of much unrest, violence and vigilantism; this continues to be the case even today. Currently, it is illegal to slaughter a cow throughout most of India. A direct result of these anti-slaughter laws is a large population of abandoned, aged or otherwise unproductive cattle housed in shelters. In a country like India, which has few spare resources for unproductive cattle, this poses a risk of inadequate space, infrastructure, skilled labour and lack of financial and veterinary support.

Clear evidence of cattle assuming a religious or reverential role can be traced back to the temples and friezes of the Mesopotamian civilization, about 3000 BCE, which in turn influenced the Harappan civilization of the Indus Valley. There is evidence that the Indus Valley people followed the Indo-Iranian Mother Goddess cult where the "Great Mother was herself the cow, giving her milk as the life bestowing agent par excellence in the process of suckling" (Lodrick 2005). The cow as a Goddess continued to feature in several millennia of South Asian religions that followed.

As the agricultural Harappan civilization declined and fragmented, the pastoral Aryans descended from the north into what is now north western and central India. By 1500 BCE, the Aryans had composed and written the Vedas, a large body of literature that is considered the original sacred scripture of modern-day Hinduism. Early Vedic writings demonstrate not only the pastoral and economic importance of cattle but, in equal measure, the ritualistic and sacrificial role of animals. In those early Vedic times, the life of the cow was not inviolable. One of the texts, for example, cites the cow as the 'all-producing, all-containing universe' and its sacrifice was considered to be essential to the very cosmic creation of the universe (Alavijeh 2014; Parel 1969). At this stage, the cow appears to be one of several religious symbols of importance, but its life was not sacred.

The end of the Vedic period saw the emergence of the concept of *ahimsa*, or non-violence. This concept developed through various interpretations, not only in Hindu philosophy but also in the Buddhist doctrines of south-east and eastern Asia. While the late Vedic concept of *ahimsa* remained passive as an attempt to do no harm, new emerging sects like Buddhism and Jainism espoused it more actively by prohibiting animal slaughter and cruelty and extending compassion and protection to all living beings. When one of the greatest leaders of the region, King Ashoka (ca 269–262 BCE), converted to Buddhism, *ahimsa* effectively became law, specifically prohibiting slaughter of animals. In fact, the first animal shelters and hospitals in the region can be traced back to these times. Over the next few hundred years, the concept of *ahimsa* unified with that of the sanctity of the cow (a common tactic to gain the allegiance of a native population, merging older traditions with concepts of a newer, introduced culture). The sacred texts and law books from this period contain several examples of this sort of amalgamation. The *Bhagavad-Gita* (ca 300 AD), which is today considered the most important holy text of Hinduism, mentions the term *ahimsa* several times, not as doctrine but more as a desirable, noble quality to aspire to. The *Manu-Smriti*, an ancient text written sometime between ca 600 BCE and 300 AD, prohibits the eating of meat by Brahmans but not other castes. The text espouses the cow's sanctity while at times casually mentioning its slaughter and the eating of meat (Lodrick 1981). By the time of the great Hindu epic, the *Mahabharata*, written in the fourth century, both concepts—that of a holy cow as well as of its being inviolable—are presented as established doctrine, and in the *Puranas*, a group of texts written between the sixth and sixteenth centuries, the cow is firmly established as sacred and inviolable. The cow, previously only a symbol of femininity, fecundity and maternalism, had become completely holy and inviolable in its own right (Alavijeh 2014; Korom 2000; Lodrick 2005).

A few hundred years later (eleventh century A.D.), the Mughals invaded the Indian sub-continent. By this time, the cow was firmly entrenched as an important symbol of Hinduism, to be worshipped, revered and protected from harm. While contacts with the Mughals started with peaceful exchange, they soon turned military in nature and eventually most of the region came under Muslim rule. The cow became a rallying point between the invading beef-eating culture and native communities. Not all interactions on the matter were negative. Muslim kings like Babar and Akbar in the fourteenth century passed edicts either forbidding cow slaughter, or heavily taxing it, in an effort to promote goodwill and reconciliation with native populations. However, other leaders like Aurangzeb slaughtered cows in deliberate disregard of the Hindu population, to assist in their subjugation (Lodrick 1981, 2005; Parel 1969). In the fifteenth century, the great Maratha leader of western India, Shivaji, used the issue as a source of inspiration and strength in the struggle of the Maratha people against Muslim rule. Shivaji is still viewed by many as a cow vigilante. Under his leadership, the sanctity of the cow became even more firmly entrenched as a core element of Hindu identity and power; to this day parts of western India remain a conservative Hindu Maratha stronghold under his banner (Korom 2000; Lodrick 1981; Parel 1969). The issue has remained a point of contention with the resident Muslim communities, who not only eat beef when available but also perform animal sacrifice during their own religious festivals, even in contemporary times.

Nationalistic sentiment continued to grow during the 200 years of the British Raj, which was officially instituted in 1858, partly as a reaction to the Revolt of 1857 (also called the Indian Mutiny) in which Indian soldiers violently revolted against British officials due to a rumour that, to undermine Hindu beliefs, the British used tallow from beef to grease their cartridges (Geaves 1996; Lahiri 2003). The first organized nationalistic cow protection movement was established in 1881, started by Swami Dayanand Saraswati, a famous saint of the time and a prototype of many such saints who followed his lead in using the cow for dual political and religious motives. Dayanand's followers called themselves '*go rakshaks*' or cow protectors, and their first task was to petition the government to ban cow slaughter. The petition failed and, as a reaction, many more cow protection organisations sprang up, including the Rashtriya Swayam Sevak Sangh in 1925, which still retains political significance in India today. These organisations made it their aim to encourage re-conversion to Hinduism, raise the status of lower castes by offering them cow protection duties, make provocative speeches on the issue of cow slaughter and distribute pamphlets containing inflammatory and emotional phrases. Another freedom fight leader of the time, Bal Gangadhar Tilak, created political festivals such as the Shivaji festival that invoked both the cow symbol and Shivaji's struggle against the Mughals (Batra 1981; Parel 1969).

Against the backdrop of these historical events, India gained independence as a secular country in 1947 and started the process of writing a Constitution for the very first time. A Constituent Assembly was established, and it took the Assembly 3 years to deliberate on the terms of the final Constitution. From records, it is evident that the issue of legally banning cow slaughter was fraught with religious and emotional sensitivities. Three factions developed in addressing the issue: Hindus, Muslims and secular thinkers. Hindus argued that cow slaughter went against the very core of their religion, while Muslims argued that any ban on cow slaughter infringed their right to earn a living as butchers, leather makers, tanners and slaughtermen. The secular faction, who included India's first prime minister, Jawaharlal Nehru, were committed to creating a truly secular Constitution that could not possibly contain prejudices or sanctions for specific religious beliefs. Nehru is quoted to have said that the agitation caused by the Hindu faction was "futile, silly and ridiculous" and he was "prepared to resign prime minister-ship" over the matter (Batra 1981; Simoons 1973; Simoons et al. 1979). So charged was the atmosphere that riots broke out intermittently during these deliberations, triggered in various places by instances of carrying beef through the village, slaughter of cows, and even the discovery of animal remains near a place of worship (Batra 1981). In 1950, when the final draft was passed, the matter was put to rest by reaching a compromise: rather than enforce the ban by law, an Article was inserted into the Directive Principles, which acted as a guide to various States in making their own enforceable laws (Bakshi and Kashyap 1982):

> The State shall endeavour to organise agriculture and animal husbandry on modern and scientific lines and shall, in particular, take steps for preserving and improving the breeds, and prohibiting the slaughter of cows and calves and other milch and draught cattle.

By the mid-1950s, many States started using this guide to write and enact laws to bring the ban into effect. Gradually as laws came into effect all over the country, several prosecutions followed, mainly against cattle buyers and sellers, hide merchants and butchers. The legislation was new, and neither the courts nor citizens had much prior precedence to use in these cases. Where offenders were prosecuted, the courts were often labelled as violating the defendant's Constitutional rights to equality, a livelihood and religious freedom. In cases where courts ruled against the Hindu conservatives, agitations and riots often followed. The first Supreme Court case in 1958 seemed to be influenced by the precarious nature of the issue: it ruled against slaughter of any cows as they were deemed sacred to Hindus (Batra 1981; Simoons 1973). Through all of these cases, the courts gradually elaborated on the vague guidelines set forth in the Constitution and defined the species, age, sex and usefulness of cattle allowed to be slaughtered, as well as legalities of slaughter-houses, beef markets and appropriate levels of punishments for the offences. Gradually, a framework of case law was built, and State laws were periodically modified based on this.

In spite of substantial case law, post-independence India continued to see periodic riots, agitations and even killings in the name of the sacred cow. In 1966, several conservative Hindu political parties such as the Rashtriya Seva Sangh, sections of the Jain community, the Arya Samaj community, the Bhartiya Jana Sangh and others joined hands to form a Committee for Cow Protection. They staged what was perhaps the largest demonstration in history before the parliament, to demand a complete nationwide ban on cow slaughter. Several people were killed, property was damaged and agitators were jailed as a result of the march. Yet the committee's activities continued: they disseminated inflammatory pamphlets, made provocative speeches and gradually secured popular support. In 1979, a well-known human rights leader, Vinoba Bhave, declared that he would perform a fast-unto-death to secure a national ban on cow slaughter. Conservative Hindu organisations from all corners of the country helped advertise the fast and they agitated and demonstrated to show support. An aggressive display of opposition was staged by Christian and Muslim communities: beef was cooked and consumed on streets and leaders made angry speeches in reaction to Vinoba Bhave's movement. Finally the fast was called off with a promise to amend the Constitution if enough parliamentary support could be garnered, which proved impossible (Batra 1981).

The contention remains to this day. In 2002, self-proclaimed cow protection agencies lynched members of the Dalits community, who traditionally hold the job of skinning cattle carcasses (Chigateri 2008). In 2004, the House of Representatives resolved to yet again seek a national ban of cow slaughter, without success (Lodrick 2005). The year 2016 witnessed alleged cattle traders and their families being stripped naked and beaten in public (Katesiya 2016). These right-winged Hindutva (strongly Hindu-biased groups that often propagate anti-secular, anti-immigrant sentiment) actions are increasingly backed by constitutional Supreme Court support (Barak-Erez 2010; Chigateri 2008).

The modern Indian political-religious-cultural landscape is still dominated by cow worship in many forms: veneration of cow dung, urine, milk, milk products and even dust from cow prints, religious affiliations of cow shelters, avoidance of beef and celebration of cow-centric festivals such as Govardhan Puja and Gopashtami (Batra 1981; Korom 2000; Lodrick 2005). Ample instances of vigilantism by self-proclaimed cow protection agencies, eagerness of the police and courts to prosecute alleged offenders and bureaucracy, blocking, cancelling and delaying of cattle slaughter and transport licenses are seen in media reports. Beliefs and opinions remain emotionally charged and do not seem to have changed relative to the pace of urbanization, modernization and increasing education or affluence levels.

The ban on killing cattle led to the practice of sheltering cattle in shelters, traditionally referred to as gaushalas, as described above. This is an ancient tradition in India, with documented evidence of their existence since the third century BC. Old, abandoned, unproductive, rescued and infirm cattle are still housed in these shelters.

These shelters exist throughout almost all of the country but are more abundant in north, west and central India. Slaughtering of cows is now banned in most states (22 of 29) (Sarkar and Sarkar 2016). Moreover, the concept of *Ahimsa*—non-violence towards all sentient beings—is the philosophy of the three main religions, Hinduism, Jainism and Buddhism, practised by 81% of the people in India (Lodrick 1981).

After independence, the institution of the gaushala was reinvigorated and consolidated as the droughts and famines of the twentieth century across the country forced villagers to send their cattle to the local cow shelters in order to pool their resources and save their cattle from death. The assistance of cow shelters during those difficult times helped farmers to sustain themselves through the active provision of material and moral support by the general public, in the form of donations of money, fodder and voluntary work.

### 14.1.2 Recent Trends: The Role of the Modern Gaushala and the Need for Cow Shelters

Between the late twentieth century and the present, massive urbanization has taken place in India, which has led to the shrinking of grazing land for cows, even in the villages. The mechanization of agricultural operations has rendered the male cattle largely without purpose, as previously they would have been used for draught power. The fragmentation of habitats has compounded the shortage of fodder for the cows and conventional smallholder dairy farming is becoming unsustainable and unprofitable. Cows are often abandoned by farmers in the streets or sent to cow shelters. The overpopulation of cows in the streets has given rise to human–animal conflicts in the form of crop raiding by cattle, traffic accidents and public health concerns. There are approximately 5.28 million stray cattle in India (DAHD 2014) and 436,727 in gaushalas (CPCB 2020). The cow shelters have been put under pressure to house an increasing number of these abandoned cattle, raising concerns about their welfare. Consequently, over a period of 30 years, the number of cow shelters increased from nearly 3000 (in 1956) to 5000 (Chakravarti 1985), of which 1837 were registered with the Animal Welfare Board of India (AWBI), the government statutory body responsible for the welfare of animals in the country (GOI 2017) and and 5964 in 2020 (CPCB 2020). However, there are unverified reports of more than 10,000 gaushalas today (SPKMCT undated). Many of the unregistered cow shelters sustain themselves with financial support from the public or businesses and are independent of financial support of the AWBI or the government. Many communities provide charity as a family tradition in the name of the holy cow. However, some of the cow shelters are run as commercial dairies masquerading as gaushalas to attract government and public funding (TOI 2018). Others sell products like cow dung, sometimes in logs for cremation of the dead as per Hindu rituals, incense sticks, distilled cow urine and vermicompost.

The gaushala network across the country is being strengthened by the central and provincial governments through the establishment of new cow shelters to meet the demand to shelter the continually increasing stray/abandoned cattle population of the country. Good welfare of the cattle in the shelters is important in order to maintain the public support for these unique traditional institutions.

### 14.1.3 The Welfare of Sheltered Cattle

The welfare of cattle in gaushalas has not been scientifically investigated until recently. Thus, what is presented in this section is a summary of our experiences and the results of our studies into the health and welfare aspects of gaushalas (Sharma et al. 2019a, b, 2020) and the findings of others.

#### 14.1.3.1 Sheltering Objectives

The objective of sheltering is to provide for the welfare of cattle, defined by the AWBI as being a life based on the five freedoms of animal welfare (FAWC 1992) until they die from natural causes. The objective of the shelters is to provide the cows with a life that has a higher level of welfare than they would experience as strays, in the streets and fields where they may be beaten while raiding crops, sustain automobile collisions, feed on garbage with a high content of plastics, suffer extremes of the weather and are sometimes inhumanely smuggled to clandestine slaughter houses. The rescue of abused cattle into shelters aims to improve their condition. The application of animal welfare science can assist the gaushalas in achieving these aims. Cattle welfare assessment protocols, for instance, can be used to assess the welfare of cattle in gaushalas (Sharma et al. 2019a). An assessment of the presence of wounds, injuries, swellings and body condition score (BCS) at the time of entry into the shelters should be carried out to assess immediate needs, and it should be followed by subsequent periodic checks. The welfare parameters assessed should include BCS, hair loss, lesions on the skin, limbs, joints, tail, eyes, horns, lameness, AD (avoidance distance), resting, lying down and getting up behaviour, general demeanour, positive interactions and agonistic behaviour.

#### 14.1.3.2 Health Management of Sheltered Cattle

The majority of the cattle sheltered in the cow shelters are old, unproductive and infertile. They have been abandoned by their owners (often dairy farmers) on the streets, sent directly to the shelters or rescued during their transportation for slaughter. These cattle often suffer from poor body condition and health due to persistent neglect by the owners, inadequate food and shelter in the streets and being transported under inhumane conditions. Restorative care and treatment of such animals in the shelter is a very important function. Veterinarians, either in-house or on-call, treat these sheltered cattle for various ailments according to the needs of the shelter management team (Sharma et al. 2020). Minor health issues are taken care of by the shelter managers, who in many places are qualified paraveterinary workers.

Adequate isolation and quarantine facilities are required after admission before mixing with other residents, to limit disease transmission and reduce fighting. Aggression by older cows towards new entrants is common, and the already weak new entrants might suffer serious, even fatal, injuries. The freshly admitted cattle may be weak and undernourished, requiring additional care and justifying their initial separation from the herd. There are guidelines for the introduction of the new animals in the shelters, which the shelter management should implement (FIAPO 2017), as well as advice from the Animal Welfare Board of India (AWBI 2018). More than half of the shelters have quarantine facilities within the shelter for the newly admitted cattle (Sharma et al. 2020), although not all are of prescribed standards. This provision of quarantine is subject to the size of the shelters and availability of labour and infrastructure. Quarantine is mostly limited to keeping the new animals in separate sheds (sometimes simply makeshift partitioning of existing sheds) for 1 or 2 weeks to observe any illness.

Many cattle admitted to the shelters suffer from infectious and contagious diseases (Ramanjeneya et al. 2019). During this period of isolation, these cattle must be vaccinated against diseases that are endemic to the region, especially foot and mouth, haemorrhagic septicaemia and black quarter disease. Most gaushala cattle are vaccinated against these endemic diseases as the government provides these vaccines free of charge and provides personnel from the state animal husbandry department to administer them. This checks the disease spread to the cattle owned by households in the vicinity of the shelter. Close surveillance must be kept on these isolated cattle to observe symptoms of these diseases. Additionally, testing of these cattle, especially for diseases like tuberculosis and brucellosis is required; sometimes these diseases are the reason for abandonment and shelters often harbour diseased cattle (Singh et al. 2004; Srinivasan et al. 2018). However, screening of sheltered cattle to check the presence of these two diseases is rarely done as the vaccine costs and veterinarian availability to monitor the test response is limited. There are some reports of screening cattle shelters against these diseases for research purposes (Sharma et al. 2015; Shringi 2004). This period of quarantine is an opportunity for the stockpersons to positively interact with the cattle so that they get used to the shelter environment. Fear of humans may initially be high due to past negative experiences with humans in the streets or elsewhere. All the cattle should be tagged for identification purposes as soon as they are admitted to the shelters, to facilitate maintenance of health and other records.

Ecto- and endoparasiticidal treatments of the cattle must be considered, following faecal examination. Tick and lice infestation in shelter cows reduce cows' welfare by blood sucking and skin irritation and are reported in both shelters and smallholder livestock communities (Chavhan et al. 2013; Singh et al. 2018). They transmit babesiosis, anaplasmosis, Lyme disease, encephalitis, borreliosis and blue tongue virus. Sheltered cattle also suffer from endoparasites (Hirani et al. 2006) and need routine deworming, with regular changing of the drugs used to avoid development of resistance. The use of a veterinarian is advised for routine health needs and tackling emergencies. If the veterinarian is in-house, it will encourage screening and segregation of cows with infectious and zoonotic diseases.

Routine foot care through the installation of footbaths and foot trimming helps in maintaining good hoof health. Some hoof enlargement and lameness has been observed in sheltered cattle but at a much lower prevalence than in Western dairies (Sharma et al. 2019a, b).

## 14.2 Housing of Sheltered Cattle and the Shelter Environment

One of the biggest welfare problems in shelters has been identified as overstocking (Sharma et al. 2019a) (Figs. 14.1 and 14.2). Sheltered cattle should have 6 to 7 $m^2$ each for movement and rest (Manoharan 2013). This will prevent injuries, disease and limit dirtiness of the animals to acceptable levels. Adequate space also prevents competition for space and food (Fregonesi et al. 2007). Cattle in shelters should have regular access to open yards outside the sheds and preferably to pasture grazing; however, with rapid urbanization, grazing lands are reducing. Guidelines regarding the construction of shelter sheds, space allowance, size of outdoor loafing areas, size of mangers and water troughs have been documented by a non-governmental organisation (FIAPO 2017) supported by AWBI.

Every shelter manager should know the shelter's carrying capacity. Separation of the cattle by sex and age will help to reduce aggression between the cattle and unwanted pregnancies. If space is limited, male cattle must be castrated. There should be a hospital area for sick cattle for palliative care and to prevent the spread of infectious diseases.

Tethering of animals is a common practice in many shelters to increase stocking density. It reduces the freedom to express normal behaviour, particularly the ability to walk (Veissier et al. 2008) (Fig. 14.3). Cattle are naturally gregarious and have a

**Fig. 14.1** A gaushala with a high stocking density, providing evidence of overcrowding

**Fig. 14.2** A gaushala with a low stocking density allowing cows to relax in an open yard

rich social behaviour repertoire (Broom 2013). Loose housing allows for ease of feeding, lying down and rising (Popescu et al. 2014). If cattle have to be tethered, the rope should not be less than 5 m long to ensure adequate movement and there should be a daily release period of 2–4 h for exercise. Tethering also reduces labour efficiency (Phillips 2018). Allowing frequent interactions between conspecifics in the shelters by avoiding tethering will reduce stress in the animals and can be reflected by the reduction in avoidance distance.

Adequate ventilation is very important to avoid accumulation of ammonia, methane, carbon dioxide and heat generated from dung in the sheds. Many are in old-fashioned buildings, without the recommended eaves height of 3.5 m (Davis et al. 2016). Flooring must be impervious to water and slurry for effective cleaning and maintenance of good health of the hooves and cow comfort when lying down (Phillips 2018). There is a lack of uniformity in flooring with combinations of earthen, brick, stone and concrete floors (Sharma et al. 2019b). Concrete floors allow regular cleaning and more resistance to wear and tear. The floor gradient should not allow water, urine and slurry to stagnate on the floor, but it should also allow cows to stand and walk without slipping and bearing weight unevenly on their claws. A minimum slope of 0.5% (1:200) is recommended (Moran 2012). The floors should also have adequate friction so that the hoof has good grip and cattle can walk freely without slipping, but not so much that the hoof horn wears excessively. Adequate floor friction reduces falls and improves cleanliness of the cows; ideally,

**Fig. 14.3** Tethered cows in a typical gaushala

the coefficient of friction should be between 0.4 and 0.6 (Sharma et al. 2019b; Telezhenko et al. 2017). If possible, tamped concrete flooring is advisable as it is more slip-resistant than grooved flooring (Albutt et al. 1990).

Shelters should not be located in areas that are noisy and busy such as industrial areas, as cattle are able to hear sounds at higher frequencies than humans. The permissible sound level limit is below 90–100 dB, but the nuisance value of sound is related more to sound attributes than just volume (Phillips 2018). Adequate lighting of shelter sheds is also important for the health and welfare of cows as they like to feed in the light, and cow reproductive physiology is affected by photoperiod, which in turn affects behaviour (Phillips and Leaver 1986; Phillips and Schofield 1989). Light intensity in the sheds should be between 161 and 215 Lux (Buyserie et al. 2001).

## 14.3 Feeding and Nutrition of Sheltered Cattle

Due to the high cost and lack of returns, good nutrition of sheltered cows is challenging. They are commonly fed on wheat and rice straw alone, which have low energy and protein characteristics. Locally available agricultural by-products, including straws, stovers, bagasse and seed hulls, can help in meeting any shortages of fodder (Figs. 14.4 and 14.5). Waste from the local vegetable markets can also be fed, but it must be checked for fungal infestation.

Provision of green roughage for the cattle is often neglected because of financial constraints. Feeding of urea/molasses-treated straw or urea-molasses mineral bricks can replace green roughage (Srinivas and Gupta 1997; Misra et al. 2006), offering an alternative nitrogen source. Urea/molasses-treated straw must be gradually introduced to avoid gastrointestinal disturbances. Like many other by-products, it is vulnerable to spoilage through fungal infestation, hence it needs careful storage.

Many shelters have vendors sitting at the gates or in surrounding roadsides selling green fodder bales of alfalfa, sweet and berseem clover. The public buy these to feed to the cows, which is perceived as a religious duty. The duty of the shelter management is to ensure that this fodder is made available uniformly to all cattle in the shelter, rather than just to those near the shelter entrance.

Many visitors to the shelters bring home-cooked bread, flour and grains for the cows. Overfeeding of these types of concentrate can lead to ruminal acidosis and other digestive disorders, sometimes with fatal consequences (Kataria and Kataria 2009). Kitchen waste, grain gleanings and floor sweepings are also provided, but shelters should check their quality before allowing them to be fed to the cattle.

**Fig. 14.4** Cows in a typical loose housing shelter with rice straw in the feed bunk

**Fig. 14.5** Cows receiving green fodder in the yard of a shelter

## 14.4 Stockpersonship of Sheltered Cattle

A positive human–animal relationship is paramount in shelters to give a positive message about the concept of sheltering and welfare to the society that venerates the cow. It can also improve the health and body condition of cows (Hemsworth 2003; Hemsworth and Coleman 2010). The distance to which humans can approach a cow at a feed bunk before she moves away (the avoidance distance) can be measured to test how good human–animal relationships are at the shelter (Mazurek et al. 2011; Windschnurer et al. 2009). A shelter survey has demonstrated that most cattle are not afraid of close contact with humans (Sharma and Phillips 2019). Despite the fact that rescued cattle might have recollections of physical abuse in the streets or illegal transport, positive human behaviour in the shelters (frequent and positive patting, no shouting, sticks or prods) can reduce the avoidance distance. The human–animal relationship in shelters can be improved by adequate training of the workers to understand the concept of humane sheltering. At all times, it must be remembered that human–animal relationships are dynamic and changes in human behaviour can improve this relationship (Waiblinger et al. 2006).

## 14.5 Disposal of Waste from Cattle Shelters

Disposal of dung and slurry generated in the shelter needs careful consideration, as flies and mosquitoes breed on this waste and cause disruptions to cattle behaviour and spread of diseases, due to being vectors for transmission of diseases like sarcocystosis, mycobacterium infections, campylobacteriosis, filariasis, brucellosis and cryptosporidiosis (Fischer et al. 2001; Gestmann et al. 2012; Markus 1980; Nichols 2005). Often, mechanised disposal is not available. Similarly, disposal of carcasses and afterbirths (placentae) must be done properly to prevent transmission of diseases through vectors to healthy cattle in the shelter. In practice, the carcases are mainly disposed of by deep burial within the shelter premises or removal by the local municipal authorities, and in a few cases just thrown into nearby ravines/crevasses (Sharma et al. 2020). Afterbirths are often just disposed of along with dung in the dung mounds piled up in one corner of the shelter.

## 14.6 Economics of the Sheltering of Cattle

The expenditure involved in running a cattle shelter depends on the size of the shelter. Shelters are usually run by private trusts owned by merchants, charitable societies, temple-based organisations, municipalities and the government. Many of the older shelters are run by trading communities as a part of their traditional philanthropic practices of donating some of their profits for cattle welfare. Those owned by temple trusts are supported by devotees' donations. The most recently established shelters are either run by the government/municipalities or local social community organisations or in a public–private partnership model. Generally, the trading community and temple trust-owned shelters are well funded, recognising that the cow is a mother goddess in religion. Government shelters are also partially reliant on public donations, given that they are established to limit the abandonment of cows to the streets and to prevent them from crop raiding, being traffic hazards and scavenging on garbage in the streets. These government/municipality run shelters suffer from overpopulation of cattle, beyond their carrying capacities, but keep accepting cows due to public pressure. The shelters run by private trusts and charitable societies restrict admission of cattle beyond their carrying capacities.

In a survey of 54 Indian shelters, we found that median annual expenditure of shelters was 50,000 USD, whereas the median income generated was just 1800 USD, mostly from the sale of dung as manure (Sharma et al. 2020). A number of other innovative means of creating income for the shelters, by selling their by-products, are evident. In some shelters, the sale of male and female calves' urine (as a biopesticide or medicine for traditional healing practices) and milk is practised. Some have installed cow urine distillation apparatus for preparation of 'Gau-Ark',—a potion used in traditional medicine. Others manufacture dung cakes for use as fuel in the last rites of Hindus, especially in areas where availability of wood is limited. A few have started making paper and incense sticks from the fodder waste and dung produced in the shelters. Vermicomposting of cow dung is used to make organic manure for sale.

## 14.7 Future Perspectives and Conclusions

Sheltering of cattle provides a method of prevention of animal wastage that contrasts with most of the dairy production activities in the world, nearly all of which are industry-oriented. Cow shelters in India are primarily a public-owned, financed and directed animal welfare initiative. They have sustained themselves through the centuries of their existence by reorienting themselves to the needs of the society. Initially, they functioned purely to fulfil the religious aspirations of the communities as retirement homes for unproductive cattle, then they contributed as rescue centres to house cattle during droughts and famines, through the pooling of resources of the communities. Later, they had roles as centres for milk production and cattle breeding, and finally they have returned to their role as retirement homes. Because of the emphasis given to establishing dairy farms in the private sector and mini dairies in small landholder farming communities, a place for cows to live when they become unproductive is required. In Western dairy systems, these cattle would be culled, but this cannot happen in India (as has been discussed above). Moreover, the number of street/abandoned cattle has increased considerably as the Indian cow population expands to meet the growing demand for milk and milk products. In this role, the gaushalas are mostly functioning as rescue homes for abandoned street cattle to manage the burgeoning population of street cows in India. It is a testament to a society in which reverence for the cow has been a hallmark of its existence.

The challenges are multifaceted, ranging from the gradual habitat fragmentation of the rural areas due to rapid urbanization, to shortages of fodder, labour and veterinary expertise. At the same time, there are increased expectations of the public as a result of improved literacy levels, awareness about animal welfare and willingness to provide financial support. However, the management of the cow shelters needs to use modern scientific principles to support good animal welfare. The future is therefore likely to be an amalgamation of religion and science, in order for sustainable institutions to emerge. A survey that we conducted of residents close to shelters suggests that the public strongly support them (Sharma et al. 2020). In the face of an increasing street cow population, cattle shelters should not just turn into cattle pounds hoarding cattle. There is enough active financial and logistic support from the current political system in the country to save this rich cultural heritage, which is a unique example of animal welfare for the rest of the world, based on the principle of non-violence towards all sentient beings. Shelters will be expected to generate some of their own resources for sustainability, through the sale of innovative products. This is where science has an important role in identifying the beneficial properties of cow urine, dung, etc. Cows are also believed to be one of the best ways to mass produce vaccines, for example against COVID-19 (Palca 2020). Technological innovation can also be used to solve the limited fodder availability to the cattle in shelters, through use of by-products and other novel feed sources. There is also scope for rehabilitating some younger female cows back to the farmers, with the assistance of veterinarians.

To ensure the sustainability of cow shelters, based on society's principles of animal welfare, welfare assessments should be used for auditing of the shelters. These assessments will provide objectivity in the routine management of shelters, feedback to all the stakeholders and methods to improve the performance of the shelters in the science-based management of cattle welfares. Cow shelters fulfil a role in the humane husbandry of cattle through appropriate care in the form of sheltering, feeding and health management, routine management and adequate funding from all stakeholders (the public, philanthropists, corporate bodies and government). Cattle shelters are evolving with the times to deal with the gradual increase of street and abandoned cattle in India, through the amalgamation of science and tradition. This model of preventing animal wastage is worth emulating in the western world for humane care of retired dairy cattle.

## References


Alavijeh AZ (2014) Representations of cows in different social, cultural, religious and literary contexts in Persia and the world. Asian J Res Soc Sci Humanit 3:215–218. http://www.ajssh.leena-luna.co.jp/AJSSHPDFs/Vol.3(1)/AJSSH2014(3.1-20).pdf

Albutt RW, Dumelow J, Cermak JP, Owen JE (1990) Slip-resistance of solid concrete floors in cattle buildings. J Agric Eng Res 45:137–147. https://doi.org/10.1016/S0021-8634(05)80145-3

Alkire S, Kanagaratnam U, Suppa N (2019) The Global Multidimensional Poverty Index (MPI) 2019. Oxford Poverty and Human Development Initiative, University of Oxford. https://ophi.org.uk/multidimensional-poverty-index/databank/global-comparisons. Accessed 1 November 2019

Animal Welfare Board of India (AWBI) (2018) Advisory to initiate necessary action for rescue and rehabilitation of stray animals. http://awbi.in/awbi-pdf/Advisory_on_stray_animals_12_7_18.pdf

Anon (2019) Livestock population in India rises 4.6 per cent to nearly 536 million, shows census. https://www.thehindubusinessline.com/economy/agri-business/livestock-population-in-india-rises-46-to-nearly-536-m-shows-census/article29708130.ece

Bakshi PM, Kashyap SC (1982) The constitution of India. Universal Law Publishing, New Delhi, India

Barak-Erez D (2010) Symbolic constitutionalism: on sacred cows and abominable pigs. Law Cult Humanit 6:420–435. https://doi.org/10.1177/1743872110374265

Batra SM (1981) Cows and cow-slaughter in India: religious, political and social aspects (No. 88). ISS Occasional papers. Erasmus University Rotterdam. http://hdl.handle.net/1765/34927

Broom DM (2007) Quality of life means welfare: how is it related to other concepts and assessed? Anim Welf 16:45–53

Broom DM (2013) Cow welfare and herd size: towards a sustainable dairy industry. Cattle Pract 21:169–173

Bruijnis MRN, Meijboom FLB, Stassen EN (2013) Longevity as an animal welfare issue applied to the case of foot disorders in dairy cattle. J Agric Environ Ethics 26:191–205. https://doi.org/10.1007/s10806-012-9376-0

Buyserie AC, Dahl GE, Gamroth MJ (2001) Managing light in dairy barns for increased milk production. Technical Report of the Oregon State University Extension Service. https://ir.library.oregonstate.edu/concern/administrative_report_or_publications/xd07gt178

Central Pollution Control Board (CPCB) (2020) Report on environmental guidelines for dairy farms and gaushalas. India Environment Portal | News, reports, documents, blogs, data, analysis on environment & development | India, South Asia. Accessed 12 May 2020

Chakravarti AK (1985) Cattle development problems and programs in India: a regional analysis. GeoJournal 10:21–45. https://doi.org/10.1007/bf00174664

Chavhan P, Maske D, Jagtap H (2013) Prevalence of arthropod parasites in bovines (cattle and buffalo) in the eastern zone of the Vidarbha region. Adv Lif Sci I2:60–61

Chigateri S (2008) 'Glory to the cow': cultural difference and social justice in the food hierarchy in India. South Asia J South Asian Stud 31:10–35. https://doi.org/10.1080/00856400701874692

Chiumia D, Chagunda MGG, Macrae AI, Roberts DJ (2012) Predisposing factors for involuntary culling in Holstein–Friesian dairy cows. J Dairy Res 80:45–50. https://doi.org/10.1017/S002202991200060X

Compton CWR, Heuer C, Thomsen PT, Carpenter TE, Phyn CVC, McDougall S (2017) A systematic literature review and meta-analysis of mortality and culling in dairy cattle. J Dairy Sci 100:1–16. https://doi.org/10.3168/jds.2016-11302

Davis R, Watts P, Stafford R (2016) Covered housing systems. In: Watts PJ, Davis RJ, Keane OB, Luttrell MM, Tucker RW, Stafford R, Janke S (eds) Feedlot design and construction. Meat and Livestock Australia, Sydney, pp 20–26

De Vries A (2017) Economic trade-offs between genetic improvement and longevity in dairy cattle. J Dairy Sci 100:4184–4192. https://doi.org/10.3168/jds.2016-11847

Department of Animal Husbandry, Fisheries and Dairying, Ministry of Agriculture, Government of India, New Delhi (2014) 19th Livestock Census of India-2012, All India Report. http://dahd.nic.in/sites/default/files/Livestock%20%205.pdf

Drewnowski A (2010) The cost of US foods as related to their nutritive value. Am J Clin Nutr 92:1181–1188. https://doi.org/10.3945/ajcn.2010.29300

Duncan IJH (2006) The changing concept of animal sentience. Appl Anim Behav Sci 100:11–19. https://doi.org/10.1016/j.applanim.2006.04.011

Esslemont R, Bailie J, Cooper M (1985) Fertility management in dairy cattle. Collins, London

Farm Animal Welfare Council (1992) FAWC updates the five freedoms. Vet Rec 131:357

Federation of Indian Animal Protection Organisations (FIAPO) (2017) Guidelines for management of gaushalas, pp 1–18. http://www.fiapo.org/fiaporg/wp-content/uploads/2018/01/GUIDELINES-FOR-MANAGEMENT-OF-GAUSHALAS.pdf

Fischer O, Mátlová L, Dvorská L, Švástová P, Bartl J, Melichárek I, Weston RT, Pavlík I (2001) Diptera as vectors of mycobacterial infections in cattle and pigs. Med Vet Entomol 15:208–211. https://doi.org/10.1046/j.1365-2915.2001.00292.x

Fishwick VC, Russell KN (1955) Dairy farming, theory and practice. Crosby, Lockwood and Sons, London, p 115

Fregonesi JA, Tucker CB, Weary DM (2007) Overstocking reduces lying time in dairy cows. J Dairy Sci 90:3349–3354. https://doi.org/10.3168/jds.2006-794

Geaves RA (1996) India 1857: a mutiny or a war of independence? A muslim perspective. Islam Stud 35:25–44. http://www.jstor.org/stable/20836926

Gestmann F, Förster M, Mehlhorn H, Sievert K, Messler S, Neuhausen N, Petersdorf S, Pfeffer K (2012) Flies as vectors of microorganisms potentially inducing severe diseases in humans and animals. In: Mehlhorn H (ed) Arthropods as vectors of emerging diseases. Springer, Berlin, pp 195–226

Government of India (2017) Basic animal husbandry and fisheries statistics. Ministry of Agriculture, Department of Animal Husbandry, Dairying & Fisheries, Government of India, New Delhi, India. Accessed from http://dadf.gov.in/sites/default/filess/Livestock%20%205_0.pdf

Hadley GL, Wolf CA, Harsh SB (2006) Dairy cattle culling patterns, explanations, and implications. J Dairy Sci 89:2286–2296. https://doi.org/10.3168/jds.S0022-0302(06)72300-1

Haskell MJ, Rennie LJ, Bowell VA, Bell MJ, Lawrence AB (2006) Housing system, milk production and zero-grazing effects on lameness and leg injury in dairy cows. J Dairy Sci 89:4259–4266

Headey DD, Alderman HH (2019) The relative caloric prices of healthy and unhealthy foods differ systematically across income levels and continents. J Nutr 149:2020–2033. https://doi.org/10.1093/jn/nxz158

Hemsworth PH (2003) Human–animal interactions in livestock production. Appl Anim Behav Sci 81:185–198. https://doi.org/10.1016/S0168-1591(02)00280-0

Hemsworth PH, Coleman GJ (2010) Human-livestock interactions: the stockperson and the productivity of intensively farmed animals. CABI, Wallingford, UK, p 208

Hirani N, Patel A, Patel P, Joshi R (2006) Prevalence of gastrointestinal parasites in cows of gaushalas in and around Anand, Gujarat, India. Indian J Field Vet 1:23–25

Jordan B (2005) Science-based assessment of animal welfare: wild and captive animals. OIE Rev Sci et Tech 24:515–528. https://doi.org/10.20506/rst.24.2.1588

Kataria A, Kataria N (2009) Diagnosis and control of outbreaks of acidosis in cattle. Indian Cow Sci Econom J 6:60–63

Katesiya G (2016) Gujarat: 7 of Dalit family beaten up for skinning dead cow. The Indian Express. http://indianexpress.com/article/india/india-news-india/gujarat-7-of-dalit-family-beaten-up-for-skinning-dead-cow-2910054/

Keeling LJ, Rushen J, Duncan IJH (2011) Understanding animal welfare. Anim Welf 2011:13–26. https://doi.org/10.1186/1751-0147-50-S1-S1

Korom FJ (2000) Holy cow! The apotheosis of zebu, or why the cow is sacred in Hinduism. Asian Folklore Stud 59:181–203. https://doi.org/10.2307/1178915

Lahiri N (2003) Commemorating and remembering 1857: the revolt in Delhi and its afterlife. World Archaeol 35:35–60. https://doi.org/10.1080/0043824032000078072

Leaver JD (1982) Milk production: science into practice. In: Handbook in agriculture. Longman Group, Harlow, UK, p 151

Lodrick DO (1981) Sacred cows, sacred places: origins and survivals of animal homes in India. University of California Press, Berkeley

Lodrick DO (2005) Symbol and sustenance: cattle in south Asian culture. Dialect Anthropol 29:61–84. https://doi.org/10.1007/s10624-005-5809-8

Manoharan T (2013) Expert system for cattle and buffalo. http://www.agritech.tnau.ac.in/expert_system/cattlebuffalo/aboutus.html

Markus MB (1980) Flies as natural transport hosts of Sarcocystis and other Coccidia. J Parasitol 66:361–362. https://doi.org/10.2307/3280842

Mazurek M, McGee M, Minchin W, Crowe MA, Earley B (2011) Is the avoidance distance test for the assessment of animals' responsiveness to humans influenced by either the dominant or flightiest animal in the group? Appl Anim Behav Sci 132:107–113. https://doi.org/10.1016/j.apPplanim.2011.03.001

McMeekan CP (1960) Grass to milk, a New Zealand philosophy. The New Zealand Dairy Exporter, Wellington, NZ, p 155

Misra A, Reddy GS, Ramakrishna Y (2006) Participatory on-farm evaluation of urea molasses mineral block as a supplement to crossbred cows for dry season feeding in rain-fed agro-ecosystem of India. Livest Res Rural Dev 18:1–11

Moran J (2012) Managing high grade dairy cows in the tropics. CSIRO Publishing, Collingwood, VIC, p 265

Moran J, Doyle R (2015) Cow talk: understanding dairy cow behaviour to improve their welfare on Asian farms. CSIRO Publishing, Clayton South, Australia, p 27

Nichols GL (2005) Fly transmission of Campylobacter. Emerg Infect Dis 11:361–364. https://doi.org/10.3201/eid1103.040460

Nix J (1989) Farm management pocketbook, 20th edn. Wye College, University of London, London

Palca J (2020) Researchers look to cattle for treatment for COVID-19: shots - health news: NPR. https://www.npr.org/sections/health-shots/2020/06/17/877844458/cows-help-with-covid-19-treatment-no-bull. Accessed 1 November 2022

Parel A (1969) The political symbolism of the cow in India. J Commonwealth Comp Polit 7:179–203. https://doi.org/10.1080/14662046908447106

Phillips CJC (2018) Principles of cattle production, 3rd edn. CABI, Oxfordshire, UK

Phillips CJC, Leaver JD (1986) The effect of forage supplementation on the behaviour of grazing dairy cows. Appl Anim Behav Sci 16:233–247. https://doi.org/10.1016/0168-1591(86)90116-4
Phillips CJC, Schofield SA (1989) The effect of supplementary light on the production and behaviour of dairy cows. Anim Sci 48:293–303. https://doi.org/10.1017/S0003356100040290
Popescu S, Borda C, Diugan EA, Niculae M, Stefan R, Sandru CD (2014) The effect of the housing system on the welfare quality of dairy cows. Ital J Anim Sci 13:2940. https://doi.org/10.4081/ijas.2014.2940
Ramanjeneya S, Sahoo SC, Pathak R, Kumar M, Vergis J, Malik SVS, Barbuddhe SB, Rawool DB (2019) Virulence potential, biofilm formation, and antibiotic susceptibility of Listeria monocytogenes isolated from cattle housed in a particular gaushala (cattle shelter) and organized farm. Foodborne Pathog Dis 16:214–220. https://doi.org/10.1089/fpd.2018.2494
RSPCA (Australia) History | RSPCA ACT (n.d.). https://www.rspca-act.org.au/about-us/history. Accessed 2 June 2020
Russell K (1974) The principles of dairy farming, 7th edn. Farming Press Limited, Ipswich, UK, p 227
Sarkar R, Sarkar A (2016) Sacred slaughter: an analysis of historical, communal, and constitutional aspects of beef bans in India. Polit Relig Ideol 17:329–351. https://doi.org/10.1080/21567689.2016.1259108
Sharma A, Phillips CJC (2019) Avoidance distance in sheltered cows and its association with other welfare parameters. Animals 9:396. https://www.mdpi.com/2076-2615/9/7/396
Sharma KK, Kalyani IH, Kshirsagar DP, Patel DR (2015) Determination of herd prevalence of brucellosis using rose Bengal plate test and indirect ELISA. J Anim Res 5:105–108. https://doi.org/10.5958/2277-940X.2015.00017.0
Sharma A, Kennedy U, Schuetze C, Phillips CJC (2019a) The welfare of cows in Indian shelters. Animals 9:172. https://www.mdpi.com/2076-2615/9/4/172
Sharma A, Kennedy U, Phillips C (2019b) A novel method of assessing floor friction in cowsheds and its association with cow health. Animals 9:120. https://www.mdpi.com/2076-2615/9/4/120
Sharma A, Schuetze C, Phillips CJC (2020) The management of cow shelters (gaushalas) in India, including the attitudes of shelter managers to cow welfare. Animals 10:211. https://doi.org/10.3390/ani10020211
Shree Pitamerdas Kewalram Maheshwari Charitable Trust (SPKMCT) (undated) Help gaushala. www.helpgaushala.com | An online platform for the purpose of Gauseva through Gaushala. Accessed 1 November 2022
Shringi BN (2004) Bovine tuberculosis in organised farms. Indian Vet J 81:1394–1395
Simoons FJ (1973) The sacred cow and the constitution of India. Ecol Food Nutr 2:281–295. https://doi.org/10.1080/03670244.1973.9990349
Simoons FJ, Batra SM, Chakravarti AK, Diener P, Ferro-Luzzi GE, Harris M et al (1979) Questions in the sacred-cow controversy [and comments and reply]. Curr Anthropol 20:467–493. https://doi.org/10.1086/202319
Singh S, Bist B, Chauhan R (2004) Seroprevalence of brucellosis in a gaushala of Mathura district and its public health significance. J Immunol Immunopathol 6:131–132
Singh KV, Singh M, Kumar A, Mukesh M (2018) Epidemiology and seasonal variation of Ixodid ticks in dairy cattle of Gangetic plains of Uttar Pradesh, India. Q Res J Plant Anim Sci 33:237–242
Srinivas B, Gupta BN (1997) Rumen fermentation, bacterial and total volatile fatty acid (TVFA) production rates in cattle fed on urea-molasses-mineral block licks supplement. Anim Feed Sci Technol 65:275–286. https://doi.org/10.1016/S0377-8401(96)01062-0
Srinivasan S, Easterling L, Rimal B, Niu XM, Conlan AJK, Dudas P, Kapur V (2018) Prevalence of bovine tuberculosis in India: a systematic review and meta-analysis. Transbound Emerg Dis 65:1627–1640. https://doi.org/10.1111/tbed.12915
Telezhenko E, Magnusson M, Bergsten C (2017) Gait of dairy cows on floors with different slipperiness. J Dairy Sci 100:6494–6503. https://doi.org/10.3168/jds.2016-12208

The Times of India (2018) Many gaushalas are dairies in disguise, FIAPO probe reveals. The Times of India September 10 2018. https://timesofindia.indiatimes.com/india/many-gaushalas-are-dairies-in-disguise-fiapo-probe-reveals/articleshowprint/65754844.cms

Veissier I, Andanson S, Dubroeucq H, Pomiès D (2008) The motivation of cows to walk as thwarted by tethering. J Anim Sci 86:2723–2729. https://doi.org/10.2527/jas.2008-1020

Waiblinger S, Boivin X, Pedersen V, Tosi M-V, Janczak AM, Visser EK, Jones RB (2006) Assessing the human–animal relationship in farmed species: a critical review. Appl Anim Behav Sci 101:185–242. https://doi.org/10.1016/j.applanim.2006.02.001

Windschnurer I, Boivin X, Waiblinger S (2009) Reliability of an avoidance distance test for the assessment of animals' responsiveness to humans and a preliminary investigation of its association with farmers' attitudes on bull fattening farms. Appl Anim Behav Sci 117:117–127. https://doi.org/10.1016/j.applanim.2008.12.013

# Cattle Welfare in Smallholder Dairy and Pastoralist Beef Systems in Sub-Saharan Africa

# 15

James Nguhiu-Mwangi

## Contents



J. Nguhiu-Mwangi (✉)
Department of Clinical Studies, Faculty of Veterinary Medicine, University of Nairobi, Nairobi, Kenya
e-mail: jnmwangi@uonbi.ac.ke



M. Haskell (ed.), *Cattle Welfare in Dairy and Beef Systems*, Animal Welfare 23,
https://doi.org/10.1007/978-3-031-21020-4_15

**Abstract**

The ability of dairy and beef cattle to comfortably interact and cope with their environment through physiological, behavioural and psychological systems, defines the state of their welfare. This is largely dependent on providing the cattle with the five universal welfare freedoms (FAWC, Farm Animal Welfare Council Press Statement. December 5th, 1979). A high percentage of sub-Saharan African countries' population depends on livestock, including cattle, for their livelihoods, hence the necessity to consider how the cattle are kept with regard to their welfare. A high percentage of dairy cattle are kept in smallholder system while the beef cattle are primarily kept in transhumance pastoralist systems. The two systems of keeping cattle in sub-Saharan Africa have, to a greater extent, a negative effect on their welfare and to a lesser extent some positive influences on their welfare. The purpose of this chapter is to describe the factors that affect the welfare of cattle negatively and positively, respectively, in both systems. The description is derived from what is documented in the literature and from personal observation during veterinary practice and social events. The main factors that influence welfare of cattle include diversified substandard zero-grazing unit designs, practices and utility elements in these zero-grazing units, stockperson practices and how they handle cattle, feeding and watering practices, limited resources, lack of knowledge, transhumance pastoralist life style, lack of policy, law and its enforcement, lack of veterinary professionals and practices, improper transportation and slaughter methods, ethnic, religious and cultural beliefs.

It can generally be concluded that there are several multifaceted factors that limit the implementation of good cattle welfare practices in smallholder and pastoralist systems. Education and information on welfare are essential for successful implementation of good animal welfare. Dialogue between advocates of cattle welfare and the communities with acknowledgement of their ethnic, religious and cultural beliefs and practices would be a positive step towards convincing these communities to embrace practices that promote animal welfare and change some of their traditional beliefs that violate animal welfare.

**Keywords**

Zero-grazing welfare · Pastoralism systems · Stockperson practices · Beef welfare

## 15.1 Introduction

Livestock production is an important activity that makes a major contribution to the livelihood of communities in sub-Saharan Africa. This has been the case for many past generations and continues to be so in present times. Both dairy and beef cattle are of great importance for these communities. Cattle can have a direct impact on their owner's livelihood through provision of meat and milk or an indirect facilitation of their livelihood through the provision of draught power and transportation. Additionally, animals may be sold occasionally providing extra revenue (Devereux 2014). In the African context, ownership of a certain threshold number of cattle commands prestige and respect among some communities. Many poor and middle-income households in sub-Saharan Africa depend on cattle for numerous resources and outputs such as milk, meat, dowry payments, payments of penalties imposed by community courts and draught power especially where oxen are used for ploughing and transportation. Other needs provided by cattle include manure for use in fertilizing the land for crop production, the dung for plastering their mud houses, the sale of skins and hides and generally for alleviation of poverty as well as giving the communities a sense of security (Kristjanson et al. 2004; Devereux 2014).

The rearing of beef cattle among small ruminants (sheep and goats) is the main engagement for pastoral communities who occasionally move from place to place in a transhumant or nomadic lifestyle. These communities move long distances in search of pasture and water, due to the fact that pastoral lands are in the arid and semi-arid regions that receive limited rainfall and have long dry seasons (Masiga and Munyua 2005; McDermott et al. 2010). Production of dairy cattle in sub-Saharan Africa is mainly in smallholder zero-grazing systems with less than 10 dairy cows (in some households fewer than 5), while pastoral beef production is mainly a large-scale system with stretches of rangeland grazing grounds being used. However, the availability of large pieces of land is diminishing at a high rate to pave way for various forms of developments and other uses (Devereux 2014). Most of this chapter will discuss the aspects of the smallholder dairy and pastoralist beef cattle systems. Additionally, in some African communities, indigenous cattle are used to provide draught power mainly for ploughing the semi-arable lands especially where agricultural machinery are unavailable, unaffordable or cannot be ploughed where the land area is too small. These ploughing cattle have their welfare contravened in various ways including long hours of work and few hours of rest, using heavy inefficient ploughs that are difficult to pull, being whipped to make them move faster in ploughing (which leaves them with wounds and scars), ploughing while hungry and thirsty for long periods, being given inadequate feed for much of the time and therefore having fair to poor body condition (Ndou et al. 2011).

All these livelihood activities involving the use of cattle do not consider the welfare of the cattle as a primary concern. In fact, many of the activities largely contravene the welfare of cattle. Some of the welfare inadequacies include the use of feeds with incomplete nutrient profiles, inadequate and unclean water, delayed or non-treatment of diseases, absence or unreliable veterinary service availability, overloading or overworking of the cattle, improper housing and living environments (Qekwara et al. 2019). Overemphasis on maximizing profits and production leads to regarding cattle solely as economic commodities, which is currently the situation in commercial agriculture, especially for dairy cattle, while ignoring their welfare conditions (Fraser 2005, 2008; Qekwara et al. 2019). These issues will be discussed in more depth in the sections below.

## 15.2 Descriptions of Typical Smallholder Dairy Production and Pastoral Beef Production Systems in Sub-Saharan Africa

The most common smallholder dairy cattle production units are on small pieces of land between 0.2 and 4 hectares with each having less than 10 head of adult cows (McDermott et al. 2010). However, there are a few that have more than 10 head of adult cows (Nguhiu-Mwangi 2007). The smallholder dairy production systems in Africa vary depending on the circumstances such as the location of the farm, availability of feed, land resources, financial ability of the farmer and the market availability for milk. They are also greatly varied in terms of the types of feeds, feeding regime and management methods (Nguhiu-Mwangi et al. 2008; Ojango et al. 2017). Smallholder dairy production systems contribute about 80% of all the milk produced in Africa (Ojango et al. 2017). In Kenya for example, milk forms an essential part of the diet for many communities, hence the high demand for milk at the markets, which promotes the keeping of dairy cattle by the smallholder farmers (Thorpe et al. 2000). The smallholder dairy production systems not only provide nutrition from milk, but also socioeconomic benefits for many households and employment to herdsmen (Moll et al. 2007). The exotic breeds of dairy cattle such as Friesian, Ayrshire, Guernsey, Jersey and the crosses between any two of these are the type of cows kept in the smallholder dairy production systems. A typical smallholder dairy unit is zero-grazing (i.e. the cows do not have access to grazing, and all feed and forage is brought to them). It should have a feeding area with feeding troughs, a lying area with individual cow cubicles, a walking area between the feeding troughs and cubicles, a separate milking area and a separate calf-pen area. Some smallholder units may have individual stalls per cow (Fig. 15.1).

Pastoralism can be viewed as a socio-cultural system in which there is a strong interaction between herders, animals and resource management (mainly land) with pastoralists involved mainly in herding cattle for grazing, which provide them with the means of livelihood. However, pastoralists have adopted other sources of supplementing their livelihood and a number have turned to agro-pastoralist systems (Nyariki and Ngugi 2002). Pastoralism can be categorized into pastoral nomadism, which involves extensive herding with much mobility and no cultivation, semi-nomadic pastoralism also called agro-pastoralism that mixes cultivation agriculture and livestock herding, and transhumance pastoralism, which involves seasonal movements from a homestead area to areas where pasture grass and water are available while mixing herding, cultivation agriculture and other forms of trade for livelihood (Nyariki and Ngugi 2002). Beef cattle in sub-Saharan Africa are herded either in community-owned, publicly owned or privately owned rangelands. Pastoralist beef cattle are herded mainly in the community or publicly owned rangelands. Occasionally the pastoralists invade privately owned rangelands and graze their cattle there illegally, especially in the dry seasons when there is a scarcity of pasture grass.

More than 70% of all the beef produced in the sub-Saharan African countries is from beef cattle raised in arid and semi-arid lands (ASALs) (Mwangi et al. 2020).

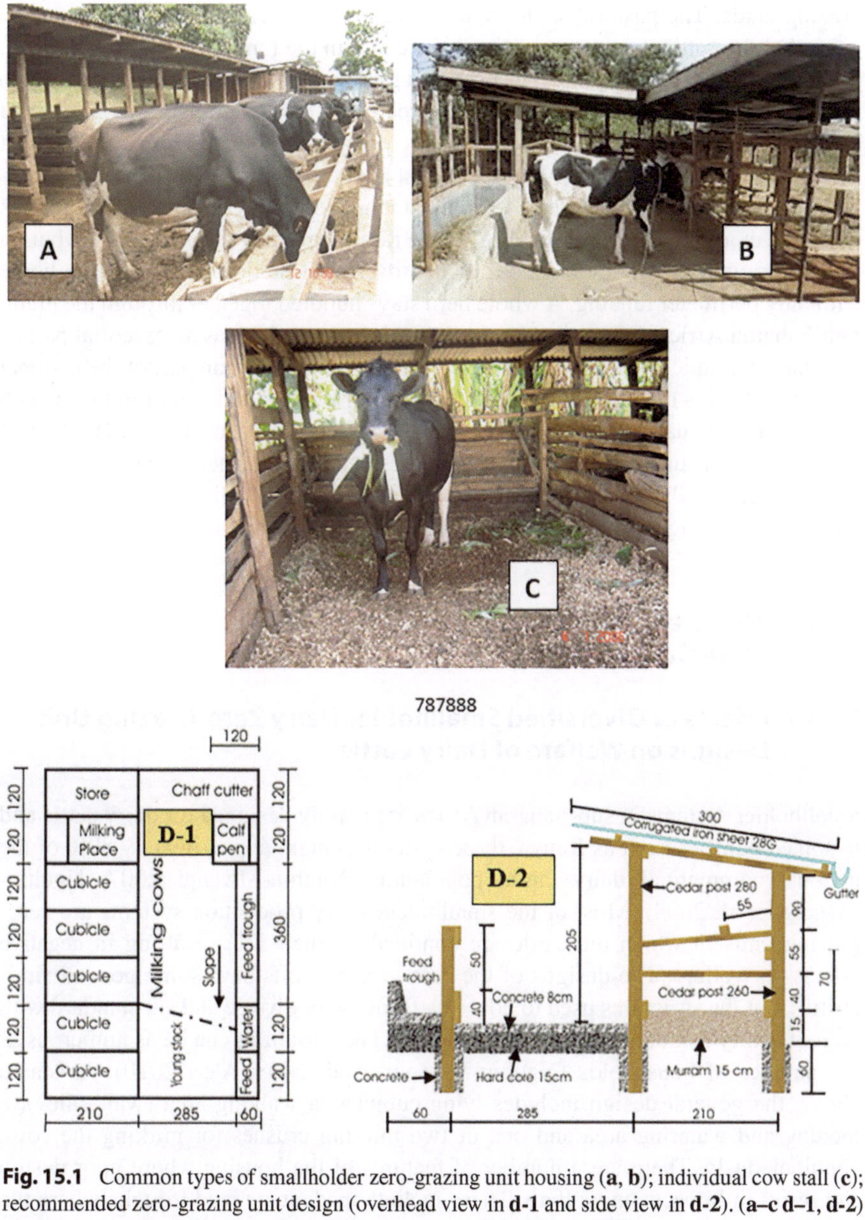


**Fig. 15.1** Common types of smallholder zero-grazing unit housing (**a**, **b**); individual cow stall (**c**); recommended zero-grazing unit design (overhead view in **d-1** and side view in **d-2**). (**a–c d–1**, **d-2**) (*Drawings courtesy of Livestock Kenya*: https://livestockkenya.com/index.php/blog/cattle/45-recommended-zero-grazing-housing-plan)

This predominance of the pastoral system is demonstrated by the fact that more than 98% of the cattle herds in Tanzania and 75% of the herds in Kenya are kept by pastoralists (Nyariki and Amwata 2019). Pastoralist beef cattle systems involve communal management, which involves high mobility of the cattle in search of pasture grazing lands. The pastoralists normally have a mixed herd with cattle, sheep and goats, but the cattle are the only animals involved in the long-distance movement in search of grazing lands, while the sheep and goats graze in pastures that are close to the home-farms. The majority of the cattle reared by the pastoralists are of the Zebu breed and cross breeds, but there are a few private commercial beef ranches that mainly have the Boran breed. The pastoralist beef cattle are sold in markets on designated days per region and they are walked many kilometres to the market centres (Mwangi et al. 2020). Pastoralist beef cattle rarely have any formal housing, but on the pastoralist farms, they have designated overnight holding sites near the home with only perimeter fencing. A whole herd stays huddled together through the night. Sub-Saharan African countries do not recognize pastoralism as an essential part of national economic development and therefore fail to a apportion part of the national budget to it. This results in the marginalization of the pastoral communities, which exacerbates a situation of persistent poverty (Nyariki and Amwata 2019). Most of the beef is consumed by urban and city residents in the homes and the hotels (Mwangi et al. 2020). The number of beef cattle kept by individual pastoralists is quite variable ranging from few tens to several hundred animals.

## 15.3 Welfare of Cattle in Smallholder Dairy Zero-Grazing Units

### 15.3.1 Effects of Diversified Smallholder Dairy Zero-Grazing Unit Designs on Welfare of Dairy Cattle

Smallholder systems in sub-Saharan Africa are mainly designed for dairy cattle and in some countries such as Kenya, these systems contain approximately 80% of the nation's commercial dairy cattle population (Nguhiu-Mwangi 2007; Nguhiu-Mwangi et al. 2008). Most of the smallholder dairy production systems are zero-grazing units in which the cattle are confined all their life, resulting in negative effects on welfare. The designs of the zero-grazing units have some general similarities, but the structures used to house cattle are very diverse and substandard with no uniformity from one household to another. The variations can be as numerous as the number of households (Nguhiu-Mwangi et al. 2008; Aleri 2010). As stated above, the general design includes lying cubicles, a walking area (walk alley), a feeding and watering area and one or two milking crushes for milking the cows simultaneously. There are a number of features of the housing where poor design can result in impacts on welfare. These include inadequate feed troughs where the available space at the troughs is smaller than the number of cows, the width of the troughs may be too narrow for the head of a cow to fit, and the neck-rail is too low (Fig. 15.2). Likewise, the walking area may be small, and the lying cubicles and milking crushes too small (Nguhiu-Mwangi 2007; Nguhiu-Mwangi et al. 2008; Aleri et al. 2012). Apart from substandard designs, inappropriate materials are used to construct cattle houses/barns, which result in poor quality finishes of floors, cubicle partitioning, feed troughs and crushes, thus increasing the likelihood of injuries,

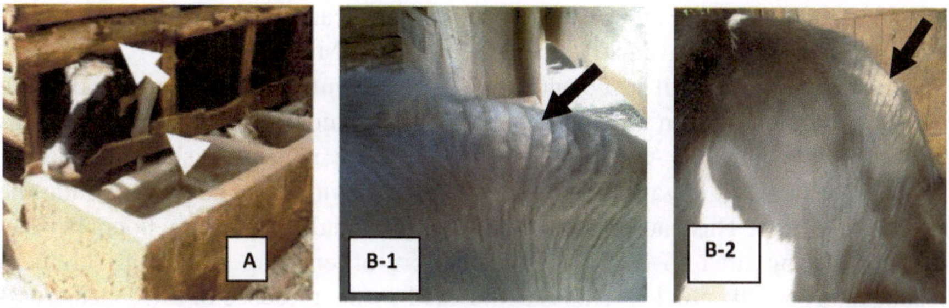


**Fig. 15.2** Low-positioned neck-rail in (**a**) shown by the bold white arrow, hardened callus-forming constantly bruised skin on the dorsal aspect of the neck in (**b-1**) and in (**b-2**) shown by the respective arrows. (*Photographs courtesy of JW Aleri*)

lameness and other detrimental effects on welfare (Nguhiu-Mwangi et al. 2012, 2013).

### 15.3.2 Elements of Housing: Feed Troughs, Passageways and Lying Areas

Poor design of the feeding troughs in the zero-grazing units leads to the cattle crowding against each other and the fixtures and competing to feed. The subordinate cattle are often not able to consume enough feed because they are displaced by the dominant animals. During these negative interactions, they are likely to injure each other. The body condition scores of these cattle are low because of this inadequate access to feed. Even when there is enough feed in the feed troughs, since some of them never get consistent access to the feed (Nguhiu-Mwangi 2007; Aleri et al. 2012; Nguhiu-Mwangi et al. 2012). The neck-rails are often fixed in a low position to prevent cattle from wasting feed, but they can cause bruises over the back of the neck of the animal, which results in the formation of thickened protective hard skin callus in the constantly bruised area. The skin in these areas never regains the normal texture but remains hardened for the life of the animal (Aleri et al. 2012; Nguhiu-Mwangi et al. 2012) (Fig. 15.2). Many farmers in the smallholder dairy zero-grazing units construct feed troughs that are excessively wide or deep. As the cow attempts to reach the feed at the widest or deepest points of the trough, the brisket area is repeatedly bruised. Subsequently, the skin over the brisket region can also harden into a callus as it does on the neck, and remains hardened for the life of the cow or as long as the trough dimensions are not corrected.

In most zero-grazing units, the walking area (walk alley) is smaller than the recommended dimensions (Livestock Kenya), which reduces the ability of cattle to move freely in the space, thus the necessary exercise needed for good blood circulation in the feet is hindered. In some of these zero-grazing units, the walking area is restricted to the extent that when the unit is fully stocked, the cattle are unable to turn around, but only walk forward and backward. When the cattle move backward, the skin over their pelvic protuberances (such as tuber coxae) is prone to injury by bumping into the cubicle poles and sharp edges of the cattle-barn structures (Nguhiu-Mwangi 2007; Aleri 2010; Aleri et al. 2012; Nguhiu-Mwangi et al. 2012). Cubicles that are narrow and not of sufficient length are not comfortable for cows, and so the cows may prefer lying in the walking areas between cubicles and the feed troughs to lying in the cubicles. Additionally, when the cows enter these short and narrow cubicles, they struggle with the process of lying down and standing up within the cubicle structure. This causes the cows to remain standing for many hours per day in the walking area to avoid the discomfort of entering and lying in these inadequately-sized cubicles. This is aggravated by the presence of a wooden cross-bar at the head-end of the cubicle, which reduces the lunging space and bob zone, thus making the act process of lying down and standing up a struggle that strains the limbs (Nguhiu-Mwangi 2007; Nguhiu-Mwangi et al. 2012) (Fig. 15.3).

The lack of provision of cubicle bedding in most zero-grazing units is of major concern with respect to the welfare of dairy cows. A number of smallholder dairy farmers do not provide any bedding and others provide scanty amounts of bedding materials in the cubicles. In such cases, the cows lie directly on bare, or scantily-bedded concrete floors, which results in excessive pressure on the body causing decubital/pressure wounds especially on bony protuberances (Nguhiu-Mwangi 2007; Aleri 2010; Nguhiu-Mwangi et al. 2013). The types of cubicle bedding materials provided by the farmers can be inappropriate for lying on and lead to discomfort or injuries to the skin when animals repeatedly lie on them. These include stone pieces (locally called hard-cores) (Fig. 15.4) as well as wet bedding. Wet bedding results from either leakage of rain-water into the cubicles or urine accumulation from the cattle due to bedding in the cubicle not being renewed regularly. The wet

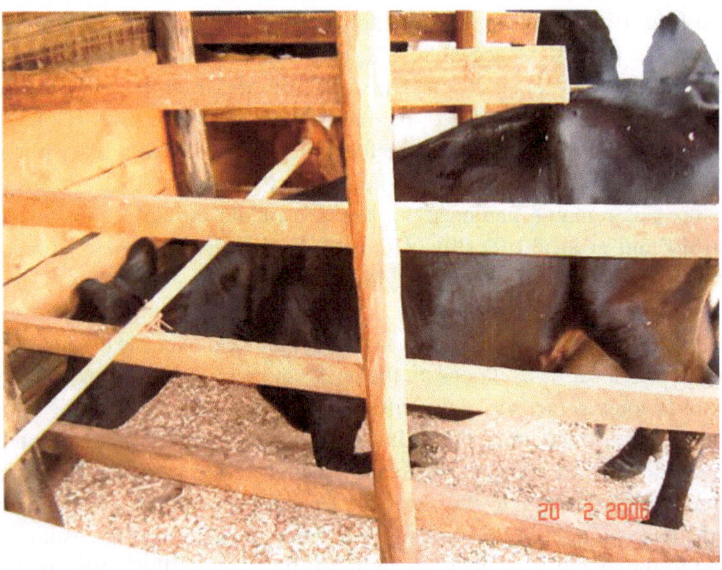


**Fig. 15.3** A cow struggling and straining to stand in a narrow small cubicle with a low-level head-end bar that reduces the lunging and bob zones

**Fig. 15.4** Example of comfortable bedding (sand) in (**a-1**) and uncomfortable bedding with small sharp stone pieces in (**a-2**) on which the cattle are reluctant to lie

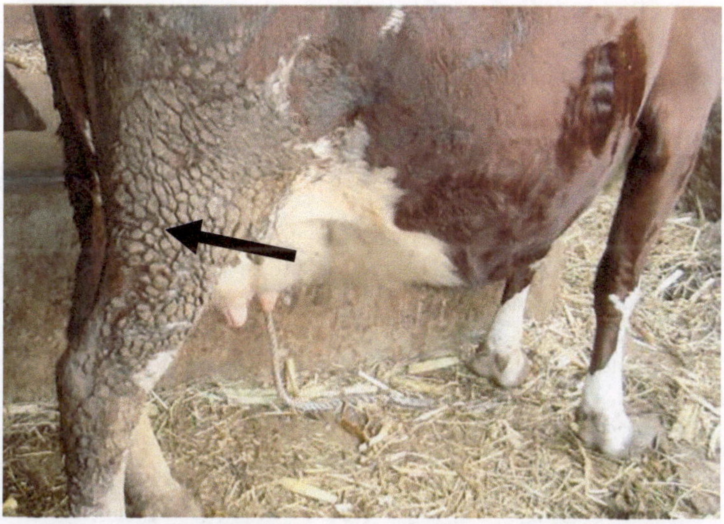

**Fig. 15.5** Moderate dried dung-matting of the skin (indicated by the arrow) from dung-wet cubicle bedding or slurry accumulation in the cow housing. (*Photograph courtesy of JW Aleri*)

bedding becomes a source of discomfort when cattle lie on it, produces excess heat that causes mild superficial scalding of the skin when the animals lie on it repeatedly. Urine-soaked and soiled bedding promotes the production of ammonia, which is an irritant to the lungs (Nguhiu-Mwangi 2007).

The type and state of floor can be an impediment to achieving good cattle welfare. The common types of floors in the smallholder dairy cattle zero-grazing units in sub-Saharan Africa include earthen, concrete and slatted wooden floors. Although an earthen floor is soft and comfortable for cattle, it is difficult to keep it hygienic because it is non-washable, hence it typically remains wet ground for most of the time. When cattle are kept on such a floor, their skin can become constantly matted with mud mixed with dung, eventually forming a layer of dry crust on the hair, which when rubbed against the floor or other objects results in the removal of the hair, leaving exposed patches of skin (Fig. 15.5). The hooves of cattle living on earthen-floor zero-grazing units become overgrown because of the lack of abrasive wear, which makes cows lame and uncomfortable (Aleri et al. 2012; Nguhiu-Mwangi et al. 2012, 2013). A large proportion of the zero-grazing units with concrete floors in the sub-Saharan Africa have features that pose a risk to welfare. These include small to large holes that often occur on old and heavily worn concrete, making the floor uneven, with pits that hold slurry and edges that cause claw disorders leading to lameness (Fig. 15.6). The occurrence of claw disorders caused by the sharp edges of the holes in the concrete is exacerbated by wetness from the accumulated slurry, which softens and weakens the horn capsule of the claw. Slippery concrete floors lead to incidents in which animals fall, resulting in limb injuries and further difficulties in movement. Excessively rough concrete finishing results in over-wearing of the horn of the sole with eventual thinning of the horn to the extent that it causes discomfort when cattle bear weight in standing or walking postures (Nguhiu-Mwangi et al. 2012; Telezhenko et al. 2017). Slatted wooden floors are uncomfortable for cattle because they become too slippery when urine and dung are present. This can lead to slipping and falling of cattle resulting in injuries. The spaces between the wooden slats become areas of discomfort for standing or walking cattle with the longitudinal slat edges forming pressure points for the sole (Telezhenko et al. 2017).

The use of improperly finished wooden materials to construct the zero-gazing units and the presence of dilapidated structures are an additional source of injury for dairy cattle. These can often be the cause of injury, from minor skin injury to severe bone injuries (Aleri 2010). The feed trough is the element of the zero-grazing unit that has the greatest potential to cause trauma and injury to the tongue and skin around the head when inappropriate materials are used in its construction. Feed

**Fig. 15.6** Old and worn out concrete with an example of an excavation defect (shown by metallic tape measure dipping) in a dairy cow zero-grazing unit. The concrete edges of the defect are sharp enough to cause claw disorders. (*Photograph courtesy of JW Aleri*)

**Fig. 15.7** Dilapidated cattle-house structures with loose hanging wood and iron sheets, which is an injury risk to any cow inside (**DS**). Feed trough (**FT**) made of timber and iron sheets. The double-sided arrow shows the width of the feed trough. Note the iron sheets are broken leaving sharp edges that can easily injure the animals. (*Photographs courtesy of JW Aleri*)

troughs made of iron sheet materials easily cut the tongue as it sweeps the trough to force the feed into the mouth (Fig. 15.7). In addition, dilapidated roofs, and patches where the roofing is absent, expose the cows to an increased likelihood of injuries but also expose them to adverse climatic factors such as wind, rain, direct sunlight heat and the cold at night (Aleri 2010; Aleri et al. 2012; Nguhiu-Mwangi et al. 2013) (Fig. 15.7).

### 15.3.3 Stockperson Practices

A number of practices by the stockpersons and farmers in the dairy zero-grazing units predispose cattle to the likelihood of injuries. One such dangerous practice is the act of chopping of fodder with a machete (locally known as a 'panga') at the feed trough as the animal is eating (Fig. 15.8). This causes accidental cut-wounds on the muzzle of the cattle. Some of the stockpersons are cruel to the cattle and whip them, which instills fear in them as evidenced by the avoidance distance when approaching the animals (Aleri 2010; Aleri et al. 2012). Sufficient feed may be given to the cattle by the farmer, but in the absence of the farmer, some stockpersons fail to give the cattle enough feed, which means that the animals are partially starved. This is a common problem in smallholder dairy herds.

### 15.3.4 Effects of Varied Feeding and Watering Practices on the Welfare of Dairy Cattle in Zero-Grazing Units

Feed types, quantities, quality and feeding practices in smallholder zero-grazing dairy units are as varied and inconsistent as the number of units. They also vary with time even within the same zero-grazing unit. Most of the fodder types include Napier grass (*Pennisetum purpureum*), mixed grass hay, maize stover (the stalk and

**Fig. 15.8** A stockperson chopping banana stem for the cow using a machete (panga) in the feed trough while the animal is simultaneously feeding. The tip edge of the machete is seen projecting from the banana stem (arrow). (*Photograph courtesy of JW Aleri*)

leaves 'left-over' after harvest of the grain) and mixed weed plants. In most cases, Napier grass, maize stover and mixed grass hay are harvested at an over-mature stage to the extent that the feed has a very high fibre content. Thus, the animals may receive a sufficient quantity of feed, which may just help to animal's rumen, but with minimal nutritional value owing to the high fibre content (Aleri 2010), thus failing to fulfil the nutrient needs of the animal, which has an impact on welfare. Napier grass is much better than grass hay and maize stover because it is fed to cattle while still green. In most African smallholder farms, grass hay and maize stover are fed to the animals in a completely dry state which means that both feeds have very low nutritional value (Pandley and Voskuil 2011). Cattle that are fed only on these dry or overgrown fodders have consistently low body condition scores. In addition, the high fibre content of overgrown dry or almost dry fodders is of low digestibility and can cause rumen impaction when cattle are fed large quantities. This is exacerbated by difficulties in getting an adequate amount of water, as cattle are often given small amounts only once per day. Rumen impaction causes discomfort and occasionally mortality in cattle (Priyanka and Dey 2018). In most cases, the smallholder dairy farmers practice agro-livestock farming, where they have pieces of land where they grow food crops for human consumption. It is from these pieces of land that they harvest invading weeds to feed their cattle when there is not enough normal fodder available. The weeds are usually in a lush phase of growth, which when fed in sufficient quantities to satisfy the animal's hunger, frequently cause frothy (foamy) ruminal tympany that is uncomfortable and could subsequently necessitate rumenotomy or occasionally result in death of cattle if not treated early (Abdisa 2018).

Supplementary feeds may be fed to the cattle and are usually concentrates that are either bought commercially or homemade. However, the cost of these feeds is high and most farmers cannot afford them regularly, hence the cattle are fed with insufficient quantities and only occasionally. Many smallholder farmers with zero-grazed dairy cows can afford concentrate feeds only at the end and beginning of each month after receiving payment for milk sold the previous month. Consequently by the second week of the month, the amount of concentrate feeding diminishes, becomes irregular in quantity and frequency, and eventually there is none until the next payment is received. Some of the farmers try to extend the period of concentrate feeding not only by reducing the quantity of feed that is fed at a time, but also by feeding it to the dairy cows on alternate days. When dietary and feed manipulation is done in a consistent way, ruminants are able to adapt and cope with the change in feeding pattern. The adaptation of the rumen is critical in this process, as it is the major organ that must adapt to the changes in feed type and timing of feed delivery. When the rumen can adapt to the change, the animal continues to perform almost optimally. The nutritional 'stress' effects of a variable concentrate feeding regime are more pronounced in high producing animals, which require more high-quality feed than low producing animals. However, the inconsistent and frequent changes in the patterns of feeding concentrates to the smallholder dairy cows do not allow them time to adapt to the new feeding mode, thus causing added nutritional stress, which is a welfare concern (Nguhiu-Mwangi 2007; Nguhiu-Mwangi et al. 2008). This problem can only be solved by providing concentrates consistently or withdrawing it permanently to enable the body to physiologically adapt either to their presence or absence (Nguhiu-Mwangi et al. 2008).

Some smallholder farmers overfeed cows with concentrates in an attempt to get a high milk yield, which results in grain/carbohydrate rumen overloading, leading to rumen acidosis and subsequent systemic lactic acidosis. These cows may die of systemic acidosis and those that recover have sequela of complications triggered by lactic acidosis such as laminitis. Other smallholder farmers feed the cattle with moderately high quantities of concentrates for long periods. These levels of concentrates do not cause rumen overload but moderate levels of lactic acid are produced in the rumen leading to subacute ruminal acidosis (SARA). This can subsequently cause subclinical laminitis, and consequently chronic laminitis with irreversible changes to the claws (Donovan et al. 2004; Vermunt 2004; Somers et al. 2005), which is a serious welfare issue (Fig. 15.9). It is important to feed cattle with the appropriate levels of concentrate, which is dependent on their metabolic needs. Low-income smallholder farmers may be unaware of the effects of this type of feeding on the health of the cattle. They are constrained by lack of sufficient money to enable them to seek early treatment for the animals, which may result in excessive damage to the claws, thus necessitating culling of the affected cows.

Provision of mineral and vitamin supplements to dairy cattle in sub-Saharan smallholder dairy systems is not a regular or consistent practice. In fact, a small percentage of the smallholder farmers do not give minerals and vitamins to their

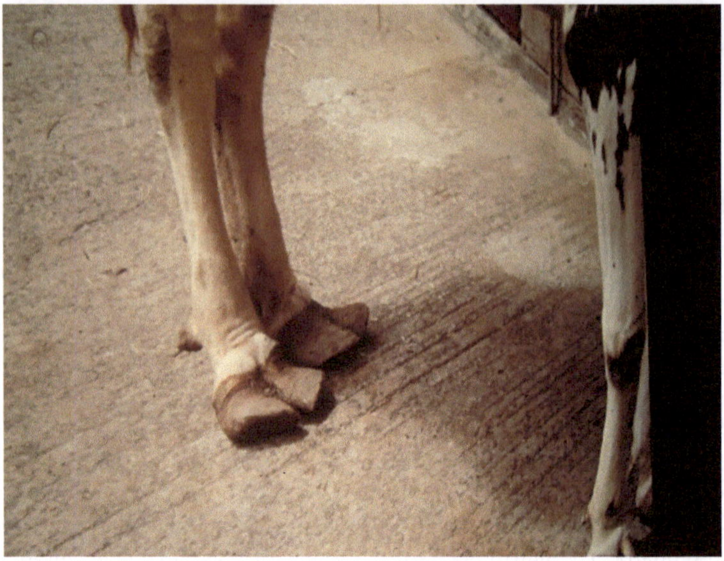

**Fig. 15.9** A cow showing chronic laminitis of the hind limb claws predisposed by subacute ruminal acidosis as an effect of long period of feeding moderately high-level amounts of concentrate

dairy cattle at all (Mapiye et al. 2009). A few of the financially-secure smallholder farmers give appropriate mineral-vitamin rations for dairy cows, which is in powder or grain form. However, a large proportion of smallholder farmers give mixtures of improper mineral and vitamin supplements, with a number just providing only occasional access to mineral blocks for cattle to lick, which does not provide sufficient quantities of minerals. Therefore, the dairy cattle in these smallholder units frequently suffer mineral and vitamin deficiencies, which results in poor body condition and may increase the likelihood of the animals succumbing to other diseases.

Provision of water to dairy cattle in the smallholder units is variable in quantity, quality and frequency. The amount of water provided ranges from ad libitum provision to only providing a small amount of water, once per day, which is the case for many smallholder units. The provision of a measured amount once per day is frequently inadequate for the cattle. The reason that water is provided only once per day is because of difficulties of getting water, long distances that must be travelled to fetch it, or due to insufficient finances to buy an adequate amount of water. The quality of the water can range from clean, soft water, through to clean, hard water or grossly contaminated water. The water and the watering troughs may become green-coloured due to the growth of algae when water is held in the trough for too long or when the trough is not washed regularly (Fig. 15.10). Water that is grossly contaminated and the water in the troughs with algal growth are unhygienic for cattle. The watering troughs are mainly made of concrete, metal or plastic. The main reason the smallholder farmers provide poor quality water is the scarcity of water in some regions as well as limited finances to afford the purchase of enough fresh water for daily replenishment. An additional issue is that some farmers use plastic or metal

**Fig. 15.10** Concrete watering trough (**a**) and smaller metallic (**b-1**) and plastic (**b-2**) containers used for providing the cattle with water under zero-grazing system. The trough is dirty with a slimy dark layer on the sides and contains unclean water at the bottom. The metallic and plastic containers are too small for the supply of sufficient water or to comfortably fit the cow's head inside. (*Photograph courtesy of JW Aleri*)

containers, which are too small to hold enough water for a cow or for the cow to comfortably fit its head into while drinking (Fig. 15.10).

### 15.3.5 Effects of Limited Availability of Resources to Smallholder Farmers

Inadequate finances and the lack of other resources are major barriers preventing farmers in sub-Saharan Africa from providing good standards of welfare for cattle in smallholder dairy units. The majority of smallholder livestock farmers in sub-Saharan Africa are poor or low-income earners. They are either employed in low-paying jobs as well as doing smallholder dairy cattle farming, or are unemployed, thus depending only on the income from the small dairy cattle units. The lack of available funds makes it difficult for smallholders to afford enough materials to construct adequately designed zero-grazing units or housing for dairy cattle to fully experience good welfare. This leads to substandard zero-grazing unit structures that cannot provide good cow comfort (Rahman et al. 2005). As discussed above, the result of this is the use of unrefined construction materials that have sharp edges and parts that pose a risk of injury to the cattle, with floors, cubicles, feeding troughs, watering troughs and walking areas that are grossly uncomfortable for the cattle. It also makes it difficult to afford feed and water in sufficient quantity and quality for good health, immunity and production to be sustained. This leads to a situation in which the freedom from hunger and thirst as well as freedom from pain, injury and disease is compromised for the cattle managed in these substandard zero-grazing units.

In the poor households experiencing such deficient financial circumstances, the resources for subsistence is shared between the needs of the people in the household and the animals. The priority is the needs of the people, while those of the animals are secondary, consequently resulting in minimal feeding and water for animals (Mapiye et al. 2009). This is the reason why the promotion of dairy cattle welfare must not ignore the economic output and the need for the alleviation of poverty of the low-income smallholder dairy cattle farmers. This is essential for these farmers to successfully embrace the practices leading to better welfare (Kristjanson et al. 2004; Lawrence and Stott 2009).

## 15.4 The Five Freedoms Perspective in Pastoralist Systems

### 15.4.1 Management in Pastoralist Systems

Generally, most of the indigenous beef cattle are reared in the arid and semi-arid lands of sub-Saharan Africa, which frequently experience periods of prolonged drought. Death is common due to starvation, thirst, predators, diseases, theft and bandit raids in some regions (Catley et al. 2014). These common disasters have a serious impact on the welfare of beef cattle in the pastoral communities.

It is important to have an overview of the key aspects of the pastoralist system of cattle production in order to understand its effects on animal welfare from the perspective of the Five Freedoms (FAWC 1979). The pastoral systems in sub-Saharan Africa keep beef cattle in medium to large rangelands in arid and semi-arid lands. These arid and semi-arid lands constitute more than 50% of the land in Africa. In these areas, the main livestock keeping system is nomadic or transhumant pastoralism (Devereux 2014). In the pastoral areas, about 80% of the household livelihood is derived from livestock keeping and the pastoralists live on milk, meat and blood as well as the sale of livestock. They also exchange their animals among themselves or with other communities for foods such as cereals using barter trade (McDermott et al. 2010).

The nutritional management of pastoral beef cattle entails the grazing of pasture in the rangelands, mostly in the semi-arid areas that the pastoralists occupy. Cattle are accompanied by herdsmen while grazing, but tethering is rarely practised. The majority of the grazing rangelands are communal but individual pastoralists or pastoralist families own a few of them. These lands are dry for most months of the year and cannot effectively be used for agro-farming. Insufficient short periods of rains in the semi-arid areas result in a scarcity of pasture grass. Therefore, the beef cattle and other livestock very often do not have sufficient feed to satisfy their hunger, which is a major welfare issue (Phillips 2002; Devereux 2014). The quality of pasture grass is also poor, overly dry, tough and of poor nutritive value. In addition, the beef cattle have to trek long distances during grazing in search of places in these rangelands with sufficient grass for them because most areas in the pasture lands are overgrazed. The movement of cattle in search of pasture may be within the same rangeland and when necessity demands it, they trek to other rangelands, which in some cases may be as far as 100 km away or more. In the dry seasons, beef cattle are grazed along the roadsides where there may be some grass left to grow before it is cut. The scarcity of pasture grass results in the cattle having poor body condition scores, which is an indicator of hunger and poor welfare. Beef cattle are rarely given mineral and vitamin supplements. Very occasionally, pastoralists may provide mineral lick blocks, but not consistently. Other than pasture, this type of cattle is never given other types of feeds such as concentrates (Ali et al. 2006; Dwyer 2009; McDermott et al. 2010).

Water is a very rare commodity in the pastoral grazing lands, particularly under drought conditions. When it is available, it may be a long distance from the grazing areas, so the animals have to trek in the heat of the sun under conditions of intense thirst to reach it. There is no formal organized provision of reliable water except for in a few ranches that are owned by the more progressive pastoralist farmers who are better enabled financially. For most of the pastoral pasture areas, available pools of water may be present alongside seasonal rivers and temporary dams. This water is usually dirty, and in many cases, people also use the same water. There may be large numbers of animals at the water sources because they are shared across a number of different herds. Animals will also stand in the water while drinking and will urinate and defaecate into it. This makes the water unhygienic for the cattle to drink as, they can contract diseases and ingest the eggs of parasites while drinking. In terms of the Five Freedoms, the scarcity of water and the long distance that must be walked to reach the water from the grazing areas, means that beef cattle are thirsty for long periods, which is a violation of their freedom from thirst (Opiyo et al. 2011).

The long distances that beef cattle must walk in search of grazing pastures and water have other consequences. The animals, including the young calves and newborns, walk while hungry and thirsty under the intense heat of the sun with very cold nights in the open. This is exhausting and stressful for the animals, which violates their freedom from thermal and physical discomfort. Due to hunger and thirst under heat stress, the weak, sick and young cattle succumb to this stress, become recumbent and eventually die as a result of the inability to move and feed anymore. Most rangeland where the cattle move about for pasture and water, has rough, rocky and hilly terrain that exacerbates the stress and discomfort especially for the weak, sick and young cattle (Coppolillo 2000; Bulitta et al. 2012; Devereux 2014), further increasing the mortality rate.

Pastoralist communities place great value on their cattle and are greatly attached to them. It is very rare to see a pastoralist striking their cattle with sticks or any form of whip. They drive them gently with care, which avoids fear and distress caused by mistreatment and poor human–animal relationships. However, fear and distress in these pastoralist beef cattle is caused by the indiscriminate mixing of different ages of cattle. This results in aggressive behaviour of adult and dominant cattle towards the weak and young cattle, which causes fear, reduces feeding and behavioural freedom in the latter group. Fighting is common among these cattle especially amongst the bulls that are herded together. The other factor that may be a negative influence on the welfare of the beef cattle is that when they are indiscriminately mixed, they also mate indiscriminately, resulting in frequent inbreeding (personal observation). Nevertheless, there are communities with elders who are knowledgeable in traditional matters of breeding. These elders select certain lineages of indigenous beef cattle and use only selected bulls from that lineage which avoids inbreeding (Mgongo et al. 2014). However, the positive aspect of free space and mixing of the cattle is the freedom to express normal behaviour without hindrance. The calves are suckled and cared for by their mothers and benefit from staying with them without any constraints, which is of great benefit for their welfare.

### 15.4.2 Transportation and Slaughter of Beef Cattle

Various methods of transportation and slaughter processes for beef cattle are practised, most of which contravene the standards required for good animal welfare. Pastoralist beef cattle are transported to slaughterhouses or abattoirs by various means that cause stress and poor welfare. One of the ways is trekking on foot for long distances without feed and water. They may take many days to reach the holding grounds of the slaughterhouse. The other mode of transport is use of inappropriately constructed trucks whose payload sections have not been compartmentalized to allow proper restraint of the cattle from strenuous uncontrolled movements while the truck is in motion. The flooring in some trucks is quite slippery. Cattle may fall accidently in the bed section of the truck, resulting in injuries, especially when the size of the truck bed is too large for the number of animals. However, in most cases, the number of cattle exceeds the capacity of the truck and the cattle are squeezed, and suffer from discomfort and distress. Some of the trucks have no roof to shield the animals from the direct heat of the sun and from rain. Other trucks are covered with canvas material and have solid metal sides, which means that the ventilation is not adequate when the lorry is overloaded during transport. The cattle can be transported long distances to get to the slaughterhouses with no provision of feed and water. As a result of transporting an excessive number of cattle in trucks without adequate ventilation, feed and water, some of them die in transit. The lack of properly designed trucks to transport cattle to slaughterhouses, especially from pastoral community regions (which are a distance from slaughterhouses) leads to significant mortalities caused by injuries from other animals as well as congestion and suffocation (Wambui et al. 2016).

At the slaughterhouses, the cattle remain in the holding ground for many hours to days without feeding or drinking water. The long holding periods may be due to the pastoralists waiting to sell the cattle to prospective individual buyers or agents as well as waiting for a slaughter time due to congested space in the slaughter-line.

During the slaughter of cattle, a captive bolt gun is used to ensure humane killing. However, the handling of beef cattle as they enter the killing line is rough and rather brutal, causing them fear and distress. There is no consideration of animal welfare in the attitudes adopted by the butchers in the slaughterhouse. The cattle are waiting in the slaughter-line while those being killed and bled are within close sight, which could be distressful for the ones that are waiting on the line, due to the smell of blood or sound of the commotion that occurs during the slaughter of the preceding animals. Occasionally, the humane killing method is not carried out accurately and the cattle fall and have their necks cut to bleed them while still sensitive to pain. They writhe in pain and there is a time-lag between neck-cutting and death, which is an animal welfare contravention. When cattle are locally slaughtered in the homes of the pastoralists, and not in the abattoir (which happens occasionally during cultural ceremonies such as marriages, circumcision and oath-taking among others), the cattle are not humanely killed but manually restrained and the neck is cut for bleeding to occur while the animal is fully conscious, resulting in tremendous suffering of the animals.

### 15.4.3 Effects of Beliefs, Religion and Culture in Pastoralist Systems

It is well recognized that cultural beliefs, perceptions and attitudes can influence the care, management and handling of animals. Specifically, those who value animals will treat them better, while others who value animals less will not pay much attention to their welfare. The latter attitude leads to diminished welfare and productivity of cattle as a result of heightened fear and stress (Hemsworth 2003).

Tribal, ethnic and community beliefs, religious practices and culture influence how all animals including cattle, are treated in this region. The improvement and implementation of animal welfare is grossly undermined by certain cultural beliefs and practices in sub-Saharan Africa (Serpell 2004). These practices are applied to pastoralist beef cattle and not to smallholder dairy cattle. Concerns about animal welfare are overridden by the view that cattle are commodities that can be used to fulfil some cultural purposes in dowry and marriage ceremonies (Maitra 2007; Bawa 2015). Tribes, ethnic and clan groups engage in conflicts that result in cattle raids. The cattle are ruthlessly handled, mistreated and injured in such conflicts (Gray et al. 2003).

Some cultures and traditional religions in sub-Saharan Africa practise inhumane animal sacrifices. These superstitions seek for blessings and help from ancestors during ceremonies such as marriages, birth of babies, funerals and taking of oaths (Qekwara et al. 2019). In these sacrifices, cattle and other animals are subjected to significant pain due to practices such as killing by way of cutting the neck without stunning and causing the animal to bleed to death, with the aim of collecting blood. Some of the ethnic groups believe that the loud bellowing of cattle after cutting the neck signifies acceptance of the sacrifice by the ancestors, yet in reality, it is due to pain. During normal traditional slaughtering of cattle, rituals that subject them to pain and suffering are also practised (Mnguni 2006).

Some ethnic groups such as the Maasai in Kenya practise traditional bleeding of cattle by puncturing the jugular vein using an arrow shot at close range in order to harvest blood for drinking in traditional ceremonies (as cited by Ndou et al. 2011). The arrow is wide, causes much pain, significant trauma and may introduce bacteria that may lead to severe infection of the neck region.

Some traditions believe that when animals, such as cattle, perform vigorous running for several hours before slaughter the meat becomes tender. Those who hold such a belief usually stab the animal to trigger running, which causes eventual exhaustion, with obvious negative effects on welfare (Andersen et al. 2005; Mnguni 2006). It is important to try and understand the reasons behind these cultural practices to be able to design effective mitigating measures that aim to improve animal welfare. This approach will encourage the communities to embrace and adopt the remedies being introduced (Ndou et al. 2011).

## 15.5 Animal Health Care Inadequacies and Practices

The 'Freedom from pain, injury and disease' concept is not always consistently applied in either the smallholder dairy cattle or the pastoralist beef cattle, despite the suffering of cattle due to diseases being of major concern in sub-Saharan Africa. There are several factors that lead to inconsistent availability of animal health care for these cattle. For both the smallholder dairy farmers and pastoralists, lack of financial security limits their ability to afford professional health care for their animals when they are sick or when they need disease prevention treatments such as regular deworming and tick control. Tick infestation is one of the secondary welfare problems that develops in the beef animals in the pastoralist systems as a result of trekking in the rangelands and along the roads in search for pastures. Tick control is not regularly or effectively practised in the pastoralist communities. Controlling ticks by the use of acaricides is difficult to implement due to the transhumance or nomadic lifestyles of the pastoralists. Additionally, these communities are unable to regularly afford acaricides for the large number of their beef cattle, sheep and goats. Therefore, the cattle end up suffering from tick-borne diseases, especially theileriosis among others.

When the cattle are sick, the farmers often seek veterinary or animal health care too late or fail to have them treated at all. Despite efforts to improve delivery of veterinary services, there is still a shortage of veterinary professionals in sub-Saharan African countries (Heffernan and Misturell 2002; Ilukor 2017). In most African countries, the main veterinary services are offered through the government. These services suffer from inadequate funding due to the prioritization of human health services over veterinary services. Veterinary services are somehow neglected during revenue allocation (Qekwara et al. 2019). The smallholder dairy farmers and the pastoralists turn to self-diagnoses, buying of medicines and self-treatment after the animals become severely sick (Lamuka et al. 2017). The medications given by the owners of the cattle are more often than not without professional advice, devoid of accurate diagnoses, the wrong medicines are used, and under-dosing, over-dosing and inconsistent dosing occur. These issues coupled with delayed treatment often result in failure of the animals to recover and subsequent death. Therefore, drug abuse is rampant in smallholder dairy cattle and pastoralist beef cattle. Failure to cure the disease may result in the development of antimicrobial drug resistance from under-dosing and inconsistent dosing, as well as development of toxicity from over-dosing (Love et al. 2011). Resistance to anthelmintics and acaricides also develops due to under-dosing or using the drugs for shorter than the recommended duration.

The routine management procedures carried out on cattle, such as castration, disbudding, dehorning and animal identification (including ear tagging and branding) are done using methods that cause pain and suffering. These procedures are done without analgesia. Closed castration is done using unconventional equipment and occasionally by use of crude methods such as crushing the spermatic cord area with a metallic or wooden mallet against a stone. Some people crush the testes within the scrotum. Disbudding is done by use of a red-hot metallic object directly over the horn bud of calves that are past the recommended age for disbudding. It is done without pain management, while dehorning is done with cutting objects also without analgesia. It has been recommended that disbudding calves immediately after birth or at least up to 2 weeks causes less pain than doing it thereafter (AABP guidelines 2019). However, most recommend disbudding before 6 to 8 weeks because from the 8 weeks, the horn bud firmly attaches to the periosteum of the frontal bone and disbudding can open into the sinuses with risk of infection (Faulkner and Weary 2000; Mainau et al. 2012; AVMA 2014). Pain sensitivity continues on a diminishing level for 9 weeks post-disbudding (Adcock and Tucker 2018). Cattle identification is done in some pastoral communities with red-hot metallic objects directly applied on the skin. Occasionally, when the red-hot metallic object is placed on the horn-bud area for a long period, it exposes the underlying surface of the frontal bone. These practices are all done in most cases without pain management, which is counter to good animal welfare practices (Githaiga 2014).

The scarcity of veterinary services is another factor that limits consistent health care for cattle in sub-Saharan Africa, particularly for the pastoralist systems. This is due to a low number of veterinarians and other animal health care workers, who are inequitably distributed in the country and who prefer setting up private veterinary clinics and services in urban and peri-urban areas, where there is the financial boost of pet or small animal practice. Rarely would a veterinarian set up practice in the pastoralist land due to the nature of transhumant or nomadic lifestyle and the tendency of pastoralists not to seek professional animal health care services. This means that the veterinarian response time to sick cattle can be prolonged, and consequently the animals suffer for longer periods than necessary. Considering that both the smallholder dairy farmers and pastoralists only seek veterinary services in the late stages of the disease when it has advanced to a severe state, the delayed response time by the veterinarians only worsens the situation and the animals are unlikely to recover when finally treated. Attempts to introduce pilot mobile veterinary clinics in the drylands of Kenya close to the pastoralists have revealed that if veterinary services were easily accessible, the pastoralists would utilize them.

Studies suggest that financial affordability of veterinary services is a problem in the pastoralist communities but the main problem is the inaccessibility of these services (Onono et al. 2013; Omondi et al. 2021).

The transhumant lifestyle of the pastoralists makes it difficult to regularly seek veterinary services or even to access the services while in transit. The pastoralists' beef cattle are subjected to prolonged suffering from untreated diseases, which is exacerbated by the long distances walked under harsh conditions. Routine vaccination against diseases such as foot-and-mouth, lumpy skin disease, anthrax and blackquarter is not done as necessary. Some of these diseases such as foot-and-mouth and lumpy skin disease occur in annual cycles, causing the death of many cattle and leaving those that recover from the diseases with prolonged painful after-effects. For example, foot-and-mouth leaves the animal with open wounds on the feet, udder, teats and sometimes oral cavity.

## 15.6 The Level of Knowledge and Its Impact

The general level of education of the low-income smallholder farmers and their knowledge of animal welfare are likely to influence their attitudes and practices regarding cattle welfare. Generally, there is a low level of knowledge about animal welfare and animal needs in most places in sub-Saharan Africa. Not only is the knowledge of animal welfare low but the knowledge of proper animal management is also generally poor. This influences the attitudes and practices that smallholder farmers adopt towards cattle. In a study done by Aleri (2010) within the smallholder units, it was found that dairy farmers and stockpersons had inadequate knowledge about animal welfare and had a negative attitude towards animals, which led to cruel animal handling. However, the attitudes of farmers and public towards animals have generally tended to improve due to campaigns and training done in the form of seminars and workshops by non-governmental organizations such as World Animal Protection (WAP) (Thornton 2010). To succeed in bringing animal welfare issues to the attention of the community, deliberate efforts to hold education and awareness seminars should be made. However, for people in sub-Saharan Africa to embrace the knowledge and practices promoting good welfare in cattle, the awareness and education campaigns must include demonstrating the production and economic benefits of good welfare provisions for the animals. A comprehensive working document entitled 'Kenya National Animal Welfare Strategy and Action Plan 2017-2022' was prepared with the collaboration of many stakeholders including animal owners. This working document has several strategic objectives that address the implementation of animal welfare through comprehensive animal welfare communication, awareness and advocacy campaigns, as well as promoting education, training, research and capacity building in animal welfare among stakeholders. These objectives include involving smallholder dairy farmer and pastoralist communities in dialogue on issues surrounding culture and beliefs, which will motivate them to embrace good animal welfare practice and hopefully abandon some of the cultural practices that cause animal suffering (KNAWSAP Draft 4 2017).

## 15.7 Status and Impact of Policy and Law

Some of the sub-Saharan African countries have animal welfare policies and laws, while others do not. Another important point of concern is whether these policies and laws are enforced where they do exist, and whether the law enforcers themselves have adequate knowledge of these policies and laws. Despite there being a large number of production animals, including cattle, in sub-Saharan Africa, most of the countries neither have laws and policies governing animal welfare nor ways of enforcing them (Asebe et al. 2016). Many sub-Saharan African countries do not have a single mention of animal welfare in their laws and constitution, and the few that have animal welfare laws are not stringent in their enforcement (Masiga and Munyua 2005). This translates to minimal support for enforcement and implementation of animal welfare laws in Africa in the face of cultural and traditional practices that conflict with animal welfare policies (Ramaswamy 1998; Ndou et al. 2011).

The World Organization for Animal Health (OIE) has demanded the inclusion of animal welfare as one of the competencies that is critical for the delivery of veterinary services. This has forced some African countries including those of sub-Saharan Africa to start implementing the consideration of animal welfare by incorporating it into their veterinary services (Bahari et al. 2006; Molomo and Mumba 2014). Laws and codes of practice are essential for a positive influence on animal welfare, especially in the way the animals are handled (Broom 2000). Some Eastern and Southern African countries have several acts of parliament that deal with welfare issues. These include the 'Prevention of Cruelty to Animals Act' that deals mainly with animal welfare, the 'Branding Act' dealing with animal identification, the 'Animal Diseases and Pest Control Act', 'Meat Control Act' as well as the 'Veterinary Surgeons and Veterinary Paraprofessionals Act' (Masiga and Munyua 2005). The non-governmental organizations concerned with promotion of animal welfare in sub-Saharan African countries concentrate on equines, dogs, abandoned, injured and homeless animals. They rarely concern themselves with cattle and other ruminants (Asebe et al. 2016). Therefore, in sub-Saharan Africa, the welfare of cattle, especially beef cattle, is largely ignored by the governments, non-governmental organizations and even by the national veterinary services.

## 15.8 Positive Aspects of Cattle Welfare in Smallholder Dairy and Pastoralist Beef Systems

Despite the many factors that negatively contribute to cattle welfare in the smallholder dairy and pastoral beef production systems, there are few positive aspects. In a significant number of smallholder dairy units, there is close interaction between the stockpersons/farmer and the cattle. The frequent presence of human in the animal unit without imposing negative interactions with the cattle improves the human–animal interaction and reduces their fear of humans. The daily presence of stockpersons in the cattle unit during cleaning, feeding, milking or changing of bedding makes the cattle used to human presence and they become less fearful. This improves animal responsiveness to humans and decreases fear and distress (Ebinghaus et al. 2018). As mentioned above, cattle in pastoralist systems have almost complete behavioural freedom. Calves in the pastoralist beef cattle system suckle their mothers for a long period, hence a strong bond between the mother and the calf develops.

## 15.9 Cattle Welfare Research in Sub-Saharan Africa

Generally, research in sub-Saharan Africa is limited by a lack of resources including funding, infrastructure, equipment, government support and community attitudes. As discussed above, governmental funding is focused on animal production. Non-governmental organizations primarily support animal welfare research in equines, dogs and cats. These organizations are devoted to securing some funding for these species, while much less attention is given to production animals. Although the concept of animal welfare is well advanced in developed countries, in sub-Saharan Africa, its introduction is at the stage of gaining gradual acceptance. However, for the majority of animal owners, especially those owning production animals such as cattle, it has not been embraced as a routine concept and translated into day-to-day practice.

There are a number of organizations that are championing campaigns on improving animal welfare practices in Africa generally, including sub-Saharan Africa. These include international organizations such as World Animal Protection (WAP), Africa Network for Animal Welfare (ANAW). There are also national bodies such as the local Veterinary Welfare Organizations and local Animal Protection Organizations. As described in the previous sections, the take-up of good animal welfare practices by communities in sub-Saharan Africa is influenced negatively by many interplaying factors including culture, religion, lack of financial resources, superstitious practices, bad attitudes towards animals and other similar mindsets. Understanding how to improve animal welfare against the background of these constraints requires a considerable research input. The scanty research that exists on the welfare of cattle is almost entirely on dairy cattle and not on beef cattle. Few postgraduate research theses have been written on the welfare of dairy cattle, mainly in research done in Kenya (Aleri 2010; Aleri et al. 2012; Nguhiu-Mwangi et al. 2013; Kathambi et al. 2018) and Ethiopia (Jerlström 2013). An analysis done on the number of research publications in the scientific literature from 1990 to 2019 on the welfare of beef cattle shows only one such publication in sub-Saharan Africa. This study was done in Kenya on the design of trucks for the transportation of beef cattle and its impact on the death of animals (Nalon et al. 2021). This suggests that research work on pastoral beef cattle in sub-Saharan Africa has scarcely been done or documented. As a way of improving advocacy, research on the effective methods of creating awareness on animal welfare needs to be carried out in earnest. This would enhance the take-up of methods that improve animal welfare by communities (Qekwara et al. 2019).

## 15.10 Conclusions

The important factor to bear in mind, as part of any conclusion about cattle welfare in this region, is that households and communities largely depend on these animals for their livelihood, thus emphasizing their significance. Due to the importance of these animals to the communities, the introduction of ways of improving cattle welfare by showing how the animals' production will benefit, is likely to encourage embracing of the practices. Practices that promote good welfare of smallholder dairy and pastoralist beef cattle in sub-Saharan Africa face several multifaceted influencing factors. The vast diversity of suboptimal smallholder dairy and semi-nomadic pastoralist beef production systems pose problems for the implementation of good welfare practices for the cattle. Suboptimal feeding practices, low financial security, failure of enforcement of existing animal welfare policies and laws, and a complete lack of animal welfare policies and laws in some countries as well as inadequate provision or non-availability of health care professionals are some of the main factors inhibiting practice of good welfare for these cattle. Other inhibiting and conflicting factors include a lack of adequate knowledge about animal welfare, cultural, religious and superstitious beliefs and generally negative attitudes towards animals, as well as a mindset of looking at animals as objects of use and trade rather than living or sentient beings. Education on matters of welfare and dialogue with communities in order to factor-in their cultural and religious beliefs and practices, is the best way to facilitate them embracing good animal welfare practices and to abandoning some of the practices that have negative impacts on animals.

**Acknowledgement** The author would like to acknowledge Dr. Joshua Wafula Aleri who was his Masters student in an animal welfare research, for allowing him to borrow some of the photographs from his MSc thesis.

Note on PhotographsUnless stated, all other photographs were provided by the author.

## References

AABP (American Association of Bovine Practitioners) Guideline (2019) Dehorning guideline. https://aabp.org/Resources/AABP_Guidelines/Dehorning-2019.pdf. Revised November 2019

Abdisa T (2018) Study on the prevalence of Bovine Frothy bloat in and around Kebele Lencha, Tokke Kutaye District, Oromia Region. Approaches Poultry Dairy Vet Sci 2(3):144–153. https://doi.org/10.31031/APDV.2018.02.000537

Adcock SJJ, Tucker CB (2018) The effect of disbudding age on healing and pain sensitivity in dairy calves. J Dairy Sci 101:10361–10373. https://doi.org/10.3168/jds.2018-14987

Aleri JW (2010) Welfare of dairy cattle in the smallholder (zero-grazing) production systems in Nairobi and its environs. MSc Thesis, University of Nairobi, Kenya. http://41.204.161.209/

Aleri JW, Nguhiu-Mwangi J, Mogoa EM, Mulei CM (2012) Welfare of dairy cattle in the smallholder (zero-grazing) production systems in Nairobi and its environs. Livest Res Rural Dev 24(9):159. http://www.lrrd.org/lrrd24/9/aler24159.htm

Ali BH, Al-Qarawi AA, Mousa HM (2006) Stress associated with road transportation in desert sheep and goats, and the effect of pretreatment with xylazine or sodium betaine. Res Vet Sci 80:343–348

Andersen H, Oksbjerg N, Young J, Therkildsen M (2005) Feeding and meat quality-a future approach. Meat Sci 70:543–554

Asebe G, Gelayenew B, Kumar A (2016) The general status of animal welfare in developing countries: the case of Ethiopia. J Vet Sci Technol 7:332. https://doi.org/10.4172/2157-7579.100033

AVMA (American Veterinary Medical Association) (2014) Literature review on the welfare implications of dehorning and disbudding cattle. https://www.avma.org/resources-tools/literature-reviews/welfare-implications-dehorning-and-disbudding-cattle

Bahari MM, Maiseli NG, Mruttu HA, Joseph W (2006) Animal welfare advocacy and veterinary profession: Tanzania experience. In: Proceedings of the 24th Tanzania veterinary association (TVA) scientific conference, Arusha, Tanzania

Bawa JAK (2015) Bride Price in Ghana: an assessment of its social representation in selected communities in northern Ghana. UDS Int J Dev 2(2):74–85

Broom DM (2000) Welfare assessment and welfare problem areas during handling and transport. In: Grandin T (ed) Livestock handling and transport livestock, 2nd edn. CABI, New York, pp 43–61

Bulitta F, Gebresenbet G, Bosona T (2012) Animal handling during supply for marketing and operations at an abattoir in developing country: the case of Gudar market and ambo abattoir, Ethiopia. J Serv Sci Manag 5:59–68

Catley A, Admassu B, Bekele G, Abebe D (2014) Livestock mortality in pastoralist herds in Ethiopia and implications for drought response. Disasters 38(3):500–516. https://doi.org/10.1111/disa.12060

Coppolillo PB (2000) The landscape ecology of pastoral herding: spatial analysis of land use and livestock production in East Africa. Hum Ecol 28(4):527–560. https://doi.org/10.1023/A:1026435714109

Devereux S (2014) Livestock and livelihoods in Africa: maximizing animal welfare and human wellbeing. IDS (Institute of Development Studies) working paper 451. ISSN: 2040-0209. ISBN: 978-1-78118-213-0

Donovan GA, Risco CA, DeChant TGM, Tran TQ, van Horn HH (2004) Influence of transition diets on occurrence of subclinical laminitis in Holstein dairy cows. J Dairy Sci 87:73–84

Dwyer CM (2009) Welfare of sheep: providing for welfare in an extensive environment. Small Rumin Res 86:14–21

Ebinghaus A, Ivemeyer S, Knierim U (2018) Human and farm influences on dairy cows' responsiveness towards humans: a cross-sectional study. PLoS One 13(12):e0209817. https://doi.org/10.1371/journal.pone.0209817

Farm Animal Welfare Council (FAWC) (1979) Farm Animal Welfare Council Press Statement. December 5th, 1979

Faulkner PM, Weary DM (2000) Reducing pain after dehorning in dairy calves. J Dairy Sci 83:2037–2041

Fraser D (2005) Animal welfare and the intensification of animal production. An alternative interpretation. FAO Readings in Ethics

Fraser D (2008) Understanding animal welfare. Acta Vet Scand 50(Suppl 1):S1. https://doi.org/10.1186/1751-0147-50-S1-S1

Githaiga M I (2014) The effect of Maasai cultural practices on the welfare of domestic animals: a case of Kajiado Central district, Kenya. Master of Arts Thesis, University of Nairobi. http://erepository.uonbi.ac.ke/bitstream/handle/11295/74177/Maina_TheeffectofMaasaiculturalpracticesonthewelfareofdomesticanimals.pdf?sequence=1

Gray S, Sundal M, Wiebusch B, Little M, Leslie P, Pike I (2003) Cattle raiding, cultural survival, and adaptability of east African pastoralists. Curr Anthropol 44(S5):S3–S30

Heffernan C, Misturell F (2002) The delivery of veterinary services to the poor: preliminary findings from Kenya. Veterinary Epidemiology and Economics Research Unit. Department of Agriculture. University of Reading RG6 6AT

Hemsworth PH (2003) Human–animal interactions in livestock production. Appl Anim Behav Sci 81(3):185–198

Ilukor J (2017) (2017) improving the delivery of veterinary services in Africa: insights from the empirical application of transaction costs theory in Uganda and Kenya. Revue Scientifique et Technique (International Office of Epizootics) 36(1):279–289

Jerlström J (2013) Animal welfare in Ethiopia: transport to and handling of cattle at markets in Addis Ababa and Ambo. Research thesis. Swedish University of Agricultural Sciences, Uppsala, Sweden

Kathambi EK, Van Leeuwen JA, Gitau GK, McKenna SL (2018) A cross-sectional study of the welfare of calves raised in smallholder dairy farms in Meru, Kenya, 2017. Vet World 11(8):1094–1101

Kenya National Animal Welfare Strategy and Action Plan (KNAWSAP) (2017–2022) Draft 4, March 29, 2017

Kristjanson P, Krishna A, Radeny M, Nindo W (2004) Pathways out of poverty in western Kenya and the role of livestock. PPLPI working paper no. 14. FAO, Rome, Italy

Lamuka PO, Njeruh FM, Gitau GC, Abey KA (2017) Camel health management and pastoralists' knowledge and information on Zoonoses and food safety risks in Isiolo County, Kenya. Pastoralism 7(20):1–10

Lawrence AB, Stott AW (2009) Profiting from animal welfare: an animal-based perspective. J R Agric Soc Engl 170:40–47

Livestock Kenya (n.d.). https://livestockkenya.com/index.php/blog/cattle/45-recommended-zero-grazing-housing-plan. Accessed 30 December 2022

Love DC, Davis MF, Bassett A, Gunther A, Nachman KE (2011) Dose imprecision and resistance: free-choice medicated feeds in industrial food animal production in the United States. Environ Health Perspect 119(3):279–283

Mainau E, Templex D, Manteca X (2012) Impact of dehorning and disbudding on the well-being of calves. The Farm Animal Welfare Fact Sheet. No. 2/August 2012. https://www.fawec.org/media/com_lazypdf/pdf/Fact_sheet_FAWEC_2_en.pdf

Maitra S (2007) Dowry and bride price. International encyclopedia of the social sciences, 2

Mapiye C, Chimonyo M, Dzama K, Raats JG, Mapekula M (2009) Opportunities for improving Nguni cattle production in the smallholder farming systems of South Africa. Livest Sci 124:196–204

Masiga WN, Munyua SJM (2005) Global perspectives on animal welfare: Africa. Rev Sci Tech 24(2):579–586

McDermott JJ, Staal SJ, Freeman HA, Herrero M, Van de Steeg JA (2010) Sustaining intensification of smallholder livestock systems in the tropics. Livest Sci 130(1–3):95–109

Mgongo FOK, Matiko MK, Batamuzi EK, Wambura RM, Karimuribo ED, Mpanduji DG, Massawe LB, Silayo RS, Kimbita E, Kiwia H (2014) Pastoral indigenous breeding practices and their impact on cattle reproduction performance: the case of Kilosa and Gairo Districts. Livestock Research for Rural Development 26(4). http://www.lrrd.org/lrrd26/4/mgon26076.htm, open access

Mnguni M E (2006) An investigation into the commercial and the Zulu traditional modes of slaughtering, butchering, culinary properties and service with special reference to sociocultural ritual behaviors in KwaZulu-Natal. M Tech Thesis, Durban University of Technology, Durban, South Africa. https://hdl.handle.net/10321/77

Moll HAJ, Staal SJ, Ibrahim MNM (2007) Smallholder dairy production and markets: a comparison of production systems in Zambia, Kenya and Sri Lanka. Agric Syst 94(2):593–603. https://doi.org/10.1016/j.agsy.2007.02.005

Molomo M, Mumba T (2014) Drivers for animal welfare policies in Africa. Rev Sci Tech 33(1):47–53. https://doi.org/10.20506/rst.33.1.2265

Mwangi V, Owuor S, Kiteme B, Giger M (2020) Beef production in the rangelands: a comparative assessment between pastoralism and large-scale ranching in Laikipia County, Kenya. Agriculture 10:0399–0414. https://doi.org/10.3390/agriculture10090399

Nalon E, Contiero B, Gottardo F, Cozzi G (2021) The welfare of beef cattle in scientific literature from 1990 to 2019: a text mining approach. Front Vet Sci 7:588749, open access. https://doi.org/10.3389/fvets.2020.588749

Ndou S P, Muchenje V, Chimonyo M (2011) Animal welfare in multipurpose cattle production systems and its implications on beef quality. Afr J Biotechnol 10(7):1049–1064. http://www.academicjournals.org/AJB, ISSN 1684–5315 © 2011

Nguhiu-Mwangi J (2007) Characteristics of laminitis and associated lesions in dairy cows in Nairobi and its environs. PhD Thesis, University of Nairobi, Kenya. http://41.204.161.209/

Nguhiu-Mwangi J, Mbithi PMF, Wabacha JK, Mbuthia PG (2008) Factors associated with the occurrence of claw disorders in dairy cows under smallholder production systems in urban and peri-urban areas of Nairobi, Kenya. Veterinarski Arhiv 78(4):343–355

Nguhiu-Mwangi J, Mbithi PMF, Wabacha JK, Mbuthia PG (2012) Risk (predisposing) factors for non-infectious claw disorders in dairy cows under varying zero-grazing systems. In: Periz Marin CC (ed) A bird's eye view of veterinary medicine. InTech Open Access Publishers, Rijeka, Croatia, pp 393–422

Nguhiu-Mwangi J, Aleri JW, Mogoa EM, Mbithi PMF (2013) Indicators of poor welfare in dairy cows within smallholder zero-grazing units in the peri-urban areas of Nairobi, Kenya. In: Payan-Carreira R (ed) Insights from veterinary medicine. InTech Open Access Publishers, Rijeka, Croatia, pp 49–88. isbn:978-953-51-1005-7

Nyariki DM, Amwata DA (2019) The value of pastoralism in Kenya: application of total economic value approach. Pastoralism Res Policy Pract 9:9. https://doi.org/10.1186/s13570-019-0144-x

Nyariki DM, Ngugi RK (2002) A review of African pastoral production systems: approaches to their understanding and development. J Hum Ecol 13(3):237–250

Ojango JMK, Mronde R, Okeyo AM, Rege JEO, Chagunda MGG, Kugonza DR (2017) Improving smallholder dairy farming in Africa. In: van Belzen N (ed) Achieving sustainable production of milk Volume 2: safety, quality and sustainability. Burleigh Dodds Science Publishing Limited, Cambridge, UK. https://doi.org/10.19103/AS.2016.0005.38

Omondi I, Baltenweck I, Kinuthia E, Kirui L, Njoroge-Wamwere G, Bett B, Munene A, Onle S, Dida D, Kiara H (2021) Mobile veterinary clinics in the drylands of Kenya: securing pastoralists' livelihoods by bringing services close. Dev Pract 31(5):1–19. https://doi.org/10.1080/09614524.2020.1863917

Onono JO, Wieland B, Rushton J (2013) Constraints to cattle production in a semiarid pastoral system in Kenya. Trop Anim Health Prod 45:1415–1422. https://doi.org/10.1007/s11250-013-0379-2

Opiyo EO, Mureithi SM, Ngugi RK (2011) The influence of water availability on pastoralist's resource use in Mwingi and Kitui District in Kenya. J Hum Ecol 35(1):43–52

Pandley GS, Voskuil GCJ (2011) Manual on improved feeding of dairy cattle by smallholder farmers. Golden Valley Agricultural Research Trust, pp 1–52. https://agriprofocus.com/upload/Dairy_manual__Feeding_(GART_Zambia,_2011)1423741222.pdf

Phillips C (2002) The welfare of cattle during transport, marketing & slaughter. In: Cattle behaviour & welfare, 2nd edn. Blackwell Science, New York, pp 38–48. https://doi.org/10.1002/9780470752418

Priyanka M, Dey S (2018) Ruminal impaction due to plastic materials - an increasing threat to ruminants and its impact on human health in developing countries. https://www.ncbi.nlm.nih.gov/pmc/articles/PMC6200578/pdf/VetWorld-11-1307.pdf

Qekwara DN, McCrindle CME, Cenci-Goga B, Grace D (2019) Animal welfare in Africa: strength of cultural traditions, challenges and perspectives. In: Hild S, Schweitzer L (eds) Proceedings: animal welfare: from science to law. La Fondation Droit Animal, Éthique et Sciences, Paris. isbn:978-2-9512167-4-7. http://www.Foundation-froit-animal.org/proceedigs-aw/

Rahman SA, Walker L, Ricketts W (2005) Global perspectives on animal welfare: Asia, the Far East and Oceania. Rev Sci Tech 24(2):597–612

Ramaswamy NS (1998) Draught animal welfare. Appl Anim Behav Sci 59:73–84

Serpell JA (2004) Factors influencing human attitudes to animals and their welfare. Anim Welf 13:S145–S151. ISSN 0962-7286

Somers JGCJ, Schouten WGP, Frankena K, Noordhuizen-Stassen EN, Metz JHM (2005) Development of claw traits and claw lesions in dairy cows kept on different floor systems. J Dairy Sci 88:110–120

Telezhenko E, Magnusson M, Bergsten C (2017) Gait of dairy cows on floors with different slipperiness. J Dairy Sci 100:6494–6503. https://doi.org/10.3168/jds.2016-12208

Thornton PK (2010) Livestock production: recent trends, future prospects. Philos Trans R Soc 365:2853–2807. https://doi.org/10.1098/rstb.2010.0134

Thorpe W, Muriuki H G, Omore A, Owango M O, Staal S (2000) Development of smallholder dairying in Eastern Africa with particular reference to Kenya. UZ/RVAU/DIAS/DANIDA-ENRECA Project Review Workshop 10th-13th January 2000, Bronte Hotel, Harare, Zimbabwe

Vermunt J (2004) Herd lameness: a review, major causal factors, and guidelines for prevention and control. In: Proceedings of the 13th International Symposium and 5th Conference on Lameness in Ruminants, 11th–15th February, Maribor, Slovenija, pp 1–15

Wambui JM, Lamuka PO, Karuri EG, Matofari JW, Abey KA (2016) Design of trucks for long distance transportation of cattle in Kenya and its effects on cattle deaths. Afr J Food Agric Nutr Dev 16(4):1–15

# Welfare and Health Challenges of 'New Entry' Dairying: a Practitioner's Perspective

16

Philip Chamberlain

## Contents




**Abstract**

In recent years, increasing household incomes and urbanisation, particularly in Asian countries, have resulted in a greater consumption of dairy products. This



P. Chamberlain (✉)
Chamberlain Veterinary Services Pty Ltd, SAMFORD, QLD, Australia

University of Queensland, Brisbane, QLD, Australia
e-mail: philip@chamberlainvet.com







M. Haskell (ed.), *Cattle Welfare in Dairy and Beef Systems*, Animal Welfare 23,
https://doi.org/10.1007/978-3-031-21020-4_16

demand has led to a major expansion in a number of small-scale dairy farms and the development of many large-scale farms. However, in some cases, the appropriate expertise was not in place, leading to a number of major cow welfare issues. This chapter focuses on welfare issues in new entry cattle farming development, especially in countries or regions that are not traditional dairy farming areas. The effect of drivers such as market pressures, implementation of development programmes and social constraints will be discussed. The cow welfare problems that occur in new large-scale and small-scale dairy systems are different. In new large dairy systems, the issues are mainly due to poor feed supply planning, poor building design and poor management of staff and resources. However, most welfare issues in new small-scale developments arise from the adoption of traditional animal housing system design and practices. Each area of animal management is discussed in detail in this chapter, with a focus on health and welfare management including housing, ventilation, feeding, watering and disease and health management across all ages of dairy cattle.

**Keywords**

Cattle welfare · Cattle nutrition · Developing country · Traditional practices · Foreign aid

## 16.1 Background

Increasing household incomes and urbanisation in recent years has driven a rise in disposable income, resulting in greater consumption of more expensive protein sources derived from animals, with milk and dairy products being important components (Fuller et al. 2005; OECD-FAO 2022). This is particularly true across the Asian countries, but with particularly marked increases in milk consumption and production in both Vietnam and China (FAO 2009; Fuller et al. 2005). To meet this demand, there has been a corresponding major increase in the number of dairy cows and milk production in these areas (FAO 2009). In some cases, the demand has been met by the construction of large-scale dairy farms, but in other cases, an increase in the number of small-scale family-owned dairies.

However, rapid increases in the number of cows and dairy farms bring many challenges, particularly in countries like China, where dairying and milk consumption in the local population was uncommon before the 1990s (Fuller et al. 2005), and so there is little experience of dairy farming. Dairy farm management is complex, with high-producing dairy cattle requiring managers to have an exceptionally good understanding of feed and nutrition, managing cow body condition, managing the effects of environmental conditions and cow husbandry, comfort and health. In the author's opinion, based on personal experiences in the dairy industries of Britain, Germany, Australia, China, India, Southeast and Central Asia, most experienced farmers are committed to ensure that good animal welfare is achieved, mainly from an animal care perspective. Good dairy animal welfare and health is associated with low stress levels in animals and farmers, and where farmers are rewarded for maintaining good welfare standards, it generally results in higher production and lower costs. However, welfare issues in new entry dairying system developments are mainly the consequence of inexperience in dairy system design, planning, nutrition and appropriate animal management for the region or at the level of production required.

A fundamental issue is that the cows in poorly designed, and managed systems are likely to experience chronic stress (Grelet et al. 2022). Moberg (2000) described chronic stress as occurring when an animal experiences a series of acute stressors whose accumulative biological cost forces the animal into a pre-pathological or pathological state. Chronic stress is very common in many new entry small-scale dairy systems that have little or no cattle experience and rely on local tradition. These stressors include poor nutrition, lack of adequate drinking water, poor staff training in cow husbandry, inadequate facilities to ensure cow comfort, poor disease management and heat stress. Factors such as heat stress and lack of availability of water will result in low voluntary feed intake (VFI) which will result in poor milk production, loss of body condition and consequently low immune responsiveness and increase susceptibility to disease (Kadzere et al. 2002). Poor human-animal interactions can also affect fertility and disease susceptibility (Dobson and Smith 2000; Ivemeyer et al. 2011).

This chapter will start by discussing some of the more global or regional issues that affect cow welfare. The specific issues affecting large- and small-scale dairies will be identified. The housing, management and health issues surrounding mature cows, calves and heifers will then be discussed.

## 16.2 General and 'Global' Issues

There are a number of wider societal and international issues that have a general influence on the way that new entry dairy farms are constructed and managed, and which have a direct or indirect effect on cow welfare.

### 16.2.1 Intensification

Intensification in agriculture can be defined as a situation where there is an increase in volume of outputs resulting from higher-grade inputs with an accompanying reduction in other inputs such as labour or time (FAO 2004). Intensification is often seen as a solution to the need to increase agricultural productivity in the face of greater consumer demand. In the dairy farming context, intensification is typically characterised by increasing animal numbers, a drive towards higher milk yields and the adoption of practices that reduce labour costs and increase farm outputs (Alvarez et al. 2008). However, the implementation of these practices often results in more stress on the cattle and people, resulting in poor welfare outcomes.

#### 16.2.1.1 Stocking Density

In an attempt to increase milk output in response to rapidly growing demand, some farms have responded by simply increasing the number of cattle in an existing unit. Increasing cattle numbers in existing farms, both in pasture-based and housed systems, often leads to poor welfare outcomes (Rushen 2017). For example, in Australia, many pasture-based farms recently purchased by one large foreign company have recently come under scrutiny for poor welfare outcomes. The animals were overstocked, underfed and suffered from poor body condition, reflecting a lack of appropriate farm and animal management. In general, when standards of management, health and welfare are good, and cows are fed well, fewer cattle can be kept and still have the same overall farm-level milk production and with lower costs, compared to a farm with large numbers of inefficient cattle.

#### 16.2.1.2 Breed Selection

Many new entrant dairy farms, particularly the large-scale farms, choose the Holstein Friesian (HF) breed of cow because they have the potential for high milk production in intensive systems in temperate climates. This choice of breed has also been made on many northern Australian dairy farms. However, in the hot and humid conditions of the tropics, many of these cows will suffer from heat stress and have poor production, fertility and health (Dairy Australia 2019; Kadzere et al. 2002). This issue is discussed at greater length in Sect. 16.4 (Welfare Challenges).

#### 16.2.1.3 Management Practices

Intensification can result in management practices that compromise cow welfare. These include tail docking to improve operator comfort and calving induction to meet market milk requirements. These were common practices in the Australian dairy industry until recently. However, these practices are now being phased out in Australia. Dehorning of adult cows is now also being phased out in favour of disbudding of young calves with less adverse welfare outcomes. Grazing management systems can also compromise cow welfare. Many farmers using an intensive pasture-based dairy farming system in New Zealand maintain their cows at a relatively low body condition score to encourage good grass utilisation. This is accepted as being normal practice in New Zealand, but many international experts consider these cows to be in chronically poor condition.

### 16.2.2 Importation and Transport

Many breeding cattle are transported internationally, to develop new dairy farms or expand current dairy systems. The international sea and land transport of live cattle, especially in tropical conditions, has sometimes resulted in poor cattle welfare outcomes during and following transit (Phillips and Santurtan 2013; Hing et al. 2021). This is often related to poor ventilation, heat stress, poor staff training and poor husbandry in sea transport (Phillips and Santurtan 2013). Poor welfare outcomes occur in land transport due to overcrowding and travelling long distances without breaks (EFSA AHAW Panel 2022). Inadequate access to feed and water is a problem in both land and sea (Hing et al. 2021). This is especially the case with sea transit of well-conditioned beef cattle and meat sheep for slaughter in the destination countries, especially when crossing the equator at sea. Poor health and welfare outcomes have also been observed with international transport of breeding dairy and beef cattle. This has sometimes resulted in deaths in transit, mainly from heat stress. Many countries also import pregnant heifers in mid to late gestation, to gain maximum genetic material (i.e. gaining both a cow and a calf), often resulting in abortion and subsequent loss of condition after arrival. Australia has implemented a monitoring system, where government veterinarians must accompany and monitor each sea shipment of live animals (Commonwealth of Australia 2021). New Zealand has now banned the sea transport of live cattle, and Australia is restricting the transport of female cattle more than 190 days pregnant (Commonwealth of Australia 2021).

### 16.2.3 Market Pressures

Achieving a reliable daily quantity of milk to supply a local liquid milk market often drives new dairy systems to utilize high-producing breeds that are unstainable in their local conditions, especially in tropical Asia (as outlined above and discussed further below). Many large-scale new entry farming systems are developed by inexperienced investors, who underestimate the amount of high-quality feed and level of staff training and experience required to achieve the desired production level in high-yielding breeds. This often results in welfare problems. This is especially evident when attempting to produce a consistent daily supply of drinking milk throughout the year, especially when feed supply is seasonally influenced. This happens, for example, when forage quality is poor in summer in the tropics or during severe winters in more temperate regions. This often results in undernutrition and poor welfare and productivity in large new farms that have not planned well for these feed supply fluctuation scenarios.

### 16.2.4 Government and Nongovernment Organisations' (NGOs) Cattle Industry Development Programs

Many of these programs have been focused on quickly improving milk supply within the host country, without a comprehensive understanding of the resource requirements and potential risks. Some have been highly successful; however, a number of these programs in Central Asia, South-eastern, Southern Asia and elsewhere have recently caused significant welfare issues and deaths of animals, due to lack of feed, heat stress and poor farmer husbandry knowledge. These issues have generally been due to lack of planning by the importing country authorities and the NGOs. It has typically involved importing high-producing North American or Western European cattle breeds into environments that are unsuitable for them. There is also often little emphasis placed on the supply and cost of appropriate feed, as well as a lack of farmer training in the level of husbandry needed for the imported cattle that have a high genetic merit for milk yield. This is usually the result of the importing governments wanting to achieve a rapid change in cattle productivity through importation, rather than focussing on improving their current cattle over time, using appropriate cross breeding through importation of semen. A recent example of a government-funded project in Sri Lanka resulted in many cattle starving to death, due to insufficient feed and poor feed availability and poor farmer understanding of appropriate animal husbandry practice.

### 16.2.5 Social Constraints

What is accepted as good welfare practice in one country or community may not be considered necessary in another. The commonly quoted example is the Indian situation where cows are considered sacred. This means that older cows cannot be culled or slaughtered, and so these animals and many unwanted male calves, find their way onto the streets and become feral (see Chap. 14). Feral cattle in India are often fed scraps by the general public and survive this way. Many are now taken into cow shelters (Gaushalas) (see Chap. 14 for more detail), which are mainly supported by charities and government funding. Euthanasia of injured cattle is often forbidden, and cattle are generally expected to die of *natural causes*. Some states now give veterinary officers permission to euthanise injured or sick cattle, but many other cattle, in the author's opinion, suffer an inhumane death. Large new dairy farms in India generally keep nonproductive cattle, and sometimes these animals are poorly fed to save money. In Southeast Asia, family status within the community is often dictated by the number of cattle the family owns. This is an incentive to retain chronically diseased or nonpregnant cattle, at the expense of space and feed that could be used to keep fewer, more efficient cattle under better welfare conditions. In Indonesia, cattle are also considered a liquid asset, so many farmers keep chronically ill cattle, until they need some funds, especially in small new farms.

## 16.3 Large-Scale vs Small-Scale Dairy Farming

The development of new dairy farms is happening in one of two ways. Large-scale farms are being developed that are owned by large organisations which employ staff to run the farm (see FAO 2009). These farms are being developed across the world, but particularly in China and Southeast Asia (e.g. USDA 2020). New small-scale farms are generally being established in areas that currently have family-run dairies (traditionally known as small-holder dairies) but which are taking on some Western concepts. There are many of these new small-scale dairy farms being developed in Southeast Asia and India. There are welfare issues that are particular to each type of farm, and this section will describe the welfare issues for each system.

### 16.3.1 Small-Holder Systems

Small-holder systems are generally family farms that contain from 2 to 20 cows (e.g. Devendra 2001), and are very common in India, Africa and Southeast Asia (www.fao.org/dairy-production-products). Most farms tether cattle by the head on cement flooring, often without matting, in low-roofed sheds with little ventilation (see Fig. 16.5). Most use the milk primarily for family consumption and then sell excess milk locally, either processed in the home or as fresh milk. Most farms have no refrigeration and deliver fresh milk daily to local cooperatives for refrigeration and further processing. Much of this milk has high bacterial loads and a short shelf life. Consequently, most milk in India is historically boiled before consumption. This results in a reduction of the risk of zoonotic disease transmission (e.g. brucellosis and tuberculosis).

General welfare issues in small-holder systems arise from the following:

(i) Economic and social drivers that compromise cow welfare. The drivers for small farmers in keeping dairy cattle are variable and include supplying the household, status, tradition in the family or region or having social/religious significance, as well as income from selling cows and some dairy products. These cattle management systems are generally based on old traditional approaches that are not focused on cow comfort or welfare.

(ii) Continuing local dairying traditions (particularly the practice of tethering). Local tradition often dictates that farmer practices in new farms remain the same when new dairy units are built in the same region, irrespective of whether they are good for cow welfare and productivity and without reference to the latest knowledge. The continued practice of tethering of cattle by the head on cement floors with no opportunity for exercise is, in the author's opinion, the greatest cattle welfare issue in the world and is only changing slowly. These are common farming practices in Southeast Asia and India. These cattle are often tethered by the head, confined on concrete, often with no matting in poorly ventilated sheds with low roofs and solid walls. There is often poor nutrition, resulting in poor body condition, and water is often only provided 1–2 times daily. The lack of opportunity to exercise also contributes to poor hoof health. Cases of lameness are often not diagnosed, as most cattle are unable to walk freely. Many of these cattle have not calved for more than a year, are not pregnant and producing less than 2 l of milk daily. In Indonesia, recent effort from trained extension officers is convincing some small farmers to improve husbandry, record production, oestrous and calving dates. This allows farmers to identify and cull nonproductive cows, feed the remainder better, allow them to exercise and have free access to water, resulting in improved profitability.

(iii) Lack of farmer knowledge. Many farmers lack knowledge of appropriate record keeping, feeding, good management, reducing animal stress and maintaining good health. In many new and older dairy industries in Central Asian countries, knowledge in areas such as cow, heifer and calf nutrition, husbandry, health, and disease management planning is often severely lacking. Production and reproduction records are often lacking, which means that culling of unproductive animals to allow herd improvement is not facilitated. This results in available feed being spread across more cows and poor body condition.

(iv) Poor resourcing. Poor resourcing is often a problem and includes lack of design and operating skills and capital and operating funds. For example, many farmers lack the ability to treat lameness, mastitis and other conditions. There is also a lack of appropriate infrastructure. This includes the lack of well-ventilated housing, continuous water supplies and access to adequate nutrition, exercise areas and appropriate bedding and flooring for cattle.

(v) Lack of support. For example, in Central Asia, there is a complete lack of diagnostic veterinary and agronomic support services, resulting in poor animal health planning, health and disease management, a lack of vaccination programmes, high levels of lameness, poor mastitis control and lack of advice on animal nutrition. There are very few veterinary animal health management services available. This results in extremely poor animal growth and health outcomes, and high prevalence of chronic disease, low productivity and high death and culling rates.

### 16.3.2 Large-Holder Systems

Large-holder systems generally involve 100 s (family owned) to 1000 s (corporate owned) of cows. In China, most farms now have more than 1000 cows, with some farms owing up to 17,000 cows (FAO 2009). These farms either process their own product or sell to large corporate processors. The farm planning concepts for new farms are generally developed by businesspeople, who engage designers and construction companies to build their systems. The final design and construction rely completely on the level of experience of the consultants, and this is sometimes lacking.

General welfare issues in large-holder systems arise from the following:

(vi) The drivers for larger farmer development often have little regard for welfare and are variable. These include existing small farm gaining economies of scale, investment of surplus funds, status, profit and return on investment. Increasing the scale of small-holder farms often results in maintaining the farm's previous poor management systems and potentiating welfare issues, including poor nutrition, record keeping, husbandry and ventilation.

(vii) Poor planning of large-scale sites. Poor planning, especially in the development of greenfield sites can very often have welfare ramifications. Commonly, issues arise due to lack of experienced input into the planning phase of building and site construction, including lack of appropriate feed supply, poor ventilation and heat stress management, inappropriate breed for the climate and lack of health management planning.

(viii) Lack of specialised veterinary training. A lack of personnel with specialised veterinary training and knowledge can result in very poor welfare outcomes. For example, many new large farms in developing countries employ local veterinarians with suboptimal training and experience in dairy, resulting in a high prevalence of metabolic disease (especially ketosis), hypocalcaemia, acute mastitis, left displaced abomasum and fatty liver disease.

(ix) Lack of a reliable supply of quality feed in the region. This is a common occurrence and is a major issue in Northern Asia in particular, where most small farms have disappeared, and new farms are generally a minimum of 1000 cows. Some of these farms have been very successful; however, many of these farms often have limited land for cropping and depend on many of the surrounding farms to supply corn for silage and grass for green-chop and hay. The quality of the fresh forage is variable, which sometimes results in the conserved forage made from it being too wet, mouldy and of poor quality, leading to digestive upsets. These feeds supplied by local farmers feeds also often contain pesticide and herbicide residues.

(x) Unsuitable housing. Many new dairy buildings in developing countries are designed by upscaling the small farm model, often resulting in buildings with low roofs, solid walls and poor ventilation. This will often result in poor air quality and can cause pneumonia in the cattle (see section below for further detail on housing and welfare issues).

(xi) Lack of staff training and management. Many large-scale farm owners prefer hiring staff with educational qualifications rather than experience. This lack of practical experience has negative consequences for cow health and welfare that is discussed further in Sect. 16.4. Also, the demise of small farming dairy systems in Northern Asia has led to a lack of experienced stock people in the countries concerned.

Additionally, many managers of new large-scale farms neglect the importance of planning and monitoring and fail to implement appropriate action planning and standard operating procedures for staff. This often results in staff not understanding their role in animal health management, especially in detection and reporting of disease and taking appropriate preventative and corrective action.

(xii) Inappropriate genetics for the climate. Many governments and companies developing new large dairy production systems are swayed by the high production rates of HF cattle from temperate climates, especially those from North America. Cattle or semen is imported from these countries, and the new farmers in Asia expect similar production levels in their hot and tropical climates. The effects on the welfare of the cattle will be considered in more depth below.

## 16.4 Welfare Challenges

There are a number of specific cow housing and management issues that are associated with poor cow welfare outcomes in new entry dairying systems. The details of the source of the problems and their impacts on cow health and welfare are discussed in the sections below. This is followed by a discussion of the specific animal health and disease issues that are problematic in new entrant dairying and a consideration of calf and heifer management.

### 16.4.1 Provision of Feed and Water

#### 16.4.1.1 Drinking Water

Important considerations in the provision of water for dairy cattle include factors that influence the amount of water needed, such as the prevailing weather conditions and the age and stage of lactation of the cows (e.g. Moran 2005). Water quality and palatability must also be considered, as well as providing enough watering points to allow all animals to access water freely, with minimal competition (www.ahdb.org.uk). Many new farm managers underestimate the water requirements, trough access space and flow rates necessary to satisfy a large herd of lactating cows. In hot weather, an individual lactating dairy cow can drink over 200 l of water/day, and most want to drink at the same time especially just after milking. Poor water intake will result in low VFI and lower production and immunity.

In small-holder farms in the tropics, cattle often do not have continuous access to water, with water being bucketed to cows twice daily, and often not sufficient for satiation. This is obviously a significant welfare concern.

In large herds, cattle must have adequate trough space to minimise the effects of competition for water and bullying by dominant cows. In these herds, cows tend to drink large volumes of water immediately following milking, and sufficient trough space to allow cows to drink undisturbed and sufficient water flow to replenish troughs quickly are important. Water quality, temperature and taste can also influence voluntary water intake, with cows preferring cool, clean and fresh water, with a low-salt content. A period of acclimatisation is sometimes required when cattle are offered water from a new source. Forages with high water content will reduce voluntary water intake, but water must always be on-offer. Cows that are injured and unable to walk to water need to be offered water often, especially in hot weather.

If the ability of a cow to access water is chronically reduced, she will have lower voluntary feed intake, reduced production, may suffer from dehydration in hot conditions and have reduced immune system responsiveness to disease (Dahl et al. 2020).

#### 16.4.1.2 Feed Supply: Quantity and Quality

Depending on the level of production and feed quality, cows will consume between 2 and 5 percent of their body weight on a dry matter basis per day (e.g. www.ahdb.org.uk). Some managers of new large-scale farms underestimate the amount of feed required or that is available on a year-round basis, for a large herd of lactating cows. Low levels of dry matter intake (DMI) can occur when the moisture content of green feed supplied is underestimated. This is especially the case in many new small-scale farms, where grass is cut and carried to the cows, resulting in high water intake in the feed. Diets low in dry matter content are not optimal as the animal is ingesting excess moisture and insufficient dry matter. Supplementary high dry matter content feed should be supplied (www.ahdb.org.uk). As for access to feed, it is important to provide sufficient trough/bunk space to allow all animals access to feed. It is not unusual in large new housed systems for too little trough space to be provided for both feed and water, often resulting in injuries, low feed intake, low body condition score and low productivity. See Fig. 16.1 for an example of good feeding system.

It is also not uncommon for large new dairy farms that have imported cattle to exhaust feed supplies quickly, resulting in low production, loss of body weight and low levels of immunity. Incidences of deaths from starvation have been recorded in poorly planned and implemented large- and small-scale dairy development programs, where large numbers of cattle are imported to a new area, with insufficient planning for continuous feed supply and staff training.

The development of a 3-year rolling feed management plan is essential in new large and small dairy systems to ensure there is adequate feed availability to meet the continuous demands of a growing and productive herd. These plans are often

**Fig. 16.1** Example of plenty of feed in front of healthy cows

poorly developed, with inadequate assurance of supply, resulting in poor production and animal health and sometimes death.

#### 16.4.1.3 Diet Formulation, Feeding Management and Body Condition Score

Dairy cow nutrition is extraordinarily complex, and details will not be given in this chapter. In summary, however, dairy cows need a diet that is balanced for energy, protein, fibre, vitamins and minerals (e.g. Moran 2005). This is often misunderstood in new farming systems. Energy (from carbohydrates) and protein are the most critical components, especially in the tropics, where the forage quality is often poor and especially lacking energy. A lack of energy in the overall diet often results from not providing concentrate feed. High-yielding cattle in peak lactation typically require to be fed concentrate feeds (e.g. www.ahdb.org.uk). Cows need energy for maintenance, activity, pregnancy, production and gaining body condition. A diet low in energy will result in low body condition, production, poor reproductive rates, weight loss and an increase in disease susceptibility (see Moran 2005). Low reproductive rates often result in overstocking on the farm, due to farmers keeping nonpregnant cows, with the attitude that 'they will eventually become pregnant', whilst these cows drain feed reserves that could have been fed to the other more productive cows. This affects welfare but also reduces farm profitability.

On many new large-scale dairies in developing countries with a high level of production, different formulations of diet are often not made on the farm to fit the requirements of the different stages of lactation. This means that late lactation cows and dry cows are often fed a diet best suited to early lactation cows, resulting in an intake of excess energy. This results in over-conditioning at calving and high rates of dystocia (difficult birth), hypocalcaemia (milk fever), ketosis, retained placenta and uterine infections. The longer-term results are often poor body condition, poor production, poor reproductive efficiency and higher death rates, as well as weakness and poor growth rates in the calves born (LeBlanc 2010; Mee 2008; Vanholder et al. 2015). Cows may become over-conditioned if they have taken an extended period to become pregnant (see also Vanholder et al. 2015).

The most useful tool in assessing the adequacy of the level of the nutrition supplied, especially energy, is body condition scoring. Body condition score (BCS) is a reasonable indicator of long-term energy balance (i.e. energy in the feed, minus energy for growth, maintenance, reproduction and production). A number of BCS scales exist worldwide, but the most common international BCS scale is 1–5, where 1 is extremely skinny, and 5 is very overweight and 2.3–3 is an average BCS (Roche et al. 2004). There are optimal average BCSs for each stage of lactation (Ishler et al. 2016). Cows that are underconditioned in late lactation are generally underconditioned at calving and then often lose further condition in early lactation. This results in low immune responsiveness, and cows often develop metritis (uterine infection), lameness, mastitis, metabolic disease and have low reproductive rates (Roche et al. 2009).

Many new large and most new small dairy systems do not have staff trained in BC scoring, resulting in high levels of dystocia, ketosis and culling from the herd.

A monthly sample of cattle from each stage of lactation in large herds, and all cattle in small herds, can be BC scored, and this is a great aid to understanding if feeding quality and quantity is appropriate or not, and dietary change can be implemented quickly. This process is simple and trains farmers and staff to observe cattle well. BC scoring is not difficult and greatly facilitates the management of the cows and is fundamental to achieving a productive and healthy herd.

#### 16.4.1.4 Forage Availability and Quality

The aspects of the forage that are important include the palatability, digestibility, nutrient content, continuous availability and quantity.

In large dairy systems, where cattle are housed and rarely have access to pasture, feed is usually supplied in the form of conserved forage (silage and hay) and concentrates (e.g. crushed grain, by-products, minerals and vitamins). This is often also supplemented by fresh cut grass. This supply of feed needs to be planned 18 months in advance, so that crops can be planted, grown, harvested and stored, or supply contracts developed for advanced purchase. Back-up supplies also need to be identified. Large new dairies, with little experience in growing or advanced purchasing of feed, especially in areas where forage growth, is seasonally dependant and sometimes exhaust their feed supply prior to next season's supply becoming available. If alternate supplies cannot be sourced quickly, this can result in poor nutrient intake, loss of condition, starvation and death, and there are examples of this occurring quite recently.

Many managers of large new farms in nontraditional dairy regions also lack experience in growing, harvesting and storing forage crops, often resulting in mouldy conserved feed being used, causing digestive upset and illness. Specific mycotoxins may also be present in this feed and can be tested for (including aflatoxins) and counteracted using antifungal additives (Galvano et al. 2001). Old harvesting equipment is also often utilized by many large new farms to reduce capital costs. However, many old harvesters drop small metal fragments into the feed, and fencing wire may also be taken in, chopped up and included in the silage or hay, resulting in 'hardware disease' where metal penetrates the stomach wall and enters the heart, causing a painful death. It is important that fencing wire scraps are not left in fields that are to be harvested for cattle feed.

In many small herds, including newly established small farms, the feed ration is often based on the manual collection of grasses by the farmers. Many farmers cut the grass when it is overmature, as this has a greater weight. However, much of the mature grass, especially in the tropics, has very low nutritive value. Farmers should be encouraged to cut the grass when it has the greatest nutritive value, rather than at the greatest weight. Also, many of these farms do not keep good farm records, resulting in an excess of nonproductive cattle being present in the herd. These should be culled, and fewer cattle fed well, including addition of concentrates in the diet. Record keeping is essential, to be able to understand individual cows' production and pregnancy status and coordinate it with the optimal grass cutting periods.

#### 16.4.1.5 Feed Management

Some large new systems buy harvesting, storage, mixing and feed-out machinery with insufficient capacity. This can result in poor mixing, with some cows receiving a slug of high-energy feed, which may cause acute rumen acidosis. Also, if urea is used as a source of nonprotein-nitrogen, this can be toxic and cause illness and death if an excess is ingested (Austin 1967). The cattle should be fed at least twice daily, with feed that moves beyond the cows' reach pushed-up between feeds. In early to mid-lactation, continuous feeding of a diet with excess energy and a lack of fibre may result in chronic acidosis and chronic laminitis (lameness). This is often subclinical, where symptoms are not observed early in the condition. However, subclinical laminitis may result in in prolonged standing time, lameness, a reduction in feed intake and consequently, poor body condition. Where laminitis is due to poor management of feed, this may affect a large percentage of the herd.

### 16.4.2 Breed/Type Suitability and Replacement Strategies

Cattle that are genetically suited to the prevailing climatic environment are much less likely to suffer from thermal stress (Santana et al. 2017), be healthier, reproduce better and require lower input costs (Polsky and von Keyserlingk 2017). Holstein Friesians (HFs) are well suited to temperate conditions. However, many large new farms and industries in the tropics are often blinded by the very high production levels possible from HF genetics derived from temperate regions of North America. Many new farmers in warmer climates assume that such production levels are achievable for them, but do not take their local conditions into account when purchasing cattle, semen or embryos to establish or grow their herds. For example, in tropical dairy industries, it is still common to utilize HF cows and attempt to keep them cool and comfortable; however, this is often not always possible without extremely good technology and management. There are often severely negative welfare outcomes when temperate breeds are used in tropical, hot and humid conditions. Many HF-type cattle will lose excessive body condition, develop severe lameness and suffer from mastitis and other diseases, despite attempts to cool them (see Fig. 16.2a). Other breeds (e.g. Jersey, Brown Swiss, Danish Red, local breeds, and cross-breeds between local breeds and a European breed) generally require lower input costs and manage hot conditions better than pure HF (see Fig. 16.2b). Under ideal temperate conditions, these alternative breeds generally produce less milk. However, under tropical conditions, whilst HF cows may have a high milk yield in their first lactation in the new situation, they may take long periods to become pregnant, so are dry for long periods of time. They are often more stressed and more susceptible to lameness, mastitis and local diseases. Consequently, under tropical and subtropical conditions, HF type cattle are often less profitable over a 3–4-year period, compared to more tropically adapted breeds (e.g. Marshall et al. 2020). For example, purebred HF in India are much more susceptible to theileriosis (a debilitating and often fatal tick-borne protozoal disease) than local breeds or crossbreds, which are well adapted to local conditions and diseases. With the use of appropriate

**Fig. 16.2** (a) Holstein Friesian-type cow in the tropics (upper photograph); (b) Holstein Friesian × Jersey crossbred cow in the tropics (lower photograph)

genetic selection and cross-breeding programmes, these breeds can contribute valuable traits for hardiness, longevity, productivity, low levels of disease and good reproductive performance, despite the lower genetic potential for milk yield.

Small-scale and large-scale farmers are also often heavily influenced by the marketing pressures of semen-selling companies. For example, in a tropical coastal SE Asian country (hot and humid all year), a new farmer with 20 cows was asked by the author which was his best cow. He identified a 6-year-old HF-type cow, on her second lactation, which was severely lame on multiple legs, and had a BCS of 1.5 (very thin—see Fig. 16.2a). She had last calved over 1 year ago, was not pregnant and producing only 5 l of milk/day. In contrast, a 6-year-old HF × Jersey crossbred cow in the same herd was on her third lactation and healthy, with a BCS of 3.0. She had last calved approximately 6 months ago, was 3 months pregnant and producing 15 l of milk/day. When asked why the HF was better than the HF × Jersey, the farmer said that 'HF are better, as they give more milk'. This statement highlights the influence of inappropriate genetics sales company marketing.

However, that is not to say that HF type cattle do not do well in any part of Asia. HF cattle are better suited to temperate regions and have been observed by the author to thrive and produce well in the sub-zero conditions in north China and

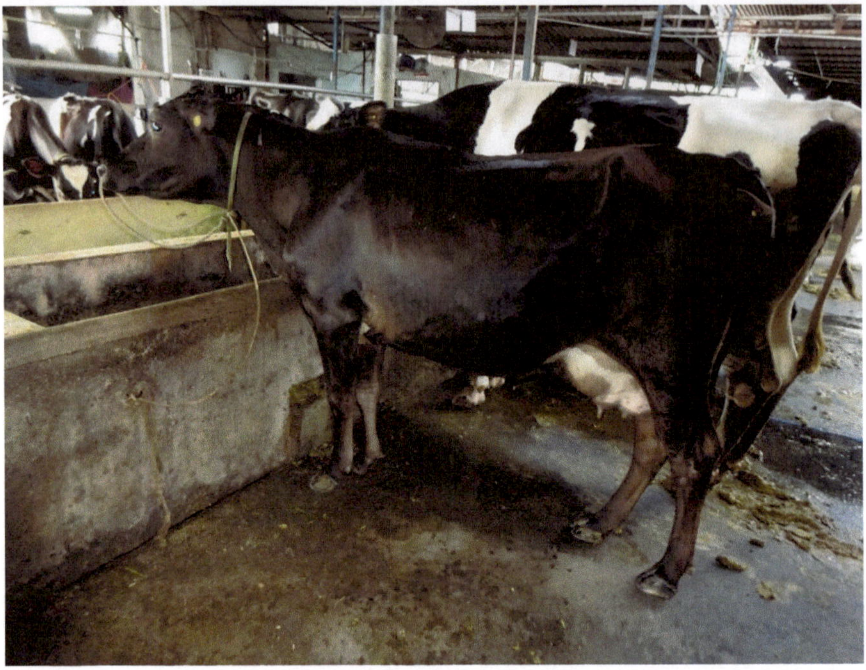

**Fig. 16.2** (continued)

Central Asia, as well as the very-high-altitude regions in Southeast Asia, if they are managed well.

Cross breeds, often utilizing local breeds or strains, have often been proven to be more robust than pure breeds, as they also possess hybrid vigour. Hybrid vigour is where the traits of the offspring are superior to the average of the parent generation (Simm et al. 2021). Some hybrid vigour can be maintained using a three-way cross in specific situations, with superior welfare and productivity outcomes being maintained in each new cross (Simm et al. 2021). An example of a three-way cross producing cow genotype suitable for use in tropical areas is, firstly, Holstein Friesian (Breed A) × Breed B (typically Jersey) with the offspring mated to Breed C (typically a Nordic Red type). Crossbred cattle are generally hardier and more productive in situations of lower management capacity.

The drive for high milk yields combined with a lack of knowledge of the cow development and longevity can also have adverse consequences for welfare. In developing countries, milk production in the first lactation is often used as a measure of animal performance or 'success' and is used to select animals for breeding. However, a good milk yield in the first lactation is not a guarantee of good production later in life, particularly in systems that pose nutritional and thermal challenges. Many animals that have a high milk yield in first lactation have poor longevity as they are culled due to poor health and reproductive performance prior to, or during their second lactation. The additional costs and consequences of early culling are often not considered when using high genetic merit (for milk yield in temperate regions) breeds in tropical climates. Where these cows do manage to produce a calf, this only serves to perpetuate a genetic line with poor resilience and longevity in the new environment. Realistic cost/benefit analyses on these farms, however, generally reveal that hardier breeds of cattle, especially crossbreds, are more profitable over a 3-year period, as they have shorter dry periods and have lower maintenance costs even though they have lower genetic merit for milk yield. These cattle generally manage nutritional and heat stress better, become pregnant faster, lose less condition, require less feed, have less disease and stay in the herd longer (Marshall et al. 2020).

In the author's opinion, crossbred cattle generally have better health and welfare outcomes in tropical environments, especially in new farms, where dairy farm management experience is often lacking. It must be pointed out that there are examples of large-scale purebred HF herds being successfully managed in the tropics, where very efficient barn cooling is achieved. However, these are very expensive systems to build and run; the level of operational management must be extremely high and continuously maintained. However, the risks for welfare are high if there is mechanical or management failure.

## 16.4.3 Housing and Environment

### 16.4.3.1 Housing and Management Principles

Dairy cow housing systems are designed to protect cows from adverse weather conditions and also to provide easy access for farmers to manage and feed the cows. Good housing facilities provide dry areas for lying, thermal comfort and good access to feed and water (e.g. Rushen 2017). Likewise, pasture systems must provide enough grass but also preserve enough clean pasture to allow cows to lie down and rest. Water and shelter must also be provided at pasture (Mee and Boyle 2020 and see Chap. 5). The concept of cow comfort is central to the provision of a living environment that is good for cow welfare. Cow comfort has many definitions, but generally occurs when the cow is at peace with her perception of the world and is thought to be suffering from minimal stress (Moran and Doyle 2015). As stated above, some stress is necessary (eustress) for learning and animal adaption (e.g. if stressed by the hot sun, cattle will learn to seek shade and cool areas). However, the effects of excessive stress (distress) are generally negative, and if continuous (chronic), poor welfare outcomes are the result (Grelet et al. 2022). In general, positive cow comfort will result in better welfare and higher productivity (e.g. Moran and Doyle 2015).

Cattle grazed on pasture are generally easier to manage, as diseases do not spread as quickly as they would in housed cows, the animals get exercise and display oestrous signs readily. These animals have the choice of what and where they graze, rest and are able to seek shade (Mee and Boyle 2020). However, they will suffer heat stress in fields with no shade available, often damaging pastures and creating muddy conditions during wet periods. Cows then often suffer a high degree of lameness and mastitis from this mud; consequently most large herds are held in housed and feedlot conditions.

The objective of housing cows and of intensive cow management is to control variables to achieve high production, with the lowest level of stress possible. This allows closer management of the animals, but farmers are then responsible for ensuring that animals' nutritional, thermal, physical and other welfare needs are met. This requires that the building design and facilities within it as well as staff training and skills are optimized. These aspects will be discussed in more detail below.

#### 16.4.3.2 Heat Stress, Ventilation and Air Quality in Housing

The major stress that housing seeks to abate is environmental stress, e.g. heat, cold, rain, mud, etc. However, housing animals in poorly designed and sited buildings is a major contributor to heat stress (Toledo et al. 2022). Heat stress in dairy systems is a major issue across Asia. Heat stress occurs when the production of heat from the body (e.g. from digestion, metabolism, pregnancy, milk production and muscular activity) is greater than the body's capacity to lose heat. Heat stress results in reduced food intake, reduced body condition, lowered reproduction and productivity, lowered immune responsiveness, poor health and higher levels of disease (Kadzere et al. 2002). The ability to lose heat is more difficult when ventilation and air exchange are poor, especially when environmental heat and humidity are high. The temperature humidity index (THI) is used as a measure and predictor of the cattle's ability to lose heat (e.g. Habeeb et al. 2018) (Figs. 16.3 and 16.4).

Many new large dairy systems in Asia have often not been designed well to combat heat stress. Sheds have been built away from prevailing winds or in valleys. In other cases, they have been built too close together, with roofs that are too low or made of heat-conducting material such as corrugated iron or with solid walls that do not allow sufficient air movement (see Fig. 16.5). These location and design factors result in poor airflow, ventilation and air exchange. Poor ventilation and air flow can result in pneumonia because of poor air quality or heat stress when air flow is low.

Increasing air flow across cattle by having open-sided buildings to take advantage of the prevailing winds and the installation and use of fans will assist in heat loss. The use of water misters to cool the air in low humidity situations, or water sprays to wet cattle in high-humidity situations, can be effective in reducing heat load. However, these cooling systems have their limitations and must be managed by experienced staff. Unlimited access to cool drinking water is also necessary in managing heat stress. Heat-stressed cattle tend to stand for longer, and this is a major cause of poor welfare in tropical Asia. Heat loss is also compounded when cattle are held closely together, radiating heat to each other and reducing air flow between cattle, so stocking density should be considered.

As an example of poor new construction, the author visited a small new farm in tropical Southeast Asia, where the farmer had built solid 2.5-meter-high concrete walls, with no windows, and bought in HF-type cattle. When asked why, that farmer explained that the solid walls were necessary for security, and he was not aware that heat stress was an issue (see Fig. 16.5 for an example) and that HF give more milk.

**Fig. 16.3** Cows on rubber mats with open walls in a shed in India

Many cows on this farm were panting excessively, even though the THI was below 70. A comparison was a new 20 cow farm in Vietnam, with the barriers around the edge of the shed consisting of rails instead of walls, built from locally sourced bamboo (Fig. 16.6). Steel mesh could be used if security is an issue.

In some cases, both heat and cold are climatic factors that need to be considered in the design and location of the housing. Some areas of northern and central Asia have very hot summers and very cold winters. When low-roofed sheds are used in these climates, the ventilation is poor causing heat stress as discussed above. However, they are usually closed-up in winter to stop water, urine and faeces freezing. The lack of ventilation results in poor air quality because of high concentrations of ammonia. This often contributes to the occurrence of pneumonia in adult cows, heifers and calves.

Holstein Friesian animals are more susceptible to heat stress because of their high metabolic heat production. This breed is used in many of the new farms across Asia. In the summer these cattle often suffer from heat stress quite quickly, stand for longer periods, resulting in weight loss and impaired production, lameness, reproductive ability and immune responses. Many large farms in the Middle East can successfully cool and maintain high production from HF cattle in hot dry conditions. However, in humid tropical conditions, especially in Southeast Asia, cooling is much more difficult due to the ambient humidity, which reduces the possibility for evaporation and its cooling effects. Sufficient cooling in tropical and subtropical conditions is often not achieved, resulting in poor welfare and productivity of

**Fig. 16.4** Cows in free stalls on sand bedding with open walls in Indonesia

temperate breeds of cattle, especially in HF cattle. Certain breeds and crossbreeds can withstand higher maximum daily THI levels. The Indian website Agri-Farming provides a good guide to dairy cow housing in tropical areas that has very good sections on ventilation (Agri-Farming India 2021).

#### 16.4.3.3 Concrete Walkways

In cow housing, large areas of cement are generally necessary for flooring and walkways. Many new large dairy systems install rough cement laneways and walkways wearing down the cow's feet and causing lameness. Newly laid concrete is also more abrasive for the feet than older concrete (McDaniel 1983). Many staff, especially in new systems, are not trained to move cattle slowly along concrete paths, resulting in damage to legs and hooves. This is especially the case when cattle are forced to turn corners on concrete. Many floors are also not cleaned appropriately, allowing for foot infections and diseases (e.g. foot rot) to spread. To maintain the quality and integrity of the flooring, regular maintenance and repair is required.

#### 16.4.3.4 Lying Areas and Bedding

A housing system that requires cows to stand for excessively long periods is not compatible with the concept of cow comfort. If the bedding/lying area is not well designed and maintained, cows will stand for excessively long periods. Cows may lie down for up to 12 h/day (Tucker et al. 2021). Standing for long periods can result

**Fig. 16.5** Poorly ventilated housing with solid walls and tie stalls in Indonesia

in poor voluntary feed intake as cows will then prioritise lying over feeding (Tucker et al. 2021), poor rumen movement, low saliva production and rumination, laminitis and lameness, rumen acidosis, indigestion, loss of body condition, poor immune response and ill health.

As discussed above, tethering cattle by the head is a major welfare issue in smallholder systems. New-build large farms are often installing free-stall (or cubicle) systems where cattle can choose a stall to lie in. However, the lying area for cattle is often not designed well and is not comfortable, with stalls being too long for the size of the average cow, resulting in defecation and urination in the bedding and a higher incidence of mastitis. Sometimes stalls are built too wide for the average size of the animal, allowing cattle to lie across the stalls, often resulting in neighbouring cows standing on and damaging teats and udders of other cows.

The bedding that is provided is also important in terms of cow comfort (Tucker et al. 2021). If the bedding is not comfortable (i.e. either hard or wet), cattle will tend to stand for longer periods. Lying surfaces should be comfortable, dry and not a source of bacteria or ammonia that can cause mastitis and pneumonia and soft enough to cushion the cow when she is lying down (Rushen 2017). In free-stall systems, a range of bedding can be used, with the most common being sand, wood dust/shaving, dry manure solids (DMS) and rubber mattresses or mats, with dry sawdust spread regularly on top. If the bedding is not comfortable (i.e. either hard or wet), cattle will tend to stand for longer periods. If cattle are soiling the beds, the

**Fig. 16.6** Small, cheap and efficient housing with free stalls in Vietnam. Note the open sides of the shed that allow good air flow

lying areas need to be cleaned of faeces regularly. Water should not be allowed to contact bedding areas, especially if DMS is used, as these wet areas become a source of bacteria. It is not uncommon to see the bedding areas thoroughly wet from water leaks or severe storms. These wet beds are a common source of infection, including mastitis in cows and navel and joint ill in newborn calves.

In open-barn systems, where cattle are held in groups in a large, covered pen, the bedding provided can also be sand, wood dust/shavings or composting bedding systems (CBS) (see Fig. 16.7). CBS (including composting DMS) systems must be tilled by machine twice daily to allow air to infiltrate and bacterial degradation to occur at lower depths. These CBS systems require constant assessment and management (Dairy Australia 2020). It is common to see these systems being attempted and failing due to poor management in large new dairy systems in Asia, resulting in very high levels of mastitis and other diseases. The main reason that CBS fail is the lack of organic fibre, excessive moisture and lack of tillage.

In conclusion, a lack of attention to issues affecting cow comfort is common in large- and small-scale new farms, especially in tropical Asia, with the major issues being poor shed design, cow poor management and the use of pure HF genetics in tropical environments. Lessons are being learned, with some of the new dairy farms being designed with health and welfare in mind. However, some large new systems are still being built without sufficient planning resulting in inappropriate cow housing and inadequate or inappropriate feeding areas, leading to poor cow comfort, welfare and productivity.

**Fig. 16.7** Cows on compost bedding (CBS) in India

### 16.4.4 Staff Hiring and Training

Many new large dairy systems in Asia have a preference for hiring staff with formal qualifications rather than experience and a positive attitude towards animals and dairying. Many new large, corporate-owned dairy systems in developing countries do not have any incentives to encourage staff loyalty. Few have staff training or career development programs in pace. This often results in a high staff turnover, staff not having the skills to be able to recognise developing animal health issues, not understanding their role or failing to implement appropriate animal management and husbandry practices. This often results in disease or poor health not being recognised promptly, poor animal husbandry and cattle movement practices and poor feeding and milking management leading to poor welfare outcomes, especially in relation to high levels of lameness, mastitis and metabolic disease. An appropriate Staff Management Plan for new farms must be based on employing staff with good attitudes, inducting them into their role, training them in animal husbandry, observation and recording of health and welfare issues and supplying them with standard operating procedures for actions to take when issues arise. In some cases, farms rely on technology, such as robotic milkers, but do not have the skills to repair this equipment themselves or the service agreements in place for maintenance, meaning that equipment remains unrepaired often resulting in poor welfare outcomes.

## 16.5 Animal Health

In large herds, good herd health and good welfare outcomes are usually a result of appropriate health management planning and must be specific to that region. Poor outcomes usually result from inadequate planning and poor staff training and attitude. Health planning has many aspects discussed below.

a. **Nutritional and feeding plan**. As discussed above, a primary driver of good health is the delivery of sufficient quantity of a well-balanced diet for each class of stock and stage of lactation. This program must be defined and planned for. As discussed above, many large new farms do not have an adequate feed-sourcing program in place. The testing of individual ingredients and of the mixed ration is generally considered necessary to determine if the diet is of sufficient quality to meet the animals' requirements. Many new large farms in Asia do not test feeds for nutrient balance, resulting in inappropriate BCSs, poor health and welfare outcomes.
b. **Health management planning**. This is often missing in large new herd development programs. This comprises health and disease risk identification for that region/area, and a plan to reduce risk through the development of written protocols, including:
   i. **Disease control, vaccination and parasite control plan.** For that region and farm. The risks of specific diseases must be assessed and control programs set in place to prevent or control these diseases. Many new farms do not have appropriate plans in place, and consequently disease and poor welfare often occur. For example, a high incidence of theileriosis was seen in a large new farm in India, due to lack of planning for control of ticks.
   ii. **Biosecurity and disease risk minimisation.** Biosecurity plans are necessary for each farm, especially for larger farms. These define that the disease testing and vaccinations required before animals are purchased and the necessary quarantine prior to introduction. The farm biosecurity plan should also define the protocol for people, vehicles and feed entering the farm, to prevent disease being imported onto the farm by these routes. A lack of biosecurity has been the cause of foot and mouth disease outbreaks in large and small farms across Asia. A similar case occurred when local cattle with *Brucellosis* were imported onto a new farm in India, causing abortions in the cows and transmission of the disease to people. Many mastitis outbreaks can also be traced back to older cows entering the farm, instead of importing healthy heifers.
   iii. **Staff training in animal husbandry.** Staff capacity is probably one of the biggest challenges and cause of animal health/welfare issues in dairy systems worldwide. A Staff Management Plan is often not in place, usually resulting in poor health and welfare outcomes. Large new dairies that involve staff in decision-making and that recognise staff for their work and feedback generally have better animal management and welfare outcomes. Many examples of staff not recognising loss of body condition, ill health or disease

early have resulted in large disease outbreaks, especially of ketosis and mastitis. This could have been controlled quickly if appropriate staff had been employed, trained and BCS and observational training had been implemented.

Staff need to be trained well to recognise what is normal behaviour and in animal body language, to be able to recognise when animals are not well. Sick animals generally have reduced appetites, are losing body condition, are often separated from the herd, have droopy ears and dull eyes and often have fever, breath quickly and are nonresponsive.

Additionally, a lack of mating records and identification of cows that are due to calve is also an issue in many new large herds, leading to cows calving unmonitored in the dry cow herd. Lameness and poor welfare are often the result of staff not understanding how to move cows quietly and at their own pace, resulting in hoof and leg injuries.

iv. **Metabolic diseases**. Besides transmissible diseases (bacterial, viral, protozoal and parasitic), metabolic (body metabolism imbalance) diseases are common in high-producing dairy cattle. Examples of these diseases include milk fever (hypocalcaemia), ketosis, displaced abomasum and fatty liver. Metabolic disease is often a consequence of inappropriate body condition and feeding, especially in late lactation and in the dry period. Metabolic disease outbreaks are common in developing countries in many large new herds, due to a lack of implementation of an appropriate Health Management Plan, Feed Management Plans, BC scoring and staff capacity building programs.

v. **Reproductive disorders**. Good reproduction outcomes result from healthy animals, being fed and managed properly. In many large new farms, cattle take long periods to become pregnant, often the result of stress, poor BCS and poor oestrous detection Many of these cattle then have long dry periods and gain excess body condition prior to calving. This often results in calving difficulties and ketosis following calving, as well as long inter-calving intervals and over-conditioning prior to the next calving.

vi. **Chronic disease and culling**. Most large dairy herds in worldwide cull at least 25% of the herd annually, to remove cattle with chronic disease (e.g. lameness, mastitis and reproductive disorders) and maintain genetic improvement. Most of these culled cattle go for slaughter. However, in India the cow is held as sacred, and slaughter of older cows and male calves is not permitted. Some of these chronically ill cattle find their way into Gaushalas (see Chap. 14); however, many remain on-farm and take up valuable space and consume valuable feed and often suffer from chronic lameness or mastitis. In many Asian cultures, social status is often linked to the number of cattle a farmer owns. Consequently, many cows with chronic mastitis or those that are not pregnant are kept in the herd, at the expense of other potentially more healthy and productive cattle. In India, the greater use of sexed semen would reduce the number of male calves born and roaming the street.

In smaller herds, the same animal health planning principles apply; however, many small new farms commonly adopt local traditional farming practices, due to a lack of contemporary knowledge based on science. This is most evident in the continuation of poor nutritional and animal husbandry practices in new farms, as mentioned previously. The use of pure HF genetics is also increasing in many of these regions, compounding the issues. Many new smaller herds also do not keep mating or production records, or understand health management planning, often resulting in significant disease and poor health/welfare outcomes.

## 16.6 Calf and Youngstock Rearing and Management

Calves in new entrant dairy systems, both large and small, are generally separated at birth from their mothers, as is the case in developed countries. Calves are notoriously susceptible to disease, often resulting in chronic illness or death. Internationally, the goal for preweaning calf mortality rates is less than 6% (Santman-Berends et al. 2019). On many new dairy farms in tropical areas, however, preweaning calf mortality rates are more likely to be 15–25% and can often be as high as 50%, which is an indication of extremely poor calf management. There are a number of management issues that contribute to this mortality and morbidity including overcrowding of calf housing, poor housing design, lack of colostrum feeding and poor observation and monitoring. Good calf management will usually result in a well-grown calf at weaning, giving the animal every opportunity to become a well grown and healthy heifer and eventually a milking cow. However, a major issue is that farm staff in these new dairy systems often have minimal calf-rearing experience. The key factors are discussed below.

### 16.6.1 Calf Management

There are a number of calf management practices used in new entrant dairying that result in poor calf health and welfare. Quite a common issue in new systems is attempting to rear too many calves, even bull calves, as they see these as a potential source of income. However, disease, death and heifers with poor growth rates are often the result. Cost-benefit analyses indicate that it is far better to rear a smaller number of high-value female calves properly. Male calf-rearing facilities are generally more successful when they are specialised units, even at the village level, that focus on calf rearing and nothing else.

#### 16.6.1.1 Health Monitoring

Good staff training and experience is critical for effective calf rearing, especially in recognising the early signs of poor health. The staff involved in calf rearing need to be well-trained, observant and preferably experienced. Calves develop disease very quickly, and early detection is therefore essential. Early signs of disease include being separated from other calves (if in groups), low responsiveness, lying down above normal levels, abnormal lying postures, having droopy ears, a rough coat, fast breathing, diarrhoea, nasal and eye discharge and increased body temperature (e.g. Bell et al. 2023). Targets need to be set for growth rate and weaning times. The most common predisposing factors to health and welfare issues observed in calf-rearing systems in new farming systems include the following:

i. **Insufficient feeding of colostrum.** Calves that receive insufficient colostrum, and/or are weaned prior to appropriate rumen development has occurred, generally lose weight following weaning. These poorly grown calves often develop chronic health conditions, especially pneumonia, often resulting in death or poorly grown heifers (Lopez and Heinrichs 2022). Colostrum is the first milk from a cow that has just calved. It is recommended to feed 3–4 l of colostrum, divided into in two feeds within the first 12 h of birth (Dairy Australia 2020). This provides high-quality nutrition but more importantly antibodies (passive immunity) for the calf. Calves are born with very few antibodies (immunoglobulins (IgG)) in their blood (Lombard et al. 2020). Calves will produce their own IgG, as part of the active immunity system, but this requires exposure to pathogens and takes time to develop. Thus, facilitating the uptake of antibodies through colostrum feeding is vitally important in providing calves with some defence against disease in the first few months of life (Lopez and Heinrichs 2022). However, a lack of adequate colostrum feedings and the consequential failure of IgG transfer often occurs in many new farming systems, both large- and small-scale, due to inexperience of the farmers, leaving the young calf highly susceptible to infectious disease. In the author's opinion, this is the major cause of poor health, chronic disease and death in calves and heifers, as well as poorly grown heifers entering the milking herd and early culling of cows in their first lactation. First calving heifers often have insufficient colostrum for their calf. Frozen colostrum banks are essential to ensure all calves have access to sufficient, high-quality colostrum.
ii. **Nutrition and weaning**. A lack of understanding of calf nutrition is also a major cause of some of the welfare issues and poorly grown heifers that are seen in many new large and small farms. Calves need to be fed milk at approximately 10% of their body weight/day, preferably in two feeds. Many inexperienced new entrants do not feed enough milk to young calves. Also, many new entrants wean the calves before they are eating enough concentrate (e.g. 800 g/day) to promote the development of the rumen. This will result in a rapid loss of body weight following weaning.
iii. **Calf Housing**. The aim of housing is to protect the calves against the elements (heat, cold, rain, snow) and to provide a clean and comfortable environment. However, many new calf-rearing systems are poorly designed and managed; with poor ventilation; lack of drinking water and overcrowding; with resulting welfare issues including heat stress, poor growth, diarrhoea and pneumonia; and consequently high death rates. Calves need to be kept clean, dry, be able to breath good quality air, be well fed, have access to water and be in an environment with a low risk of disease transmission. Calves can be held in group pens

**Fig. 16.8** Clean individual young calf-rearing boxes, with access to water in India

from the first day of life, but disease often results from this close contact, requiring exceptionally good management. Most farms keep calves separated for the first 4 weeks of life, to reduce disease transmission risk, as well as being separated from their faeces. This can be achieved in calf hutches, raised slatted floor pens, or on straw, that is cleaned daily and renewed often. Individual calf hutches that are kept clean and separated by an airspace are generally successfully used now by many large herds in new entrant systems (see Fig. 16.8). Recent research indicates that rearing calves in pairs allows for more appropriate social skills to develop (Van Os 2020).

#### 16.6.1.2 Common Calf Diseases

i. **Joint Ill (Navel Ill).** In the first few days following birth, calves are very susceptible to bacterial infections via the navel, usually causing a swollen navel. The infection can develop into a bacteraemia (blood infection), with bacteria infecting the leg joints (infectious arthritis). This arthritis often becomes chronic, causing chronic pain and joint swelling, weight loss and sometimes death. If noticed early on, antibiotics may control the condition, but this is not always the case, with chronic navel and joint infection often resulting (see also www.nadis.org.uk). The risk is reduced by calving cows in clean and dry conditions, preferably on new straw. The navel of the calf should not be touched or tied-off, as this will introduce bacteria. Dipping or spraying the navel with 7% tincture of iodine, 2–3 times in the first 48 h, is recommended. Navel ill and joint ill is commonly seen in many new farming systems, where cows give birth in unhygienic conditions and staff are inexperienced in calf rearing.

ii. **Diarrhoea** (scours) often begins with nutritional scours and then progresses to viral and/or bacterial scours that spreads between calves. Calves dehydrate rapidly. The most common precursors include a lack of IgG transfer from colostrum, unhygienic housing environment and too many calves in a small area (Medrano-Galarza et al. 2018). Treatment includes fluid replacement, but antibiotics are sometimes necessary. However, prevention is best. Bloody diarrhoea in older calves (2 months and older) is often caused by coccidia from calves accessing older animal faeces. Chronic diarrhoea is common on new farms and will often result in poor growth and death.

iii. **Pneumonia.** This lung condition often results in calf death. Calves that recover often have reduced lung capacity and do not grow to their full potential (Bach 2011 and see Fig. 16.9). The most common precursors include a lack of IgG transfer from colostrum and poor air quality (especially high levels of ammonia and humidity) (Mahendran et al. 2017; Bonizzi et al. 2022). Calves must always breath fresh air, even if it is cold, but without being directly affected by draughts (Lago et al. 2006; Lorenz et al. 2011). Urine in bedding and poor ventilation quickly result in ammonia production, contributing to pneumonia. Treatment includes the use of antibiotics and good nursing, but as for calf diarrhoea, prevention measures are best.

Many cases of pneumonia, due to ammonia build up, are commonly seen in new farm calf-rearing systems that have poor ventilation. This often occurs when there are walls (particularly low walls) blocking air flow, especially if the system is overcrowded. Calves will tolerate cold conditions, especially if lying in fresh and clean straw.

## 16.6.2 Heifer Rearing and Management

The same issues apply with young heifers as they do for calves, with many new farmers rearing too many heifers poorly, resulting in deaths and small heifers entering the milking herd. Poorly grown (small) heifers often experience dystocia and are often culled from the herd in the first or second lactation, due to loss of condition, bullying by older cows, disease and failure to conceive again following the first calving. These effects are due to their low immune response status and their experience of chronic stress.

### 16.6.2.1 Welfare Outcomes of Poor Heifer Management

Calves that have recovered from diarrhoea or pneumonia often have chronically damaged lungs and gastrointestinal tracts (GIT) and grow poorly as heifers (Bach 2011). Poorly grown heifers tend to become pregnant much later than their healthy counterparts, and therefore are often over-conditioned at calving. This leads to a higher incidence of dystocia, and the associated pain and distress when they calve.

**Fig. 16.9** Heifer with chronic pneumonia in North Asia. Note the poor coat and body condition, open mouth and extended neck

The result is weight loss in these newly calved heifers and higher rate of culling during their first lactation, due to disease and infertility. This confirms the need to rear a smaller number of heifer replacements well, rather than a large number poorly.

#### 16.6.2.2 Health, Growth and Reproductive Management and Planning in Heifers

This is often overlooked in new dairy systems, both large and small. A regionally specific vaccination and parasite control program should be developed and implemented in new entrant dairying areas. Internal parasites are of particular concern if heifers have access to grazing or chopped green forage, especially if effluent is used to irrigate these forage crops. Tick control is also essential where tick-borne diseases (e.g. theileriosis, babesiosis, anaplasmosis) are endemic.

#### 16.6.2.3 Targets

Each farm needs to set targets for mortality (death) rate and weight/age at mating and calving. Heart girth measurements can also be used as an indicator of weight (e.g. Sherwin et al. 2021). This, combined with body condition scoring, will give an indication of the effectiveness of feeding and health management programs. Heifers should be mated at a target weight, rather than age, to ensure they are well grown at calving. Well-grown heifers become pregnant faster, are healthier, generally produce more milk and stay in the herd longer. Many new dairy systems do not have adequate targets in place or ability to measure against these targets, resulting in inadequate growth, health and poor welfare.

## 16.7 Conclusion

In conclusion, the most important aspects to consider when implementing a new dairy system, include the development of the following:

- An appropriate business plan
- A feed management plan for sufficient feed supply and nutrition
- A housing system that allows adequate ventilation, cow cooling, cow comfort and efficient effluent collection and management
- Use of an appropriate breed to manage the local environmental conditions
- An appropriate animal health management and biosecurity program
- A calf and heifer rearing and management plan
- Appropriate staff training program and standard operating procedures
- Appropriate record collection program and use
- Appropriate monitoring to ensure cow health and welfare, especially using observation and BCS

In general, good animal welfare is central to a productive and sustainable dairy farm and dairy industry, and in general, farmers care greatly about their animals. New entrant or rapidly developing dairying farming systems, whether they be small-holder systems or large intensive farms, face many challenges. Many new entrants are very successful in achieving good health and welfare outcomes; however, many are not as successful.

Poor welfare outcomes in new or developing diary systems can be summed up as follows. In small-holder systems in developing countries, most new dairy farmers adopt the traditional farming practices from the region and often don't seek expert advice. These traditional practices often include cows being held by the head, under low roofs with poor ventilation, lying areas based on cement often with no matting, feeding of poor-quality forage, given water only 1–2 times daily. Problematic practices extend to calves and typically consist of poor colostrum feeding and poor general calf/heifer rearing management practices. There is usually a lack of understanding of the most appropriate breed for the region; records are not kept of key information on calving date, pregnancy status and production. There is also a general reluctance to cull animals. This commonly results in high rates of disease and death of calves, chronically ill and undergrown heifers entering the dairy herd and many nonpregnant and nonproductive cows consuming a high percentage of the available feed resulting in low body condition scores and poor health of the productive animals. Recent training programs, especially in Indonesia, are having some influence in changing current farmer attitude, with a focus on improving management practices and farm profitability, also resulting in improved welfare outcomes.

There are many examples of large new dairy systems being developed and successfully managed. However, especially in warmer climates, the welfare issues that become evident in new large-holder systems are generally a result of poor understanding of what resources are required and poor planning, prior to construction. This is often related to poor infrastructure and design, poor planning of feed supplies (affected quantity and quality), poor breed suitability and staff sourcing, poor staff training and management, improper health planning and veterinary care availability. This can result in lack of feed, heat stress, poor air quality, loss of body condition, poor health, poor husbandry, high rates of disease and very poor welfare outcomes. These issues are most commonly seen when corporates build large-scale systems, without adequate due diligence and planning. Farmers do not want their animals to suffer; however, poor welfare outcomes can be the result of a lack of understanding of the complexities and what is required to build, expand and manage dairy cattle farming systems.

**Notes on Photographs** All the photographs were supplied by the author.

## References

AgriFarming, India (2021) Dairy Housing Systems – Types, Designs. https://www.agrifarming.in/dairy-housing-systems-types-design-layout

Alvarez A, del Corral J, Solis D, Peréz JA (2008) Does intensification improve the economic efficiency of dairy farms? J Dairy Sci 91:3693–3698. https://doi.org/10.3168/jds.2008-1123

Austin J (1967) Urea toxicity and its prevention. In: Briggs MH (ed) Urea as a protein supplement. Pergamon Press, Oxford, pp 173–184

Bach A (2011) Associations between several aspects of heifer development and dairy cow survivability to second lactation. J Dairy Sci 94:1052–1057. https://doi.org/10.3168/jds.2010-3633

Bell DJ, Vigors B, Duthie C-A et al (2023) Developing a tool to assess the health-related quality of life in calves with respiratory disease: content validation. Animal 17:100702. https://doi.org/10.1016/j.animal.2022.100702

Bonizzi S, Gislon G, Brasca M et al (2022) Air quality, management practices and calf health in Italian dairy cattle farms. Animals 12:2286. https://doi.org/10.3390/ani12172286

Commonwealth Government of Australia (2021) Australian Standards for the Export of Livestock 3.2. Department of Agriculture, Water and the Environment, Canberra. https://awe.gov.au/biosecuritytrade/export/controlled-goods/liveanimals/livestock/australian-standards-livestock

Dahl G, Tao S, Laporta J (2020) Heat stress impacts immune status in cows across the life cycle. Front Vet Sci 7:116. https://pubmed.ncbi.nlm.nih.gov/32211430/

Dairy Australia (2019) Cool cows: strategies for managing heat stress in dairy cows. Dairy Australia Limited. Dairy Australia Limited ABN 60 105 227 987, 40 City Road, Southbank Victoria 3006 Australia

Dairy Australia (2020). Rearing healthy calves manual – second edition. Dairy Australia. https://www.dairyaustralia.com.au/resource-repository/2020/07/09/rearing-healthy-calves-manual%2D%2Dsecond-edition#.YVlBqDFBzIU

Devendra C (2001) Smallholder dairy production systems in developing countries: characteristics, potential and opportunities for improvement – review. Asian-Aust Sci 14(1):104–113

Dobson H, Smith RF (2000) What is stress, and how does it affect reproduction? Anim Reprod Sci 60-61:743–752

EFSA AHAW Panel (EFSA Panel on Animal Health and Welfare): Nielsen SS, Alvarez J, Bicout DJ et al (2022) Welfare of cattle during transport. EFSA J 20(9):7442. https://doi.org/10.2903/j.efsa.2022.7442

FAO (2004) The ethics of sustainable agricultural intensification. FAO Ethics Series 3. Food and Agriculture Organization of the United Nations. Rome

FAO (2009) Smallholder dairy development: Lessons learned in Asia. Food and Agriculture Organization of the United Nations. Regional Office for Asia and the Pacific. Bangkok. ISBN 978-92-5-106187-9 (https://www.fao.org/3/i0588e/I0588E00.htm#Contents)

Fuller FH, Huang J, Ma H Rozelle S (2005) The rapid rise of China's dairy sector: factors behind the growth in demand and supply. Working Paper 05-WP 394. May 2005. Center for Agricultural and Rural Development, Iowa State University, Ames, Iowa, United States. 50011-1070

Galvano F, Piva A, Ritieni A et al (2001) Dietary strategies to counteract the effects of mycotoxins: a review. J Food Prot 64:120–131

Grelet C, Vanden Dries V, LeBlois J et al (2022) Identification of chronic stress biomarkers in dairy cows. Animal 16:100502. https://doi.org/10.1016/j.animal.2022.100502

Habeeb AA, Gad AE, Atta MA (2018) Temperature-humidity indices as indicators to heat stress of climatic conditions with relation to production and reproduction of farm animals. Int J Biotechnol Recent Adv 1(1):35–50. https://doi.org/10.18689/ijbr-1000107

Hing S, Foster S, Evans D (2021) Animal welfare risks in live cattle export from Australia to China by sea. Animals 11:2862. https://doi.org/10.3390/ani11102862

Ishler VA, Heinrichs J, Jones CM (2016) Body condition scoring as a tool for dairy herd management. https://extension.psu.edu/body-condition-scoring-as-a-tool-for-dairy-herd-management

Ivemeyer S, Knierim U, Waiblinger S (2011) Effect of human-animal relationship and management on udder health in Swiss dairy herds. J Dairy Sci 94:5890–5902

Kadzere CT, Murphy MR, Silankove N, Maltz E (2002) Heat stress in lactating dairy cows: a review. Livest Prod Sci 77:59–91

Lago A, McGuirk SM, Bennett TB et al (2006) Calf respiratory disease and pen microenvironments in naturally ventilated calf barns in winter. J Dairy Sci 89:4014–4025

LeBlanc S (2010) Monitoring metabolic health of dairy cattle in the transition period. J Reprod Dev 56:S29–S35. https://doi.org/10.1262/jrd.1056S29

Lombard J, Urie N, Garry F et al (2020) Consensus recommendations on calf- and herd-level passive immunity in dairy calves in the United States. J Dairy Sci 103:7611–7624

Lopez AJ, Heinrichs AJ (2022) The importance of colostrum in the newborn dairy calf. J Dairy Sci 105:2733–2749. https://doi.org/10.3168/jds.2020-20114

Lorenz I, Earley B, Gilmore J et al (2011) Calf health from birth to weaning. III. Housing and management of calf pneumonia. Ir Vet J 64:14. https://doi.org/10.1186/2046-0481-64-14

Mahendran S, Booth R, Beekhuis L et al (2017) Assessing the effects of weekly preweaning health scores on dairy calf mortality and productivity parameters: cohort study. Vet Rec 181:196. https://doi.org/10.1136/vr.104197

Marshall K, Salmon GR, Tebug S et al (2020) Net benefits of smallholder dairy cattle farms in Senegal can be significantly increased through the use of better dairy cattle breeds and improved management practices. J Dairy Sci 103:8197–8217. https://doi.org/10.3168/jds.2019-17334

McDaniel BT (1983) Management and housing factors affecting feet and leg soundness in dairy cattle. Proc AABP 14:41–49

Medrano-Galarza C, Leblanc SJ, Jones-Bitton A et al (2018) Associations between management practices and within-pen prevalence of calf diarrhea and respiratory disease on dairy farms using automated milk feeders. J Dairy Sci 101:2293–2308

Mee J, Boyle L (2020) Assessing whether dairy cow welfare is "better" in pasture-based than in confinement-based management systems. NZ Vet J 68(3):168–177. https://doi.org/10.1080/00480169.2020.1721034

Mee JF (2008) Prevalence and risk factors for dystocia in dairy cattle: a review. Vet J 176:93–101

Moberg GP (2000) Biological response to stress: implications for animal welfare. In: Moberg GP, Mench JA (eds) The biology of animal stress: basic principles and implications for animal welfare. CABI Publishing, New York, pp 1–21. https://doi.org/10.1079/9780851993591.0001
Moran J (2005) Tropical dairy farming: feeding management for small holder dairy farmers in the humid tropics. Landlinks Press, p 312
Moran J, Doyle R (2015) Cow talk: understanding dairy cow behaviour to improve their welfare on Asian farms. CSIRO Publishing, E-book platform https://doi.org/10.1071/9781486301621
OECD/FAO (2022) OECD-FAO Agricultural Outlook 2022–2031. OECD Publishing, Paris. https://doi.org/10.1787/f1b0b29c-en
Phillips CJC, Santurtan E (2013) The welfare of livestock transported by ship. Vet J 196:309–314
Polsky L, von Keyserlingk MAG (2017) Effects of heat stress on dairy cattle welfare. J Dairy Sci 100:8645–8657. https://doi.org/10.3168/jds.2017-12651
Roche JR, Dillon PG, Stockdale CR et al (2004) Relationships among international body condition scoring systems. J Dairy Sci 87:3076–3079
Roche JR, Friggens NC, Kay JK et al (2009) Body condition score and its association with dairy cow productivity, health and welfare. J Dairy Sci 92:5769–5801
Rushen J (2017) Housing and the welfare of dairy cattle. In: Webster J (ed) Achieving sustainable production of milk. Volume 3: dairy herd management and welfare. Burleigh Dodds, UK. https://doi.org/10.19103/AS.2016.0006.03
Santana ML Jr, Bignardi AB, Pereira RJ et al (2017) Genetics of heat tolerance for milk yield and quality in Holsteins. Animal 11:4–14. https://doi.org/10.1017/S1751731116001725
Santman-Berends IMGA, Schukken YH, van Schaik G (2019) Quantifying calf mortality on dairy farms: challenges and solutions. J Dairy Sci 102(7):6404–6417. https://doi.org/10.3168/jds.2019-16381
Sherwin V, Hyde R, Green M et al (2021) Accuracy of heart girth tapes in the estimation of weights of pre-weaned calves. Vet Rec Open e16. https://doi.org/10.1002/vro2.16
Simm G, Pollott G, Mrode R et al (2021) Genetic improvement of farmed animals. CABI. Wallingford, Oxfordshire, UK. ISBN: 9781789241723
Toledo IM, Dahl GE, De Vries A (2022) Dairy cattle management and housing for warm environments. Livest Sci 255:104802. https://doi.org/10.1016/j.livsci.2021.104802
Tucker CB, Jensen MB, de Passillé AM et al (2021) Lying time and the welfare of dairy cows. J Dairy Sci 104:20–46. https://doi.org/10.3168/jds.2019-18074
USDA (2020) China: Dairy and Products Annual. Foreign Agricultural Service, U.S. Department of Agriculture. https://www.fas.usda.gov/data/china-dairy-and-products-annual-3
Van Os J (2020) Pair or group housing of dairy calves. Two heads are better than one: A starter guide to pairing dairy calves Extension article Animal Welfare Science @ Uw-Madison. https://animalwelfare.cals.wisc.edu/calf_pairing/
Vanholder T, Papen J, Bemers R et al (2015) Risk factors for subclinical and clinical ketosis and association with production parameters in dairy cows in The Netherlands. J Dairy Sci 98:880–888. https://doi.org/10.3168/jds.2014-8362